PHYSIOLOGICAL AND REGULATORY FUNCTIONS OF ADENOSINE AND ADENINE NUCLEOTIDES

Physiological and Regulatory Functions of Adenosine and Adenine Nucleotides

Editors

Hans P. Baer, Ph.D.
Professor of Pharmacology
Department of Pharmacology
University of Alberta
Edmonton, Alberta, Canada

George I. Drummond, Ph.D.
Killam Memorial Professor
Biochemistry Group
Department of Chemistry
University of Calgary
Calgary, Alberta, Canada

Raven Press ■ New York

Raven Press, 1140 Avenue of the Americas, New York, New York 10036

Made in the United States of America

Library of Congress Cataloging in Publication Data
Main entry under title:

Physiological and regulatory functions of adenosine and adenine nucleotides.

Based on proceedings of a conference held in Banff, Alberta, Canada, June 11-16, 1978.
Includes bibliographical references and index.
1. Adenosine--Physiological effect--Congresses. 2. Adenine--Physiological effect--Congresses. 3. Metabolic regulation--Congresses. 4. Nucleotide metabolism--Congresses. I. Baer, Hans P. II. Drummond, George I. [DNLM: 1. Adenosine--Physiology--Congresses. 2. Adenosine--Metabolism--Congresses. 3. Adenine nucleotides--Physiology--Congresses. 4. Adenine nucleotides--Metabolism--Congresses. QV185 P578 1978]
QP951.A25P45 574.1'88 78-55809
ISBN 0-89004-305-1

Preface

Interest in the regulatory functions of adenosine and adenine nucleotides arose primarily from studies by Berne and his associates which suggested that the nucleoside might function as a physiological regulator of coronary blood flow. More recently the proposed nerve transmitter function of ATP in smooth muscle, the discovery of adenosine receptors linked to adenylate cyclase in brain, and the association of defects in purine metabolism with certain immunodeficiency diseases has led to an enormous increase in studies dealing with possible regulatory functions of adenosine and adenine nucleotides.

In view of this rapid development it seemed appropriate and timely that the various scientific groups working in this area—physiologists, pharmacologists, and biochemists—be brought together to describe their current work and to discuss new possibilities for future research. This book is based on the proceedings of such a meeting which was held in Banff, Alberta, Canada, June 11–16, 1978.

The volume contains a variety of studies dealing with the nature of "purinergic" nerves in smooth muscle and a detailed historical review of the development of the purinergic nerve hypothesis is included. Several chapters deal with the role of adenosine as a mediator of vasodilation in heart, skeletal muscle, and brain and with the presence and nature of receptors for adenosine and adenine nucleotides in a variety of tissues. Particular attention is given to possible roles of the nucleoside in regulating heart, smooth muscle, and neuronal function. The possible involvement of cyclic AMP in adenosine action and the coupling of adenosine receptors with adenylate cyclase in broken cell preparations especially from brain, heart, adipose tissue, and platelets is dealt with at length. Adenosine transport into cells and metabolism of adenosine and adenine nucleotides in several cell types is described. Abnormalities in purine metabolism in human immunodeficiency disease is a particularly exciting area dealt with, as are several studies relating to possible clinical applications of synthetic adenosine analogs. In short, this volume touches almost every aspect of this rapidly developing field.

This book should be a valuable and up-to-date reference source for the established investigator and the student alike. Above all, it should demonstrate the exciting possibilities that lie ahead for those who choose to probe the amazing diversity of function which nature seems to have bestowed on adenosine and adenine nucleotides.

Hans P. Baer, Ph.D.
George I. Drummond, Ph.D.

Acknowledgements

The conference, on which this volume is based, was supported by generous financial contributions from the following agencies and firms:

The Canadian Heart Foundation
The Alberta Heart Foundation
The Dean of Medicine, University of Alberta
University of Alberta Conference Committee
University of Calgary Conference Fund
Sandoz Ltd.
Byk Gulden Lomberg Chemische Fabrik GmbH
Dr. Karl Thomae GmbH
Hoffmann-La Roche, Inc.
The Upjohn Company
Burroughs Wellcome Company
Merck, Sharp & Dohme
Smith, Kline & French Laboratories
Louisiana State University Foundation
CIBA-GEIGY Corporation

Contents

"Purinergic" Nerves and Isolated Smooth Muscle Studies

3 Past and Current Evidence for the Purinergic Nerve Hypothesis
G. Burnstock

33 Specificity of Adenine Nucleotide Receptor Sites: Inhibition of the Guinea Pig Taenia Coli by Adenine Nucleotide Analogs
M. Helen Maguire and D. G. Satchell

45 Intramural Inhibition in Rabbit and Guinea Pig Intestine
R. C. Small and A. H. Weston

61 Distribution of Nonadrenergic, Inhibitory Nerves in the Intestine, their Structural Identification, and the Role of Prostaglandins in their Function
E. E. Daniel

69 Presynaptic Inhibition of Adrenergic Neurotransmission by Adenine Nucleotides and Adenosine
David M. Paton

79 Multiple Adenine Derivative Receptors in Rat Ileum and Electrical Degradation of Purine Drugs
K. Kakatsu and V. Bartlett

85 Effects of 2-2′-Pyridylisatogen on Responses to Adenosine 5′-Triphosphate in Isolated Smooth Muscle Preparations
M. Hooper, M. Spedding, A. J. Sweetman, and D. F. Weetman

95 Action of Adenosine on Isolated Vascular Smooth Muscle
Jeremiah T. Herlihy

103 Structure-Activity Relationship Studies of Purinergic Agonists: Some Speculations on the Source and Identity of the Purinergic Mediator
M. A. Cook, J. T. Hamilton, and F. K. Okwuasaba

Cardiovascular Actions

117 Role of Adenosine as a Mediator of Metabolic Vasodilation in the Heart and the Brain: A Brief Overview and Update

Robert M. Berne, Duane H. Foley, William P. Watkinson, Wayne L. Miller, H. Richard Winn, and Rafael Rubio

127 Sites and Mode of Action of Adenosine in the Heart. I. Coronary System
E. Gerlach, J. Schrader, and S. Nees

137 Sites and Mode of Action of Adenosine in the Heart. II. Ventricular Myocardium
J. Schrader, E. Gerlach, and G. Baumann

145 Adenosine Production in Skeletal Muscles of Different Fiber Types
Emma L. Bockman and Jack E. McKenzie

155 Cardiovascular and Metabolic Actions of Adenosine and Adenosine Analogs
Gerhard Raberger

167 Cardiac Adenosine: Electrophysiologic Effects, Possible Significance in Cell Function, and Mechanisms Controlling Its Release
Rafael Rubio, Luiz Belardinelli, Carl I. Thompson, and Robert M. Berne

183 Cardiovascular Effects of Adenosine-5′-Carboxylates and Carboxamides
Herman H. Stein and Raj N. Prasad

189 Adenosine Receptors of the Blood Platelet: Interaction with Adenylate Cyclase
Richard J. Haslam, Monica M. L. Davidson, Byron W. G. Lemmex, Janet V. Desjardins, and Brain E. McCarry

205 Evidence for an Adenylate Cyclase-Linked Adenosine Receptor Mediating Coronary Relaxation
Walter R. Kukovetz, Annelies Wurm, Sigrid Holzmann, Gerald Pöch, and Ingo Rinner

Cyclic AMP and Adenylate Cyclase Involvement, Structure-Activity Studies

217 Regulation of Cyclic Adenosine Monophosophate Accumulation in Brain Tissue: Interactions of Adenosine with Other Agonists
Theodore W. Rall

229 Adenosine and Cyclic Adenosine Monophosphate-Generating Systems in Brain Tissue
John W. Daly

243 Biochemical Characterization of Adenosine Receptors in the Brain
Hirotoshi Shimizu

249 Regulation of Mouse Neuroblastoma Adenylate Cyclase by Adenosine
Arthur J. Blume

259 Effects of Adenosine and Adenylate Cyclase in Broken Cell Preparations of Central Nervous System
Joël Premont, Jean-Pol Tassin, Gérard Blanc, and Joël Bockaert

271 Actions of Adenosine on Adenylate Cyclase
Constantine Londos, J. Wolff, and Dermot M. F. Cooper

283 Conformational Component of Structure-Activity Relationships in Nucleosides
Douglas L. Miles, Daniel W. Miles, and Henry Eyring

Transport and Metabolism of Adenosine

297 Adenosine-Binding Proteins in Dog Heart
R. A. Olsson, R. B. Vomacka, and D. G. Nixon

305 Adenosine Transport
Alan R. P. Paterson

315 Regulation of Adenosine Metabolism
J. Frank Henderson

323 Ecto-Enzymes Concerned with Nucleotide Metabolism
J. F. Manery and E. E. Dryden

Electrophysiological Studies

343 Role of Adenosine and Adenine Nucleotides in Central Nervous Function
J. W. Phyllis, J. K. Kostopoulos, J. P. Edstrom, and S. W. Ellis

361 Adenosine and Its Mononucleotides as Regulatory and Adaptive Signals in Brain
Henry McIlwain

369 Effects of Purinergic Compounds on Excitable Membranes
L. P. Jager

Fat Cells, Immune System, and Prostaglandin Involvement

379 Adenosine Effects in Adipose Tissue: Effect of Adenosine on the Cyclic Adenosine Monophosphate System and Lipolysis of Fat Cells
Thomas Trost and Klaus Stock

395 Abnormalities of Purine Metabolism in Human Immunodeficiency Diseases
J. Edwin Seegmiller

409 Interrelationship Between Adenine Nucleotides and Prostaglandins
T. Kevin Sweeney and Philip Needleman

417 Analysis of Adenosine and Adenine Nucleotides in Severe Combined Immunodeficiency Disease
J. Arly Nelson, John F. Kuttesch, Randall Goldblum, Armond S. Goldman, and Frank C. Schmalstieg

429 *Subject Index*

Contributors

V. Bartlett
Department of Pharmacology
Faculty of Medicine
Queen's University
Kingston, Ontario, K7L·3N6, Canada

G. Baumann
Department of Physiology
University of Munich
D-8000 Munich, Germany

Luiz Belardinelli
The Department of Physiology
School of Medicine
University of Virginia
Charlottesville, Virginia 22901

Robert M. Berne
Department of Physiology
School of Medicine
University of Virginia
Charlottesville, Virginia 22901

Gérard Blanc
College de France
Laboratoires de Physiologie Cellulaire
75231 Paris Cedex 05, France

Arthur J. Blume
Department of Physiological Chemistry and Pharmacology
Roche Institute of Molecular Biology
Nutley, New Jersey 07110

Joël Bockaert
College de France
Laboratoires de Physiologie Cellulaire
75231 Paris Cedex 05, France

Emma L. Bockman
Department of Physiology
Louisiana State University
School of Medicine
New Orleans, Louisiana 70112

G. Burnstock
Department of Anatomy and Embryology
University College London
London WC1E 6BT, England

M. A. Cook
Department of Pharmacology
University of Western Ontario
London, Ontario N6A 5C1, Canada

Dermot M. F. Cooper
National Institute of Arthritis, Metabolism, and Digestive Diseases
National Institutes of Health
Bethesda, Maryland 20014

John W. Daly
National Institute of Arthritis, Metabolism, and Digestive Diseases
National Institutes of Health
Bethesda, Maryland 20014

E. E. Daniel
Department of Neurosciences
McMaster University
Hamilton, Ontario L8S 4J9, Canada

Monica M. L. Davidson
Department of Pathology
McMaster University
Hamilton, Ontario L8S·4J9, Canada

Janet V. Desjardins
Department of Biochemistry
McMaster University
Hamilton, Ontario L8S 4J9, Canada

E. E. Dryden
Department of Biochemistry
University of Toronto
Toronto, Ontario M5S 1A8, Canada

J. P. Edstrom
Department of Physiology
College of Medicine
University of Saskatchewan
Saskatoon, Saskatchewan S7N 0W0, Canada

S. W. Ellis
Department of Physiology
College of Medicine
University of Saskatchewan
Saskatoon, Saskatchewan S7N 0W0, Canada

Henry Eyring
Department of Chemistry
University of Utah
Salt Lake City, Utah 84112

Duane H. Foley
Department of Physiology
University of Virginia
School of Medicine
Charlottesville, Virginia 22901

E. Gerlach
Department of Physiology
University of Munich
D-8000 Munich, Germany

Randall M. Goldblum
Department of Pharmacology and Toxicology and Department of Pediatrics
University of Texas Medical Branch
Galveston, Texas 77550

Armond S. Goldman
Department of Pharmacology and Toxicology and Department of Pediatrics
University of Texas Medical Branch
Galveston, Texas 77550

J. T. Hamilton
Department of Pharmacology
University of Western Ontario
London, Ontario N6A 5C1, Canada

Richard J. Haslam
Department of Pathology
McMaster University
Hamilton, Ontario L8S 4J9, Canada

J. Frank Henderson
Cancer Research Unit
McEachern Laboratory
University of Alberta
Edmonton, Alberta T6G·2H7, Canada

Jeremiah T. Herlihy
Department of Physiology
University of Texas Health Science Center
San Antonio, Texas 78284

Sigrid Holzmann
Institut für Pharmakodynamik und Toxikologie
A-8010 Graz, Austria

M. Hooper
School of Pharmacy
Sunderland Polytechnic
Sunderland Tyne & Wear
England

L. P. Jager
Department of Pharmacology
State University Utrecht
Utrecht 2501, The Netherlands

G. K. Kostopoulos
Department of Physiology
College of Medicine
University of Saskatchewan
Saskatoon, Saskatchewan S7N·0W0, Canada

Walter R. Kukovetz
Institut für Pharmakodynamik und Toxikologie
A-8010 Graz, Austria

John F. Kuttesch
Department of Pharmacology and Toxicology
University of Texas Medical Branch
Galveston, Texas 77550

Byron W. G. Lemmex
Department of Pathology
McMaster University
Hamilton, Ontario L8S 4J9, Canada

Constantine Londos
National Institute of Arthritis, Metabolism, and Digestive Diseases
National Institutes of Health
Bethesda, Maryland 20014

M. Helen Maguire
Department of Pharmacology
University of Kansas Medical Center
Kansas City, Kansas 66103

J. F. Manery
Department of Biochemistry
University of Toronto
Toronto, Ontario M5S·1A8, Canada

Brian E. McCarry
Department of Chemistry
McMaster University
Hamilton, Ontario L8S 4J9, Canada

Henry McIlwain
Institute of Psychiatry
University of London
London SE5 8AF, England

Jack E. McKenzie
Department of Physiology
Louisiana State University
School of Medicine
New Orleans, Louisiana 70112

Daniel W. Miles
Department of Chemistry
University of Utah
Salt Lake City, Utah 84112

Douglas L. Miles
Department of Chemistry
University of Utah
Salt Lake City, Utah 84112

Wayne L. Miller
Department of Physiology
University of Virginia
School of Medicine
Charlottesville, Virginia 22901

K. Nakatsu
Department of Pharmacology
Faculty of Medicine
Queen's University
Kingston, Ontario K7L 3N6, Canada

Philip Needleman
Department of Pharmacology
Washington University Medical School
St. Louis, Missouri 63110

S. Nees
Department of Physiology
University of Munich
D-8000 Munich, Germany

J. Arly Nelson
Department of Pharmacology and Toxicology
University of Texas Medical Branch
Galveston, Texas 77550

D. G. Nixon
Suncoast Heart Association Cardiovascular Research Laboratory
Department of Internal Medicine
University of South Florida
College of Medicine
Tampa, Florida 33620

F. K. Okwuasaba
Department of Pharmacology
University of Alberta
Edmonton, Alberta T6G·2H7, Canada

R. A. Olsson
Suncoast Heart Association Cardiovascular Research Laboratory
Department of Internal Medicine
University of South Florida
College of Medicine
Tampa, Florida 33620

Alan R. P. Paterson
Cancer Research Unit
McEachern Laboratory
University of Alberta
Edmonton, Alberta T6G·2H7, Canada

David M. Paton
Department of Pharmacology and Clinical Pharmacology
School of Medicine
University of Auckland
Auckland, New Zealand

J. W. Phillis
Department of Physiology
College of Medicine
University of Saskatchewan
Saskatoon, Saskatchewan S7N 0W0, Canada

Gerald Pöch
Institut für Pharmakodynamik und Toxikologie
A-8010 Graz, Austria

Raj N. Prasad
Abbott Laboratories, Limited
Montreal, Quebec H4P 1A5, Canada

Joël Premont
College de France
Laboratoires de Physiologie Cellulaire
75231 Paris Cedex 05, France

Gerhard Raberger
Pharmakologisches Institut der Universität Wien
1090 Wien, Austria

Theodore W. Rall
Department of Pharmacology
University of Virginia
Charlottesville, Virginia 22901

Ingo Rinner
Institut für Pharmakodynamik und Toxikologie
A-8010 Graz, Austria

Rafael Rubio
Department of Physiology
University of Virginia
School of Medicine
Charlottesville, Virginia 22901

D. G. Satchell
Department of Zoology
University of Melbourne
Parkville, Victoria 3052, Australia

Frank C. Schmalstieg
Department of Pharmacology and Toxicology and Department of Pediatrics
University of Texas Medical Branch
Galveston, Texas 77550

J. Schrader
Department of Physiology
University of Munich,
D-8000 Munich, Germany

J. Edwin Seegmiller
Department of Medicine
University of California San Diego
La Jolla, California 92093

Hirotoshi Shimizu
Nippon Roche Research Center
Kajiwara, Kamakura, Japan 247

R. C. Small
Medical School
University of Manchester
Manchester M13 9PT, England

M. Spedding
Centre de Recherche Merrell International
67084 Strasbourg Cedex, France

Herman H. Stein
Abbott Laboratories
North Chicago, Illinois 60064

Klaus Stock
Institut für Pharmakologie und Toxikologie
Medizinische Hochschule Hannover
D-3000 Hannover 61, Germany

T. Kevin Sweeney
Department of Pharmacology
Washington University Medical School
St. Louis, Missouri 63110

A. J. Sweetman
School of Health and Applied Sciences
Leeds Polytechnic
Leeds, England

Jean-Paul Tassin
College de France
Laboratoires de Physiologie Cellulaire
75231 Paris Cedex 05, France

Carl I. Thompson
The Department of Physiology
School of Medicine
University of Virginia
Charlottesville, Virginia 22908

Thomas Trost
Institut für Pharmakologie und Toxikologie
Medizinische Hochschule Hannover
D-3000 Hanover 61, Germany

R. B. Vomacka
Suncoast Heart Association Cardiovascular Research Laboratory
Department of Internal Medicine
University of South Florida
College of Medicine
Tampa, Florida 33620

William P. Watkinson
Department of Physiology
University of Virginia
School of Medicine
Charlottesville, Virginia 22901

D. F. Weetman
School of Pharmacy
Sunderland Polytechnic
Sunderland Tyne & Wear, England

A. H. Weston
Department of Pharmacology
University of Manchester
Manchester M13 9PT, England

H. Richard Winn
Department of Physiology
University of Virginia
School of Medicine
Charlottesville, Virginia 22901

J. Wolff
National Institute of Arthritis, Metabolism and Digestive Diseases
National Institutes of Health
Bethesda, Maryland 20014

Annelies Wurm
Institut für Pharmakodynamik und Toxikologie
A-8010 Graz, Austria

"Purinergic" Nerves and Isolated Smooth Muscle Studies

Physiological and Regulatory Functions of Adenosine and Adenine Nucleotides, edited by H. P. Baer and G. I. Drummond. Raven Press, New York © 1979.

Past and Current Evidence for the Purinergic Nerve Hypothesis

G. Burnstock

Department of Anatomy and Embryology, University College London, London WC1E 6BT, England

The "purinergic nerve hypothesis" was developed in detail in 1971/72 (37,38), proposing that autonomic efferent nerves that are neither adrenergic nor cholinergic release a purine nucleotide or nucleoside as the principal transmitter. Early evidence supporting the hypothesis included the observations that adenosine triphosphate (ATP) was synthesized, stored, and released from nonadrenergic, noncholinergic inhibitory and excitatory nerves supplying the guinea pig taenia coli, stomach, and bladder and that it mimicked the effects of nerve stimulation on these small muscles (56,62,63,267,275,277).

Independent of the development of the purinergic nerve hypothesis, there was growing recognition, first by Drury and Szent-Gyorgyi (109) and later by Green and Stoner (148) for the sensitivity of many tissues to adenine nucleotides and nucleosides (see ref. 41). Particularly important was the suggestion of Berne (22) that adenosine was a physiological regulator of blood flow in the heart. This concept was extended later to regulation of the vascular beds of kidney (151) and skeletal muscle (29). Holton and Holton (168,169) claimed that ATP released by antidromic stimulation of sensory nerves was responsible for vasodilatation of the rabbit ear artery, and Hughes and Vane (176) noted that the responses of the rabbit portal vein to ATP resembled the inhibitory responses to nonadrenergic, noncholinergic intramural nerve stimulation. Adenosine triphosphate was recognized as a major component in storage vesicles in many endocrine and neural tissues (see ref. 301); and Hyden (179) examined nucleotide metabolism in nerve cells and noted three to five times more nucleotide in resting anterior horn cells compared with cells following stimulation. It had also been known for some time that the application of ATP or adenosine to peripheral ganglia (117) and to various regions of the brain and spinal cord (11,32,118,131,279) produced electrophysiological, biochemical, or behavioral changes. Finally, the effects of ATP administration in man were explored widely in the 1940s and 1950s on an empirical basis, apparently with some success, especially in geriatric patients with coronary insufficiency and angina, chronic ulcers caused by poor circulation, postischemic renal failure, asthenopia, diabetes, and shock (see ref. 27).

Since 1972 considerable additional evidence in support of the purinergic nerve hypothesis has been published. Detailed reviews on different aspects of this rapidly growing field are available (39–41,43–51,66,224).

EARLY EVIDENCE FOR NONADRENERGIC, NONCHOLINERGIC AUTONOMIC NERVES

Pharmacology

In 1898, Langley noted that stimulation of the vagus nerves could produce relaxation of the stomach after blockade of excitatory cholinergic fibers with atropine (see also 111,223,226,288,319). Relaxations of the intestine by the ganglion stimulant nicotine in the presence of atropine revealed that these inhibitory responses were mediated by intramural neurons, but it was assumed that they were adrenergic (5). It was later shown that the inhibitory responses of the gut to either vagal or intramural nerve stimulation were not abolished by adrenergic neuron blocking agents (53,222,245). Whereas Paton and Vane (245) at that time favored the explanation that the inhibitory responses were caused by adrenergic nerves being stimulated peripherally to the site of action of the blocking drug, Burnstock et al (53) and Martinson and Muren (222) considered this as evidence for the existence of nonadrenergic, noncholinergic inhibitory nerves, and indeed it was shown that the inhibitory responses persisted after sympathectomy (38), and that nerve-mediated inhibition was demonstrated in developing fetal rabbit intestine before adrenergic nerves appeared (133). Details followed from the pharmacological characteristics of the nonadrenergic, noncholinergic response in the intestine (22,55,98,266), and the stomach (15,33,67,187,221,237).

Langley and Anderson (215) were the first to show that contractions of the urinary bladder in response to pelvic nerve stimulation were "atropine resistant." This led to the suggestion that there was a separate, noncholinergic, nonadrenergic innervation of the bladder (8,62,72,160).

Electrophysiology

Inhibitory junction potentials (IJPs) in response to nonadrenergic, noncholinergic nerve stimulation were first recorded in smooth muscle cells of the guinea pig taenia coli by Burnstock and colleagues (20,53,54) and were later demonstrated in smooth muscle cells of the stomach in response to stimulation of both intramural and vagus nerves (15,16,206), in the longitudinal muscle of guinea pig ileum (162,207), in the guinea pig and rabbit colon (127,128,191), and in chicken and pigeon gizzard (17).

Electronmicroscopy

Electronmicroscopic studies of axon profiles in toad lung innervated by nonadrenergic, noncholinergic nerves led Robinson et al. (267) to propose that they were characterized by a predominance of "large opaque vesicles" (LOVs) with diameters of 80 to 200 nm compared with 60 to 120 nm for "large granular vesicles" found in small numbers in both adrenergic and cholinergic nerves. LOVs have a less prominent halo between the granular core and vesicle membrane and were unaffected by 6-hydroxydopamine, which destroys adrenergic nerve terminals. Nerve profiles of this kind have also been described in other tissues in which there was pharmacological evidence for nonadrenergic, noncholinergic nerves (see ref. 38).

Distribution

Burnstock (66) reviewed the evidence for nonadrenergic, noncholinergic nerves in the autonomic nervous system of vertebrates. There is nonadrenergic inhibitory nervous control of the stomach via preganglionic vagal cholinergic nerves in fish, amphibians, reptiles, and birds as well as in mammals. The intestine of mammals also contains intramural, nonadrenergic inhibitory neurons controlled largely by cholinergic interneurons and in the distal rectum by preganglionic sacral parasympathetic fibers. Noncholinergic, nonadrenergic *excitatory* nerves are a prominent feature in both the small and large intestines of amphibians, reptiles, and birds.

Nonadrenergic, noncholinergic nerves have also been described in lung, bladder, seminal vesicles, esophagus, uterus, eye, trachea, and parts of the cardiovascular and central nervous systems (40,66,68,129).

EARLY EVIDENCE THAT A PURINE NUCLEOTIDE OR NUCLEOSIDE IS A NEUROTRANSMITTER IN NONADRENERGIC, NONCHOLINERGIC NERVES

Five criteria are generally regarded as necessary for establishing a substance as a neurotransmitter, namely, (a) synthesis and storage of transmitter in nerve terminals; (b) release of transmitter during nerve stimulation; (c) postjunctional responses to exogenous transmitter that mimic responses to nerve stimulation; (d) enzymes that inactivate the transmitter and/or an uptake system for the transmitter or its breakdown products; and (e) drugs that produce parallel blocking or potentiating effects on the responses to both exogenous transmitter and nerve stimulation.

In the early experiments carried out to try to identify the transmitter in nonadrenergic, noncholinergic nerves supplying the guinea pig taenia coli, ATP appeared to best satisfy these criteria (Fig. 1) (38). Other substances were explored as possible candidates, including catecholamines, 5-hydroxytryptamine, cyclic adenosine monophosphate (cAMP), histamine, prostaglandins, amino acids includ-

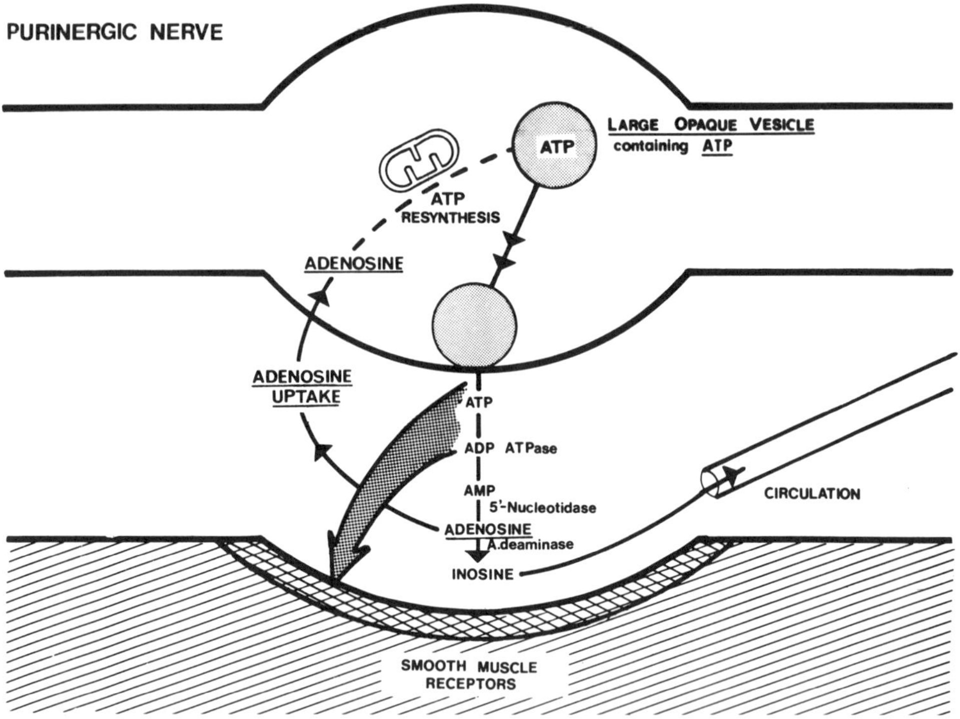

FIG. 1. Schematic representation of synthesis, storage, release, and inactivation of autonomic neurotransmitters at purinergic neuromuscular junctions. (From ref. 38, with permission.)

ing glycine and γ-aminobutyric acid (GABA), as well as the polypeptides bradykinin and substance P. However, these substances were rejected as contenders by most workers on the grounds that they were either inactive or did not mimic the nerve-mediated responses, that specific blocking drugs for these substances did not affect the nerve-mediated response, or that their action was by stimulation of nerves and not by direct action on smooth muscle.

Storage

Both ATP and the enzyme systems that synthesize ATP occur ubiquitously in cells, so that it was not contentious that nonadrenergic inhibitory nerves were also able to produce and store ATP. Tritium-labeled adenosine was taken up by preparations of stomach and intestine and rapidly converted largely to [^{3}H]ATP (305). Analysis of the radioactivity in serial frozen sections showed that when the tissue was incubated in low concentrations of [^{3}H]adenosine, most of the label was stored as [^{3}H]ATP in nerves.

It is interesting too, that the unicellular parasite *(Trypanosoma cruzi)* responsible for Chagas's disease is unable to synthesize its own adenine (119) and the LOVs in nerve profiles were damaged in the intestine of infected patients (310).

Release

Early evidence for release of ATP during stimulation of purinergic nerves was as follows: (a) venous efflux of adenosine and inosine from the stomach of both guinea pigs and toads occurred on stimulation of the vagus nerves. These compounds were likely to be the breakdown products of adenine nucleotides, because ATP introduced into the same perfusing system was quickly broken down into comparable proportions of adenosine and inosine; (b) radioactive compounds were released from guinea pig taenia coli previously incubated with [^{3}H]adenosine on stimulation of its intramural nerves.

The possibility that the purine nucleotides or nucleosides released from nerves were not neurotransmitter substances but appeared for other reasons was considered. For example, it was known that ATP was released from nerve membranes during propagation of an action potential (1). However, the amount of nucleosides collected during stimulation of nonadrenergic inhibitory nerves was calculated to be at least 1,000-fold greater than that released as a direct result of the process of axon membrane activation during impulse propagation. Release from muscle was considered unlikely, since stimulation of portions of Auerbach's plexus from turkey gizzard (heavily innervated by nonadrenergic inhibitory nerves), dissected free of the underlying muscle, still resulted in efflux of purine nucleotides, in this case mostly adenosine monophosphate (AMP). The possibility that ATP was released as a result of antidromic stimulation of enteric sensory nerves was ruled out when it was shown that the nonadrenergic inhibitory response to stimulation of the vagal nerves supplying the rabbit stomach was abolished after degeneration of the efferent, but not the afferent component, achieved by vagal section above the nodose ganglion 8 days previously.

Receptor Activation

The form and time course of the responses to exogenously applied ATP closely mimicked those of nonadrenergic inhibitory nerve stimulation. Typically, the relaxation produced by ATP and nerve stimulation rapidly reached a maximum that declined quickly; this was in contrast to the relaxations to noradrenaline (NA) and sympathetic nerve stimulation, which reached a maximum more slowly and were maintained for a longer time. The transmitter released by nonadrenergic inhibitory nerves produced IJPs, and ATP also caused rapid hyperpolarization of smooth muscle cells. The possibility that ATP might be causing relaxation by initiating action potentials in nonadrenergic inhibitory nerves was negated by the finding that tetrodotoxin abolished the inhibitory responses of atropinized intestinal muscle to stimulation of either perivascular or intramural nerves but did not affect the relaxation produced by ATP.

In the intestine the most potent inhibitory purine compounds were ATP and adenosine diphosphate (ADP), which were about equipotent, with a threshold concentration for relaxation of the taenia coli of about 0.1 μM. Adenosine monophosphate and adenosine had about 1% of the potency of ATP. No effect was found with the following related compounds up to 1 mM: Adenine, inosine, inosine monophosphate (IMP), uridine, and uridine monophosphate (UMP). The sensitivity of the gut to cAMP was also low. Of ATP and ADP, ATP was considered to be the most likely transmitter, since [^{3}H]adenosine taken up into the taenia coli was shown to be rapidly converted and stored largely as [^{3}H]ATP, with only traces of [^{3}H]ADP detectable.

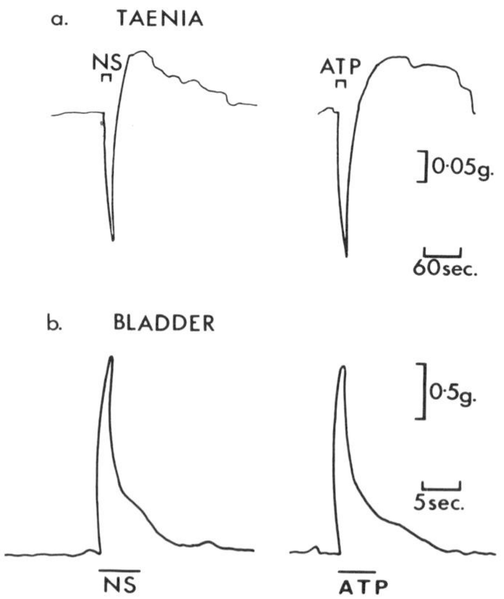

Fig. 2. A comparison of the responses to exogenous adenosine triphosphate (ATP) and intramural nerve stimulation (NS) in the presence of atropine and guanethidine. (a) inhibitory response of guinea pig taenia coli (NS: 1 Hz, 0.5 msec pulse duration, supramaximal voltage; ATP 2×10^{-6}M) (From ref. 64, with permission). (b) excitatory response of guinea pig bladder strip; (NS: 5 Hz, 0.2 msec pulse duration, supramaximal voltage; ATP: 8.5 μM) (From ref. 58, with permission).

Although the predominant response of mammalian gut to ATP was relaxation, excitatory responses of two kinds were also recognized. One consisted of contractions after washout of ATP, which were comparable to the "rebound excitation" that followed nonadrenergic inhibitory responses. The second occurred while the nucleotide was still present in the organ bath and was considered to be a direct excitatory effect of ATP. In a variety of preparations, responses to nonadrenergic, noncholinergic nerve stimulation were closely mimicked by ATP (Fig. 2.): relaxation in most mammalian gut preparations, contraction in the intestine of lower vertebrates and mammalian bladder, and biphasic responses in guinea pig ileum and toad stomach.

Inactivation

The rapid recovery of smooth muscle after application of ATP or stimulation of nonadrenergic inhibitory nerves and the absence of long-lasting action despite continued stimulation indicated an efficient inactivation mechanism. By analogy with other neuroeffector systems, it was argued that if the nonadrenergic inhibitory nerves act on the gut by releasing ATP, the action would be terminated by uptake into nerves or smooth muscle and/or breakdown of ATP by enzymes into compounds with greatly reduced potency.

When ATP was added to a perfusion fluid recycled through the vasculature of the stomach, very little ATP was recovered, but the perfusate contained substantially increased amounts of adenosine and inosine as well as some ADP and AMP. While it was recognized that there was no direct evidence for the breakdown of ATP by enzymic activity, it was pointed out that the gut was known to contain 5′-nucleotidase and adenosine deaminase. Mg^{2+}-activated ATPase has been localized in micropinocytotic vesicles in smooth muscle membranes closely adjacent (20 nm) to nonadrenergic, noncholinergic profiles in the intestine, and also in synaptic clefts in the rat cerebral cortex (163).

An uptake mechanism for adenosine, but not nucleotides, was demonstrated suggesting that ATP released from nerves was broken down to adenosine before uptake occurs, in a comparable manner to uptake of choline after breakdown of acetylcholine (ACh) released from cholinergic nerves. When the adenosine moiety of ATP was labeled with tritium and the phosphate moiety labeled with ^{32}P, the rate of uptake of ^{3}H into taenia coli was considerably greater than that of ^{32}P. Electric field stimulation at low physiological frequencies (5 Hz) showed little overflow, whereas stimulation at 30 Hz led to substantial overflow. In the presence of low concentrations of the adenosine uptake inhibitor dipyridamole, overflow at low stimulation frequencies was comparable to that at high frequencies.

Drugs

Several groups of drugs were known to block the responses to either adenine nucleotides or nucleosides in a wide variety of preparations. These included antimalarial drugs such as mepacrine, quinine, and quinidine, and methylxanthines such as caffeine and theophylline. Quinidine blocked nonadrenergic, noncholinergic nerve-mediated responses in the gut and bladder, but only in concentrations that gave little confidence in its specificity.

Low concentrations of the adenosine-uptake inhibitor, dipyridamole, were shown to potentiate the responses of the guinea pig taenia coli to both nonadrenergic inhibitory nerve stimulation and ATP, while reducing the responses to adrenergic nerve stimulation and NA (277).

Indirect support for the purinergic hypothesis came from experiments utilizing the tachyphylactic action of ATP on some preparations. For example, when tachyphylaxis to ATP was produced in the rabbit ileum, there was a consistent depression of the inhibitory responses to nonadrenergic inhibitory nerve stimulation, but not to adrenergic nerve stimulation.

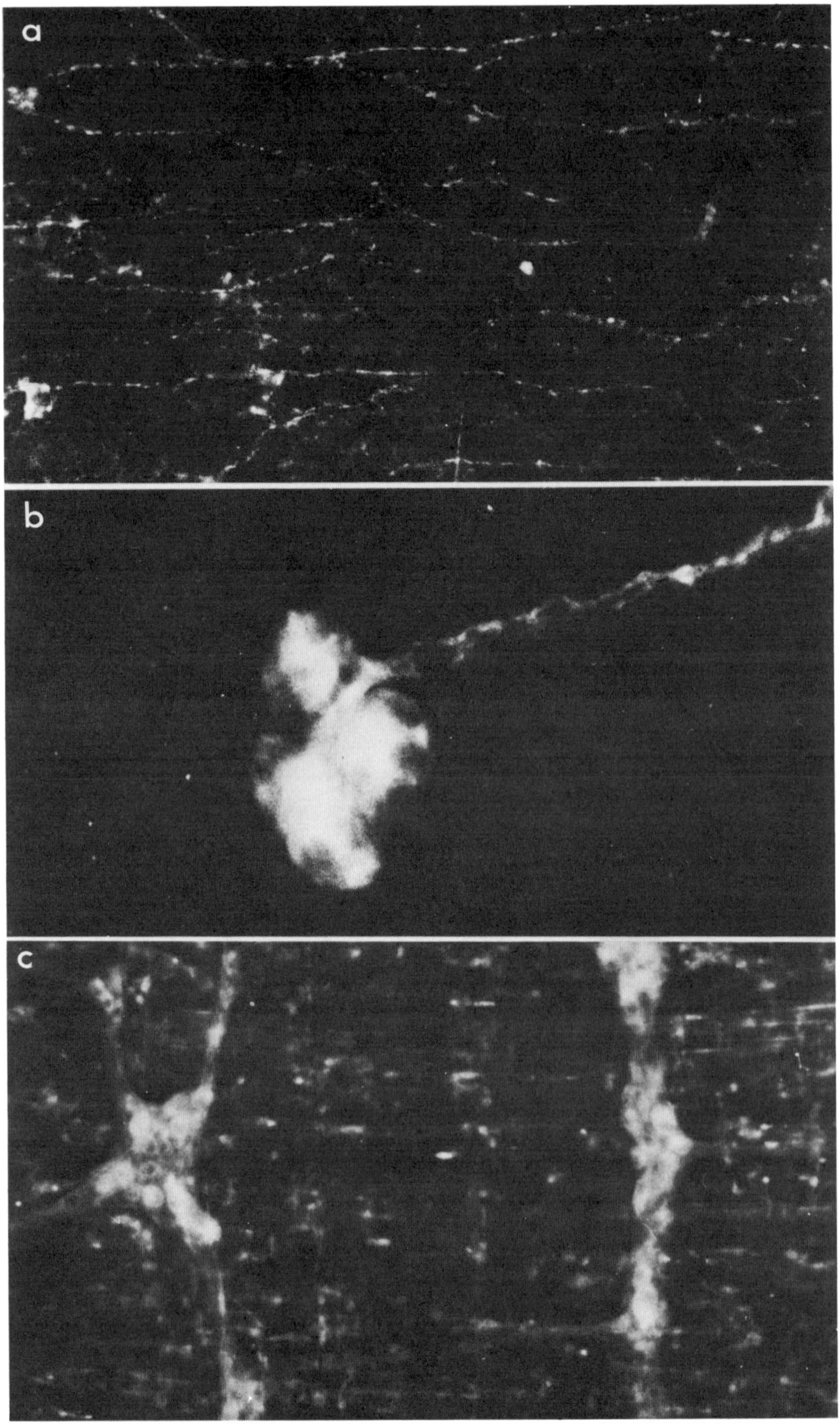
a
b
c

RECENT EVIDENCE FOR PURINERGIC NERVES

Since the purinergic hypothesis was first proposed, over 2,000 papers have been published that are either directly or indirectly related to the hypothesis; it is possible to review only a limited number of these.

Storage, Release, and Inactivation of ATP from Purinergic Nerves

Quinacrine Fluorescence Histochemistry

A fluorescence histochemical method for localizing quinacrine was introduced by Olson et al (241). This method may be selective for purinergic nerves. It is known to bind to ATP (180) and also to regions of DNA rich in adenine and thymine (112). It gives positive staining of adrenal medullary cells and blood platelets, known to contain high levels of ATP (218). Microsomal fractions obtained by differential and sucrose-density gradient centrifugation of homogenates of purinergically innervated preparations (taenia coli and bladder) preloaded with [^{3}H]adenosine and [^{14}C]quinacrine showed peaks of ^{14}C that corresponded to peaks of [^{3}H]ATP (57).

Quinacrine-positive fibers have been demonstrated in gut, gall bladder, urinary bladder, and portal vein (Fig. 3.), where pharmacological evidence for purinergic transmission has been presented recently by Burnstock and co-workers (57,58,65,81,95), but are absent from the iris, which contains abundant adrenergic and cholinergic, but not purinergic nerves.

Electronmicroscopic Autoradiography

Selective accumulation of silver grains over nerve endings has been demonstrated following exposure of brain [^{3}H]adenosine (281; and H. Shimizu, *this volume*). Preliminary studies in our laboratory (M. E. Dennison and G. Burnstock, *personal communication*) suggest that after exposure of intestine to low concentrations of [^{3}H]adenosine for short periods (45 sec) there is selective accumulation of silver grains over nerve profiles containing LOVs.

Ca^{2+}-Dependent Release from Nerve Rather than Muscle

The problem of whether ATP released during stimulation of purinergic nerves comes from nerve or secondarily from muscle has been resolved recently in our laboratory (58,59). It was shown that whereas there was a two- to sixfold increase in ATP release from the guinea pig taenia coli or bladder during isometric responses to

FIG. 3. Fluorescence histochemical localization of quinacrine in whole mount stretch preparations of: **(a)** Rabbit portal vein showing nerve plexus (× 200), **(b)** Rabbit urinary bladder showing ganglion cells associated with fine nerve trunk (× 250), and **(c)** Guinea pig ileum. Fluorescent nerve cells of Auerbach's plexus can be seen in the foreground (× 200). (By courtesy of Rahima Crowe, Anatomy Department, University College, London.)

purinergic nerve stimulation, there was no significant release of ATP during comparable responses elicited by direct muscle stimulation (Fig. 4). Further, the nerve-mediated release of ATP was Ca^{2+} dependent; a 10-fold reduction in Ca^{2+} concentration resulted in an 80 to 90% reduction in both the mechanical response and the ATP release (58). This finding is consistent with that of Holman and Weinrich (167) who showed that the amplitude of IJPs in the taenia coli was reduced by 80% when the external Ca^{2+} was reduced 10-fold.

Axonal Transport and Transynaptic Release of Nucleotides in the Brain

Schubert and Kreutzburg (281) were the first to demonstrate that after injection of [^{3}H]adenosine into the cortex there was selective uptake by certain neurons, axonal transport to their terminals, followed by release into the synaptic cleft. Release from dendrites as well as axon terminals has now been demonstrated in various regions of the CNS (178,204,268,282,283,285,286,334,335).

Storage and Release of ATP from Synaptosome and Vesicle Preparations

Synaptosome preparations have been isolated from guinea pig and rat brain for studies of ATP storage and release (12,210,258,332). Release of ATP from guinea

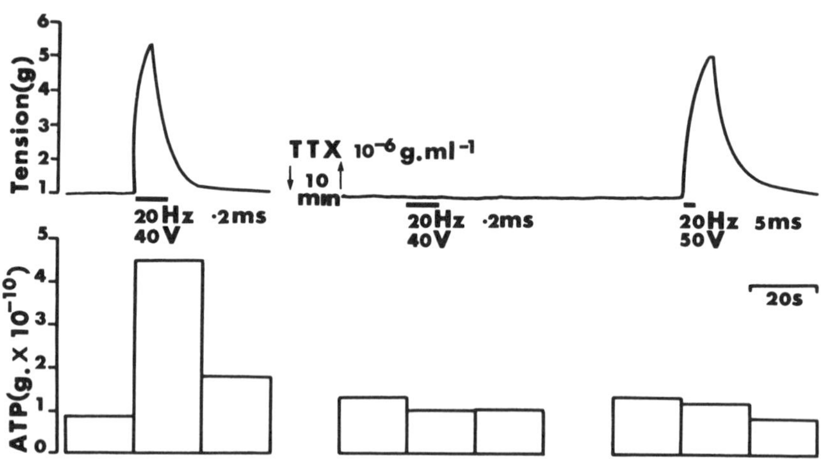

FIG. 4. Adenosine triphosphate release in response to nerve stimulation of guinea pig bladder strip (20 Hz, 0.2-msec pulse duration, and supramaximal voltage for 10 sec). Following nerve paralysis with tetrodotoxin (10^{-6} g/ml), the increase in release of ATP accompanying stimulation was abolished. No increase in ATP release was seen as a result of contraction caused by direct muscle stimulation (20 Hz, 5-msec pulse duration, and supramaximal voltage for 2 to 3 sec). The period of direct muscle stimulation (2 to 3 sec) was chosen so that the resultant contraction was similar in both magnitude and time course to the contraction elicited by nerve stimulation. Atropine (10^{-7} g/ml) and guanethidine (10^{-6} g/ml) were present. Upper ordinate: tension (g); lower ordinate: ATP (g/ml $\times$ 10^{10}). (From ref. 59, with permission.)

pig neocortex synaptosomes increased about threefold during electrical stimulation (12). Release of ATP from the preparations from rat brain was induced by high potassium or by veratridine (332). K^+-induced release of ATP was Ca^{2+} dependent, closely resembling the exocytotic release of putative transmitters from nerve terminals.

Inactivation

Some new data have appeared on the ectoenzymes involved in breakdown of extracellular ATP (159,200,201,308). Adenylate kinase was claimed to be a key enzyme in synaptosomal degradation of ATP to adenosine (200), and extracellular localization of Mg^{2+}-activated ATPase and 5′-nucleotidase in smooth muscle, mast cells, and synapses in the brain was demonstrated (3,199,220,242). The catabolism of ATP and high- and low-affinity uptake of adenosine into neuronal and non-neuronal cells was described (104,122), including evidence that endothelial cells are mainly responsible for the removal of circulating adenosine (247,284).

Purinergic Receptors

The many papers that have been published since 1972 about the potent extracellular actions of purine nucleotides and nucleosides on a wide variety of tissues have been reviewed in detail by Burnstock (41,49). These studies deal with a number of different aspects including those described below.

Structure-Activity Relationships

The structure-activity studies of a wide variety of analogs on visceral and vascular smooth muscle, cardiac muscle, and cortical neurons have been described (see in particular refs. 4,79,125,149,175,211,240,244,251,252,259,278,280,316). For most tissues the potency series is ATP $\geq$ ADP $>$ AMP $\geq$ adenosine. However, for some tissues (e.g., trachea, some areas of the brain, and astrocytoma cells) the reverse is true (see in particular refs. 78,116,211,253).

Conductance Changes

The early demonstration (19) that the IJP resulting from stimulation of nonadrenergic inhibitory nerves and the subsequent relaxation was the result of a selective increase in K^+ permeability was confirmed and extended to show that the hyperpolarization resulting from exogenous application of ATP was also the result of a selective increase in K^+ permeability (99,312). For the mechanism of *contraction* to ATP it has been suggested that ATP, by complexing Mg, enhances Ca transport across muscle membranes (94,145,174).

Synthesis of Prostaglandins

The discovery that ATP is a potent inducer of prostaglandin synthesis (231) led Burnstock et al (60) to suggest that ATP released from purinergic nerves may be linked with prostaglandins in peristalsis and possibly other physiological regulatory mechanisms. The authors showed that the prostaglandin synthesis inhibitor indomethacin blocked the "rebound contractions" that follow the inhibitory responses of the guinea pig taenia coli to purinergic nerve stimulation and ATP. Since that time, there have been several examples of such a link in the esophagus and trachea (192), bladder (189), uterus (107), and heart (13).

Blocking Agents

New blocking agents for ATP have been developed. These include the 2-substituted imidazoline compounds such as antazoline and phentolamine (276), and 2,2′-pyridylisatogen (171,195,298). Unfortunately, these drugs are not specific for ATP. In a recent report, however, it was claimed that very low concentrations of apamin (10 nM), a constituent of bee venom, selectively block both the IJPs in response to purinergic nerve stimulation in the intestine and the hyperpolarizations produced by ATP (323).

Ultraviolet Light

Ultraviolet (UV) light between 340 and 380 nm produced responses in the guinea pig taenia coli and rabbit portal vein that mimicked precisely the inhibitory response to purinergic nerve stimulation and ATP (64). Ultraviolet light did not initiate impulses in purinergic nerves since its action was unaffected by tetrodotoxin, nor did it release ATP from nerve terminals. Agents that alter postjunctional responses to ATP and purinergic nerve stimulation also altered the responses to UV light. Ultraviolet irradiation had no action on smooth muscle that was not innervated by inhibitory purinergic nerves, so that it was suggested that UV light may be acting on some part of the purinergic receptor complex and may provide a means of investigating the chemistry of purinergic receptors.

Adenylate Cyclase and cAMP

There are many reports that follow up the earlier observations of Sattin and Rall (279) that purine nucleosides act on receptors that contain adenylate cyclase and that this leads to changes in cAMP accumulation in both neuronal and nonneuronal cells, both increases (26,77,78,93,149,152,175,211,248,249,287,336) and decreases (115,216,225,325,339) having been observed. The methylxanthines competitively block these actions.

TABLE 1. *Four criteria for distinguishing two types of purinergic receptor*

Purinergic receptor	Antagonists	Agonist potencies	Changes in cAMP production	Induction of prostaglandin synthesis
P_1	Methylxanthines	AD≥AMP>ADP≥ATP	Yes	No
P_2	Quinidine Imidazolines 2′2-Pyridylisatogen Apamin	ATP≥ADP>AMP≥AD	No	Yes

Classification of Purinergic Receptors

Burnstock (49) has recently suggested a basis for distinguishing two types of purinergic receptor according to four criteria: relative potencies of agonists, competitive antagonists, changes in levels of cAMP and induction of prostaglandin synthesis (Table 1). Thus P_1 purinoceptors are most sensitive to adenosine, are blocked by methylxanthines, and occupation leads to changes in cAMP accumulation. P_2 purinoceptors are most sensitive to ATP, are blocked (although not competitively) by high concentrations of quinidine, 2-substituted imidazolines, 2′,2-pyridylisatogen and possibly by low concentrations of apamin (see above), and occupation leads to production of prostaglandin.

Evidence has been presented for two adenosine receptor sites coupled to adenylate cyclase (e.g., in liver, adrenal tumors, thyroid, platelets), one leading to stimulation and the other to inhibition of the enzyme (217). It has also been proposed that there are two ATP binding sites (255).

Distribution and Physiological Roles

Alimentary Canal

A schematic representation of the innervation of the mammalian gastrointestinal tract by purinergic nerves is shown in Fig. 5. Intramural nonadrenergic inhibitory neurons have now been demonstrated in the mammalian esophagus (100,143,260), in the stomach, where they are concerned with "receptive relaxation" (2,40,69, 78,155,184,186,240,309), small intestine (10,17,18,25,129,133,164,165,337), colon where they are involved in "descending inhibition" during peristalsis (14,61,89,120,142,190,235), in the lower esophageal (72,74,101,102,146,147, 314,326), pyloric (10,227), ileocaecal (85, but see ref. 243), and internal anal (88,132) sphincters, and the anococcygeus (57,90,91,134,135,137–139,274), and rectococcygeus muscles (81,197,198).

Evidence has been presented that these nerves are purinergic in the esophagus (82), stomach (240,297), small intestine (238,318), sphincter of Oddi (250), anococcygeus muscle (57), and rectococcygeus muscle (81).

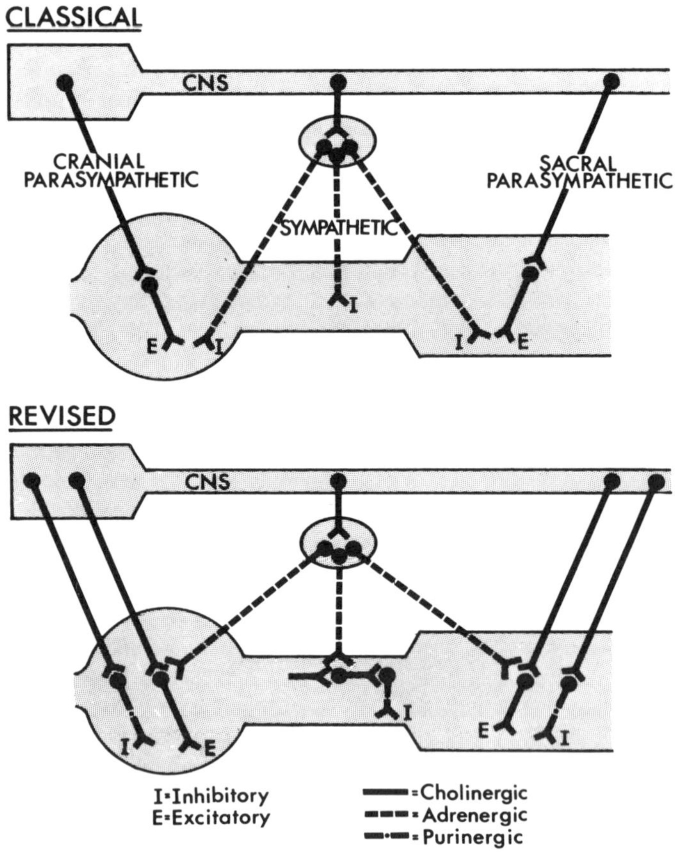

FIG. 5. Diagrammatic representation of the autonomic innervation of the gut.

Evidence for purinergic innervation of the gall bladder has also been presented (95,96), including the demonstration *in vivo* of nonadrenergic relaxation during stimulation of the cervical vagus nerve, the mimicking action of ATP, and the presence of quinacrine fluorescence in intramural neurons and fibers that persist after sympathectomy. There are some indications that purinergic nerves may also supply the pancreas. Transient hyperpolarizations occur in acinar cells on stimulation of vagal nerves (97). Adenosine triphosphate has been shown to stimulate glucagon and insulin secretion in isolated perfused pancreas (219) and LOVs have been described in nerve profiles in the endocrine pancreas of the catfish (30).

Urinogenital System

There is further evidence that the nonadrenergic, noncholinergic excitatory innervation of the urinary bladder is purinergic. Downie and Dean (106) showed that concentrations of theophylline and 2′,2-pyridylisatogen that antagonized ATP, but not carbachol contractions, of rabbit detrusor strips, depressed the nerve-mediated frequency-response curves; desensitization of the bladder to ATP also depressed the nerve-mediated responses. Hoyes and his colleagues (172,173) and Wendelarr et al. (328) demonstrated nerve profiles containing LOVs in the rat and fish bladder, respectively. Weetman and Turner (324) showed that quinidine reduced nerve-mediated contractions without affecting those in response to histamine. In a recent study of the guinea pig urinary bladder (58), a combination of evidence for purinergic innervation was presented: Fluorescence histochemical localization of quinacrine, Ca^{2+}-dependent release of ATP during stimulation of intramural nerves, and reduction of the nerve-mediated responses following desensitization to ATP.

Morphological (161,299) and pharmacological evidence (166) for nonadrenergic, noncholinergic innervation of the uterus of cat and rat has been presented, and the possibility considered that these nerves are purinergic (107).

Nonadrenergic, noncholinergic excitatory responses of the seminal vesicle to stimulation of the hypogastric nerve are closely mimicked by ATP (233).

Respiratory System

Evidence for purinergic inhibitory innervation of the guinea pig and dog trachea has been considered by several groups in recent years (66,80,83,84,116,193, 265). Adenosine triphosphate mimics the nerve-mediated response, which is unaffected by sympathectomy but potentiated by dipyridamole. Numerous intramural neurons are located in the peritracheal and peribronchial tissue. It is possible that the transmitter in these nerves is adenosine rather than ATP (49).

Richardson and Beland (264) presented compelling evidence that nonadrenergic, noncholinergic nerves are the principal inhibitory system for the smooth muscle of human airways and it was further claimed (263) that little or no nonadrenergic inhibition was observed in patients with severe obstructive airway disease.

Blood Vessels

The abundant literature on the potent actions of purine nucleotides and nucleosides on many vessels and the release of ATP or its breakdown products during nonadrenergic noncholinergic vasomotor responses of coronary, pulmonary, renal, intestinal, skeletal muscle, skin, and cerebral vascular beds, to hypoxia, ischemia, and exercise, have been reviewed in detail recently (23,50,51,234,270,303). It was suggested by Burnstock (50) that the rapid actions of ATP released from purinergic nerves acting on P_2 purinoceptors may be superimposed in some vascular beds on the slower, longer-lasting effects of adenosine released as a result of local metabolic activity during hyperactivity, acting on P_1 purinoceptors (see also ref. 289).

Pharmacological evidence that ATP might be involved in the mechanism of vasodilatation of skin vessels was considered (47,71,196) and LOVs have been identified in nerve profiles related to arteriovenous anastamoses in the skin (144,229).

Following up the work of Hughes and Vane (176) on the rabbit portal vein, the possibility that ATP is the transmitter released from the nonadrenergic, noncholinergic nerves has been considered. Intramural nerve stimulation of the rabbit portal vein produced an approximately 13-fold increase in release of [^{3}H]adenine compounds; after preloading the vein with [^{3}H]adenosine, about 50% of this release was blocked by guanethidine and all by tetrodotoxin (65,302). A quinacrine-positive perivascular plexus was demonstrated. Both this plexus and the guanethidine-resistant release of ATP remained after sympathectomy. The presence of quinacrine-positive nerve cell bodies within the wall of the rabbit portal vein is of interest because there are few reports of intramural ganglion cells in blood vessels (see ref. 170), and also because it is consistent with the observations of Hughes and Vane (177) that the ganglion stimulant nicotine produces nonadrenergic relaxation and that hexamethonium reduces the response to transmural stimulation. Finally, indirect support for purinergic innervation of the rabbit portal vein was obtained from comparative studies of the guinea pig portal vein, in which (a) a quinacrine-positive plexus was absent; (b) the release of ATP in response to intramural nerves was abolished by guanethidine or sympathectomy; (c) and in which ATP did not produce a mimicking relaxation (65). These experiments also indicated that ATP is probably released together with noradrenaline from the perivascular sympathetic nerves (see section below titled "Adenosine Triphosphate or Adenosine as a Cotransmitter and/or Neuromodulator"), although in smaller quantities since there are about three to seven times as many adrenergic as purinergic nerve varicosities in the rabbit portal vein.

Peripheral Ganglia

Interest is returning to the potent actions of purine and pyrimidine nucleotides on peripheral ganglia following the early work of Feldburg and Hebb (117). Low concentrations of ATP and uridine triphosphate (UTP) produce depolarization of bullfrog sympathetic neurons that resembles the atropine-resistant late slow excitatory postsynaptic potential (EPSP) (292). Adenosine triphosphate and adenosine also act on the superior cervical ganglion and on pelvic ganglion cells supplying the urinary bladder (311; and R.J. Theobold and W.C. Degroat, *personal communication.*)

Central Nervous System

Since 1972, there has been considerable interest in exploring the possibility that purinergic nerves are present in parts of the central as well as peripheral systems. This has been reviewed by Burnstock (46) and McIlwain (224) and includes (a) release during electrical stimulation of adenine nucleotides and their derivatives from intact brain and brain slices (158,256,257,306) or synaptosomal preparations (12,210,332); (b) depression of excitability of certain central neurons by iontophoretic application of adenine nucleotides and nucleosides (202,203,239); (c) the presence of a specific adenosine-sensitive adenylate cyclase in the brain leading to production of cAMP (93,290,291) [adenosine has also been shown to elevate brain cyclic guanosine monophosphate (cGMP) (236,272)]; (d) uptake of [^{3}H]adenosine by certain neurons, conversion largely to [^{3}H]ATP, axonal transport, and release from nerve terminals (178,204,268,282,283,285,286,334,335); and (e) the presence of LOVs and high levels of 5′-nucleotidase in the substantia gelatinosa of the spinal cord (110,307).

ADENOSINE TRIPHOSPHATE OR ADENOSINE AS A COTRANSMITTER AND/OR NEUROMODULATOR

Cotransmitter

The concept that each nerve cell synthesizes and releases only one nerve transmitter (widely known as Dale's Principle) has been reexamined and the possibility raised that some nerves may employ more than one transmitter (42,52).

Adrenergic Nerves

It is well established that ATP is released together with catecholamines from adrenal medullary granules (103,300,317). Evidence is now accumulating that ATP is also released together with NA from adrenergic nerves. Su et al. (305) showed release of radioactive compounds from guinea pig taenia coli preincubated in [^{3}H]adenosine in response to periarterial sympathetic nerve stimulation and further, that this release was blocked by guanethidine. It has been suggested that ATP may be a cotransmitter with NA in sympathetic nerves supplying the guinea pig seminal vesicle (232) and the dog heart (140). Langer and Pinto (213) considered that the substantial residual response of the cat nictitating membrane to sympathetic nerve stimulation, following depletion of their catecholamines with reserpine, might be caused by ATP released as a second transmitter from adrenergic nerves.

^{3}H compounds were released during nerve stimulation of the guinea pig vas deferens mechanically inactivated by hypertonicity with sucrose (329). The authors suggest that this provides evidence that ATP may serve as a cotransmitter with NA and that this may account for some of the current confusion about sympathetic innervation of this organ (8,114,188,295). A similar explanation may also apply to the nonadrenergic component of the response of the retractor penis to sympathetic nerve stimulation (6). An adenosine-like material was shown to be released from isolated perfused dog adipose tissue following sympathetic nerve stimulation (124). This is of particular interest in view of the evidence that adenosine is a potent inhibitor of catecholamine-induced lipolysis (125,315). Evidence has also been presented that cGMP is released together with NA from sympathetic nerves in the rat pineal (338).

Cholinergic Nerves

Adenosine triphosphate is stored together with acetylcholine (ACh) in the synaptic terminals of nerves supplying the electrical organ of torpedine rays (28,105, 183,333,341,342) and ATP has been shown to be released together with ACh from phrenic nerves in the rat diaphragm (293,294). Ewald (115) claims that the ATP released from phrenic nerves facilitates activation of cholinergic receptors as well as having a presynaptic neuromodulatory role.

Zimmerman (340) has shown that stimulation of isolated synaptosomes does not cause a fall in vesicle numbers but an approximately 50% loss of total vesicular nucleotides in parallel with a comparable presynaptic loss of ACh. Further, he showed that if [^{3}H]adenosine is perfused through the tissue, 85% is recovered in the isolated vesicle fraction as ATP. Following stimulation, a new population of vesicles appear that have a high turnover rate for both ATP and ACh. He quotes Dowdall (104) as providing evidence for a saturable uptake system for adenosine into isolated nerve terminals of Torpedo with a K_m value identical with the high affinity choline uptake system. This is taken as the first suggestion that nucleotides released into the cleft have mechanisms of reuptake, resynthesis, and vesicular reincorporation that are about as effective as those for ACh.

Sensory Nerves

The possibility that ATP might be released as a cotransmitter with substance P from sensory nerves during the axonal reflex leading to skin vasodilatation has been considered (47,70*a*).

Neurons in the Brain

Following injections into the cortex or the eye, [^{3}H]adenosine is taken up by certain neurons, transported by axonal flow to the terminals, and released into the synaptic cleft (for references, see section above titled "Storage, Release, and Inactivation of ATP from Purinergic Nerves"). Whether adenine compounds are released in addition to the principal transmitter (268,285,286) or as the principal transmitter (46,209,210,254) is not yet clear.

APUD System

It is postulated that paraneuronal cells in the gastro-entero-pancreatic endocrine system contain ATP along with various polypeptides, amines, and carrier proteins within their secretory granules (126). These cells are part of the APUD (246) or "neurocrine" system, but this raises the question of whether or not ATP may be stored with polypeptides in some enteric neurons.

Neuromodulation

Neuromodulation, defined as the regulation of release of transmitter from nerves by the action of neurohumoral agents on prejunctional receptors, is a relatively recent concept that is rapidly becoming recognized to be of physiological significance (212,296,321,330). Apart from inhibition of transmitter output by negative feedback mediated by NA through prejunctional α-adrenoceptors or by dopamine mediated by dopamine receptors, some inhibition of the activity of adrenergic nerves appears to be caused by the prejunctional actions of ACh, prostaglandin, ATP, and adenosine. Modulation of cholinergic nerve activity by the prejunctional action of catecholamines and adenine nucleotides and nucleosides also occurs. The presynaptic actions of adenine nucleotides and nucleosides on both adrenergic and cholinergic nerve terminals are blocked by methylxanthines, indicating that P_1 purinoceptors are involved and that ATP is degraded to adenosine before acting (49,153).

Modulation of Adrenergic Activity by ATP and Adenosine

Adenine nucleotides and nucleosides have been shown to inhibit NA release from adrenergic nerves. Intraarterial infusion of ATP into isolated canine subcutaneous adipose tissue inhibits NA release in response to sympathetic nerve stimulation (123). Adenosine triphosphate and its metabolites inhibit the release of NA from adrenergic nerves in the rat vas deferens (76,244) and portal vein (113). Purinergic inhibition of adrenergic transmission has also been demonstrated in ear and pulmonary arteries and portal and saphenous veins of the rabbit (304). Adenosine significantly and reversibly depresses, in a concentration-dependent manner, NA release evoked by adrenergic nerve stimulation to the rabbit kidney *(in vitro* and *in situ),* canine subcutaneous adipose tissue, guinea pig vas deferens, and probably to the guinea pig terminal ileum (157,194). In kidney, adenosine causes a concentration-dependent increase in vascular resistance and in the vasoconstrictor responses to both adrenergic nerve stimulation and NA. It was concluded from these and other experiments that adenosine acts by two discrete mechanisms, prejunctional inhibition and postjunctional enhancement, and that in this situation the latter mechanism dominates.

Adenosine triphosphate, ADP, and adenosine inhibit adrenergic vasoconstrictor transmission to isolated cutaneous (saphenous) veins of the dog by reducing NA release as well as producing postjunctional vasodilatation (320). It has been suggested that ATP or a related compound may function in the portal vein as an inhibitory modulator of adrenergic nerves as well as being the vasodilator transmitter in nonadrenergic nerves (302).

Adenosine triphosphate induction of prostaglandin synthesis has been described (231) and the possibility has been raised earlier in this chapter that purinergic nerves are linked with prostaglandin in physiological regulatory mechanisms. Thus it is of some interest that prostaglandins of the E series can also act as neuromodulators, reducing release of NA from adrenergic nerves (156,273,295).

Modulation of Cholinergic Nerve Activity by ATP and Adenosine

There is considerable evidence for a prejunctional site of action for adenosine on cholinergic terminals. Adenosine reduces ACh release from phrenic nerves in the diaphragm (141), from frog motor nerves in sartorius muscle (261), and from excitatory nerves in the intestine (150,154,230,322). Further, theophylline, which blocks adenosine, enhances the release of ACh from both intestine and cortical slices (280,322).

It has been proposed that depression of spontaneous activity of neurons in cerebral and cerebellar cortices by ATP may be largely presynaptic (251). Both in the intestine and in the cerebral cortex, the source of adenine nucleotides involved in prejunctional inhibition of ACh release may be purinergic nerves, or cholinergic nerves which either release ATP together with ACh or lead to postsynaptic release of ATP (182,322).

Prejunctional modulation of ACh output by cyclic nucleotides (108) and the purine nucleotide, coenzyme A, released from cholinergic terminals (86) have also been postulated.

OBJECTIONS RAISED AND ANSWERED

Nonadrenergic, Noncholinergic Nerves

Two main objections were raised when the existence of nonadrenergic, noncholinergic responses of the intestine were first proposed in the early 1960s.

Muscle Responses

An early objection was that the IJPs recorded during transmural stimulation of the taenia coli might be some form of muscle-propagated response (34). However, this objection was withdrawn after the authors had shown that tetrodotoxin blocked the response without affecting smooth muscle excitability (35).

Guanethidine-Resistant Adrenergic Nerves

Another objection that lingered on into the 1970s, was that there were inhibitory responses of the intestine that were actually adrenergic, but were resistant to blockade by adrenergic neuron or receptor blocking drugs (245,327). The demonstration that inhibitory responses persisted after sympathectomy (38) removed any doubts about the existence of nonadrenergic, noncholinergic nerves.

Purinergic Nerves

Low Potency to ATP

Several authors have been worried about the low potency of ATP in some preparations supplied by nonadrenergic, noncholinergic nerves. For example, regarding the rat bladder, recent experiments in our laboratory (31) have suggested that rapid breakdown of ATP (which contracts the bladder) to AMP and adenosine, both of which produce relaxation via P_1 purinoceptors, accounts for the low potency of exogenously applied ATP (7). This is supported by the finding that β-γ-methylene ATP, which is more resistant to degradation to adenosine, is about 100-fold more potent than ATP and precisely mimics the nonadrenergic, nerve-mediated contractions.

Dipyridamole

Low concentrations of dipyridamole potentiate the relaxations of the intestine in response to both purinergic nerve stimulation and ATP, largely because of its action in preventing neuronal and extraneuronal uptake of the breakdown product adenosine (277).

Ambache et al. (7) showed that dipyridamole failed to potentiate the excitatory responses of the guinea pig urinary bladder to either nonadrenergic nerves or ATP. They argued that this was evidence against purinergic motor transmission. However, the bladder in inhibited by adenosine, so that prevention of uptake of adenosine by dipyridamole would reduce the contractile responses (58).

Release of Tritium-Labeled Compounds

Stimulation of the guinea pig taenia preincubated in [^{3}H]adenosine led to release of labeled compounds into the perfusate. However, whereas Su et al. (305) believed that most of the label was released from nerves, Kuchii et al. (205) concluded that the release was from extraneuronal sources. This problem was reexamined recently by Rutherford and Burnstock (271) and it was shown that ATP released from the muscle pool can be clearly distinguished from that released from nerve by applying tetrodotoxin. Incubation of taenia coli with low concentrations of [^{3}H]adenosine leads to preferential uptake into nerves, so that release of label on transmural stimulation is largely blocked by tetrodotoxin, whereas release of label after incubation in high concentrations of [^{3}H]adenosine is largely tetrodotoxin resistant, since it originates mostly from the heavily loaded muscle. Furthermore, there is substantial release of tritium from taenia coli during stimulation of purinergic nerves when no relaxation occurs, the tone of the preparation having been previously lowered with papaverine (B.M. Paddle and G. Burnstock, *unpublished results*). Release of ATP from nonadrenergic, noncholinergic nerves rather than from muscle using the firefly

ATP assay method is described above in the section titled "Storage, Release, and Inactivation of ATP from Purinergic Nerves."

Tachyphylaxis to ATP and Responses to Purinergic Nerves

Although tachyphylaxis to ATP in the rabbit ileum (56) and rabbit bladder (106) was claimed to produce depression of the response to purinergic nerve stimulation, this was not found by Weston (331) following desensitization of the guinea pig ileum, or by Ambache and Zar (8) in the guinea pig urinary bladder. Recent studies in our laboratory (58) have shown that the tone of the guinea pig bladder is increased by induction of prostaglandin synthesis during ATP tachyphylaxis and that this leads to substantial potentiation of the nerve-mediated responses. If the experiment is repeated in the presence of the prostaglandin-synthesis inhibitor indomethacin, there is no increase in tone, and the response to nerve stimulation is virtually abolished following desensitization to ATP. Okwuasaba et al. (240) have also shown that following tachyphylaxis of guinea pig stomach fundic strips to ATP, the nonadrenergic inhibitory nerve responses are abolished, and they regarded this as strong support for purinergic nerve hypothesis.

Lack of Mimicking Action of ATP

Gillespie and his colleagues have shown that the anococcygeus muscle of cat, rabbit, and rat receives, in addition to excitatory nerves, a nonadrenergic, noncholinergic innervation (90,91,134,137–139). Although these workers showed that ATP mimicked the nerve-mediated inhibition of the anococcygeus in the cat and rabbit, they argued that is was unlikely to be the transmitter since it caused a contraction of the rat anococcygeus and the transmitter was likely to be the same in all three species. However, in a recent study in our laboratory (57), it was shown that, in the the presence of low concentrations of the prostaglandin synthesis inhibitor indomethacin, the ATP-induced contraction of the rat anococcygeus was converted to a relaxation. Since ATP induces prostaglandin synthesis (see section above titled "Purinergic Receptors") it was argued that this could explain the anomalous result. It was further shown that release of ATP increased three to six times during stimulation of the nonadrenergic inhibitory nerves and quinacrine fluorescence, which was unaffected by sympathectomy, was localized in ganglion cells and fibers in both rat and rabbit preparations.

FUTURE DEVELOPMENTS

There is still a great deal to be done to establish and extend knowledge of purinergic nerves. Promising areas for advance include (a) quantitative studies of release, storage, and inactivation of transmitter from intact and isolated purinergic nerve terminals and vesicles; (b) affinity labeling of purinergic receptors (see refs. 87,92,136,269); (c) light and electronmicroscopic studies of the distribution of purinergic nerves using autoradiographic, quinacrine, and immunofluorescence methods and the uranaffin reaction (262); (d) the development of specific competitive blocking drugs for both P_1 and P_2 purinoceptors; (e) examination of the role of ATP released as a cotransmitter from adrenergic, cholinergic, and possibly tryptaminergic and peptidergic nerves, and the neuromodulatory activity of its breakdown product adenosine; (f) studies of the physiological roles of purinergic nerves in different organ systems and peripheral ganglia; and (g) investigations into the possible clinical implications of purinergic nerves and development of therapeutic agents in relation to diseases such as achalasia of the esophagus (70,120,121,313), irritable colon, asthma, coughing, vomiting, angina, renal ischemia, shock, skin inflammation, migraine, bladder incontinence, and urine retention.

CONCLUSIONS

It is some 15 years since the beginnings of our work in this field and eight years since the "purinergic nerve hypothesis" was formulated. During this time, two important ideas have become widely accepted: (a) that nonadrenergic, noncholinergic nerves are widely distributed in the autonomic nervous system, and (b) that there are potent extracellular actions of purine nucleotides and nucleosides, many of which appear to be involved in physiological regulatory mechanisms. Most of the criticisms that have been raised about the purinergic nerve hypothesis have been answered and a good deal of new supporting evidence has accumulated, especially in the past two to three years. Apart from its role as a principal transmitter, ATP appears to be a cotransmitter in some adrenergic and cholinergic nerves. Following release, ATP or its breakdown product adenosine may act on pre- and/or postjunctional purinergic receptors. With the rapid development of knowledge about the intraneuronal location of various polypeptides in recent times, some workers have claimed that nonadrenergic, noncholinergic nerves are "peptidergic" rather than "purinergic." Although there is some support for the presence of polypeptides in sensory nerves and interneurons, there is little evidence that they mimic the actions of the transmitter released from nonadrenergic, noncholinergic efferent nerves supplying various visceral and vascular smooth muscles.

ACKNOWLEDGMENTS

Part of this work was supported by grants from the Medical Research Council and the Wellcome Trust.

REFERENCES

1. Abood, L. G., Koketsu, K., and Miyamoto, S. (1962): *A. J. Physiol.*, 202:469–474.
2. Abrahamsson, H. (1973): *Acta Physiol. Scand.*, 89:406–414.
3. Akisaka, T., and Oda, M. (1977): *Arch. Hist. J.*, 40:63–72.
4. Ally, A. I., and Nakatsu, K. (1976): *J. Pharmacol. Exp.*, 199:208–215.

5. Ambache, N. (1951): *Br. J. Pharmacol.,* 6:51–64.
6. Ambache, N., and Killick, S. W. (1978): *Br. J. Pharmacol.,* 63:25–34.
7. Ambache, N., Killick, S. W., and Woodley, J. P. (1977): *Br. J. Pharmacol.,* 61:464.
8. Ambache, N., and Zar, M. A. (1970): *J. Physiol. (Lond.),* 210:761–783.
9. Anuras, S., Christensen, J., and Cooke, A. R. (1977): *Am. J. Physiol.,* 233:E28–E31.
10. Anuras, S., Cooke, A. R., and Christensen, J. (1974): *J. Clin. Invest.,* 54:529–535.
11. Babskii, E. B., and Malkiman, I. (1950): *C. R. Acad. Sci. U.S.S.R.,* 74:1135–1137.
12. Barberis, C., and McIlwain, H. (1976): *J. Neurochem.,* 26:1015–1021.
13. Bass, S. W., Ramuez, M. A., and Aviado, D. M. (1972): *Toxicol. Appl. Pharmacol.,* 21:464–481.
14. Bayliss, W. M., and Starling, E. H. (1900): *J. Physiol. (Lond.),* 26:107–118.
15. Beani, L., Bianchi, C., and Crema, A. (1971): *J. Physiol. (Lond.),* 217:259–279.
16. Beck, C. S., and Osa, T. (1971): *Am. J. Physiol.,* 220:1397–1403.
17. Bennett, A., and Stockley, H. L. (1973): *J. Physiol. (Lond.),* 233:34P–35P.
18. Bennett, A., and Stockley, H. L. (1977): *Br. J. Pharmacol.,* 61:573–578.
19. Bennett, M., Burnstock, G., and Holman, M. E. (1963): *J. Physiol. (Lond.),* 169:33–34P.
20. Bennett, M., Burnstock, G., and Holman, M. E. (1966): *J. Physiol. (Lond.),* 182:541–558.
21. Bennett, T., (1969): *Br. J. Pharmacol.,* 37:585–594.
22. Berne, R. M. (1963): *Am. J. Physiol.,* 204:317–322.
23. Berne, R. M., Rubio, R., and Curnish, R. R. (1974): *Circ. Res.,* 35:262–271.
24. Bianchi, C., Beani, L., Frigo, G. M., and Crema, C. (1968): *Eur. J. Pharmacol.,* 4:51–61.
25. Biber, B., and Fara, J. (1973): *Experientia,* 29:551–552.
26. Blume, A. J., Dalton, C., and Sheppard, H. (1973): *Proc. Natl. Acad. Sci. U.S.A.,* 11:3099–3102.
27. Boettge, K., Jaeger, K. H., and Mittenzwei, H. (1957): *Arzneim. Forsch.,* 7:24 59.
28. Bohan, T. P., Boyne, A. F., Guth, P. S., Narayanan, Y., and Williams, T. H. (1973): *Nature,* 244:32–34.
29. Boyd, I. A., and Forrester, T. (1968): *J. Physiol. (Lond.),* 199:115–135.
30. Brinn, J. E. (1975): *Cell Tissue Res.,* 162:357–365.
31. Brown, C., Burnstock, G., and Cocks, T. (1979): *Br. J. Pharmacol.,* 65:97–102.
32. Buchthal, F., Engbaek, L., Sten-Knudsen, O., and Thomasen, E. (1947): *J. Physiol. (Lond.),* 106:3–4P.
33. Bülbring, E., and Gershon, M. D. (1967): *J. Physiol. (Lond.),* 192:823–846.
34. Bülbring, E., Kuriyama, K., and Tomita, T. (1965): *J. Physiol. (Lond.),* 181:8–10P.
35. Bülbring, E., and Tomita, T. (1967): *J. Physiol. (Lond.),* 189:299–316.
36. Burnstock, G. (1969): *Pharmacol. Rev.,* 21:247–324.
37. Burnstock, G. (1971): *Nature,* 229:282–283.
38. Burnstock, G. (1972): *Pharmacol. Rev.,* 24:509–581.
39. Burnstock, G. (1975): In: *Handbook of Psychopharmacology,* vol. 5, edited by L. L. Iversen, S. D. Iversen, and S. H. Snyder, pp. 131–194. Plenum Press, New York.
40. Burnstock, G. (1975): *J. Exp. Zool.,* 194:103–133.
41. Burnstock, G. (1976): *J. Theor. Biol.,* 62:491–503.
42. Burnstock, G. (1976): *Neuroscience,* 1:239–248.
43. Burnstock, G. (1977): In: *Purine and Pyrimidine Metabolism,* pp. 295–314. Ciba Foundation Symposium No. 48, Elsevier/Excerpta Medica/North Holland, Amsterdam.
44. Burnstock, G. (1977): *Fed. Proc.,* 36:130–137.
45. Burnstock, G. (1977): In: *Comparative Pharmacology of Synaptic Receptors,* pp. 26–33. Nauka Science Publications, Leningrad.
46. Burnstock, G. (1977): In: *Neuroregulators and Psychiatric Disorders,* edited by E. Usdin, D. A. Hamburg, and J. D. Barchas, pp. 470–477. Oxford University Press, New York.
47. Burnstock, G. (1977): *J. Invest. Dermatol.,* 69:47–57.
48. Burnstock, G. (1977): In: Wenner-Gren Center International Symposium series, vol. 30, *Neurogenic Control of Brain Circulation,* edited by C. Owman and L. Edvinsson, pp. 3–14, Pergamon Press, Oxford and New York.
49. Burnstock, G. (1978): In: *Cell Membrane Receptors for Drugs and Hormones: A Multidisciplinary Approach,* edited by L. Bolis and R. W. Straub, pp. 107–118. Raven Press, New York.
50. Burnstock, G. (1978): In: *Mechanisms of Vasodilation,* edited by P. M. Vanhoutte, Karger, Basel.
51. Burnstock, G. (1978): In: *Handbook of Physiology (Vascular Smooth Muscle),* edited by D. Bohr, A. D. Somlyo, and H. V. Sparks. American Physiological Society, Williams &. Wilkins, Baltimore *(in press).*

52. Burnstock, G. (1978): *Prog. Neurobiol.,* 11:205–222.
53. Burnstock, G., Campbell, G., Bennett, M., and Holman, M. E. (1963): *Nature,* 200:581–582.
54. Burnstock, G., Campbell, G., Bennett, M., and Holman, M. E. (1964): *Int. J. Neuropharmacol.,* 3:163–166.
55. Burnstock, G., Campbell, G., and Rand, M. J. (1966): *J. Physiol. (Lond.),* 182:504–526.
56. Burnstock, G., Campbell, G., Satchell, D. G., and Smythe, A. (1970): *Br. J. Pharmacol.,* 40:668–688.
57. Burnstock, G., Cocks, T., and Crowe, R. (1978): *Br. J. Pharmacol.,* 64:13–20.
58. Burnstock, G., Cocks, T., Crowe, R., and Kasakov, L. (1978): *Br. J. Pharmacol.,* 63:125–138.
59. Burnstock, G., Cocks, T., Kasakov, L., and Wong, H. K. (1978): *Eur. J. Pharmacol.,* 49:145–149.
60. Burnstock, G., Cocks, T., Paddle, B. M., and Staszewska-Barczak, J. (1975): *Eur. J. Pharmacol.,* 31:360–362.
61. Burnstock, G., and Costa, M. (1973): *Gastroenterology,* 64:141–144.
62. Burnstock, G., Dumsday, B. H., and Smythe, A. (1972): *Br. J. Pharmacol.,* 44:451–461.
63. Burnstock, G., Satchell, D. G., and Smythe, A. (1972): *Br. J. Pharmacol.,* 46:234–242.
64. Burnstock, G., and Wong, H. (1978): *Br. J. Pharmacol.,* 62:293–302.
65. Burnstock, G., Wong, H. K., and Crowe, R. (1978). *(Submitted for publication.)*
66. Cameron, A. R., and Kirkpatrick, C. T. (1977): *J. Physiol. (Lond.),* 270:733P.
67. Campbell, G. (1966): *J. Physiol. (Lond.),* 185:600–612.
68. Campbell, G. (1970): In: *Smooth Muscle,* edited by E. Bulbring, A. Brading, A. Jones, and T. Tomita, pp. 451–495. Edward Arnold, London.
69. Campbell, G. (1975): *Comp. Biochem.,* 50:169–170.
70. Castell, D. O. (1975): *Ann. Int. Med.,* 83:390–401.
70*a*. Chahl, L. A. (1977): *Eur. J. Pharmacol.,* 44:45–49.
71. Chapman, L. F. (1977): *J. Invest. Dermatol.,* 69:88–97.
72. Chesher, G. B., and James, B. (1966): *J. Pharm. Pharmacol.,* 18:417–423.
73. Christensen, J. (1976): *Clin. Gastroenterol.,* 5:15–27.
74. Christensen, J., Conklin, J. L., and Freeman, B. W. (1973): *Am. J. Physiol.,* 225:1265–1270.
75. Christensen, J., and Torres, E. I. (1975): *Gastroenterology,* 69:641–648.
76. Clanachan, A. S., Johns, A., and Paton, D. M. (1977): *Neuroscience,* 2:597–602.
77. Clark, R. B., Gross, R., Su, Y. F., and Perkins, J. P. (1974): *J. Biol. Chem.,* 249:5296–5303.
78. Clark, P. B., and Seney, M. N. (1976): *J. Biol. Chem.,* 251:4239–4246.
79. Cobbin, L. B., Einstein, R., and Maguire, M. H. (1974): *Br. J. Pharmacol.,* 50:25–33.
80. Coburn, R. F., and Tomita, T. (1973): *Am. J. Physiol.,* 224:1072–1080.
81. Cocks, T., Burnstock, G., and Crowe, R. (1979): *Eur. J. Pharmacol., (in press).*
82. Cohen, S. (1974): *Proc. Soc. Exp. Biol. Med.,* 145:1004–1007.
83. Coleman, R. A. (1976): *Br. J. Pharmacol.,* 57:51–57.
84. Coleman, R. A., and Levy, G. P. (1974): *Br. J. Pharmacol.,* 52:167–174.
85. Conklin, J. L., and Christensen, J. (1975): *Am. J. Physiol.,* 228:1075–1081.
86. Cook, M. A., Hamilton, J. T., and Okwuasaba, F. K. (1978): *Nature,* 271:768–771.
87. Cooperman, B. S., and Brunswick, D. J. (1973): *Biochemistry,* 12:4079–4084.
88. Costa, M., Furness, J. B. (1973): *Rend. Gastroenterol.,* 5:37–38.
89. Costa, M., and Furness, J. B. (1976): *Naunyn Schmiedebergs Arch. Pharmacol.,* 294:47–60.
90. Creed, K. E., and Gillespie, J. S. (1977): *J. Physiol. (Lond.),* 273:137–153.
91. Creed, K. E., Gillespie, J. S., and McCaffery, H. (1977): J. Physiol. (Lond.), 273:121–135.
92. Cusack, N. J. (1977): *Etro. Inserm.,* Paris Meeting on Platelet Membranes, p. 38.
93. Daly, J. (1975): In: *Handbook of Psychopharmacology,* vol. 5, edited by L. L. Iversen, S. D. Iversen, and S. H. Snyder, pp. 47–107. Plenum Press, New York.
94. Daniel, E. E., and Irwin, J. (1965): *Can. J. Physiol. Pharmacol.,* 43:89–136.
95. Davison, J. S., Al-Hassani, M., Crowe, R., and Burnstock, G. (1978): *Pflug. Arch.,* 377:43–49.
96. Davison, J. S., and Fösel, S. (1975): *Digestion,* 13:251–254.
97. Davison, J. S., and Ueda, N. (1977): *J. Physiol. (Lond.),* 266:39P–40P.
98. Day, M. D., and Warren, P. R. (1968): *Br. J. Pharmacol. Chemother.,* 32:227–240.
99. Denhertog, A., and Jager, L. P. (1975): *J. Physiol. (Lond.),* 250:681–691.
100. Diamant, N. E. (1973): *Rend. Gastroenterol.,* 5:26–27.
101. Diamant, N. E., and Elsharka, T. Y. (1977): *Gastroenterology,* 72:546–556.
102. Dimarino, A. J., and Cohen, S. (1975): *Proc. Soc. Exp. Biol. Med.,* 148:1265–1269.
103. Douglas, W. W., and Poisner, A. M. (1966): *J. Physiol. (Lond.),* 183:249–256.

104. Dowdall, M. J. (1978): Quoted in Zimmerman, (ref. 340).
105. Dowdall, M. J., Boyne, A. F., and Whittaker, V. P. (1974): *Biochem. J.*, 140:1–12.
106. Downie, J. W., and Dean, D. M. (1977): *J. Pharmacol. Exp. Ther.*, 203:417–425.
107. Dozi-Vassiliades, J., Tsiamitas, C., and Kokolis, N. (1976): *Prostaglandins*, 12:515–524.
108. Dretchen, K. L., Standaer, F. G., Skirboll, L. R. and Morgenro, V. H. (1976): *Nature*, 264:79–81.
109. Drury, A. N., and Szent-Györgyi, A. (1929): *J. Physiol. (Lond.)*, 68:213–237.
110. Duncan, D., and Morales, R. (1973): *Am. J. Anat.*, 138:139–144.
111. Eliasson, S. (1952): *Acta. Physiol. Scand.*, 26 (Suppl. 95):1–70.
112. Ellison, J. R., and Barr, H. J. (1972): *Chromosoma*, 36:375–390.
113. Enero, M. A., and Saidman, B. Q. (1977): *Naunyn Schmiedebergs Arch. Pharmacol.*, 297:39–46.
114. Euler, U. S. Von, and Hedquist, P. (1975): *Med. Hypotheses*, 1:214–216.
115. Ewald, D. A. (1976): *J. Membr. Biol.*, 29:67–79.
115*a*. Fain, J. N., Pointer, R. H., and Ward, W. F. (1972): *J. Biol. Chem.*, 247:6866–6872.
116. Farmer, J. B., and Farrar, D. G. (1976): *J. Pharm. Pharmacol.*, 28:748–752.
117. Feldberg, W., and Hebb, C. (1948): *J. Physiol. (Lond.)*, 107:210–221.
118. Feldberg, W., and Sherwood, S. L. (1954): *J. Physiol. (Lond.)*, 123:148–167.
119. Fernandes, J. F., and Castellani, O. (1958): *Exp. Parasitol.*, 7:224–235.
120. Fisher, R. S., and Cohen, S. (1975): *Annu. Rev. Med.*, 26:373–390.
121. Fisher, R. S., and Cohen, S. (1976): *Clin. Gastroenterol.*, 5:29–47.
122. Fox, I. H. (1974): *Nutr. Metab.*, 16:65–78.
123. Fredholm, B. B. (1974): *Acta. Physiol. Scand.*, 90:226–236.
124. Fredholm, B. B. (1976): *Acta. Physiol. Scand.*, 96:422–430.
125. Fredholm, B. B. (1978): *Acta. Physiol. Scand.*, 102.191–198.
126. Fujita, T., and Kobayashi, S. (1978): In: *Gut Hormones*, pp. 414–422.
127. Furness, J. B. (1969): *J. Physiol.*, 205:549–562.
128. Furness, J. B. (1969): *Eur. J. Pharmacol.*, 6:349–352.
129. Furness, J. B., and Costa, M. (1973): In: *Recent Developments in Vertebrate Smooth Muscle Physiology*, pp. 123–133. Philos. Trans. Roy. Soc. Lond. [Biol.], 265.
130. Furness, J. B., and Costa, M. (1977): *Clin. Exp. Pharmacol.*, 4:37–41.
131. Galindo, A., Krnfevic, K., and Schwartz, S. (1967): *J. Physiol. (Lond.)*, 192:359–377.
132. Garrett, J. R., and Howard, E. R. (1972): *J. Physiol. (Lond.)*, 222:90P–91P.
133. Gershon, M. D., and Thompson, E. B. (1973): *J. Physiol. (Lond.)*, 234:257–277.
134. Gibson, A., and Gillespie, J. S. (1973): *Br. J. Pharmacol.*, 47:261–267.
135. Gibson, A., and James, T. A. (1977): *Br. J. Pharmacol.*, 60:141–145.
136. Giessner-Pretre, C., and Pullman, B. (1977): *J. Theor. Biol.*, 65:171–201.
137. Gillespie, J. S. (1972): *Br. J. Pharmacol.*, 45:404–416.
138. Gillespie, J. S., and McGrath, J. C. (1974): *Br. J. Pharmacol.*, 50:109–118.
139. Gillespie, J. S., and McKnight, A. T. (1978): *Br. J. Pharmacol.*, 62:267–274.
140. Gillis, R. A., Pearle, D. L., and Hoekman, T. (1974): *Science*, 185:70–72.
141. Ginsborg, B. L., and Hirst, G. D. S. (1971): *Br. J. Pharmacol.*, 43:432–433.
142. Gonella, J., and Gardette, B. (1974): *J. Physiol. (Paris)*, 68:395–413.
143. Gonella, J., Niel, J. P., and Roman, C. (1977): *J. Physiol. (Lond.)*, 273:647–664.
144. Gorgas, K., Bock, P., Tischend, F., and Curri, S. B. (1977): *Anat. Embryol.*, 150:269–289.
145. Goto, M., Yatani, A., and Tsuda, Y. (1977): *Jpn. J. Physiol.*, 27:81–94.
146. Goyal, R. K., and Rattan, S. (1975): *J. Clin. Invest.*, 55:1119–1126.
147. Goyal, R. K., and Rattan, S. (1978): *Gastroenterology*, 74:598–619.
148. Green, H. N., and Stoner, H. B. (1950): In: *Biological Actions of the Adenine Nucleotides*, Lewis, London.
149. Green, R. D. and Stanberry, L. R. (1977): *Biochem. Pharmacol.*, 26:37–43.
150. Gustafsson, L., Hedqvist, P., and Fredholm, B. B. (1977): *Acta. Pharmacol.*, (S4)49.
151. Haddy, F. J., and Scott, J. B. (1968): *Physiol. Rev.*, 48:688–707.
152. Haslam, R. J., and Rossen, G. M. (1975): *Mol. Pharmacol.*, 11:528–544.
153. Hayashi, E., Mori, M., Yamada, S., and Kunimoto, M. (1978): *Eur. J. Pharmacol.*, 48:297–307.
154. Hayashi, E., Yamada, S., and Mori, M. (1976): *Jpn. J. Smooth Muscle. Res.*, 12:257–258.
155. Heazell, M. A. (1977): *Arch. Int. Pharmacol.*, 226:109–117.
156. Hedqvist, P. (1976): In: *Prostaglandins—Physiological, Pharmacological and Pathological Aspects*, edited by S. H. H. Karim, pp. 37–61. Lancaster MTP.

157. Hedqvist, P., and Fredholm, B. B. (1976): *Naunyn Schmiedebergs Arch. Pharmacol.*, 293:217–223.
158. Heller, I. H., and McIlwain, H. (1973): *Brain Res.*, 53:105–116.
159. Henderson, J. F., Battell, M. L. and Zombor, G. (1977): *Cancer Res.*, 37:3434–3441.
160. Henderson, V. E., and Roepke, M. H. (1934): *J. Pharmacol. Exp. Ther.*, 51:97–111.
161. Hervonen, A. (1973): *Acta. Physiol. Scand.*, 88:430–432.
162. Hidaka, T., and Kuriyama, H. (1969): *J. Gen. Physiol.*, 53:471–486.
163. Hirano, H., Saito, T., and Ogawa, K. (1968): *Exp. Cell Res.*, 51:259–267.
164. Hirst, G. D. S., and McKirdy, H. C. (1974): *J. Physiol. (Lond.)*, 238:129–143.
165. Hirst, G. D., and McKirdy, H. C. (1975): *J. Physiol. (Lond.)*, 249:369–385.
166. Hollingsworth, M. (1974): *Br. J. Pharmacol.*, 52:539–547.
167. Holman, M. E., and Weinrich, J. P. (1975): *Pfluegers Arch.*, 360:109–119.
168. Holton, F. A., and Holton, P. (1954): *J. Physiol. (Lond.)*, 126:124–140.
169. Holton, P. (1959): *J. Physiol. (Lond.)*, 145:494–504.
170. Honig, C. R., and Frierson, J. L. (1976): *Am. J. Physiol.*, 230:493–507.
171. Hooper, M., Spedding, M., Sweetman, A. J., and Weetman, D. F. (1974): *Br. J. Pharmacol.*, 50:458P–459P.
172. Hoyes, A. D., Barber, P., and Martin, B. G. H. (1975): *Cell Tissue Res.*, 164:133–144.
173. Hoyes, A. D., Bourne, R., and Martin, B. G. H. (1976): *Br. J. Urol.*, 48:43–53.
174. Hrdina, P. D., Bonaccorsi, A., and Benvenuti, F. (1970): *Eur. J. Pharmacol.*, 12:249–252.
175. Huang, M., and Drummond, G. I. (1976): *Biochem. Pharmacol.*, 25:2713–2719.
176. Hughes, J., and Vane, J. R. (1967): *Br. J. Pharmacol. Chemother.*, 30:46–66.
177. Hughes, J., and Vane, J. R. (1970): *Br. J. Pharmacol.*, 39:476–489.
178. Hunt, S. P., and Kunzle, H. (1976): *Brain Res.*, 112:127–132.
179. Hydén, H. (1947): *Symp. Soc. Exp. Biol.*, 1:152–162.
180. Irvin, J. L. and Irvin, E. M. (1954): *J. Biol. Chem.*, 210:45–56.
181. Iso, T. (1974): *Jpn. J. Pharmacol.*, 24:641–642.
182. Israel, M., Lesbats, B., Manaranc, R., Marsal, J., Mastourf, P., and Meunier, F. M. (1977): *J. Neurochem.*, 28:1259–1267.
183. Israel, M., Lesbats, B., Massal, J., and Meunier, F. M. (1975): *C. R. Acad. Sci. [D](Paris)*, Serie D, 280:905–908.
184. Ito, Y., and Kuriyama, H. (1975): *Jpn. J. Physiol.*, 25:333–344.
185. Iverson, L. L., Mitchell, J. F., and Srinivasan, V. (1971): *J. Physiol. (Lond.)*, 212:519–534.
186. Jahnberg, T. (1977): *Scand. J. Gastroenterol.*, 12(S46):5–32.
187. Jansson, G. (1969): *Acta. Physiol. Scand.*, 75:245–252.
188. Jenkins, D. A., Marshall, I., and Nasmyth, P. A. (1977): *Br. J. Pharmacol.*, 61:649–655.
189. Johns, A., and Paton, D. M. (1977): *Prostaglandins*, 13:245–254.
190. Jule, Y. (1975): *J. Physiol. (Paris)*, 70:5–26.
191. Jule, Y., and Gonella, J. (1972): *J. Physiol. (Paris)*, 64:599–621.
192. Kamikawa, Y., Serizawa, K., and Shimo, Y. (1977): *Eur. J. Pharmacol.*, 45:199–203.
193. Kamikawa, Y., and Shimo, Y. (1976): *J. Pharm. Pharmacol.*, 28:294–297.
194. Kazic, T., and Milosavljevic, D. (1976): *Arch. Int. Pharmacodyn.*, 223, 187–195.
195. Kazic, T. and Milosavljevic, D. (1977): *J. Pharm. Pharmacol.*, 29:542–545.
196. Kiernan, J. A. (1975): *Q. J. Exp. Physiol.*, 60:123–130.
197. King, B. F., McKirdy, H. C., and Wai, S. S. (1977): *J. Physiol. (Lond.)*, 264:P607.
198. King, B. F., and Muir, T. C. (1977): *J. Physiol. (Lond.)*, 273:48–49P.
199. Klaushofer, K., and Pavelka, M. (1975): *Histochemistry*, 43:373–385.
200. Kluge, H., Fischer, H. D., Schwarzenfeld, I. U., Haubenreiser, J., Zahlten, W., and Hartmann, W. (1975): *Acta Biol. Med. Ger.*, 34:1279–1282.
201. Kolassa, N., Stengg, R., and Turnheim, K. (1977): *Can. J. Physiol.*, 55:1039–1044.
202. Kostopoulos, G. K., Limacher, J. J., and Phillis, J. W. (1975): *Brain Res.*, 88:162–165.
203. Kostopoulos, G. K., and Phillis, J. W. (1977): *Exp. Neurol.*, 55:719–724.
204. Kruger, L., and Saporta, S. (1977): *Brain Res.*, 122:132–136.
205. Kuchii, M., Miyahara, J. T., and Shibata, S. (1973): *Br. J. Pharmacol.*, 49:642–650.
206. Kuriyama, H., Osa, T., and Tasaki, H. (1970): *J. Gen. Physiol.*, 55:48–62.
207. Kuriyama, H., Osa, T., and Toida, N. (1967): *J. Physiol., (Lond.)*, 191:239–256.
208. Kuriyama, H., and Tomita, T. (1965): *J. Physiol. (Lond.)*, 178:270–289.
209. Kuroda, Y, and Kobayashi, K. (1975): *Proc. Jpn. Acad.*, 51:495–500.

210. Kuroda, Y., and McIlwain, H. (1974): *J. Neurochem.,* 22:691–699.
211. Kuroda, Y., Saito, M., and Kobayashi, K. (1976): *Brain Res.,* 109:196–201.
212. Langer, S. Z. (1974): *Biochem. Pharmacol.,* 23:1793–1800.
213. Langer, S. Z., and Pinto, J. E. B. (1976): *J. Pharmacol. Exp. Ther.,* 196:697–713.
214. Langley, J. N. (1898): *J. Physiol. (Lond.),* 23:407–414.
215. Langley, J. N., and Anderson, H. K. (1895): *J. Physiol. (Lond.),* 19:122–130.
216. Londos, C., and Preston, M. S. (1977): *J. Biol. Chem.,* 252:5951–5956.
217. Londos, C., and Wolfe, J. (1977): *Proc. Natl. Acad. Sci. U.S.A.,* 74:5482–5486.
218. Lorez, H. P., DaPrada, M., and Launay, J. M. (1977): In: *Fluorescence Microscopy of 5-HT Organelles in Normal and Storage Pool Deficient Blood Platelets (P),* Abstract, 9th Annual Meeting, Unions of Swiss Societies of Experimental Biology, Zurich.
219. Loubatieres-Mariani, M. M., Loubatieres, A. L., Chapal, J., and Valette, G. (1976): *C. R. Soc. Biol. (Paris),* 170:833–836.
220. Magro, A. M. (1977): *Clin. Exp. Immunol.,* 30:160–167.
221. Martinson, J. (1965): *Acta. Physiol. Scand.,* 65 (Suppl. 255):1–23.
222. Martinson, J., and Muren, A. (1963): *Acta. Physiol. Scand.,* 57:309–316.
223. May, W. P. (1904): *J. Physiol. (Lond.),* 31:260–271.
224. McIlwain, H. (1977): *Neuroscience,* 2:357–372.
225. McKenzie, S. G., Frew, R., and Bar, H. P. (1977): *Eur. J. Pharmacol.,* 41:193–203.
226. McSwiney, B. A., and Robson, J. H. (1929): *J. Physiol. (Lond.),* 68:124–131.
227. Mir, S. S., Mason, G. R., and Ormsbee, H. S. (1977): *Gastroenterology,* 73:432–434.
228. Mitchell, G. W., and Wood, J. D. (1976): *Eur. J. Pharmacol.,* 40:63–66.
229. Molyneux, G. S. (1976): *J. Anat.,* 121:420P.
230. Mori, M., Yamada, S., Takamura, S., and Hayashi, E. (1973): *J. Jpn. Pharmacol.,* (Suppl. 124) 23:137.
231. Needleman, P., Minkes, M. S., and Douglas, J. R. (1974): *Circ. Res.,* 34:455–460.
232. Nakanishi, H., and Takeda, H. (1972): *Jpn. J. Pharmacol.,* 22:269–270.
233. Nakanishi, H., and Takeda, H. (1973): *Jpn. J. Pharmacol.,* 23:479–490.
234. Oberg, B. (1976): *Annu. Rev. Physiol.,* 38:537–570.
235. Ohashi, H., Naito, K., Takewaki, I., and Okada, T. (1977): *Jpn. J. Pharmacol.,* 379–387.
236. Ohga, Y., and Daly, J. W. (1977): *Biochim. Biophys. Acta,* 498:46–60.
237. Ohga, A., Nakazato, Y., and Saito, K. (1969): *J. Physiol. Soc. Jpn.,* 31:92–93.
238. Ohkawa, H. (1974): *Bull. Yamaguchi Med. School,* 21:31–45.
239. Okada, Y., and Kuroda, Y. (1975): *Proc. Jpn. Acad.,* 51:491–494.
240. Okwuasaba, F. K., Hamilton, J. T., and Cook, M. A. (1977): *Eur. J. Pharmacol.,* 46:181–198.
241. Olson, L., Alund, M., and Norberg, K. (1976): *Cell Tissue Res.,* 171:407–423.
242. Ovtscharoff, W., and Manolov, S. (1977): *Acta Histochem. (Jena),* 60:329–338.
243. Pahlin, P. E., and Kewenter, J. (1976): *Am. J. Physiol.,* 231:296–305.
244. Paton, D. M., Bar, H. P., Clanachan, A. S., and Lauzon, P. A. (1978): *Neuroscience,* 3:65–70.
245. Paton, W. D. M., and Vane, J. R. (1963): *J. Physiol. (Lond.),* 165:10–46.
246. Pearse, A. G. (1968): *Proc. R. Soc. Lond. [Biol.],* 170:71–80.
247. Pearson, J. D., Carleton, J. S., Hutching, A., and Gordon, J. L. (1978): *Biochem. J.,* 170:265–271.
248. Peck, W. A., Carpenter, J. G., and Schuster, R. J. (1976): *Endocrinology,* 99:901–909.
249. Penit, J., Cantau, B., Huot, J., and Jard, S. (1977): *Proc. Natl. Acad. Sci. U.S.A.,* 74:1575–1579.
250. Persson, C. G. A. (1976): *Br. J. Pharmacol.,* 58:479–482.
251. Phillis, J. W., and Edstrom, J. P. (1976): *Life Sci.,* 19:1041–1053.
252. Phillis, J. W., and Kostopoulos, G. K. (1975): *Life Sci.,* 17:1085–1094.
253. Phillis, J. W., Kostopoulos, G. K., and Limacher, J. J. (1974): *Can. J. Physiol.,* 52:1226–1229.
254. Phillis, J. W., Kostopoulos, G. K., and Limacher, J. J. (1975): *Eur. J. Pharmacol.,* 30:125–129.
255. Powers, S. G., Griffith, O. W., and Meister, A. (1977): *J. Biol. Chem.,* 252:3558–3560.
256. Pull, I., and McIlwain, H. (1972): *Biochem. J.,* 130:975–981.
254. Pull, I., and McIlwain, H. (1973): *Biochem. J.,* 136:893–901.
258. Pull, I., and McIlwain, H. (1974): *Biochem. J.,* 144:37–41.
259. Raberger, G., Schutz, W., and Kraupp, O. (1977): *Arch. Int. Pharmacodyn. Ther.,* 230:140–149.
260. Rattan, S., and Goyal, R. K. (1975): *Gastroenterology,* 69:154–159.
261. Ribeiro, J. A., and Walker, J. (1975): *Br. J. Pharmacol.* 54:213–218.
262. Richards, J. G., and DaPrada, M. (1977): *J. Histochem. Cytochem.,* 25:1322–1336.
263. Richardson, J. B. (1977): *Proc. Int. Union Phys. Sci.,* vol. XII, Paris.

264. Richardson, J., and Beland, J. (1976): *J. Appl. Physiol.*, 41:764–771.
265. Richardson, J. B., and Bouchard, T. (1975): *J. Allergy Clin. Immunol.*, 56:473–480.
266. Rikimaru, A., and Suzuki, T. (1971): Tohoku. *J. Exp. Med.*, 103:303–315.
267. Robinson, P. M., McLean, J. R., and Burnstock, G. (1971): *J. Pharmacol. Exp. Ther.*, 179:149–160.
268. Rose, G., and Schubert, P. (1977): *Brain Res.*, 121:353–357.
269. Rosenblit, P. D., and Levy, D. (1977): *J. Supramol. Struct.*, 127:45–51.
270. Rubio, R., Berne, R. M., Bockman, E. L., and Curnish, R. (1975): *Am. J. Physiol.*, 228:1896–1902.
271. Rutherford, A., and Burnstock, G. (1978): *Eur. J. Pharmacol.*, 48:195–202.
272. Saito, M. (1977): *Biochim. Biophys. Acta*, 498:316–324.
273. Sakato, M., and Shimo, Y. (1976): *Eur. J. Pharmacol.*, 40:209–214.
274. Sanwai, S., and Coupar, I. M. (1976): *Clin. Exp. Pharmacol. Physiol.*, 3:141–145.
275. Satchell, D. G., and Burnstock, G. (1971): *Biochem. Pharmacol.*, 20:1694–1697.
276. Satchell, D., Burnstock, G., and Dann, P. (1973): *Eur. J. Pharmacol.*, 23:264–269.
277. Satchell, D. G., Lynch, A., Bourke, P. M., and Burnstock, G. (1972): *Eur. J. Pharmacol.*, 19:343–350.
278. Satchell, D. G., and Maguire, M. M. (1975): *J. Pharmacol. Exp. Ther.*, 195:540–548.
279. Sattin, A., and Rall, T. W (1970): *Mol. Pharmacol.*, 6:13–23.
280. Sawynok, J., and Jhmandas, K. H. (1976): *J. Pharmacol. Exp. Ther.*, 197:379—390.
281. Schubert, P., and Kreutzberg, G. W. (1974): *Brain Res.*, 76:526–530.
282. Schubert, P., and Kreutzberg, G. W. (1975): In: *Physiology and Pathology of Dendrites: Advances in Neurology*, vol. 12, edited by G. W. Kreutzberg, pp. 255–268.
283. Schubert, P., and Kreutzberg, G. W. (1975): *Brain Res.*, 85:317–319.
284. Schubert, P., and Kreutzberg, G. W. (1976): In: *The Cerebral Vessel Wall*, edited by J. Cervos-Navarro, et al., pp. 207–213. Raven Press, New York.
285. Schubert, P., Lee, K., West, M., Deadwyle, S., and Lynch, G. (1976): *Nature*, 260:541–542.
286. Schubert, P., Rose, G., Lee, K., Lynch, G., and Kreutzberg, G. W. (1977): *Brain Res.*, 134:347–352.
287. Schultz, J., and Daly, J. W. (1973): *J. Biol. Chem.*, 248:853–859.
288. Semba, T., Fujii, K., and Kimura, N. (1964): *Jpn. J. Physiol.*, 14:319–327.
289. Sen, A. K., Sunahara, F. A., Talesnik, J., and Endrenyl, L. (1977): *Br. J. Pharmacol.*, 61:381–393.
290. Shimizu, H. (1973): *Proc. 4th Int. Soc. Neurochem.*, Tokyo, pp. 88–89.
291. Shimizu, H., and Okayama, H. (1973): *J. Neurochem.*, 20:1279–1283.
292. Siggins, G. R., Gruol, D. L., Padjen, A. L., and Forman, D. S. (1977): *Nature*, 270:263–265.
293. Silinsky, E. M. (1975): *J. Physiol. (Lond.)*, 247:145–162.
294. Silinsky, E. M., and Hubbard, J. I. (1973): *Nature*, 243:404–405.
295. Stjarne, L. (1972): *Nature*, 241:190–191.
296. Stjarne, L. (1975): *Med. Biol.*, 52:320–326.
297. Spedding, M. (1977): *Br. J. Pharmacol.*, 155–156P.
298. Spedding, M., and Weetman, D. F. (1976): *Br. J. Pharmacol.*, 57:305–310.
299. Sporrong, B., Chase, L., Owman, C., and Sjöberg, N. O. (1977): *Am. J. Obstet.*, 127:811–817.
300. Stevens, P., Robinson, R. L., Van Dyke, K., and Stitzel, R. (1972): *J. Pharmacol. Exp. Ther.*, 181:463–471.
301. Stormorken, H. (1969): *Scand. J. Haematol. Suppl.*, 9:24 pp.
302. Su, C. (1975): *J. Pharmacol. Exp. Ther.*, 195:159–166.
303. Su, C. (1977): In: *Modes of Vasoconstrictor and Vasodilator Neurotransmission. Blood Vessels (in press)*.
304. Su, C. (1978): *Blood Vessels*, 15:183–189.
305. Su, C., Bevan, J. A., and Burnstock, G. (1971): *Science*, 173:337–339.
306. Sulakhe, P. V, and Phillis, J. W. (1975): *Life Sci.*, 17:551–555.
307. Suran, A. A. (1974): *J. Histochem. Cytochem.*, 22:302–311.
308. Suzuki, S., and Nakahara, A. (1975): *Bioorganic Chem.*, 4:250–258.
309. Syromiatnikov, A. V., Groisman, S. D., and Skok, V. I. (1977): In: *Physiology and Pharmacology of Smooth Muscle*, edited by M. Papasova and E. Atanassova, pp. 108–112.
310. Tafuri, L., and Maria, A. (1970): *Rev. Inst. Med. Trop. Sao Paulo*, 12:298–309.
311. Theobald, R. J., and Degroat, W. C. (1978): *J. Am. Osteop.*, 77:396P.
312. Tomita, T., and Watanabe, H. (1973): *J. Physiol. (Lond.)*, 231:167–177.

313. Tournut, R. (1976): *Arch. Fr. Mal. App. Dig.,* 65:409–420.
314. Tuch, A., and Cohen, S. (1973): *J. Clin. Invest.,* 52:14–20.
315. Turpin, P., Duckworth, W. C., and Solomon, S. S. (1977): *J. Clin. Invest.,* 60:442–448.
316. Urquilla, P. R., Vandyke, K., and Trush, M. (1978): *J. Pharm. Pharmacol.,* 30:189–190.
317. Vandyke, K., Robinson, R., Urquilla, P., Smith, D., Taylor, M., Trush, M., and Wilson, M. (1977): *Pharmacology,* 15:377–391.
318. Vannuete, J. M., Fountaine, J., Helsen, L., and Janssen, P. A. J. (1977): *Arch. Int. Pharmacodyn. Ther.,* 227:168–170.
319. Veach, H. O. (1925): *Am. J. Physiol.,* 71:229–264.
320. Verhaeghe, R. H., Vanhoutte, P. M., and Shepherd, J. T. (1977): *Circ. Res.,* 40:208–215.
321. Vizi, E. S. (1976): In: *Physiology of Smooth Muscle,* edited by E. Bulbring and M. F. Shuba, pp. 357–367. Raven Press, New York.
322. Vizi, E. S., and Knoll, J. (1976): *Neuroscience,* 1:391–398.
323. Vladimirova, A. I., and Shuba, M. F. (1978): *Neurophysiologia,* 10:295–299.
324. Weetman, D. F., and Turner, N. (1977): *Arch. Int. Pharmacodyn. Ther.,* 228:10–14.
325. Weinryb, I., and Miclel, I. M. (1974): *Biochim. Biophys. Acta,* 334:218–225.
326. Weisbrodt, N. W., and Christensen, J. (1972): *Gastroenterology,* 62:1159–1166.
327. Weisenthal, L. M., Hug, C. C., Weisbrodt, N. W., and Brass, P. (1971): *J. Pharmacol. Exp. Ther.,* 178:497–508.
328. Wendelarr, S. E., Greven, J. A., and Keenhuis, M. (1977): *J. Morphol.,* 153:225–243.
329. Westfall, D. P., Stitzel, R. E., and Rowe, J. N. (1978): *Eur. J. Pharmacol. (in press).*
330. Westfall, T. C. (1977): *Physiol. Rev.,* 57:659–728.
331. Weston, A. H. (1973): *Br. J. Pharmacol.,* 47:606–608.
332. White, T. D. (1978): *J. Neurochem.,* 30:329–336.
333. Whittaker, V. P., Dowdall, M. J., and Boyne, A. F. (1972): *Biochem. Soc. Symp.,* 36:49–68.
334. Wise, S. P., and Jones, E. G. (1976): *Brain Res.,* 107:127–131.
335. Wise, S. P., Jones, E. G., and Berman, N. (1978): *Brain Res.,* 139:197–217.
336. Wolff, J., and Cook, G. H. (1977): *J. Biol. Chem.,* 252:687–693.
337. Wood, J. D., Rose, B. A., and Jackson, M. H. (1976): *J. Pharmacol. Exp. Ther.,* 196:71–79.
338. Zatz, M., and O'Dea, R. (1977): *Science,* 197:174–176.
339. Zenser, T. V. (1976): *Proc. Soc. Exp. Biol. Med.,* 152:126–129.
340. Zimmerman, B. G. (1978): *Neuroscience (in press).*
341. Zimmerman, H., and Denston, C. R. (1976): *Brain Res.,* III:365–376.
342. Zimmerman, H., and Whittaker, V. P. (1974): *J. Neurochem.,* 22:435–450.

Physiological and Regulatory Functions of Adenosine and Adenine Nucleotides, edited by H. P. Baer and G. I. Drummond. Raven Press, New York © 1979.

Specificity of Adenine Nucleotide Receptor Sites: Inhibition of the Guinea Pig Taenia Coli by Adenine Nucleotide Analogs

*M. Helen Maguire and **D. G. Satchell

**Department of Pharmacology and Ralph L. Smith Mental Retardation Research Center, University of Kansas Medical Center, Kansas City, Kansas 66103, and **Department of Zoology, University of Melbourne, Parkville, Victoria 3052, Australia*

Adenosine and its 5′-mono- (AMP), di- (ADP), and triphosphates (ATP) are known to have potent inhibitory effects on mammalian gut (4,8,9,13). Recently, ATP and ADP were shown to be equipotent in inhibiting the guinea pig taenia coli preparation, acting at a threshold of 0.1 μM, and to be 100 times more potent than AMP and adenosine, which were of similar potency (3,6). The inhibitory response elicited by ATP on the taenia coli rapidly reaches a maximum and regains tone on or before washout. The transient nature of the action of ATP has been explained in terms of its rapid degradation to the less potent adenosine and the concomitant inactivation of adenosine by tissue uptake and deamination (5,28).

We have studied the effects of certain adenine nucleotide analogs on the isolated guinea pig taenia coli in order to elucidate structural requirements for their action at the ATP receptor site, to determine if adenosine acts at the same site as ATP, and to establish whether or not metabolism of ATP to adenosine is a component of its inhibitory response. The analogs include 2-chloro– and 2-alkylthio–substituted derivatives (16) and phosphate chain–modified nucleotides, and a combination of both. The 2-chloro– and 2-alkylthio–substituted adenosine derivatives have previously been used to demonstrate the different selectivity of the adenosine receptor systems associated with negative inotropic action and coronary dilation in the heart (1,10,22). Further, the 2-chloro and the 2-alkylthio substituents greatly enhanced the platelet aggregating action of ADP (15,21), while differential effects obtained with 2-substituted derivatives of adenosine and AMP as inhibitors of ADP-induced platelet aggregation (14,17,21,23) suggested different sites of action for adenosine and AMP on platelets (20).

The following modified nucleotides were used in this study: adenosine 5′-α,β-methylenetetraphosphate (AOPCPOPOP); adenosine 5′-α,β-methylenetriphosphate (AOPCPOP) (see Fig. 1); 5′-adenylyl methylenediphosphonate (AOPOPCP); adenosine 5′-methylenediphosphonate (AOPCP); 6′-deoxyhomoadenosine 6′-phosphonyldiphosphate (ACPOPOP); 6′-deoxyhomoadenosine 6′-phosphonate (ACP); 5′-adenylylimidodiphosphate (AOPOPNP); 2-chloro-5′-adenylylmethylenediphosphonate (2-chloro-AOPOPCP); 2-chloro-adenosine 5′-methylenediphosphonate (2-

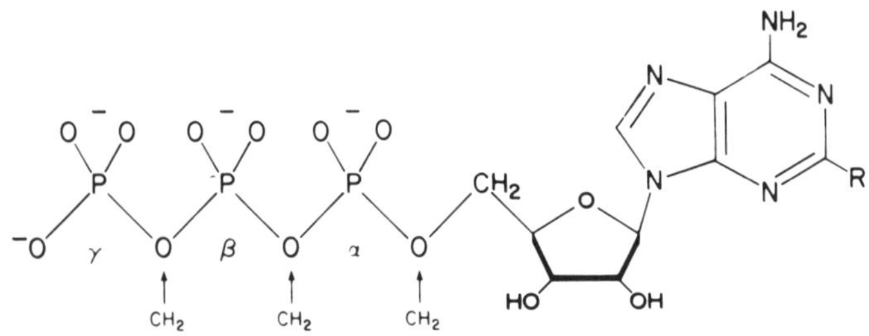

FIG. 1. Structure of ATP and 2-substituted ATP derivatives with the positions of replacement of oxygen atoms by methylene groups in the triphosphate-modified ATP analogs indicated. R = H, Cl, CH_3S.

chloro-AOPCP); 2-chloro-adenosine 5′-methylenediphosphate (2-chloro-AOPCP); 2-methylthio-5′-adenyl methylenediphosphonate (2-methylthio-AOPOPCP). Further, arabinofuranosyl-adenine triphosphate (ara-ATP) and 8-bromo-ATP were included, as similar structural modifications to adenosine resulted in abolition of its coronary vasodilator activity (7; and M. H. Maguire, *unpublished observations*).

METHODOLOGY

The effects of the analogs were investigated using guinea pig taenia coli preparations suspended in a modified Krebs solution containing hyoscine with isometric recording. Concentration-effect curves to ATP were obtained on each preparation. Adenosine triphosphate was applied in the organ bath for 30 sec before overflow washout; analogs were generally applied for 60 sec. Further experimental details have been published in the references cited below.

SOURCES OF ANALOGS

The methylene isosteres, AOPCP, AOPCPOP, AOPOPCP, and AOPCPOPOP were obtained from PL Biochemicals, Milwaukee, Wisc., who also supplied ara-ATP, 8-bromo-ATP, and AOPOPNP. Adenosine tetraphosphate (AtetraP) was obtained from Sigma Chemical Co. 6′-Deoxyhomoadenosine 6′-phosphonyldiphosphate and ACP were gifts from Dr. J. G. Moffatt, Syntex Research, Palo Alto, Calif. All other analogs were synthesized as reported (15–17). None of the analogs were contaminated with adenosine, AMP, ADP, or ATP as indicated by the synthetic routes used or by paper chromatography (15).

2-SUBSTITUTED ANALOGS OF ATP, ADP, AMP, AND ADENOSINE

The concentration-effect curves for unmodified ATP, ADP, AMP, and adenosine are shown in Fig. 2. Adenosine triphosphate and ADP gave identical curves as did AMP and adenosine, but the curves for AMP and adenosine were displaced

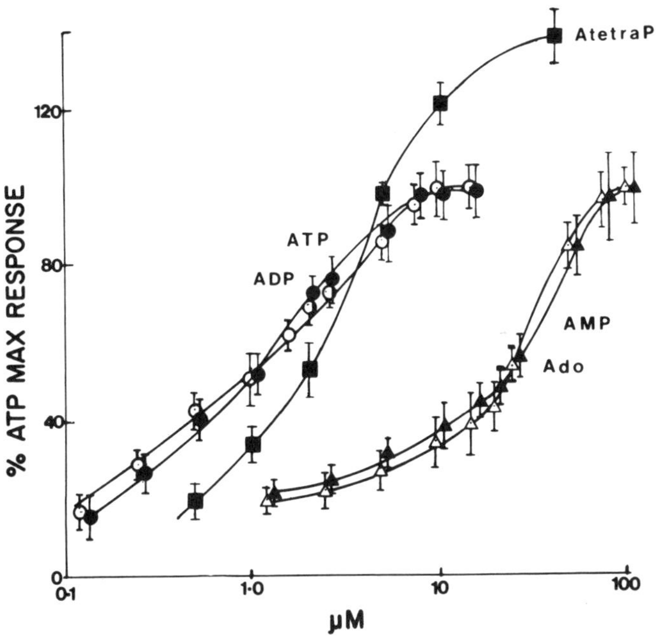

FIG. 2. Concentration-effect curves for ATP, ADP, AtetraP, AMP, and adenosine (Ado) in causing the relaxation of the isolated guinea pig taenia coli preparation. The ordinate gives the amplitude of relaxation expressed as a percentage of the maximum response to ATP. Each experimental point is the mean ± SEM of at least four values obtained from at least two preparations.

considerably to the right of those for the ATP-ADP pair and were not parallel to those of ATP and ADP. Responses to adenosine and AMP reached the same maximal relaxation as responses to ATP or ADP but required almost three times longer for maximal response (Table 1).

The 2-chloro–substituted analogs of ATP and ADP also relaxed the taenia and showed concentration-effect curves parallel to that of ATP, with similar maximum responses and times to maximum relaxation (Fig. 3, Table 1), but with a higher potency compared to ATP or ADP. 2-Chloroadenosine monophosphate and 2-chloroadenosine also elicited maximum responses similar to ATP but their dose-effect curves were steeper. 2-Chloroadenosine monophosphate had greater relative activity than 2-chloroadenosine, but both took three times as long as ATP to reach maximum relaxation (Table 1).

2-Methylthioadenosine triphosphate and 2-methylthio-ADP also inhibited the taenia coli but were 50 and 30 times more potent than ATP, respectively. The diphosphate was significantly less potent than the triphosphate analog, but both elicited the same maximum response as ATP and took somewhat longer than ATP to reach maximum relaxation (Fig. 4, Table 1). The pD_2 values, the relative activities, and the maximum responses of the 2-methylthio–substituted derivatives of AMP and adenosine show that these analogs caused inhibition of the taenia coli only at

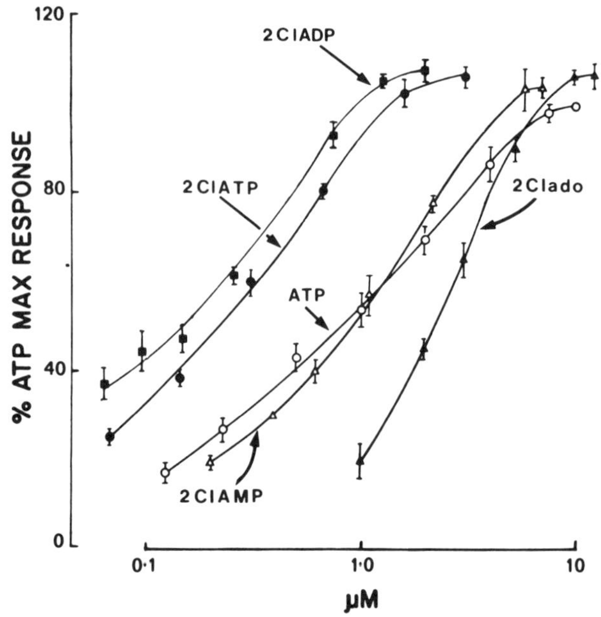

FIG. 3. Concentration-effect curves for the relaxation of the isolated guinea pig taenia coli preparation by 2-chloro-ATP (2ClATP), 2-chloro-ADP (2ClADP), ATP, 2-chloro-AMP (2ClAMP), and 2-chloroadenosine (2Clado). Details of the ordinate and experimental points are the same as in the legend to Fig. 2.

high concentrations and that they behaved as partial agonists (Fig. 4, Table 1). Responses to these derivatives took three times as long as the ATP response to reach maximum relaxation, thus resembling the responses to AMP and adenosine and their 2-chloro derivatives.

The differences in the characteristics of the inhibitory responses induced by the 2-substituted adenosine and AMP analogs and those elicited by their di- and triphosphate derivatives accord with the differences observed between the responses to ATP and ADP on the one hand and adenosine and AMP on the other and suggest that different receptor populations are involved—one receptor for ATP and ADP, and one receptor for adenosine and AMP. Moreover, no reduction of the ATP response could be achieved by the partial agonists 2-methylthioadenosine and 2-methylthio-AMP at concentrations up to 100 μM, substantiating the existence of separate sites for ATP and adenosine.

PHOSPHATE-MODIFIED ADENINE NUCLEOTIDE ANALOGS

The methylene isosteres of the adenosine polyphosphates and AtetraP, and AtetraP itself, all caused a profound degree of relaxation of the guinea pig taenia coli preparation, with maximum responses 22 to 50% greater than that to ATP (Table 1). Concentration-effect curves for these derivatives were characteristically

TABLE 1. *Inhibitory effects of adenine nucleotide analogs, adenosine analogs, adenine nucleotides, and adenosine on the isolated guinea pig taenia coli*

Compound	pD_2	Relative activity[a]	Maximum response (%)	Time to maximum[b] (sec)
ATP	6.12 ± 0.06	1.0	100 ± 11	11 ± 0.5
ADP	6.09 ± 0.08	0.93	99 ± 14	12 ± 0.08
AMP	4.66 ± 0.07[c]	0.034	100 ± 10	28 ± 2.1[c]
Adenosine	4.68 ± 0.10[c]	0.036	99 ± 18	30 ± 2.1[c]
2-Chloro-ATP	6.62 ± 0.04[c]	3.1	106 ± 5	11 ± 1.7
2-Chloro-ADP	6.74 ± 0.15[c,d]	4.0	108 ± 6	11 ± 1.4
2-Chloro-AMP	6.02 ± 0.04[d]	0.79	104 ± 5	33 ± 4.2[c]
2-Chloroadenosine	5.62 ± 0.11[c,d]	0.31	107 ± 6	39 ± 2.9[c,d]
2-Methylthio-AMP	7.82 ± 0.09[c]	50	106 ± 6	14 ± 1.1[c]
2-Methylthio-ADP	7.60 ± 0.07[c,d]	30	100 ± 4	14 ± 1.3[c]
2-Methylthio-AMP	4.33 ± 0.11[c,d]	0.016	66 ± 5	31 ± 2.4[c]
2-Methylthioadenosine	4.41 ± 0.16[c]	0.019	53 ± 5	33 ± 4.8[c]
AtetraP	5.46 ± 0.02	0.22	139 ± 9	10 ± 1.2
AOPCPOPOP	5.28 ± 0.13[c]	0.29	143 ± 7[c]	28 ± 2.3[c]
AOPCPOP	5.60 ± 0.09	0.30	133 ± 5[c]	21 ± 1.9[c]
AOPCP	5.98 ± 0.08	0.72	150 ± 12[c]	53 ± 4.2[c]
AOPOPCP	4.58 ± 0.07[c]	0.04	122 ± 11[c]	15 ± 1.9[c]
ACPOPOP	7.80 ± 0.24[c]	70.8	124 ± 2.5[c]	14 ± 1.6[c]
ACP	—	Inactive	—	—
2-Chloro-AOPCP	6.72 ± 0.06[c,d]	3.99	132 ± 9[c]	18 ± 1.2[c,d]
2-Chloro-AOPOPCP	6.92 ± 0.07[c,d]	6.3	138 ± 13[c]	10 ± 1.0[d]

[a] Antilog of the difference between the pD_2 of each compound and that of ATP.
[b] Measured at pD_2 concentrations.
[c] $p < 0.05$ against ATP.
[d] $p < 0.05$ against the compound without substitution on the purine ring.

steeper than that for ATP (Figs. 2 and 5), and the relative activities of the compounds that were not ring substituted ranged from 0.04 for AOPOPCP to 70 for ACPOPOP. 6′-Deoxyhomoadenosine 6′-phosphonate was inactive in doses up to 100 μM. 2-Chloro–substitution of AOPOPCP and AOPCP shifted their dose-effect curves to the left, increasing their relative activities 158-fold and fivefold, respectively, but did not alter the slopes of the dose-effect curves or the maximum responses. The phosphate-modified nucleotides varied in the times taken to reach maximum relaxation, e.g., AtetraP and 2-chloro-AOPOPCP were as quick to effect relaxation as ATP, whereas ACPOPOP and AOPOPCP took slightly longer, and AOPCPOP, AOPCPOPOP, and 2-chloro-AOPCP took two to three times as long as ATP. Adenosine 5′-methylenediphosphonate was particularly slow to achieve maximum relaxation, and 2-chloro–substitution resulted in a threefold reduction in this parameter.

The similar slopes of the dose-response curves for the phosphate-modified nucleotides (Figs. 2 and 5) indicate that they act at the same site, but the lack of parallelism between their slopes and that of ATP suggests that these analogs act at a

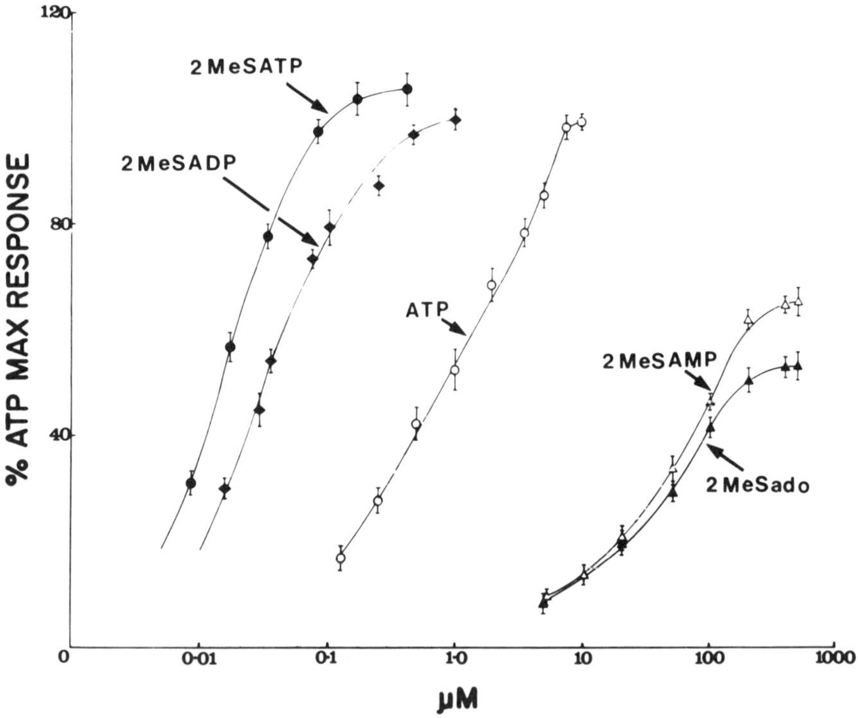

FIG. 4. Concentration-effect curves for the relaxation of the isolated guinea pig taenia coli preparation by 2-methylthio-ATP (2MeSATP), 2-methylthio-ADP (2MeSADP), ATP, 2-methylthio-AMP (2MeSAMP), and 2-methylthioadenosine (2MeSado). Details of the ordinate and experimental points are the same as in the legend to Fig. 2.

different site from that of ATP and that, *ipso facto,* a second inhibitory nucleotide site exists on the taenia coli smooth muscle. However, the possibility that the phosphate-modified analogs act at the ATP site but elicit an altered response as a result of their greater metabolic stability should be considered. Moreover, the structural modifications incorporated in the polyphosphate chains of these analogs can be expected to result in differences in their physical and chemical properties as well as in their susceptibility to enzyme action. These differences may contribute to the steep dose-effect relationships characteristic of these compounds and are considered in turn.

Since the methylene isosteres are aliphatic phosphonic acid derivatives they will be weaker acids than the analogous phosphates (12). The secondary phosphonyl dissociation contants of AOPOPCP and AOPCP, pK_as 8.4 to 8.5 and 8.0, respectively, are higher than the secondary phosphoryl dissociation constants of ATP and ADP, pK_as 7.1 and 7.0, respectively (24,25). Consequently at the pH of the Krebs medium, 7.4, the proportion of ionized to unionized species of these analogs and their ring-substituted derivatives will differ from those obtained with ATP, ADP, and their ring-substituted derivatives. Adenosine tetraphosphate and AOPCPOPOP

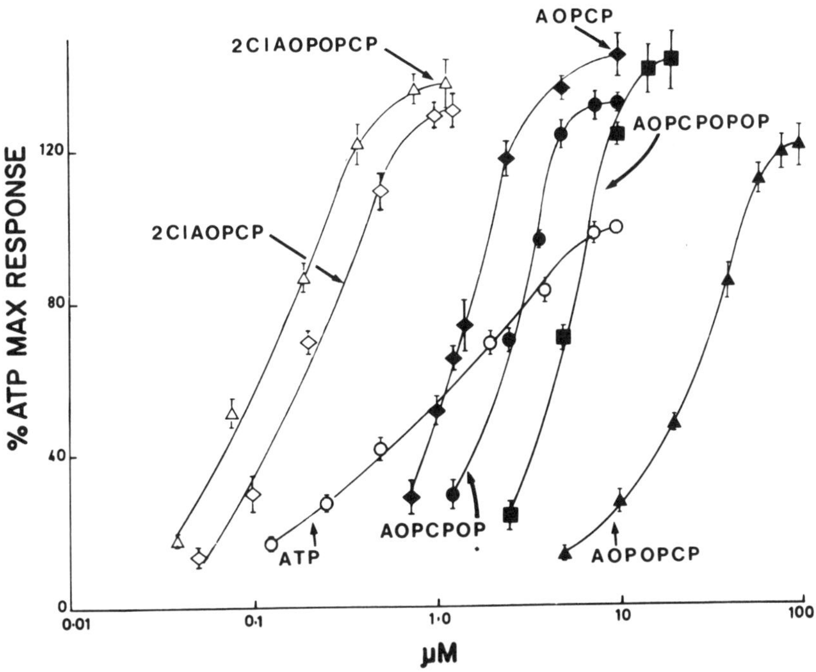

FIG. 5. Concentration-effect curves for the relaxation of the isolated guinea pig taenia coli preparation by 2-chloro-AOPOPCP (2ClAOPOPCP), 2-chloro-AOPCP (2ClAOPCP), ATP, AOPCP, AOPCPOP, AOPCPOPOP, and AOPOPCP. Details of the ordinate and the experimental points are the same as in the legend to Fig. 2.

each have an extra ionizing phosphate group to contribute an additional charge at pH 7.4.

The space-filling characteristics of the modified polyphosphate chains differ from those of the simple di- and triphosphates. Adenosine tetraphosphate and AOPCPOPOP have longer side chains than ATP. Some steric distortion is associated with the methylene isosteric analogs because the phosphorous-carbon-phosphorous bond angle is larger than the anhydride bond angle and the length of the phosphorous-carbon bond is greater than that of the phosphorous-oxygen link (19). The different steric and charge characteristics of the modified phosphate chains in these derivatives may directly influence their binding at the site at which ATP acts.

The complexation of calcium ions by AOPOPCP is greater than that by ATP, the apparent binding constants of the two nucleotides at pH 7.4 being 4,800 and 1,800 M^{-1}, respectively (32), and as the stability of the complexes formed by binding of Ca^{2+} to the phosphate groups of nucleotides increases with increase in the number of phosphate residues in the polyphosphate chain (18,30), AtetraP and AOPCPOPOP may be expected to bind Ca^{2+} better than ATP. If complexation of a membrane-associated Ca^{2+} pool is involved in the inhibitory actions of ATP and

ADP on the taenia coli the enhanced Ca^{2+}-binding capacities of the altered polyphosphate chains may contribute to their distinctive dose-response relationships.

The carbon-phosphorous bond is stable and any possible enzyme-catalyzed breakdown of the methylene isosteres can only proceed via cleavage of phosphorous-oxygen links. Accordingly ACPOPOP could give rise to ACP but not to adenosine, and the α,β-methylene isosteres AOPCPOP and AOPCP could not yield AMP, but could give rise to adenosine as a result of hydrolytic cleavage at the ester link. Degradation of the β,γ-methylene isosteres could likewise occur by splitting of the α,β-anhydride linkage, giving AMP and, subsequently, adenosine. Adenosine tetraphosphate has only anhydride bonds but even so was resistant to the action of ATPase in homogenates of rat tissues (31). Thus, although certain routes of metabolic degradation of the phosphate-modified analogs are feasible, it is likely that they will not be metabolized as readily by the intact taenia coli as ATP and ADP.

EFFECTS OF DIPYRIDAMOLE ON THE INHIBITORY RESPONSES TO NUCLEOTIDE ANALOGS

Satchell and Burnstock (27) have proposed that the inhibitory potencies of ATP and ADP are influenced by their enzymatic breakdown. It appears possible that the steep concentration-effect curves and greater maximum effects of the phosphate-modified nucleotides may be a consequence of their resistance to rapid catabolism. The inhibitory responses elicited by ACPOPOP, AOPCP, and AOPCPOP were not changed in the presence of dipyridamole, an adenosine uptake blocker. However, the concentration-effect curve of AOPOPCP was shifted to the left without any alteration in slope and with a sixfold increase in relative activity and a 16% increase in maximum relaxation. The potentiation of ATP was associated with a fivefold increase in time to maximum relaxation, whereas no such change was observed in the potentiation of AOPOPCP.

5′-Adenylyl methylenediphosphonate, with a relative activity of 0.04, was the least active of the methylene isosteres. Demonstration of a sixfold potentiation of AOPOPCP action by dipyridamole strongly suggests that it is degraded to adenosine. Conversely, as AOPOPCP has the same steep log dose-effect curves as the other phosphate-modified analogs that were not potentiated by dipyridamole, resistance to metabolic degradation would not appear to be the basis of the distinctive dose-effect relationships of these derivatives. Degradation of AOPOPCP is not likely to occur via hydrolytic cleavage at the ester linkage since AOPCPOP was not potentiated in the presence of dipyridamole. The action of a nucleoside triphosphate pyrophosphohydrolase could yield AMP, which would be dephosphorylated by 5′-nucleotidase to give adenosine. A Ca^{2+}-dependent ATP pyrophosphohydrolase has recently been demonstrated in rat liver plasma membrane (11), but not yet in the taenia coli smooth muscle.

OTHER ATP ANALOGS

Preliminary studies with the phosphate-modified derivatives AOPOPNP and 2-methylthio-AOPOPCP showed them to be weak inhibitors of the taenia coli with relative activities of 0.06 and 0.008, respectively. Both analogs had steeper concentration-effect curves than ATP and induced supramaximal relaxation. The maximum response of 2-methylthio-AOPOPCP was particularly noteworthy, since at high concentrations of 1 to 2 mM it caused twice the relaxation evoked by 10 μM ATP (Fig. 6). Arabinofuranosyl adenine triphosphate induced a transient inhibitory response, similar to that of ATP, with a relative activity of 0.8. 8-Bromo-adenosine triphosphate also relaxed the muscle and had a relative activity of 0.07. The activity of ara-ATP contrasts with the fact that ara-adenosine adenine is inactive as a coronary dilator (7).

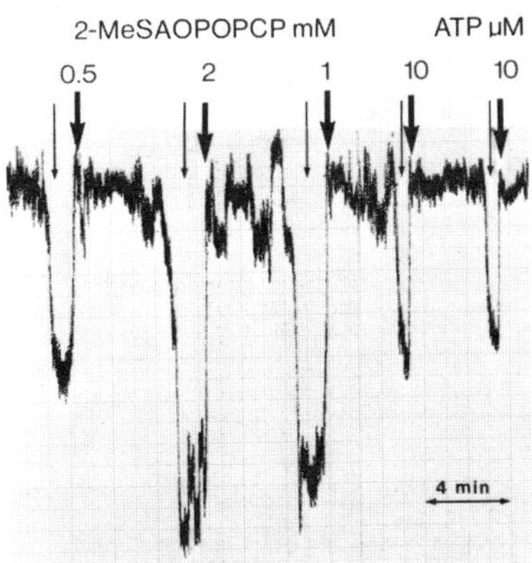

FIG. 6. Isometric recording of the relaxation of the isolated guinea pig taenia coli preparation in response to 10 μM ATP and 0.5, 1, and 2 mM 2-methylthio-AOPOPCP (2MeSAOPOPCP). Nucleotides were introduced as indicated by the heavy arrows. Washout, indicated by the light arrow, was at 30 sec for ATP and 60 sec for 2-methylthio-AOPOPCP.

CONCLUSIONS

Alterations in the purine ring, sugar moiety, and triphosphate chain of ATP resulted in modification, but not abolition, of inhibitory activity. Substantial loss of relative activity was observed on 8-substitution of the purine ring, whereas only a modest reduction in activity resulted from epimerization of the 2′-hydroxyl and some loss in activity was observed in derivatives with altered triphosphate functions. On the other hand, several structural alterations greatly enhanced the inhibitory response evoked by ATP, as in ACPOPOP and the 2-chloro and 2-methylthio derivatives of ATP and AOPOPCP. Substitution of the chloro and methylthio residues in the 2 position of ADP and ATP increased the agonist activity of the parent nucleotides without affecting other characteristics of their inhibitory responses. Similar substitution in AMP and adenosine resulted in decreases in relative activity in the 2-methylthio derivatives, and increases in the 2-chloro derivatives, although the nature of the inhibitory responses remained like those of the parent nucleotide and nucleoside. The difference in log concentration-effect curves between ATP, 2-methylthio-ATP, and 2-chloro-ATP on the one hand, and adenosine, 2-chloroadenosine, and 2-methylthioadenosine on the other, are interpreted to indicate that a separate receptor for adenosine exists on the taenia coli smooth muscle, with binding requirements differing from those of the ATP receptor. Adenosine monophosphate and its 2-substituted derivatives may act as such at the postulated adenosine receptor, or may act after dephosphorylation by 5′-nucleotidase.

Optimal inhibitory activity was associated only with those adenine nucleotide derivatives having a di- or triphosphate side chain, or isosteric equivalent, observations that indicate the important contribution of the ionized side chain to binding to the ATP receptor site.

These studies have shown that metabolic degradation of ATP to adenosine is not necessary per se for eliciting an inhibitory response. 6′-Deoxyhomoadenosine 6′-phosphonyldiphosphate was the most active analog investigated, and its metabolism could produce inactive ACP. Potentiation effects observed with dipyridamole suggest that the concomitant action of adenosine may be part of the inhibitory response to ATP and those nucleotides that may be metabolized to yield adenosine, e.g., ADP and AOPOPCP, although the slow time to maximum response typical of adenosine is not in evidence in the fast development of maximum relaxation characteristic of these nucleotides.

A rapid achievement of maximum relaxation was associated with all the nucleotide derivatives that had triphosphate side chains or equivalents, with the exception of the α,β-methylene isosteres. The significance of the differences in rapidity of development of maximum relaxation is not presently apparent, but it is not related to ease of degradation of the nucleotide to adenosine, since ACPOPOP was almost as rapid in time to maximum as ATP.

No explanation can be given at the present time for the distinctive dose-effect curves and supramaximal responses of the polyphosphate-modified adenine nucleotides. Although the difference between the slopes of their dose-effect curves and that of ATP suggests that the analogs act at a separate specific nucleotide receptor site, this point remains unresolved. In measuring the relaxation of the guinea pig taenia coli induced by the adenine nucleotides and their analogs, we have measured a response that is the end result of drug binding, receptor activation, and subsequent permeability changes, depolarization, and ion flux (26). Definition of the site of action of these analogs may have to await measurement of an intermediate event, such as Ca^{2+} flux, and its relationship to analog concentration.

REFERENCES

1. Angus, J., Cobbin, L. B., Einstein, R., and Maguire, M. H. (1971): *Br. J. Pharmacol.*, 41:592–599.
2. Ariens, E. J., and Van Rossum, J. M. (1957): *Arch. Int. Pharmacodyn. Ther.*, 110:275–299.
3. Axelsson, J., and Holmberg, B. (1969): *Acta Physiol. Scand.*, 75:149–156.
4. Bärsoum, G. S., and Gaddum, J. H. (1935): *J. Physiol. (Lond.)*, 85:1–14.
5. Burnstock, G. (1972): *Pharmacol. Rev.*, 24:509–581.

6. Burnstock, G., Campbell, G., Satchell, D. G., and Smythe, A. (1970): *Br. J. Pharmacol.*, 40:668–688.
7. Cobbin, L. B., Einstein, R., and Maguire, M. H. (1974): *Br. J. Pharmacol.*, 50:25–33.
8. Drury, A. N. (1936): *Physiol. Rev.*, 16:292–325.
9. Drury, A. N., and Szeng-Gyorgi, A. (1929): *J. Physiol. (Lond.)*, 68:213–237.
10. Einstein, R., Angus, J. A., Cobbin, L. B., and Maguire, M. H. (1972): *Eur. J. Pharmacol.*, 19:246–250.
11. Flodgaard, H., and Torp-Pedersen, C. (1978): *Biochem. J.*, 171:817–820.
12. Freedman, L. D., and Doak, G. O. (1957): *Chem. Rev.*, 57:479–523.
13. Gillespie, J. H. (1934): *J. Physiol. (Lond.)*, 80:345–359.
14. Gough, G., Maguire, M. H., and Michal, F. (1969): *J. Med. Chem.*, 12:494–498.
15. Gough, G., Maguire, M. H., and Penglis, F. (1972): *Mol. Pharmacol.*, 8:170–177.
16. Gough, G., Maguire, M. H., and Satchell, D. G. (1973): *J. Med. Chem.*, 16:1188–1190.
17. Gough, G. R., Nobbs, D. M., Middleton, J. C., Penglis-Caredes, F., and Maguire, M. H. (1978): J. Med. Chem., 21:520–525.
18. Kahn, M. M. T., and Martell, A. E. (1962): *J. Am. Chem. Soc.*, 84:3037–3041.
19. Larsen, M., Willett, R., and Yount, R. G. (1969): *Science*, 166:1510–1511.
20. Maguire, M. H., and De, A. (1978): *Fed. Proc. Abst.*, 37:547.
21. Maguire, M. H., and Michal, F. (1968): *Nature*, 217:571–573.
22. Maguire, M. H., Nobbs, D. M., Einstein, R., and Middleton, J. C. (1971): *J. Med. Chem.*, 14:415–420.
23. Michal, F., Maguire, M. H., and Gough, G. (1969): *Nature*, 222:1073–1074.
24. Myers, T. C., Nakamura, K., and Danielzadeh, A. B. (1965): *J. Org. Chem.*, 30:1517–1520.
25. Myers, T. C., Nakamura, K., and Flesher, J. W. (1963): *J. Am. Chem. Soc.*, 85:3292–3295.
26. Rang, H. P. (1971): *Nature*, 231:91–96.
27. Satchell, D. G., and Burnstock, G. (1975): *Eur. J. Pharmacol.*, 32:324–328.
28. Satchell, D. G., Lynch, A., Bourke, P. M., and Burnstock, G. (1972): *Eur. J. Pharmacol.*, 19:343–350.
29. Satchell, D. G., and Maguire, M. H. (1975): *J. Pharmacol. Exp. Ther.*, 195:540–548.
30. Smith, R. M., and Alberty, R. A. (1956): *J. Am. Chem. Soc.*, 78:2376–2380.
31. Winand-Devigne, J., Hamdir, G., and Liébecq, C. (1967): *Eur. J. Biochem.*, 1:29–32.
32. Yount, R. G., Babock, D., Ballantyne, W., and Ojala, D. (1971): *Biochemistry*, 10:2484–2489.

Physiological and Regulatory Functions of Adenosine and Adenine Nucleotides, edited by H. P. Baer and G. I. Drummond. Raven Press, New York © 1979.

Intramural Inhibition in Rabbit and Guinea Pig Intestine

R. C. Small and A. H. Weston

Department of Pharmacology, Materia Medica and Therapeutics, Medical School, University of Manchester, Manchester M13 9PT, England

The existence of a nonadrenergic, noncholinergic inhibitory neuronal system in the mammalian gastrointestinal tract was proposed in 1966 (8). The evidence that adenosine triphosphate (ATP) might be the transmitter liberated from the endings of these inhibitory nerves has been reviewed (15) and the characteristics of "purinergic" transmission have recently been described (16). However, although many workers agree that the inhibitory process in question is neither cholinergic nor adrenergic, the proposal that ATP is the inhibitory transmitter (15) is not universally accepted and is often the subject of active debate (10,43,55).

This chapter is an attempt to present a review of the characteristics of the intramural inhibitory processes in the rabbit and guinea pig gastrointestinal tracts, together with a discussion of the possible role of ATP in this inhibitory process.

CHARACTERISTICS OF THE INTRAMURAL INHIBITORY RESPONSE

Under appropriate conditions, isolated segments of intestine will exhibit neurogenic inhibition in response to electrical, chemical, or mechanical stimuli.

Electrical Stimulation

Many workers have used transmural or field stimulation (pulse width 0.5 msec or less) to activate the inhibitory neurons. If the tone of the tissue is high or if it has been treated with a muscarinic cholinoceptor antagonist, a single pulse evokes a relaxation that is rapid in onset and offset.

The smooth muscle membrane potential change that accompanies this relaxation is a discrete inhibitory postjunctional potential (IJP). The IJP follows the stimulus after a latency in the range 38 msec (41) to 160 msec (14). This latency is temperature dependent (17) but independent of the stimulus strength (41,53,64). The amplitude of the IJP is dependent on the stimulus strength (14,25,35,64) and can be as great as 25 mV (15). Thus it is probably dependent on the number of nerve fibers stimulated or the amount of neurotransmitter released. The observation that IJP amplitude is related to Ca^{2+} concentration (37) suggests that, as is known for acetylcholine and noradrenaline, transmitter release may be a Ca^{2+}-dependent process.

The effects of repetitive stimulation of intrinsic inhibitory neurons are dependent on both the tissue examined and the pulse frequency used. At pulse frequencies of 0.05 to 1.0 Hz, the first two or three IJPs elicited in the guinea pig colon generally exhibit facilitation (25). The IJPs of the longitudinal muscle of guinea pig jejunum (53) and of the circular muscle of rabbit caecum (64) exhibit depression. In these two tissues, the first IJP in a train is larger than all subsequent IJPs. In rabbit caecal muscle, this IJP depression is evident at pulse frequencies greater than 0.01 Hz and the degree of depression increases with pulse frequencies up to 0.5 Hz (Fig. 1). In the guinea pig taenia caeci, successive IJPs may vary in amplitude if the resting membrane potential of the cell fluctuates markedly, but if the resting potential is stable, then successive IJPs in a train are of constant amplitude (Fig. 1). The mechanisms that govern whether successive IJPs are of constant amplitude or exhibit either facilitation or depression have yet to be determined.

At pulse frequencies greater than 1 Hz, the IJPs of rabbit and guinea pig intestinal muscle summate and hence a greater overall hyperpolarization results (17,26,53, 62). The mechanical response to increases in pulse frequency is a progressively

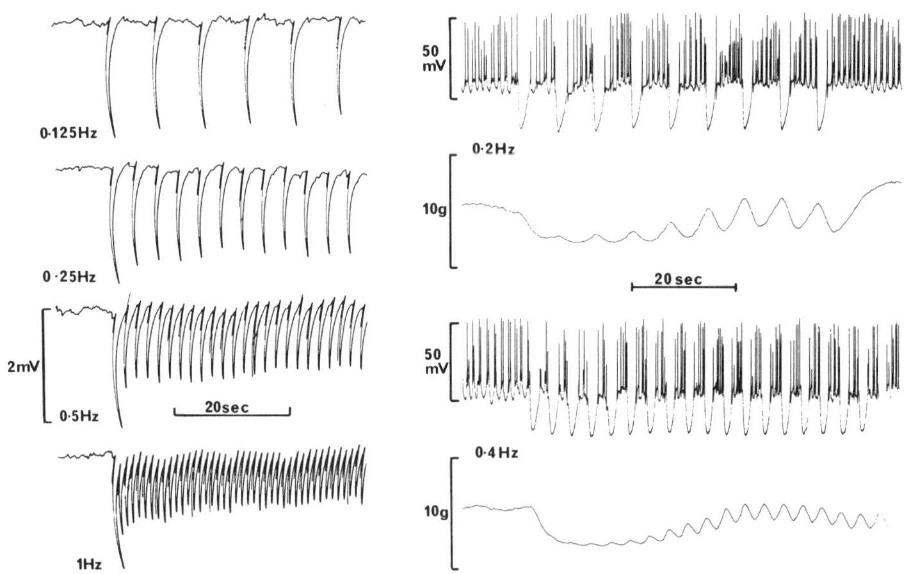

FIG. 1. Left panel: Sucrose gap recording of electrical activity of circular muscle of rabbit caecum. Atropine (170 nM) was present throughout. Inhibitory postjunctional potentials were elicited by field stimulation using trains of 0.5-msec pulses at the frequencies indicated. Note the depression that follows the first IJP in each train. (From ref. 64, with permission). **Right panel:** Simultaneous intracellular recording of membrane potential *(upper trace)* and mechanical activity *(lower trace)* in the guinea pig taenia caeci. Hyoscine 0.3 μM was present throughout. Inhibitory postjunctional potentials were elicited by field stimulation using trains of 0.5-msec pulses at the frequencies indicated. Note the relative constancy of IJP amplitude within each train. From R. C. Small, *unpublished observations.)*

greater relaxation and this is near-maximal at pulse frequencies in the range 20 to 32 Hz (Fig. 2) (58,72).

The primary inhibition evoked by intramural nerve activity is often followed by a secondary or "rebound" excitation. This is manifested as depolarization, increased spike frequency, and a rise in muscle tone above the preexisting level.

In the circular muscle of rabbit caecum, the likelihood of rebound excitation is greatly increased if triethylcholine is used instead of atropine to prevent primary cholinergic excitatory responses (62). This phenomenon may be attributable to the membrane-labilizing effect of the onium ion of the triethylcholine molecule (Fig. 3).

Other factors that predispose toward rebound excitation include the induction of inhibition just prior to an expected spontaneous wave of tension and stimulation of the inhibitory neuron for protracted periods or at high pulse frequencies (Fig. 2).

Several authors have proposed that rebound excitation is a myogenic reaction to the withdrawal of an inhibitory stimulus (11,29). The demonstration that indomethacin and polyphloretin phosphate can suppress rebound excitation in a variety of tissues has led to the suggestion that the release of prostaglandins may be responsible for this phenomenon (21,44).

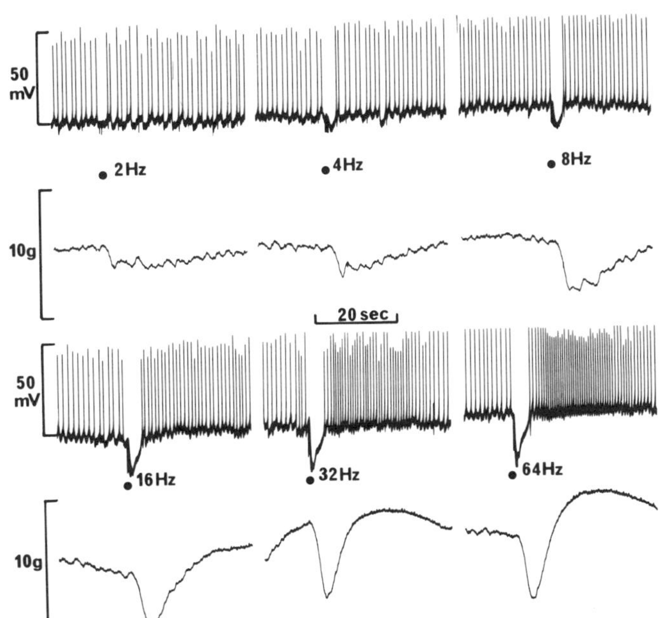

FIG. 2. Simultaneous intracellular recording of membrane potential *(upper trace)* and mechanical activity *(lower trace)* in guinea pig taenia caeci. The electrical trace is a continuous recording from a single cell. Atropine (170 nM) was present throughout. (•), Field stimulation using 0.5-msec pulses, at frequencies indicated, in trains of 1-sec duration. Note the greater overall hyperpolarization, greater relaxation, and greater rebound increase in spike frequency as stimulus pulse frequency is raised. (From R. C. Small, *unpublished observations.)*

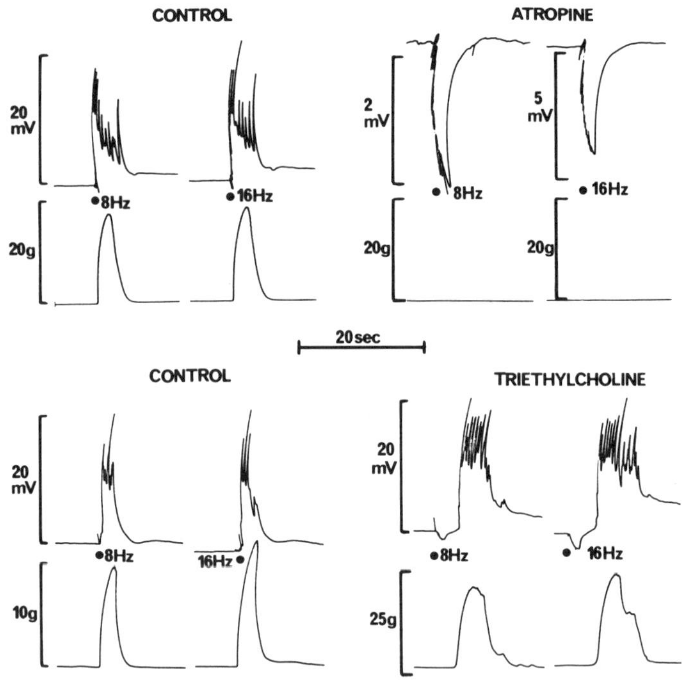

FIG. 3. Sucrose gap recording of the electrical activity *(upper trace)* and mechanical activity *(lower trace)* of circular muscle from rabbit caecum. Records from two tissues are shown. Upper records show responses obtained before (controls) and after treatment with atropine (170 nM). Control responses to field stimulation (0.5-msec pulses applied in trains of 1-sec duration at frequencies indicated) comprise a cholinergic excitatory postjunctional potential that triggers spike activity and a tension wave. After atropine, field stimulation evokes hyperpolarization with little evidence of rebound excitation. Lower records show responses obtained before (control) and after a period of repetitive stimulation in 14 mM triethylcholine. After triethylcholine, field stimulation evokes hyperpolarization followed by rebound spike firing and tension development. (From ref. 64., with permission.)

Chemical Stimulation

The relaxations induced by nicotine in the stomach and taenia caeci of the guinea pig and in the large intestine of the rabbit are resistant to sympatholytic drugs and have therefore been attributed to the activation of intrinsic inhibitory neurons (19,22,66). In the circular muscle of the rabbit caecum, the relaxant effect of nicotine is accompanied by a marked hyperpolarization of the tissue (Fig. 4) indicating that this drug can mimic both the electrical and mechanical responses associated with discharge of the inhibitory neurons. Since the action of nicotine is antagonized by hexamethonium and almost abolished by tetrodotoxin (63), it is mediated by neural nicotinic cholinoceptors and involves the discharge of nerve action potentials.

The simplest explanation for the action of nicotine is that the inhibitory neurons themselves possess nicotinic cholinoceptors (13,30). However, experimental evi-

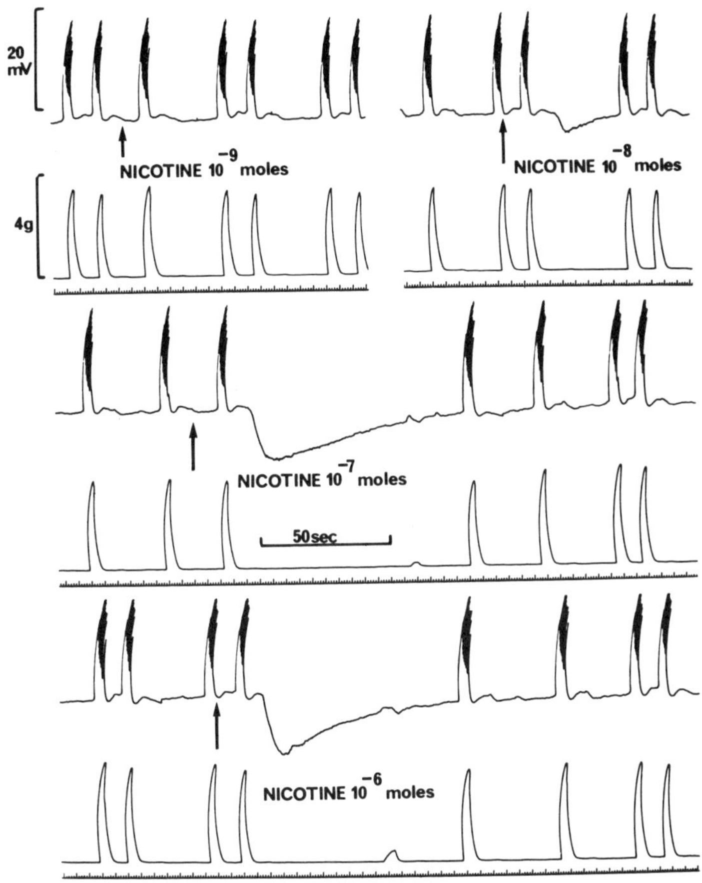

FIG. 4. Sucrose gap recording of the electrical activity *(upper trace)* and mechanical activity *(lower trace)* of circular muscle of the rabbit caecum. The bursts of spike activity and tension waves were induced by 1 μM methacholine which was present throughout. Nicotine relaxes this tissue by evoking discharge of intramural inhibitory neurons. Note that the spasmolytic action of nicotine is accompanied by abolition of the spike bursts and by marked hyperpolarization of the tissue. (From ref. 64, with permission.)

dence currently available cannot rule out the possibility that the site of action of nicotine may be on interneurons that provide the inhibitory fibers with an excitatory input (27,35).

Mechanical Stimulation

Stretching the intestinal wall may be the physiological stimulus that initiates discharge of intramural inhibitory neurons.

When the intraluminal pressure of isolated segments of guinea pig ileum is raised by a few mm of water, the longitudinal muscle contracts. This event is closely followed by contraction of the circular muscle and a simultaneous active relaxation of the longitudinal muscle (71). This active relaxation of the longitudinal muscle has been attributed to the discharge of nonadrenergic intramural neurons (47,49). When an intraluminal balloon is used to stretch a segment of guinea pig ileum, IJPs are generated in the circular muscle layer of an adjacent but aboral segment (36).

These and other observations are consistent with the suggestion that intrinsic inhibitory neurons form the final efferent link in a cascade of descending reflex arcs in the intestinal wall. Such reflex arcs are activated in response to stretch caused by a food bolus and their function is to dilate the alimentary canal in front of the advancing bolus (28).

PHARMACOLOGICAL ANALYSIS

Role of Noradrenergic Neurons

Since field or transmural stimulation probably results in the activation of all neural structures within effective range of the cathode, the possibility exists that the inhibitory response of atropinized intestine may be mediated by noradrenaline released from postganglionic sympathetic neurons. In order to assess the role of noradrenaline in this inhibitory response, many workers have compared the response with that obtained on perivascular stimulation of the extrinsic sympathetic nerves. The results of most experiments of this kind indicate that noradrenaline release contributes little, if at all, to the response to transmural stimulation.

In the guinea pig taenia caeci, fatigue of the inhibitory response to stimulation of the extrinsic sympathetic nerve supply caused by repetitive stimulation does not evoke fatigue of the response to transmural stimulation. The converse has also been demonstrated—that fatigue of the response to transmural stimulation does not evoke fatigue of the response to stimulation of the extrinsic sympathetic nerve supply (18). These observations suggest that different mechanisms underly the two inhibitory responses but it is difficult to explain why repetitive transmural stimulation should not have activated sympathetic nerve terminals and hence fatigued the response to extrinsic stimulation.

Further evidence that stimulation of the extrinsic sympathetic nerve trunks and transmural stimulation may excite different neural structures and thereby evoke intestinal relaxation has been provided by electrophysiological studies. In contrast to transmural stimulation, a single pulse applied to the extrinsic sympathetic fibers does not evoke an IJP in rabbit colon (32) or in the taenia caeci, colon, and stomach of the guinea pig (6,8,25). Repetitive stimulation of extrinsic sympathetic neurons supplying these tissues yields a relatively weak (10 to 15 mV) and slowly developing hyperpolarization together with a decline in tension and spike frequency. The failure of extrinsic sympathetic nerve stimulation to evoke an IJP in response to a single pulse may be related to the distribution of noradrenergic neurons in the wall of the gut. In nonsphincteric regions most adrenergic neurons terminate in the myenteric plexus and in the walls of blood vessels whereas relatively few fibers supply the muscle layers (29). Under such circumstances it seems reasonable to assume that most intestinal muscle cells cannot be influenced directly by noradrenaline released from sympathetic nerve terminals in response to a single stimulating pulse.

Sympatholytic drugs clearly differentiate between inhibitory responses to transmural stimulation and to stimulation of extrinsic sympathetic neurons. Antagonists at α- and β-adrenoceptors markedly suppress responses elicited by stimulation of extrinsic sympathetic nerves (2,9,12,54,72) whereas, in contrast, responses to transmural stimulation are either unaffected (9,48,62,72) or, at high stimulation frequencies, only slightly reduced (Fig. 5) (12,23,72).

The adrenergic neuron blocking agents (e.g., xylocholine, bretylium, and guanethidine readily abolish the inhibitory responses of intestinal muscle to stimulation of extrinsic sympathetic nerves (2,6,8,9,12,18,19,23,25,64,72). However, these agents either do not modify responses to transmural stimulation (8,9, 25,48,62,72) or, at high stimulation frequencies, only slightly reduce them (6,12, 19,23).

Pretreatment of animals with reserpine or 6-hydroxydopamine can be used as a means of removing the functional integrity of noradrenergic neurons. In order that sympathetic neuroeffector transmission be significantly impaired by reserpine, the noradrenaline content of a tissue must be reduced by more than 50% (56). Effective pretreatment with reserpine reduces the responses of intestinal muscle to extrinsic sympathetic nerve stimulation but does not affect responses to vagal or transmural stimulation (6,9,64,72).

Similarly, an effective 6-hydroxydopamine pretreatment schedule reduces tissue noradrenaline content below 50% of control levels, removes tissue responsiveness to tyramine, and reduces the numbers of catecholamine-fluorescent fibers in tissue sections. Such pretreatment reduces the response of intestinal muscle to extrinsic sympathetic nerve stimulation but not to field stimulation (24,64). Thus combined electrophysiological and pharmacological evidence suggests that the inhibitory response to transmural stimulation of atropinized intestine at low pulse frequencies is not mediated by noradrenaline release, although at high pulse frequencies this may contribute to the response in a minor way. Furthermore, studies of fetal and early postnatal intestine from rabbits and mice have revealed a distinct developmental separation between noradrenergic and nonadrenergic inhibitory nechanisms (31).

Role of Adenine Nucleotides

Release of ATP Following Transmural Stimulation

Several attempts have been made to demonstrate radiolabeled nucleotide release following transmural stimulation (51,52,60,68). Su and his co-workers (68) showed that it was possible to detect ^{3}H release following transmural stimulation in guinea pig taenia coli previously incubated in [^{3}H]adenosine. However, in the same study, ^{3}H release was also detected following adrenergic nerve stimulation both in the taenia and in rabbit aorta showing that ^{3}H release following exposure of tissues to

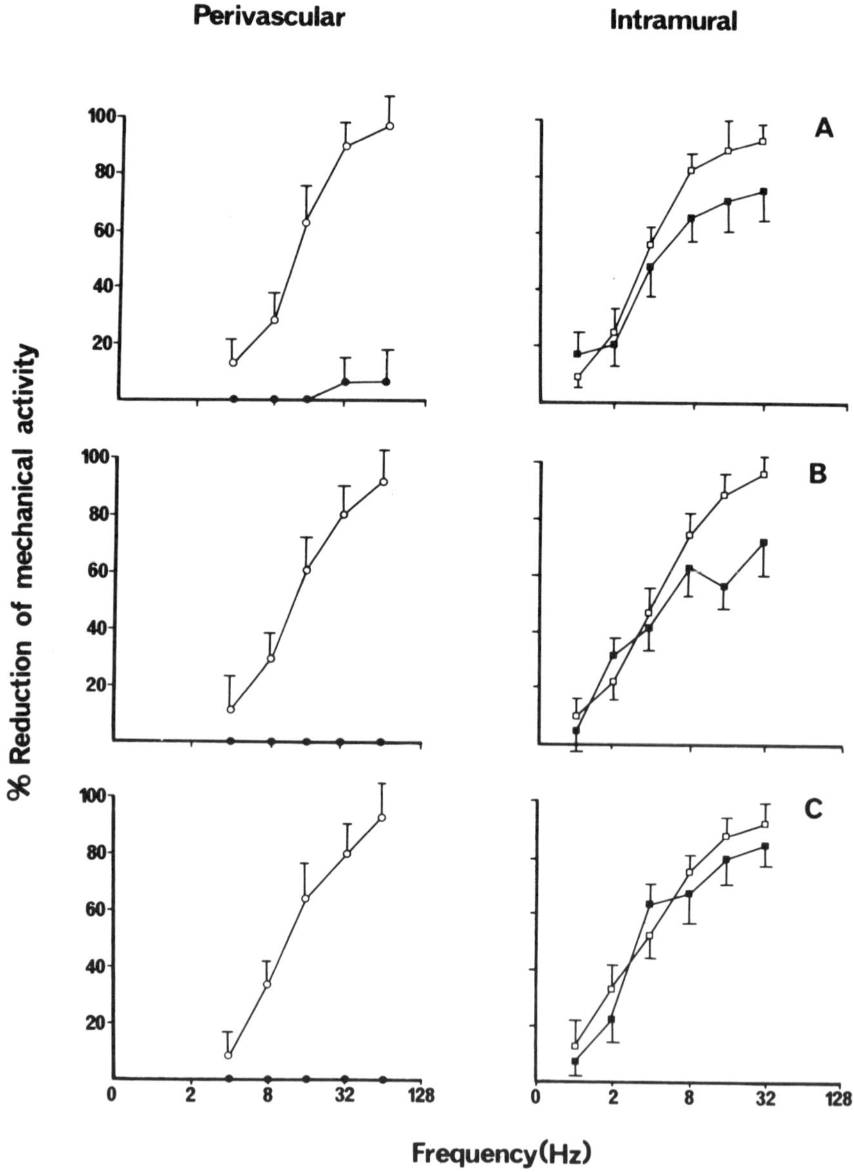

FIG. 5. Effect of drugs on the frequency-response curves to perivascular and intramural stimulation in the longitudinal muscle strip of rabbit duodenum. **A:** Effect of 2.5 μM phentolamine +4.75 μM propranolol. **B:** Effect of 10 μM guanethidine. **C:** Effect of reserpine (2 mg/kg i.p.). Responses to perivascular stimulation before (○) and after (●) drug exposure. Responses to intramural stimulation before (□) and after (■) drug exposure. (Adapted from ref. 72, with permission.)

[^{3}H]adenosine was not a unique property of the transmurally activated nerves of the taenia.

Subsequently, the results of two detailed studies using [^{3}H]adenosine in the taenia employing a variety of procedures to distinguish between neuronal and nonneuronal sources of ^{3}H release strongly suggested that the isotope originated largely from a nonneuronal source (51,52). More recently, a further attempt to resolve the situation concluded that a component of the ^{3}H release in taenia coli originated from neuronal sources but this study did not distinguish among release from adrenergic, cholinergic, or the proposed purinergic nerves of the tissue (60).

Mimicry by ATP

If, as suggested by Burnstock and his colleagues (20), ATP is the transmitter of intramural inhibitory nerves, then nervous activity should be mimicked by ATP. Ideally, the putative transmitter should be administered either by close intraarterial injection or by iontophoresis at the neuroeffector junction to obviate distributional problems arising from its simple addition to the bathing fluid. The latter method has been used most widely and has been shown to evoke rapid, primary relaxation accompanied by rebound contraction on washout in the circular muscle of the stomach, the taenia caeci, and longitudinal muscle of the colon in the guinea pig (20,27,65), and in the longitudinal muscle of the small intestine and colon of the rabbit (20,72). However, in some intestinal tissues from the same species, mimicry of inhibitory nerve activity is not observed since ATP evokes contractions of a primary nature (28).

Unless ATP can evoke neurotransmitter release by a mechanism not involving the firing of nerve action potentials, its tetrodotoxin-resistant effects imply a direct action on smooth muscle cells (58,64,72). The relaxant effects of ATP, like those of intramural nerve stimulation, are resistant to concentrations of phentolamine that antagonize noradrenaline (64,72). It is unlikely, therefore, that the inhibitory effects of ATP are mediated by α-adrenoceptors.

Among the adenine nucleotides, the relatively high potency of ATP as a smooth muscle relaxant has been advanced as evidence in favor of its transmitter role (20). However, when ATP is compared, on a molar basis, with the established neurotransmitter noradrenaline, it appears as a relatively impotent relaxant in the gut. For example, in the circular muscle of rabbit caecum and in guinea pig taenia caeci, ATP is approximately 1,000 times less potent than noradrenaline (64; and R. C. Small and A. H. Weston, *unpublished observations*). However, to claim that high molar potency as measured in the tissue bath is an essential feature of the action of a neurotransmitter and hence to deny such a function for ATP in these tissues may be unwise. It may be that the operation of very efficient inactivation mechanisms masks a high molar potency of ATP at the receptor level.

That ATP can mimic the hyperpolarization of intestinal muscle evoked by intramural nerve stimulation has been demonstrated in the guinea pig taenia caeci (42,65,70) and in the circular muscle of rabbit caecum (Figs. 6 and 7) (63). The

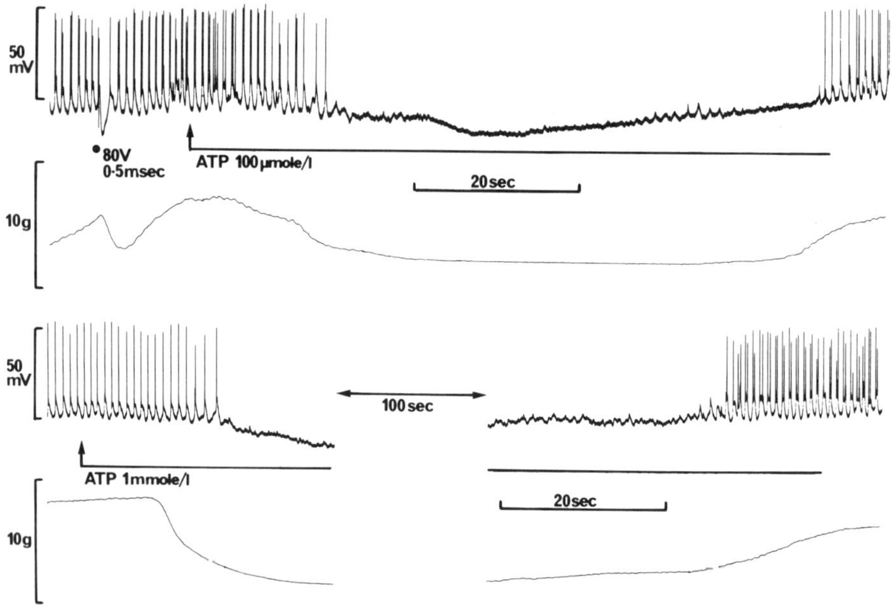

FIG. 6. Simultaneous intracellular recordings of membrane potential *(upper trace)* and mechanical activity *(lower trace)* in guinea pig taenia caeci. Hyoscine (0.3 μM) was present throughout. The upper records show an IJP and relaxation evoked by field stimulation (•) and the inhibitory effect of 100 μM ATP. The lower records show the inhibitory effect of 1 mM ATP. Note that the relaxant effect of ATP is accompanied by abolition of spike activity and by hyperpolarization of the tissue. (From ref. 65, with permission.)

threshold concentration of ATP required to hyperpolarize these tissues is 10^{-4} mole/liter. Like the IJP, ATP-induced hyperpolarization of these tissues is resistant to phentolamine.

In the taenia caeci, the hyperpolarization evoked by ATP is associated with a decrease in membrane resistance and its lack of dependence on extracellular Cl^- concentration implies an underlying selective increase in membrane K^+ conductance (42,70). A similar conductance change is believed to underly the IJP (42,69,70).

Potentiators of ATP and Intramural Nerve Stimulation

In his model of purinergic transmission, Burnstock (15) proposed that the transmitter action of ATP may be terminated first by its metabolism to adenosine and subsequently by the uptake of adenosine into nerve terminals. Although it is often less potent in intestinal muscle, adenosine shares the postjunctional activity of ATP. Hence, provided that the uptake process is efficient enough to govern the concentration of adenosine in the junctional cleft, drugs that inhibit adenosine uptake might be expected to potentiate both exogenous ATP and the stimulation of the intramural nerves. The drugs most commonly used for this purpose are dipyridamole,

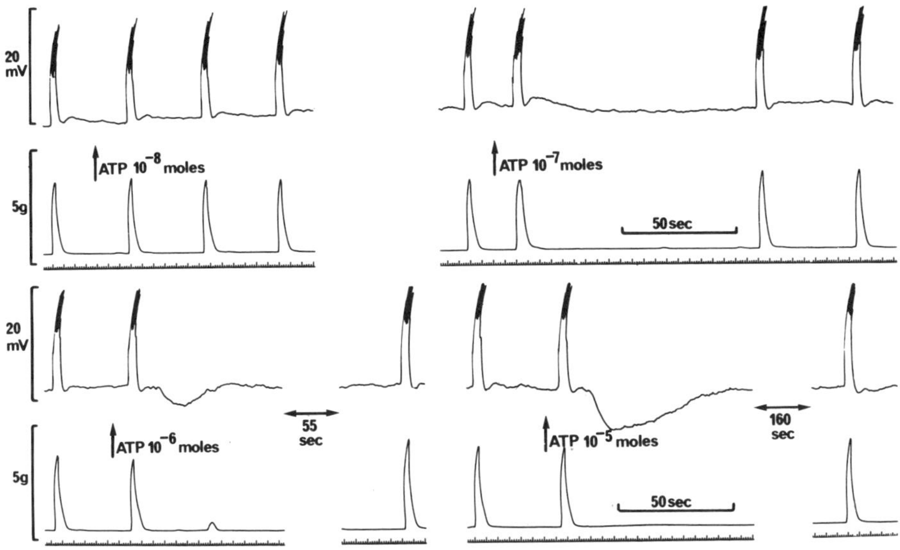

FIG. 7. Sucrose gap recordings of the electrical activity *(upper trace)* and mechanical activity *(lower trace)* of circular muscle of the rabbit caecum. Note the bursts of spike activity and tension waves induced by 1 μM methacholine which was present throughout. Note that the spasmolytic action of ATP was accompanied by abolition of spike activity and marked hyperpolarization of the tissue. (From ref. 64, with permission.)

hexobendine, and lidoflazine, all of which have been reported to potentiate the actions of adenosine on the heart and to reduce the accumulation of adenosine in this tissue (1,38,46,50).

In rabbit intestine, dipyridamole has been reported to potentiate the effects of adenosine (67) and, together with hexobendine, to potentiate the effects of intramural inhibitory stimulation and responses to adenosine and ATP in guinea pig taenia caeci (61). In studies with isotopically labeled adenosine, dipyridamole, hexobendine, and lidoflazine were each able to reduce the accumulation of radioactive nucleotides by rabbit duodenum (39). However, concentrations of these agents that produce this effect caused no significant potentiation of responses to ATP or to intramural nerve stimulation (40). More recently, other groups of workers have reported the inability of dipyridamole to enhance the inhibition produced by intramural nerve stimulation in rat stomach (34), guinea pig taenia caeci (5), and pig stomach (57). Attempts to support previous findings (61) by quantitative data and also to show potentiation of the inhibitory responses using 6-(2-hydroxy-5-nitro)-benzylthioguanosine, a more selective inhibitor of adenosine uptake than dipyridamole, have also been unsuccessful (5).

ATP Antagonists

Several compounds including quinidine (11) and imidazole (59) have been shown to antagonize the actions of ATP and related nucleotides and their effects have been

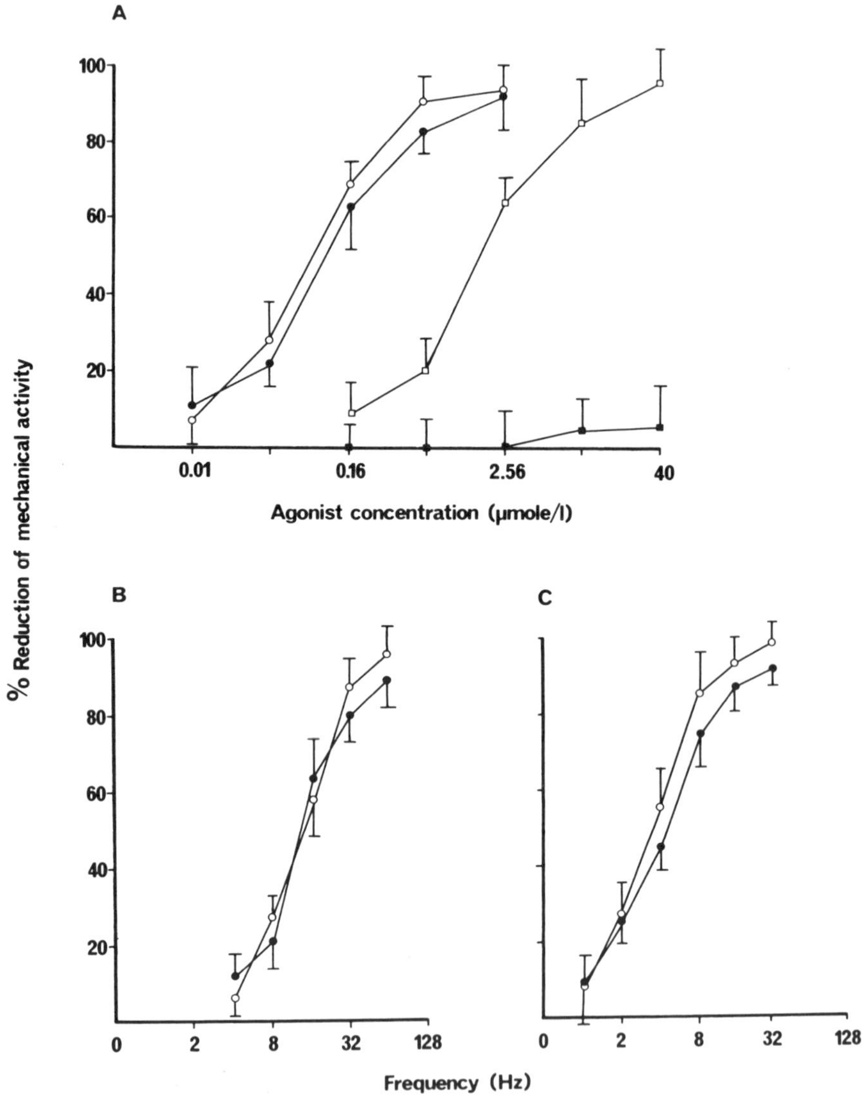

FIG. 8. Effects of desensitization with 100 μM ATP in the longitudinal muscle strip of rabbit duodenum. **(A):** Responses to phenylephrine before (○) and after (●) desensitization; responses to ATP before (□) and after (■) desensitization. **(B):** Responses to perivascular stimulation and **(C):** responses to intramural stimulation before (○) and after (●) desensitization. (From ref. 72, with permission.)

reviewed by Burnstock (16). A more recently developed antagonist is 2,2′-pyridylisatogen (PIT) (66) whose pharmacology is discussed elsewhere in this volume.

Theophylline has been reported to antagonize the effects of adenosine on rabbit duodenum (3) and, on guinea pig stomach fundus, the effects of both ATP and of intramural nerve stimulation (58). However, similar concentrations of theophylline have no such effects in guinea pig taenia caeci or in the isolated longitudinal muscle of rabbit duodenum (R. C. Small and A. H. Weston, *unpublished observations)*.

Desentization to ATP

An agent with no direct effect on smooth muscle tone that selectively antagonizes the inhibitory action of ATP in a wide variety of intestinal tissues has yet to be found. In the absence of such an agent, an unequivocal assessment of the role of ATP in mediating the intramural inhibitory response is not possible. However, attempts have been made to desensitize tissues to ATP by exposing them to high ATP concentrations and examining the responses to intramural stimulation before and after this procedure.

In rabbit ileum, Burnstock and his co-workers (20) reported that, following desensitization to ATP, responses to intramural nerve stimulation were also abolished whereas the effects of adrenergic stimulation were unaffected. In an attempt to confirm this finding, Weston (72) examined the effects of desensitization to ATP on responses elicited by stimulation of the intramural inhibitory nerves over a wide range of stimulation frequencies. In ATP-desensitized preparations, there was no change in the responses of the ileum to either intramural or adrenergic nerve stimulation (Fig. 8). The extent of this discrepancy cannot be assessed since detailed quantitative results were not given by Burnstock and his colleagues (20) and this again serves to emphasize the need for a quantitative approach to this and similar problems.

The procedure of desensitization to ATP has recently been used by other groups of workers. In guinea pig stomach fundus, Okwuasaba and co-workers (58), in a detailed quantitative study, showed that, following desensitization to ATP, the responses to intramural inhibitory nerve stimulation were much reduced whereas responses to periarterial nerve stimulation were unaffected. However, in pig stomach, desensitization to ATP has no effect on responses to nonadrenergic inhibitory stimulation (57).

DISCUSSION

The evidence in favor of a noncholinergic, nonadrenergic inhibitory neural pathway in guinea pig and rabbit intestine seems beyond dispute. However, the nature of the inhibitory transmitter remains controversial. Certainly, ATP and related nucleotides have inhibitory actions but, when applied exogenously, they are up to 1,000 times less potent on a molar basis than established transmitters such as acetylcholine and noradrenaline.

There is evidence that an increase in ATP output can be detected when intramural nerves are stimulated *and* a change in mechanical activity is produced. However, although it has been argued that this release is from the intramural nerves themselves (68), other studies suggest that the ATP is derived from the muscle cells of the effector, rather than from the inhibitory nerves (51,52), or possibly from cholinergic or adrenergic nerves (60).

Attempts have been made to analyze the intramurally mediated inhibitory response using antagonists to the action of ATP or agents that modify the processes by which ATP is inactivated. The usefulness of currently available antagonists is limited by their nonselectivity, their inherent spasmolytic activity, their failure to antagonize the inhibitory actions of ATP in all tissues, or a combination of these factors.

Dipyridamole is the drug most widely used to modify the postulated ATP inactivation process. The use of this compound as a tool for the analysis of intramural inhibitory responses illustrates a problem common to many pharmacological fields. The use of a drug who mode and specificity of action in intestinal muscle have not been firmly established means that the experimental results should be evaluated with caution. This point is illustrated by the observations that, in rabbit intestine at least, adenosine uptake is not directly affected by dipyridamole (39) and that the drug possesses actions unrelated to an interaction with adenosine (7,45). It is now clear that dipyridamole, and perhaps other more specific nucleotide transport inhibitors are of limited or no use in the analysis of intramural inhibitory responses (5).

Specific desensitization to the effects of ATP would seem to be potentially one of the most useful tools in the analysis of its possible role in the intramural inhibitory response. However, the results obtained following desensitization are equivocal. Some groups of workers have obtained evidence consistent with a transmitter role for ATP (20,58) whereas others have reached the opposite conclusion (57,72).

No doubt the quality of the pharmacological evidence either for or against a transmitter role for ATP will improve with better understanding of the mechanisms and specificity of the analytical tools employed. In the meantime, attempts to isolate the transmitter (4,33) may yield more direct evidence of its chemical nature. The inhibitory transmitter, in bovine retractor penis at least, does not seem to be a prostaglandin (4; and J. S. Gillespie, *personal communication)* or a peptide (J. S. Gillespie, *personal communication)* and the outcome of these studies is awaited with interest.

ACKNOWLEDGMENTS

We wish to thank the Alexander von Humboldt-Stiftung and the Smith Kline and French Foundation for equipment grants, and Boehringer Ingelheim, I.C.I. Ltd., and the Wellcome Trust for contributions toward traveling expenses. We are grateful to Miss J. Bamber and Mr. J. Taylor for expert technical assistance.

REFERENCES

1. Afonso, S., O'Brien, G. S., and Crumpton, Ch. W. (1968): *Circ. Res.,* 22:43–48.
2. Akubue, P. I. (1966): *J. Pharm. Pharmacol.,* 18:390–395.

3. Ally, A. I., and Nakatsu, K. (1976): *J. Pharmacol. Exp. Ther.*, 199:208–215.
4. Ambache, N., Killick, S. W., and Zar, M. Aboo (1975): *Br. J. Pharmacol.*, 54:409–410.
5. Baer, H. P., Frew, R., and Burnstock, G. (1977): *Can. J. Physiol. Pharmacol.*, 55:394–398.
6. Beani, L., Bianchi, C., and Crema, A. (1971): *J. Physiol. (Lond.)*, 217:259–279.
7. Bell, C. (1974): *Br. J. Pharmacol.*, 52:13–17.
8. Bennett, M. R., Burnstock, G., and Holman, M. E. (1966): *J. Physiol. (Lond.)*, 182:527–540.
9. Bianchi, C. P., Beani, L., Frigo, G. M., and Crema, A. (1968): *Eur. J. Pharmacol.*, 4:51–61.
10. Bloom, S. R., and Polak, J. M. (1978): *Lancet*, i:93.
11. Bowman, W. C., and Hall, M. T. (1970): *Br. J. Pharmacol.*, 38:399–415.
12. Bucknell, A. (1964): *J. Physiol. (Lond.)*, 177:58–59P.
13. Bülbring, E., and Gershon, M. D. (1967): *J. Physiol. (Lond.)*, 192:823–846.
14. Bülbring, E., and Tomita, T. (1967): *J. Physiol. (Lond.)*, 189:299–315.
15. Burnstock, G. (1972): *Pharmacol. Rev.*, 24:509–581.
16. Burnstock, G. (1975): In: *Handbook of Psychopharmacology*, vol. 5, edited by L. L. Iversen, S. D. Iversen, and S. H. Snyder, pp. 131–194. Plenum, New York.
17. Burnstock, G., Campbell, G., Bennett, M. R., and Holman, M. E. (1963): *Nature*, 200:581–582.
18. Burnstock, G., Campbell, G., Bennett, M. R., and Holman, M. E. (1964): *Int. J. Neuropharmacol.*, 3:163–166.
19. Burnstock, G., Campbell, G., and Rand, M. J. (1966): *J. Physiol. (Lond.)*, 182:504–526.
20. Burnstock, G., Campbell, G., Satchell, D., and Smythe, A. (1970): *Br. J. Pharmacol.*, 40:668–688.
21. Burnstock, G., Cocks, T., Paddle, B., and Staszewska-Barcyak, J. (1975): *Eur. J. Pharmacol.*, 31:360.
22. Bursztajn, S., and Gershon, M. D. (1977): *J. Physiol. (Lond.)*, 269:17–31.
23. Day, M. D., and Warren, P. R. (1968): *Br. J. Pharmacol.*, 32:227–240.
24. Fozard, J. R., Kelly, M. J., and Small, R. C. (1973): *Br. J. Pharmacol.*, 49:182–183P.
25. Furness, J. B. (1969): *J. Physiol. (Lond.)*, 205:549–562.
26. Furness, J. B. (1970): *J. Physiol. (Lond.)*, 207:803–821.
27. Furness, J. B. (1971): *Br. J. Pharmacol.*, 41:213–226.
28. Furness, J. B., and Costa, M. (1973): *Philos. Trans. R. Soc. Lond. [Biol.]*, 265:123–133.
29. Furness, J. B., and Costa, M. (1974): *Ergeb. Exp. Pharmakol.*, 69:1–51.
30. Gershon, M. D. (1967): *Br. J. Pharmacol. Chemother.*, 29:259–279.
31. Gershon, M. D., and Barnes-Thompson, E. (1973): *J. Physiol. (Lond.)*, 234:257–277.
32. Gillespie, J. S. (1962): *J. Physiol. (Lond.)*, 102:54–75.
33. Gillespie, J. S., and Martin, W. (1978): *J. Physiol. (Lond.)*, 280:45–46P.
34. Heazell, M. A. (1975): *Br. J. Pharmacol.*, 55:285–286P.
35. Hidaka, T., and Kuriyama, H. (1969): *J. Gen Physiol.*, 53:471–486.
36. Hirst, G. D. S., and McKirdy, H. C.(1974): *J. Physiol. (Lond.)*, 238:129–143.
37. Holman, M. E., and Weinrich, J. P. (1975): *Pfluegers Arch.*, 360:109–119.
38. Hopkins, S. V., and Goldie, R. G. (1971): *Biochem. Pharmacol.*, 20:3359–3365.
39. Hulme, M. E., and Weston, A. H. (1974): *Br. J. Pharmacol.*, 50:569–574.
40. Hulme, M. E., and Weston, A. H. (1974): *Br. J. Pharmacol.*, 50:609–611.
41. Ito, Y., and Kuriyama, H. (1973): *J. Physiol. (Lond.)*, 231:455–470.
42. Jager, L. P. (1974): *Eur. J. Pharmacol.*, 25:372–382.
43. Jones, C. J. (1978): *Lancet*, i:272–273.
44. Kamikawa, Y., Serizawa, K., and Shimo, Y. (1977): *Eur. J. Pharmacol.*, 45:199–203.
45. Kessel, D., and Dodd, D. C. (1972): *Biochim. Biophys. Acta*, 288:190–194.
46. Kolassa, N., Pfleger, K., and Rummel, W. (1970): *Eur. J. Pharmacol.*, 9:265–268.
47. Kosterlitz, H. W. (1967): *Am. J. Dig. Dis.*, 12:245–254.
48. Kosterlitz, H. W., and Lydon, R. J. (1969): *J. Physiol. (Lond.)*, 200:126–128P.
49. Kottegoda, S. R. (1970): In: *Smooth Muscle*, edited by E. Bülbring, A. F. Brading, A. W. Jones, and T. Tomita, Arnold, London.
50. Kraupp, O. von, Wolner, E., Adler-Kastner, L., Chirikdjian, J. J., Plosczanski, B., and Tuish, E. (1966): *Arzneim. Forsch.*, 16:692–696.
51. Kuchii, M., Miyahara, J. T., and Shibata, S. (1973): *Br. J. Pharmacol.*, 49:258–267.
52. Kuchii, M., Miyahara, J. T., and Shibata, S. (1973): *Br. J. Pharmacol.*, 49:642–650.
53. Kuriyama, H., Osa. T., and Toida, N. (1967): *J. Physiol. (Lond.)*, 191:257–270.
54. Lawson, W. R., and Mackenna, B. R. (1969): *J. Physiol. (Lond.)*, 202:35–36P.
55. Leading Article (1977): *Lancet* ii:1331–1332.

56. Lee, F. L. (1967): *J. Pharmacol. Exp. Ther.,* 156:137–141.
57. Ohga, A., and Taneika, T. (1977): *Br. J. Pharmacol.,* 60:221–231.
58. Okwuasaba, F. K., Hamilton, J. T., and Cook, M. A. (1977): *Eur. J. Pharmacol.,* 46:181–198.
59. Rikimaru, A., Fukushi, Y., and Suzuki, T. (1971): *Tohoku J. Exp. Med.,* 105:199–200.
60. Rutherford, A., and Burnstock, G. (1978): *Eur. J. Pharmacol.,* 48:195–202.
61. Satchell, D. G., Lynch, A., Bourke, P. M., and Burnstock, G. (1972): *Eur. J. Pharmacol.,* 19:343–350.
62. Small, R. C. (1972): *Br. J. Pharmacol.,* 45:149–150P.
63. Small, R. C. (1974): *Br. J. Pharmacol.,* 50:456P.
64. Small, R. C. (1975): Ph.D. Thesis, University of Manchester.
65. Small, R. C., and Spedding, M. (1978): *Br. J. Pharmacol. (in press).*
66. Spedding, M., Sweetman, A. J., and Weetman, D. F. (1975): *Br. J. Pharmacol.,* 53:575–583.
67. Stafford, A. (1966): *Br. J. Pharmacol. Chemother.,* 28:218–227.
68. Su, C., Bevan, J. A., and Burnstock, G. (1971): *Science,* 173:336–339.
69. Tomita, T. (1972): *J. Physiol. (Lond.),* 225:693–703.
70. Tomita, T., and Watanabe, H. (1973): *J. Physiol. (Lond.),* 231:167–177.
71. Weston, A. H. (1973): *Br. J. Pharmacol.,* 47:606–608.
72. Weston, A. H. (1973): *Br. J. Pharmacol.,* 48:302–308.

Physiological and Regulatory Functions of Adenosine and Adenine Nucleotides, edited by H. P. Baer and G. I. Drummond. Raven Press, New York © 1979.

Distribution of Nonadrenergic Inhibitory Nerves in the Intestine, Their Structural Identification and the Role of Prostaglandins in Their Function

E. E. Daniel

Department of Neurosciences, McMaster University, Hamilton, Ontario, L8S 4J9, Canada

The purposes of this chapter are (a) to summarize evidence we have obtained relating the functional distribution of nonadrenergic, noncholinergic inhibitory (NAI) nerves of the intestine to the nervous structures present, (b) to question if adenosine triphosphate (ATP) is the mediator of NAI nerves in esophageal sphincters and (c) to consider if any evidence relates prostaglandin (PG) synthesis to function of NAI nerves in gut muscle.

I conclude that there is evidence from a number of intestinal tissues that indicates a selective high density of innervation of circular muscle by NAI nerves, but within this muscle layer no distinctive class of nerve structures has been identified as NAI nerves. Indeed, most of the nerves found appear to be structurally similar if not identical to cholinergic nerves. A variety of evidence suggests that purines or other substances may be released from the same nerves that release acetylcholine but other alternatives including release from nonnervous cells need consideration. However, for the lower esophageal sphincter of the opossum, I argue that current evidence does not support ATP as the mediator for NAI nerves. Nor is there evidence from *in vivo* studies that PGs play any necessary role in NAI transmission in this tissue.

DISTRIBUTION OF FUNCTIONAL NAI NERVES

We have studied the occurrence of relaxations and/or inhibitory junction potentials (IJPs) in response to field stimulation in a variety of tissues. In rabbit jejunum and duodenum (5,12) we have found that large (5 to 20 nm) IJPs from NAI nerves occur in nearly every circular muscle cell, but they are rare in longitudinal muscle cells and small when they occur. Excitatory junction potentials in response to field stimulation were rare and nearly always found in longitudinal muscle cells. Similarly in the esophagus of the brush-tailed Australian opossum, the circular muscle responded to field stimulation by producing IJPs and relaxations (12,13). After cessation of stimulation there were "off" contractions and depolarizations that could be reduced or prevented by atropine. Furthermore, in some strips of circular muscle, atropine-sensitive "on" contractions could be produced with certain parameters of field stimulation. Thus there appeared to be a predominant NAI innervation of esophageal circular muscle, but some cholinergic innervation as well. In the North American opossum (6,7,9,11) there is virtually no evidence of functional cholinergic innervation of circular muscle. Field stimulation produces only relaxation in the lower esophageal sphincter (LES) where tone is present and atropine-insensitive "off" contractions there and in the body. Longitudinal muscle *in vivo* responds to vagal stimulation by atropine-sensitive contraction whereas circular muscle shows only relaxation (LES) or inhibition without relaxation (body) and atropine-insensitive "off" contractions.

In none of these structures was there evidence of adrenergic innervation of either muscle layer in response to field stimulation. Thus there was no contraction sensitive to phentolamine or guanethidine and no relaxation sensitive to propranolol or guanethidine.

In the small intestine of this species other investigators have provided evidence (18) that circular muscle is predominantly innervated by NAI nerves with little cholinergic or adrenergic component.

A possible deficiency in all our studies exists because we lack a reliable, potent, selective inhibitor of the NAI nerve mediator. Without one, we cannot exclude that a cholinergic mediator may have been released but its effects prevented by the increased conductance resulting from concomitant release of NAI mediator. However, this does not affect the conclusion that in a variety of intestinal regions, circular muscle is heavily innervated by NAI nerves.

DISTRIBUTION OF NERVE VARICOSITIES IN RELATION TO FUNCTION

If tissues are functionally innervated predominantly by NAI nerves, we expect that their structural characteristics will be visible. Burnstock (1,3) has associated NAI nerve function with nerves containing predominantly large (80 to 160 nm diameter) granular vesicles, usually large opaque vesicles. To test this association, we have examined the ultrastructure of nerve varicosities associated with the muscle layers of the intestine in the regions mentioned above. The classification used is shown in Table 1. One aspect of our approach may require defense. We have insisted on classifying vesicles by type and on counting numbers of vesicles of each

TABLE 1. *Definitions of types of varicosities*

1. No profile counted unless at least three vesicles and few or no neurotubules or filaments.
2. If any small (30 to 60 nm) vesicles present with dense cores, designate as varicosity with small granular vesicles.
3. If majority of vesicles are large (60 to 200 nm) granular vesicles, designate as varicosity with large granular vesicles. May be subdivided into those with large dense-cored vesicles or those with large opaque vesicles.
4. If majority of vesicles are small (30 to 60 nm) agranular vesicles, designate as varicosity with small agranular vesicles.

type as defined in Table 1 to ultimately classify varicosities. This insistence arises from the realization that all types of nerve varicosities contain a variable number of large granular vesicles and these are not randomly distributed (12,21). Thus it is not sufficient to rely on a subjective impression about "predominance" of one or other vesicle type. We have defined predominance in terms of the type of the majority of vesicles present. Even if it were argued, for example, that one large granular vesicle should be equivalent to two small agranular vesicles, the predominance of varicosities with small agranular vesicles shown in Table 2 would not be reversed or drastically altered.

Table 2 clearly demonstrates that in three intestinal regions, the circular muscle layer with a predominance of functional NAI innervation has a predominance of varicosities in which small agranular vesicles are in the majority. This strongly suggests that NAI mediator emerges from nerve varicosities with these characteristics. There are fundamentally four possible explanations: (a) the cholinergic and NAI nerves are independent but structurally similar; (b) the two mediators are contained within the same nerve, possibly in different vesicles, and released simultaneously or differentially; (c) the same mediator (acetylcholine) is released from all varicosities with a majority of small agranular vesicles and the mediator of NAI nerves is released secondarily from other structures; and (d) acetylcholine is released from gut nerves, but NAI mediator is released independently from other structures by a tetrodotoxin (TTX)-sensitive or sometimes -insensitive (9) mechanism.

Unfortunately there is no procedure analogous to the Falck technique (17) for catecholamines or immunofluorescence for peptides, etc., that enables NAI nerves to be positively identified by light microscopy. Furthermore, there appears to be no evidence excluding the existence of acetylcholinesterase in NAI nerves, so they cannot be differentiated from cholinergic nerves at the level of light microscopy. Some acetylcholinesterase activity is certainly associated with noncholinergic

TABLE 2. *Types of varicosities in intestinal muscle* [a]

Species	SAV (no.)	LDV (no.)	LDV (%)	Response to field stimulation
Rabbit jejunum				
Longitudinal muscle	60	5	7.6	Few small IJPs
Circular muscle	189	25	11.5	Many large IJPs
Plexus	895	96	10.7	Not studied
Australian opossum				
Circular muscle	712	14	1.9	Many IJPs, mostly nonadrenergic, noncholinergic relaxations
North American opossum				
Circular muscle	403	34	7.7	Exclusively nonadrenergic, noncholinergic relaxations

SAV, small agranular vesicles; LDV, large dense vesicles.
[a] Nerves near blood vessels with small granular vesicles (adrenergic) excluded.

TABLE 3. *Mediators claimed to be in large granular vesicles*

1. Peptides
2. Purines
3. 5-Hydroxytryptamine
4. Catecholamines
5. Amino acids

nerves (4,24). The possibility also cannot be excluded that NAI and cholinergic nerves are independent but structurally similar entities; however, such a hypothesis is not easily tested and thus I prefer one of the other three mentioned above.

The fact that the same nerve contains more than one type of vesicle in its varicosities has been explained in a variety of ways, e.g., that one vesicle type represents a site of synthesis, storage, or transport whereas the other is a site of release. An alternate possibility is that one nerve contains more than one mediator and each mediator is stored in a different vesicle (4). There is good evidence, too, that small agranular (cholinergic) vesicles contain and can release ATP and its breakdown products (4,15,25). Coenzyme A, a product of the acetylation of choline by choline acetylase in cholinergic nerve endings, may also be released from these nerves and act like ATP or adenosine. However, the question remains of whether or not there is any evidence that large granular vesicles store and release ATP or any mediator of NAI nerves.

A growing number of mediators have been assigned to large granular vesicles (Table 3) and recently it has been proposed (23) that the various peptides of gut neurons (e.g., VIP, substance P, etc.) are localized to these vesicles. There is convincing evidence for localization of substance P in such vesicles in brain (22). The kinds of experiments that have been used to localize putative mediators to particular vesicles are summarized in Table 4. Aside from data reported here, none of the evidence has been applied to large granular vesicles of the mammalian intestine and thus there is no direct evidence associating these vesicles in the gut with ATP, VIP, substance P, or any other proposed mediator. Therefore, there is no evidence that any of these substances is released from large granular vesicles. Since a variety of putative transmitters have been claimed to occur in such vesicles of gut nerves, and since they occur in a variety of nerves with different mediators, we have no basis for associating any given large granular vesicle with any given mediator.

Indeed, it should be noted that there is no evidence that shows conclusively that

TABLE 4. *Evidence localizing a given mediator to a particular vesicle type*

1. Isolation of vesicle type and chemical characterization
2. Selective depletion or exhaustion of mediator and concomitant alteration of vesicle type
3. Selective repletion or labeling of mediator and concomitant localization of uptake to vesicle type
4. Functional localization of mediator release and comparison to vesicles in that locus

NAI mediator or ATP or its degradation products, released during field stimulation, actually emerges exclusively or primarily from nerves rather than from an associated structure such as a Schwann cell or interstitial cell. There exist a variety of close structural contacts between these structures and gut nerves and muscles (8,9,16). Field stimulation could directly or indirectly affect structures other than nerves, and TTX could prevent these responses to field stimulation if they involve Na conductance changes in other than nerve cells or changes (muscle contractions) that are consequent to nerve activity. Thus we cannot decide among possibilities (b), (c), and (d); i.e., we do not know whether NAI mediator is released along with acetylcholine from the same nerve ending, or sequentially from another locus following release of acetylcholine, or independently from another structure by a TTX-sensitive mechanism.

Although this may seem a negative summation of our state of knowledge, it is necessary at this stage to clarify what is established and what is not. Only then will we be able to ask appropriate questions and assemble a valid description of events related to function of NAI nerves.

IS ATP THE MEDIATOR OF RELAXATION OF THE LES?

In the course of studies of the control over relaxation of LES in the North American opossum, we tested the effects of ATP. It affected the LES only in high doses of 0.1 to 1.0 mg/ml. Sometimes it caused a simple contraction (24 of 52 strips) and in other instances it caused a brief relaxation followed by a contraction (28 of 52 strips). The extent of ATP-induced relaxation was not related to the preexistent active tension. Furthermore, theophylline had no selective effect on ATP-induced effects or on relaxation to field stimulation. Thus we considered it unlikely that ATP was the mediator of relaxation or ''off'' contractions in LES. Burnstock et al. (2) suggested that PGs might be involved in the ''off'' contractions of gut muscle since indomethacin (IDM) inhibited them selectively. This was not confirmed by Mitchell and Wood (20). We examined this matter further and considered if any PG might be involved in any aspect of NAI-mediated relaxation.

ROLE OF PGs IN NAI-MEDIATED RESPONSES OF OPOSSUM LES

We examined the effects of a variety of PGs on LES muscle *in vitro*. $PGF_{2\alpha}$, PGE_1, PGE_2, PGA_2, stable endoperoxides [U 44069 (15S)-hydroxy-9α, 11α-(epoxymethano)-prosta-5Z, 13E-dienoic acid and U 46619 (15S)-hydroxy-11α, 9α-(epoxymethano)prosta-5Z, 13E-dienoic acid], thromboxane B_2 (TXB_2), and prostacyclin (PGI_2) were studied in doses of 1 ng/ml to 10 μg/ml. Only PGI_2 was an effective relaxant and was active over the dose range of 10^{-8} to 10^{-6} g/ml. However, it reduced active tension only about 10%. Its lack of efficacy may have been related to its short half-life. Arachidonic acid (18 to 174 μg/ml) caused contractions that could be inhibited by 10 μg/ml of IDM or of 5,8,11,14-eicosatetraynoic acid (ETA). Indomethacin and ETA did not significantly affect

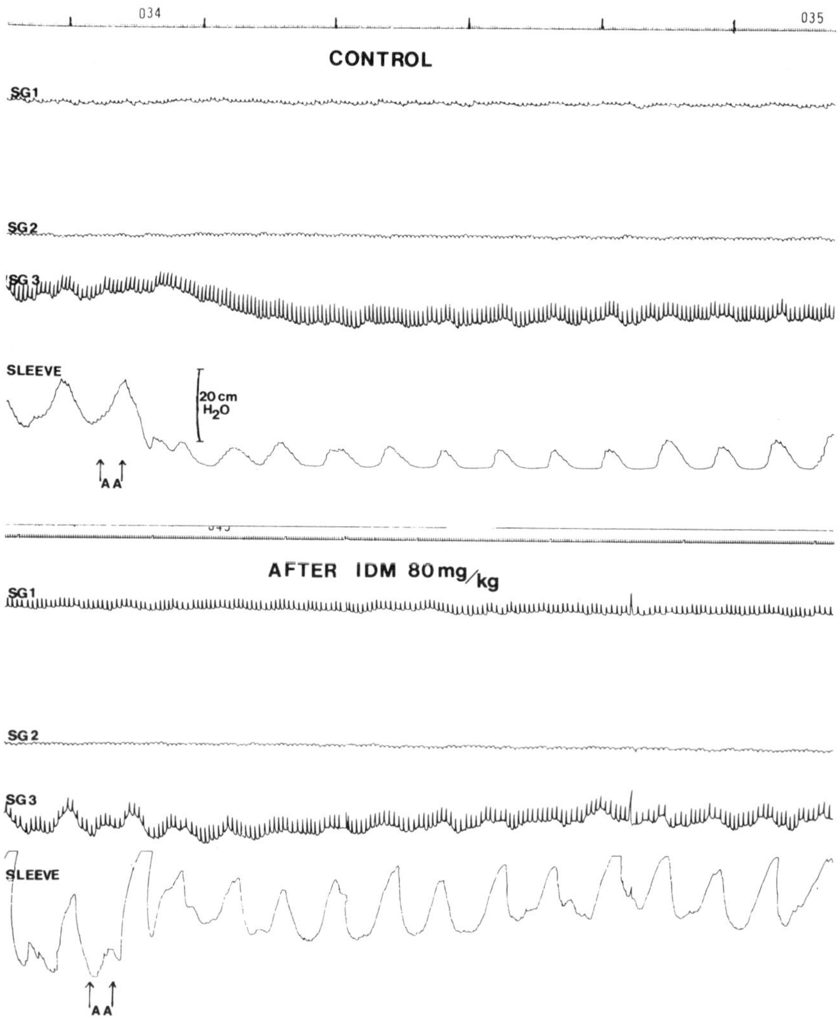

FIG. 1. Responses to 37 μg arachidonic acid (AA) (given intraarterially to opossum LES) before and after IDM, 80 mg/kg, i.v. Strain gauges 1 (SG_1) and 2 (SG_2) were 4 and 2 cm proximal to the LES. Strain gauge 3 (SG_3) was on the LES. A "sleeve" (14) was inserted through the LES. Note that the prolonged decrease in LES pressure to AA was reversed by IDM to an increase. Strain gauge calibrations are meaningless when oriented on the circumference of esophagus, but the sleeve pressure calibration is shown.

responses to PGs mentioned below. The most potent contracting substances were stable endoperoxides active in doses of 10^{-9} g/ml upward. Thromboxane B_2 was effective in the dose range 10^{-8} to 10^{-5} g/ml. Prostaglandin F_2 or PGA_2 were effective contracting agents only in the range of 0.5 to 10 μg/ml. Prostaglandin E_1 or PGE_2 were also effective only in the dose range of 0.5 to 10 μg/ml and caused weak relaxation in some strips and weak contractions in others. Thus it is clear that the PG receptors exist on LES muscle. Conceivably then, PGs must be involved in

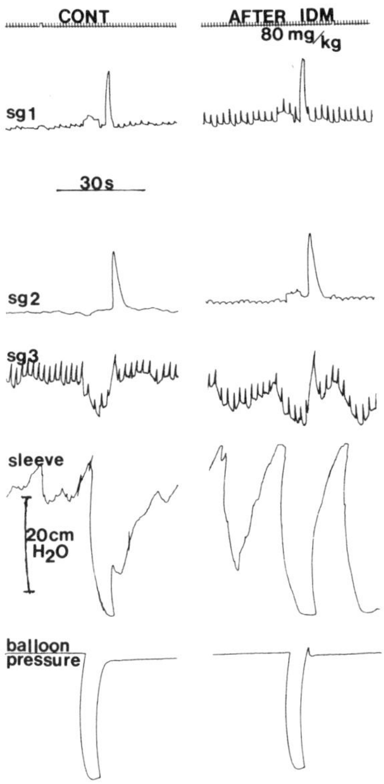

FIG. 2. Responses to balloon distention (with 15 ml air, 7.5 ml proximal to LES) on opossum esophagus. Same preparation as in Fig. 1. Responses are shown before and after IDM. The response of esophageal body (SG_1 and SG_2) is an "off" contraction, whereas LES relaxes to balloon distention. Both responses are retained after IDM. Note the much larger spontaneous oscillations of LES pressure that followed IDM treatment and the delayed recovery of LES tone following relaxation after IDM treatment. Calibrations as in Fig. 1.

mediating NAI-type responses, PGI_2 as a relaxant and endoperoxides as contractile agents.

We studied this possibility *in vitro* and *in vivo*. Our *in vitro* data are presented in detail elsewhere (9–11). Briefly, they showed that the TTX-sensitive relaxations and "off" contractions to field stimulation with 0.5-msec pulses were not readily inhibited by doses of IDM or ETA that blocked responses to arachidonic acid. There was evidence of PG involvement in TTX-insensitive relaxations to field stimulation with 5-msec pulses. We also found that 0.5 to 10 μl/ml of ethyl alcohol markedly and selectively inhibits "off" contractions and this may account for effects of agents such as IDM or PGs which are sometimes dissolved in ethanol.

In vivo, in contrast to the response *in vitro,* intraarterial arachidonic acid (infused via the left gastric artery), caused prolonged LES relaxation, possibly by releasing PGs from the mucosa removed *in vitro*. This could be inhibited or reversed by large doses of IDM (30 to 80 mg/kg) dissolved in 1% Na_2CO_3 (Fig. 1). However, these doses of IDM had no effect on either relaxations of LES or "off" contractions of body or LES (Fig. 2). Thus there is no reason to conclude that PGs play any essential role in NAI responses.

Our conclusions then are as follows:

1. The varicosities in regions with a predominant nonadrenergic inhibitory in-

nervation do not contain a large class of nerve varicosities with large granular vesicles.

2. Most nerves that appear to be associated with the function of non-adrenergic inhibitory nerves resemble "cholinergic" nerves.

3. In LES, there is little evidence to associate nerve-mediated relaxation with a purine transmitter.

4. There is no evidence that PGs are necessary for either relaxation responses or "off" responses of nonadrenergic inhibitory nerves *in vivo*.

ACKNOWLEDGMENTS

This work was supported by the Medical Research Council of Canada and carried out in collaboration with Dr. S. Sarna, Dr. W. Waterfall, Ms. Jean Crankshaw, Dr. G. Taylor, and Professor Mollie E. Holman.

REFERENCES

1. Burnstock, G. (1972): *Pharmacol. Rev.*, 24:509–581.
2. Burnstock, G., Cocks, T., Paddle, B, and Staszewska-Barcjak, J. (1975): *Eur. J. Pharmacol.*, 31:360–362.
3. Burnstock, G. (1975): *J. Exp. Zool., (Suppl.)* 194:108–134.
4. Burnstock, G. (1976): *Neuroscience*, 1:239–248.
5. Cheung, D. W., and Daniel, E. E. (1978): *Gastroenterology*, 74:1019 (abst.).
6. Christensen, J. (1970): *Gastroenterology*, 59:909–916.
7. Christensen, J., and Lund, G. F. (1969): *J. Clin. Invest.*, 48:408–419.
8. Daniel, E. E. (1977): In: *Nerves and the Gut*, edited by F. P. Brooks and P. W. Evers, pp. 154–196. Charles B. Slack, Inc., Thorofare, N.J.
9. Daniel, E. E., Crankshaw, J., and Sarna, S. (1978): *Am. J. Physiol. (in press)*.
10. Daniel, E. E., Crankshaw, J., and Sarna, S. (1978): *Int. Symp. of the Esophagus and Gastroesophageal Junction*, pp. 14–20. held Mar. 19–25, Ixtapa, Mexico. Marine Laboratories, Kansas City.
11. Daniel, E. E., Sarna, S. K., and Crankshaw, J. (1977): *Proc. Sixth Int. Symp. on Gastrointestinal Motility*, edited by H. Duthie, held Sept. 12–16, Edinburgh, Scotland. MTP Press Ltd, Lancaster, England.
12. Daniel, E. E., Taylor, G. S., Daniel, V. P., and Holman, M. E. (1977): *Can. J. Physiol. Pharmacol.*, 55:243–250.
13. Daniel, E. E., Taylor, G. S., and Holman, M. E. (1976): *Gastroenterology*, 70(5).
14. Dent, J., and Chu, B. (1976): *Gastroenterology*, 71:263–267.
15. Dowdall, M. J., Boyne, A. F., and Whittaker, V. P. (1974): *Biochem. J.*, 140:1–12.
16. Duchon, G., Henderson, R., and Daniel E. E. (1974): *Proc. Fourth Int. Symp. on Gastrointestinal Motility*, edited by E. E. Daniel, pp. 635–646. Mitchell Press, Vancouver, B.C.
17. Falck, B. (1962): *Acta Physiol Scand. (Suppl.)*, 56:1–25.
18. Hirst, G. D. S., and McKirdy, H. C. (1974): *J. Physiol. (Lond.)*, 238:129–144.
19. Kuroda, Y., and McIlwain, H. (1974): *J. Neurochem.*, 22:691–699.
20. Mitchell, G-W., and Wood, J. D. (1976): *Eur. J. Pharmacol.*, 40:63–66.
21. Oki, M., and Daniel, E. E. (1977): *Gastroenterology*, 73:1029–1040.
22. Pickel, V. M., Reis, D. J., and Leeman, S. E. (1977): *Brain Res.*, 122:534–540.
23. Polak, J. M., Bishop, A. E., and Bloom, S. R. (1978): *Gastroenterology*, 74:1080 (abst.).
24. Robinson, P. (1971): *Prog. Brain Res.*, 34:371–376.
25. White, T. D. (1977): *Nature*, 267:67–68.
26. Zimmerman, H., and Denston, C. R. (1976): *Brain Res.*, 11:365–376.
27. Zimmerman, H., and Whittaker, V. P. (1974): *J. Neurochem.*, 22:435–450.

Physiological and Regulatory Functions of Adenosine and Adenine Nucleotides, edited by H. P. Baer and G. I. Drummond. Raven Press, New York © 1979.

Presynaptic Inhibition of Adrenergic Neurotransmission by Adenine Nucleotides and Adenosine

David M. Paton

Department of Pharmacology and Clinical Pharmacology, School of Medicine, University of Auckland, Auckland, New Zealand

In the last decade it has become apparent that, in addition to the postsynaptic alpha- and beta-adrenoceptors located on the effector organ, there are presynaptic alpha-adrenoceptors located on the terminals of adrenergic nerves (11,24,29). Activation of these presynaptic receptors causes inhibition of transmitter release and appears to be involved in the physiological regulation of release. There are also presynaptic beta-adrenoceptors. Activation of these receptors increases release, particularly at low frequencies of stimulation (11).

Evidence has also been obtained for the presence of specific inhibitory presynaptic receptors for muscarinic agonists, dopamine, prostaglandins, and opiates, and for specific facilitatory presynaptic receptors for nicotinic agonists and angiotensin (11,24,29). It should be noted, however, that these latter presynaptic receptors are not present in all peripheral adrenergic nerves and their physiological roles, if any, are uncertain (11,24,29).

It is only recently that the effects of adenine nucleotides and adenosine on adrenergic transmission have been studied. Such studies have been stimulated by a belief that adenosine 5′-triphosphate (ATP) is released together with noradrenaline during adrenergic nerve stimulation (4) and by the demonstration that ATP and adenosine inhibit cholinergic transmission (20,28).

EVIDENCE FOR AN INHIBITORY EFFECT OF ADENINE NUCLEOTIDES AND ADENOSINE ON ADRENERGIC TRANSMISSION

Two approaches have been used in studies designed to determine the effect of adenine nucleotides and adenosine on adrenergic transmission. In the first, the effects of these agents on responses to nerve stimulation and exogenous noradrenaline have been compared. In the second, their effects on the release of [^{3}H]noradrenaline have been determined.

In the isolated rat portal vein, ATP and adenosine 5′-diphosphate (ADP) were 10 to 15 times more potent inhibitors of responses to transmural stimulation than of responses to noradrenaline (26). Qualitatively similar responses were obtained with

ATP in rabbit ear artery, saphenous vein, and thoracic aorta (26). By comparison, papaverine was an equipotent inhibitor of responses to transmural stimulation and noradrenaline (26).

Adenosine and ATP were also more potent inhibitors of responses to adrenergic nerve stimulation than of responses to exogenous noradrenaline in rat vas deferens (5) and canine blood vessels (27).

By contrast, in the perfused rabbit kidney, adenosine increased the perfusion pressure and potentiated responses to both adrenergic nerve stimulation and to noradrenaline (8).

Adenosine and adenine nucleotides have been found to reduce the release of [^{3}H]noradrenaline during adrenergic nerve stimulation in a number of blood vessels (7,8,26,27) and in guinea pig and rat vas deferens (5,8). These latter findings suggest that the reduced response to nerve stimulation produced by adenosine and adenine nucleotides results from a reduction in release of transmitter.

The time course of these inhibitory effects has also been studied. The inhibitory action of the purines has been found to rapidly reach a maximum and to be readily reversible.

MECHANISM OF INHIBITORY ACTION ON ADRENERGIC TRANSMISSION

Studies performed to date have indicated that the presynaptic inhibitory effects of adenosine and adenine nucleotides are not the result of an action on presynaptic alpha- or beta-adrenoceptors, muscarinic receptors, or dopaminergic receptors since they were not antagonized by phenoxybenzamine (5,7,26), propranolol (26), atropine (5,26), or haloperidol (26).

Adenine nucleotides can stimulate the biosynthesis of prostaglandins (15) and prostaglandins of the E series inhibit adrenergic transmission presynaptically in many tissues (11,24,29). However, it seems unlikely that stimulation of prostaglandin biosynthesis can account for the inhibitory presynaptic actions of adenosine and adenine nucleotides since indomethacin did not greatly alter their inhibitory actions in vas deferens (5). In rabbit ear artery, however, indomethacin inconsistently reduced the action of ATP (26).

Theophylline has been extensively employed as an antagonist of ATP and adenosine at postsynaptic sites (3). Theophylline was a weak antagonist of the presynaptic inhibitory actions of adenosine and ATP in vas deferens (5) and blood vessels (26,27) and it also antagonized the presynaptic inhibitory action of adenosine and adenine nucleotides on cholinergic transmission (20).

In taenia coli, 2,2′-pyridylisatogen has been reported to selectively antagonize the postsynaptic actions of ATP and ADP but not those of adenosine 5′-monophosphate (AMP) or adenosine (23). However, in rat vas deferens (5) and rabbit saphenous vein (26), 2,2′-pyridylisatogen failed to antagonize the presynaptic inhibitory action of ATP.

These studies suggest that adenosine and adenine nucleotides inhibit peripheral adrenergic transmission by interacting with a specific presynaptic purinergic receptor.

Adenosine 5′-triphosphate, ADP, and adenosine were all more potent inhibitors of adrenergic transmission at low stimulation frequencies (27). In this respect purinergic presynaptic inhibition is similar to alpha-adrenoceptor presynaptic inhibition (11,24,29). In canine blood vessels, adenosine inhibited release produced by electrical field stimulation and by elevated potassium (30 mM), but had no effect on release produced by tyramine (27). The first two procedures cause calcium-dependent release whereas tyramine does not. These findings suggest that adenine nucleotides and adenosine may inhibit adrenergic transmission by an action on calcium fluxes. It should be noted, however, that ATP did not greatly alter release induced by either elevated potassium (70 mM) or tyramine (10 μM) in rabbit aorta (26).

RELATIVE POTENCIES OF ADENOSINE AND ADENINE NUCLEOTIDES

In rat portal vein transmurally stimulated at 2 Hz, ATP was the most potent inhibitor of adrenergic transmission (26). In this tissue it had a threshold concentration of about 0.1 μM and an ED_{50} of about 1 μM. Adenosine 5′-diphosphate was about half as potent, whereas AMP and adenosine were about one-third as potent. By comparison in rat vas deferens, adenosine was the most potent with a threshold concentration of about 3 μM (5).

The true potencies of adenosine and adenine nucleotides as presynaptic inhibitors of adrenergic transmission are, however, probably masked by the effect of uptake of adenosine by tissue. The evidence for this is that in rat vas deferens the inhibitor of adenosine uptake 6-(2-hydroxy-5-nitro)-benzylthioguanosine (HNBTG) (18), potentiated the inhibitory effects not only of adenosine but also of ATP, ADP and AMP (5), and it was suggested that the adenine nucleotides must first be hydrolyzed by extracellular phosphorylytic enzymes to adenosine in order to cause presynaptic inhibition (5). Such rapid hydrolysis is known to occur in several tissues (for example, see ref. 2).

STRUCTURE-ACTIVITY RELATIONSHIPS FOR PRESYNAPTIC INHIBITION BY ADENOSINE

The structure-activity relationships for the presynaptic inhibition of adrenergic transmission have been examined extensively in the rat vas deferens (19). In this study HNBTG was used to block the possible uptake of adenosine and analogs.

The effect of substituents to the adenine moiety of adenosine are shown in Table 1. Adenosine and adenosine-N^1-oxide were the most potent inhibitors tested. 6-Methylaminopurine riboside and 6-hydroxylaminopurine riboside were slightly less potent than adenosine. 6-Δ^2-Isopentenylaminopurine riboside had weak activity. The other analogs tested had little or no activity.

TABLE 1. *Effect of changes to the purine portion of adenosine on presynaptic inhibition of transmission in the rat vas deferens.*

Compound	Inhibition of twitch response (%)
DMSO	5 ± 1
Adenosine and derivatives	
Adenosine	65 ± 3
Inosine	6 ± 1
Adenosine-N^1-oxide	60 ± 13
8-Bromoadenosine	8 ± 1
Purine riboside	13 ± 3
Adenine	9 ± 1
6-Substituted purine ribosides	
6-Methylamino-	49 ± 9
6-Dimethylamino-	10 ± 2
6-Hydroxylamino-	51 ± 7
6-Δ^2-Isopentenylamino-	23 ± 6
6-Methyl-	12 ± 3
6-Methoxy-	9 ± 1
6-Mercapto-	8 ± 1
6-Chloro-	9 ± 2

The effect of each compound was determined at 100 μM. Mean ± SEM.
From ref. 19, with permission.

The effect of substituents or modifications to the ribose moiety of ribose are shown in Table 2. 5′-Deoxyadenosine and 5′-nitroadenosine were slightly less potent than adenosine. 3′-Deoxyadenosine also produced significant inhibition. All the other analogs had little or no activity.

The effect of substituents at C_2 of the purine ring has also been examined in rat vas deferens in the absence of inhibition of the uptake of adenosine (15).

TABLE 2. *Effect of changes to the sugar moiety of adenosine on presynaptic inhibition of transmission in the rat vas deferens*

Compound	Inhibition of twitch response (%)
DMSO	5 ± 1
Adenosine	65 ± 3
2′-Deoxyadenosine	10 ± 1
3′-Deoxyadenosine	24 ± 5
5′-Deoxyadenosine	45 ± 12
Adenine xylofuranoside	10 ± 1
Adenine arabinoside	7 ± 1
2′,3′-Diacetyladenosine	9 ± 2
2′,3′-Isopropylidene adenosine	8 ± 2

The effect of each compound was determined at 100 μM. Mean ± SEM. From ref. 19, with permission.

2-Chloroadenosine, 2-bromoadenosine, and 2-hydroxyadenosine were all more potent and longer acting than adenosine. 2-Fluoroadenosine was also more potent but had a similar duration of action.

These studies have shown that the structural requirements for activity of adenosine analogs at the presynaptic purinergic receptor in rat vas deferens include a primary or secondary amine function at C_6 of the purine ring with little tolerance for major steric changes or substituents on the sugar moiety. A variety of substituents at C_2 of the purine ring can also be tolerated.

EFFECT OF UPTAKE AND DEAMINATION ON THE POTENCY OF ADENOSINE

As indicated previously, in untreated rat vas deferens, 2-chloroadenosine is a more potent inhibitor of transmission than adenosine (15). However, whereas the inhibitory action of adenosine was potentiated by inhibitors of adenosine uptake (i.e., dipyridamole and HNBTG) and by an inhibitor of adenosine deaminase (2′-deoxycoformycin), that of 2-chloroadenosine was not (15). When both adenosine uptake and deamination were inhibited, the inhibitory potencies of adenosine and 2-chloroadenosine did not differ significantly.

These findings indicate that the effective concentration of adenosine at the presynaptic purinergic receptor is reduced by uptake into tissues (smooth muscle ?) and by deamination. Consequently, it is necessary to inhibit these processes before any conclusions can be drawn about the true relative potencies of agonists at this receptor.

RELATIONSHIP TO OTHER PURINERGIC RECEPTOR SYSTEMS

The presynaptic inhibitory effect of adenosine in the rat vas deferens (19) shows structure-activity requirements similar to those found postsynaptically in the rabbit intestine (13) and in the coronary artery (1,6,17). A notable difference is adenosine-N^1-oxide which is a potent agonist in the vas and intestine but not in the coronary artery. 6-Δ^2-Isopentenylaminopurine ribose is clearly active in the intestine but marginally active in the other two tissues.

In the guinea pig ventricle, adenosine causes an increase in adenosine 3′:5′-monophosphate (cyclic AMP) (9). In this system 6-methoxypurine riboside, 2′-deoxyadenosine and 2′,3′-isopropylidene adenosine are agonists, but are inactive in the vas. Adenosine also causes an increase in cyclic AMP levels in brain slices (12). In this system a number of adenosine derivatives inhibited the action of adenosine. This effect was not observed in the vas. Finally, the structure-activity requirements for the inhibition of adenyl cyclase in intestinal smooth muscle (14) differ markedly from those in the vas deferens. These features are summarized in Table 3.

These considerations suggest that there may be different types of purinergic receptors present in different tissues. However, elucidation of this possibility will depend on more detailed studies of the relative order of potency of agonists in

TABLE 3. *Comparison of structure-activity relationships for different purinergic receptor systems*

Compound	Presynaptic inhibition of adrenergic transmission	Coronary flow	Rabbit intestinal relaxation	Cyclic AMP in brain slices	Cyclic AMP in ventricle	Adenylate cyclase inhibition
Adenosine	+	+	+	+	+	+
Adenosine-N¹-oxide	+	−	+		+	−
Purine ribosides						
6-Methylamino-	+	−	+		+	−
6-Hydroxylamino-	+	+	+		+	+
6-Δ^2-Isopentylamino-	+	−	+			−
6-Methyl-	−	−	−	+		−
6-Methoxy-	−		−		+	−
6-Mercapto-	−	−	−		−	+
Sugar modifications						
2-Deoxyadenosine	−	−	−	inhi	+	+
3-Deoxyadenosine	−		−	inhi	−	
5-Deoxyadenosine	+	+		inhi	inhi	
2′,3′-Isopropylidene adenosine	−		−		+	−

Data on presynaptic inhibition of adrenergic neurotransmission from ref. 19; on coronary flow from refs. 1, 6, and 17; on rabbit intestinal relaxation from ref. 13; on cyclic AMP in brain slices from ref. 12; on cyclic AMP in ventricle from ref. 9; and on inhibition of adenylate cyclase in small intestine from ref. 14.

Symbols: (+), significant effect; (−), no significant effect; (inhi), inhibition of adenosine-induced response by the compound.

different tissues and the use of more potent receptor antagonists when these become available.

PHYSIOLOGICAL ROLE OF PRESYNAPTIC INHIBITION OF ADRENERGIC TRANSMISSION BY ADENINE NUCLEOTIDES AND ADENOSINE

Is ATP available in adrenergic vesicles for release? The actual amount of ATP in the large vesicles is low (10) whereas the molar ratio of noradrenaline to ATP is about 50 in splenic nerve terminals (31). Because there is no convincing evidence that the small vesicles contain ATP, Smith (22) has concluded that caution is required in any discussion about the possible release of ATP from adrenergic nerves. Other potential sources of ATP are purinergic nerves (3) and cholinergic nerves (21).

Is ATP released by nerve stimulation? Nerve stimulation has been shown to cause the release of ATP from cholinergic nerves to the rat diaphragm (21) whereas exposure of rat brain synaptosomes to elevated potassium or veratridine caused the release of ATP (30). It seems likely that the synaptosomal preparation contained adrenergic nerve terminals.

After exposure of rabbit portal vein to [^{3}H]adenosine, subsequent transmural stimulation caused the release of tritiated adenine nucleotides and adenosine (25). This release of ATP was significantly reduced by guanethidine and abolished by tetrodotoxin. Su (25) concluded that the purine nucleotides were released from both adrenergic and nonadrenergic nerves.

Is it likely that the concentration of ATP would be sufficient to cause presynaptic inhibition? Su (25) considered this question. Based on assumptions about the molar ratios of ATP to noradrenaline in vesicles and the amount of noradrenaline released, he concluded that ATP might attain a concentration of about 1 μM in the rat portal vein during nerve stimulation at 16 Hz (26). However, if the molar ratio of noradrenaline to ATP is indeed greater than he assumed and if the frequency of nerve stimulation is considerably lower than 16 Hz as is likely under physiological circumstances, then the concentration of ATP is likely to be much lower. It should also be noted that 300 μM ATP was required to inhibit adrenergic transmission in the rat portal vein (7).

Is adrenergic transmission potentiated by receptor antagonists for adenine nucleotides and adenosine, and inhibited by drugs that block the uptake and/or deamination of adenosine? In the rat portal vein (7), noradrenaline release was significantly reduced in the presence of an adenosine uptake inhibitor (i.e., dilazep) and an adenosine deaminase inhibitor (i.e., erythro-9[2-hydroxy-3-nonyl]adenine), in keeping with a physiological role for a purinergic negative feedback on release. However, in rat vas deferens, responses to transmural stimulation were unaffected by theophylline and were reduced by 2,2′-pyridylisatogen (5).

It is clear that a great deal of additional information is required before it can be finally established whether or not the presynaptic release of a purine compound during adrenergic transmission has a physiological role.

CONCLUSIONS

Adenosine, ATP, ADP, and AMP inhibit peripheral adrenergic transmission apparently by an action on a presynaptic purinergic receptor. The true potency of adenosine at this site is masked by uptake and deamination. The structural requirements for activity of adenosine analogs include a primary or secondary amine function at C_6 of the purine ring with little tolerance for major steric changes or substituents on the sugar moiety. A variety of substituents at C_2 of the purine ring can also be tolerated.

Outstanding questions for future work include: How does adenosine produce presynaptic inhibition? Do all peripheral adrenergic fibers possess this purinergic system? Is it also present on central adrenergic neurons? And what is the physiological importance of this form of presynaptic inhibition?

ACKNOWLEDGMENTS

Original investigations were supported by grants from the Medical Research Council of Canada and the Alberta Heart Foundation. I wish to acknowledge the work of my collaborators at the University of Alberta who made this study possible: Professor H. P. Baer, Dr. A. S. Clanachan, Dr. A. Johns, Mrs. P. Lauzon, and Mr. M. J. Muller.

REFERENCES

1. Augers, J. A., Cobbin, L. B., Einstein, R., and Maguire, M. H. (1971): *Br. J. Pharmacol. Chemother.*, 41:592–599.
2. Baer, H. P., and Drummond, G. I. (1968): *Proc. Soc. Exp. Biol. Med.*, 127:33–36.
3. Burnstock, G. (1975): In: *Handbook of Psychopharmacology,* vol. 5, edited by L. I. Iverson, S. D. Iverson, and S. H. Snyder, pp. 131–194. Plenum Press, New York.
4. Burnstock, G. (1976): *Neuroscience,* 1:239–248.
5. Clanachan, A. S., Johns, A., and Paton, D. M. (1977): *Neuroscience,* 2:597–602.
6. Cobbin, L. B., Einstein, R., and Maguire, M. H. (1974): *Br. J. Pharmacol. Chemother.*, 50:23–33.
7. Eneroz, M. A., and Saidman, B. Q. (1977): *Naunyn Schmiedebergs Arch. Pharmacol.*, 297:39–46.
8. Hedqvist, P., and Fredholm, B. B. (1976): *Naunyn Schmiedebergs Arch. Exp. Pathol. Pharmacol.*, 293:217–223.
9. Huang, M., and Drummond, G. I. (1976): *Biochem. Pharmacol.*, 25:2713–2719.
10. Lagercrantz, H. (1976): *Neuroscience,* 1:81–92.
11. Langer, S. Z. (1977): *Br. J. Pharmacol.*, 60:559–568.
12. Mah, H. D., and Daly, J. W. (1976): *Pharmacol. Res. Commun.*, 8:65–79.
13. McKenzie, S. G., Frew, R., and Bär, H.-P. (1977): *Eur. J. Pharmacol.*, 41:183–192.
14. McKenzie, S. G., Frew, R., and Bär, H.-P. (1977): *Eur. J. Pharmacol.*, 41:193–203.
15. Muller, M. J., and Paton, D. M. (1978): *Proc. Can. Fed. Biol. Soc.*, 21:144.
16. Needleman, P., Minkes, M. S., and Douglas, J. R. (1974): *Circ. Res.*, 34:455–460.
17. Olsson, R. A., and Patterson, R. E. (1976): *Prog. Biophys. Mol. Biol.*, 4:227–248.
18. Paterson, A. R. P., Kim, S. C., Bernard, O., and Cass, C. E. (1975): *Ann. N.Y. Acad. Sci.*, 255:402–411.
19. Paton, D. M., Bär, H.-P., Clanachan, A. S., and Lauzon, P. A. (1978): *Neuroscience,* 3:65–70.
20. Sawynok, J., and Jhamandas, K. H. (1976): *J. Pharmacol. Exp. Ther.*, 197:379–390.
21. Silinsky, E. M. (1975): *J. Physiol.*, 247:145–162.
22. Smith, A. D. (1978): In: *The Release of Catecholamines from Adrenergic Neurons,* edited by D. M. Paton. Pergamon Press, Oxford *(in press).*

23. Spedding, M., and Weetman, D. F. (1976): *Br. J. Pharmacol. Chemother.*, 57:305–310.
24. Starke, K. (1977): *Rev. Physiol. Biochem. Pharmacol.*, 77:1–124.
25. Su, C. (1975): *J. Pharmacol. Exp. Ther.*, 195:159–166.
26. Su, C. (1978): *J. Pharmacol. Exp. Ther.*, 204:351–361.
27. Verhaeghe, R. H., Vanhoutte, P. M., and Shepherd, J. T. (1977): *Circ. Res.*, 40:208–215.
28. Vizi, E. S., and Knoll, J. (1976): *Neuroscience,* 1:391–398.
29. Westfall, T. C. (1977): *Physiol. Rev.*, 57:659–728.
30. White, T. D. (1978): *J. Neurochem.*, 30:329–336.
31. Winkler, H. (1976): *Neuroscience,* 1:65–80.

Physiological and Regulatory Functions of Adenosine and Adenine Nucleotides, edited by H. P. Baer and G. I. Drummond. Raven Press, New York © 1979.

Multiple Adenine Derivative Receptors in Rat Ileum and Electrical Degradation of Purine Drugs

K. Nakatsu and V. Bartlett

Department of Pharmacology, Faculty of Medicine, Queen's University, Kingston, Ontario, K7L 3N6, Canada

Although it has been 80 years since a noncholinergic, nonadrenergic component of the autonomic nervous system was alluded to by Langley (3), the identity of the transmitter still remains to be established. Adenosine and its nucleotides, adenosine monophosphate (AMP), adenosine diphosphate (ADP), and adenosine triphosphate (ATP), have been suggested to act as neurotransmitters or as mediators of other types of intercellular communication (see ref. 4). Currently ATP is the most intensely investigated candidate for the transmitter of noncholinergic, nonadrenergic nerves. As Burnstock (2) pointed out in his 1972 review, the usual criteria required of a putative transmitter are not particularly useful when considering a substance such as ATP. This is because its presence, release, synthesis, and degradation are not unique to nonadrenergic, noncholinergic nerves (6). However, drugs that either block or enhance the effects of ATP would be of particular importance in establishing its possible transmitter function.

Previous work from this laboratory (1) indicated that one might be able to exploit theophylline because of its ability to block adenosine (and ATP) receptors in the isolated rabbit ileum. In this tissue adenosine induced inhibition of spontaneous contractions and the adenosine receptors were found in the muscle cells (most likely on the cell surface) as opposed to nerves. Theophylline antagonized adenosine in a manner that appeared to be competitive. The adenine nucleotides also inhibited spontaneous activity of the ileum and this effect was similarly antagonized by theophylline. A series of experiments was then undertaken in an attempt to determine if the electrically induced inhibition of rabbit intestine might be inhibited by methylxanthines and thus be attributable to an adenine nucleotide.

Intestinal sections were mounted in conventional organ baths, and exposed to drugs and electrical stimulation. An example of typical tracings is shown in Fig. 1. Electrical stimulation rapidly inhibited the spontaneous activity and this inhibition was not blocked by atropine or antiadrenergic drugs. When theophylline was added to the baths and the electrical stimulation repeated, the results were highly inconsistent. In only about 30% of tissues theophylline blockade of the electrically induced response was obtained as shown in Fig. 1. In tissues in which theophylline blocked the effects of electrical stimulation, an unusual feature was the slow onset of action.

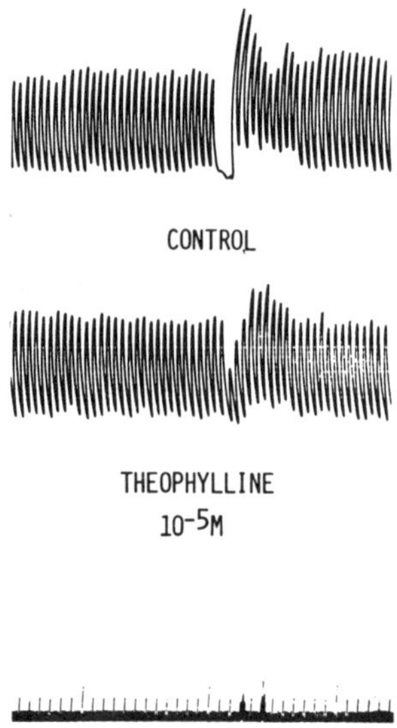

FIG. 1. Inhibition of rabbit ileum. The upper tracing shows the inhibitory effect of electrical stimulation on the isolated rabbit ileum, 30 Hz, 0.5 msec. The lower tracing shows the effect of theophylline (10 μM) on the electrically induced response.

The maximum blockade by theophylline was not observed until 15 to 30 min after the drug was added to the bath. In our study, the greatest nonadrenergic, noncholinergic inhibitory effect was observed at a stimulation frequency of about 30 Hz and the greatest effect of theophylline was at frequencies above 100 Hz. The interanimal inconsistency persisted even when different strains of rabbits were used. Because of the difficulties encountered above it was decided to repeat these experiments in another species and to investigate the effects of electrical stimulation on the agonists used.

RECEPTOR STUDIES

Rat ileum was prepared as a classic organ bath preparation so that isometric contractile activity of longitudinal muscle could be monitored. Figure 2 shows typical traces of spontaneous activity of the rat ileum and its inhibition by ATP. Similar inhibition was induced by AMP, ADP, and adenosine. However, marked differences in potency exist as shown in Fig. 3. Clearly ADP and ATP are much more potent than adenosine in inducing inhibitory responses. It should also be noted that the curves for ATP and ADP are somewhat bell shaped. This is because high concentrations of ATP and ADP caused contractions of the tissue. At high concentrations neither adenosine nor AMP appeared to cause contractions. These contractions appear to be receptor mediated and not the result of chelation of ions by the phosphate moieties because the effects were not reproduced with either pyrophosphate or tripolyphosphate.

Another difference in the response to agonists is revealed by the effects of theophylline as shown in Fig. 4. It can be seen that the dose-response curves for adenosine were shifted to the right whereas those for ATP were not affected. This

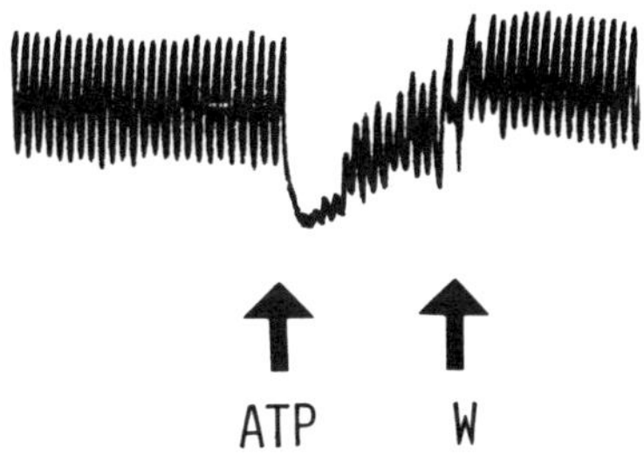

FIG. 2. Inhibition of rat ileum by ATP. Rat ileum was bathed in an organ bath containing Krebs bicarbonate solution gassed with 95% O_2 and 5% CO_2. The temperature was 37° and the base-line tension was 1.0 g. The dose of ATP giving this inhibition was 3 μM.

theophylline blockade of adenosine action and lack of blockade of ATP is consistent with the existence of two receptors; the contractile action of ATP may indicate that a third type of ATP or nucleotide receptor exists in the rat ileum. Experiments using dipyridamole also support such an interpretation. In this preparation dipyridamole did not increase the effectiveness of exogenous adenosine but antagonized the effect of the nucleoside and shifted the dose-response curve to the right. In comparison, dipyridamole had no detectable effect on the response to ATP.

Results from another series of experiments also support the concept of distinct receptors for ATP and adenosine. Tissues were exposed to ATP for 30 min and then tested for their responsiveness to ATP and adenosine. These tissues responded only poorly to ATP whereas adenosine was still effective in producing inhibition. Other tissues were exposed to adenosine for 30 min and found to be less sensitive to adenosine but still sensitive to ATP. In these experiments ADP mimicked ATP and the effects of AMP most closely resembled those of adenosine.

ELECTROCHEMICAL STUDIES

In addition to tissue variability, another factor that could contribute to the inconsistency of our pharmacological studies is the chemical instability of agonists

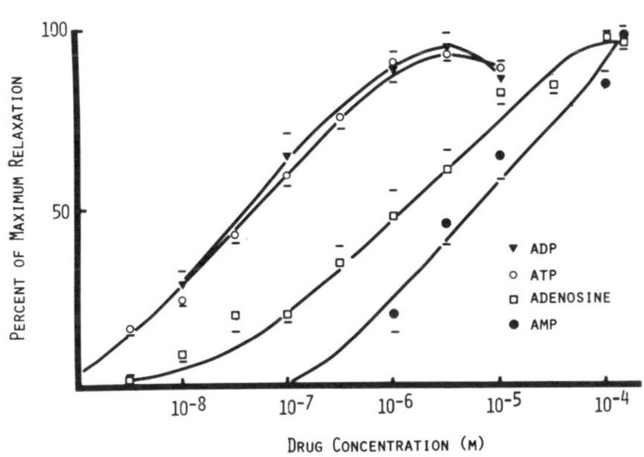

FIG. 3. Rat ileum dose-response curves for ATP, ADP, AMP, and adenosine. Each point represents the mean ± SEM and N = 7 or 8.

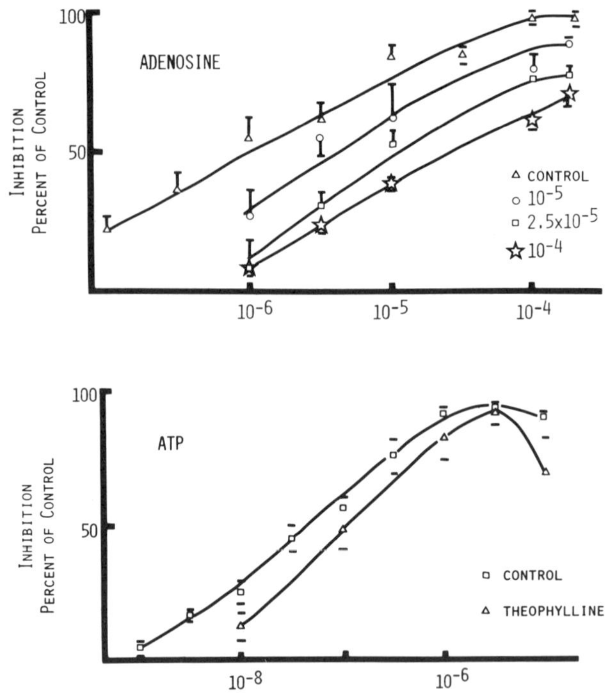

FIG. 4. Effect of theophylline on responses to adenosine and ATP. The upper panel shows that theophylline shifted the dose-response curve for adenosine far to the right. Theophylline concentration (*upper panel*): ▲, control; ○, 10 μM; □, 25 μM; ☆, 100 μM. Adenosine triphosphate (*lower panel*): □, no theophylline, Δ, theophylline 100 μM.

used. Therefore we investigated the possible effects of electrical stimulation on theophylline, adenosine, and adenine nucleotides in physiological buffers.

From an electrochemical point of view one would expect that at least some alteration of drug molecules would take place because the voltages used in biological experiments exceed their redox potentials. To test the sensitivity of these drugs to electrochemical oxidation and reduction, drug solutions were subjected to pulsatile direct current in rudimentary half-cells. The compounds were dissolved in either phosphate buffer (10 mM pH 7.4 with 100 mM NaCl) or Tyrode's solution. Experimental conditions are given in the legend to Fig. 5, which shows the effect of electrical stimulation on ATP, as monitored by a change in the ultraviolet (UV) spectrum. It can be seen clearly that after 10 min ATP was altered at the anode because the absorbance peak at 260 nm disappeared. The cathodic effect was a diminution in peak size without any shift in the position of the peak; this would be consistent with partial electrochemical reduction. Similar changes were observed for the adenosine antagonist, theophylline; anodic current caused a loss of the peak at 273 nm (Fig. 6). Adenosine monophosphate, ADP, and adenosine were also tested and the results were essentially the same as for ATP. Other agents examined

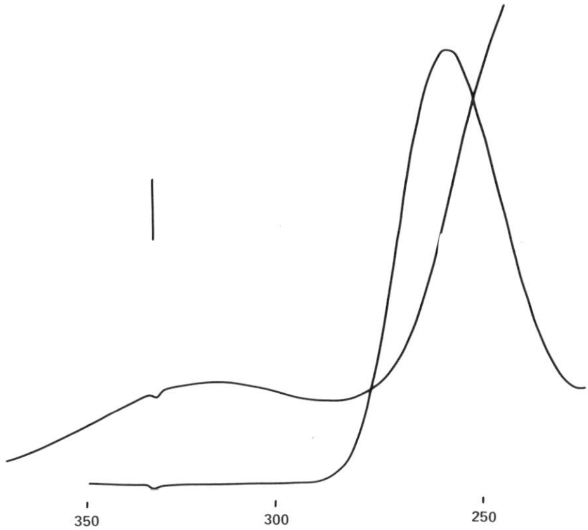

FIG. 5. Electrically induced degradation of ATP. Adenosine triphosphate was diluted in Tyrode's solution to make a final concentration of 90 μM. Aliquots (1.2 ml) were placed in each of two half-cells that were made by filling the bottom of a U-tube with 1% agar and 100 mM NaCl. The solutions were stimulated through platinum wire electrodes using a Grass model SD 9 stimulator; 1 Hz, 200 msec, 100 V nominal. Vertical bar represents 0.05 AU.

were epinephrine, norepinephrine, indomethacin, and aspirin. All were indeed sensitive to electrical current. Epinephrine at the anode lost its absorbance peak whereas the absorbance spectra for aspirin and indomethacin at both anode and cathode were altered. Indomethacin has absorbance peaks at 260 nm and at 319 nm, and the most marked effect of anodic current was elimination of the 260 nm peak. Cathodic current resulted in replacement of both peaks with a broader peak at 266 nm. The data indicated that electrical stimulation was capable of altering these compounds in physiological media. It should be noted that the endpoint monitored

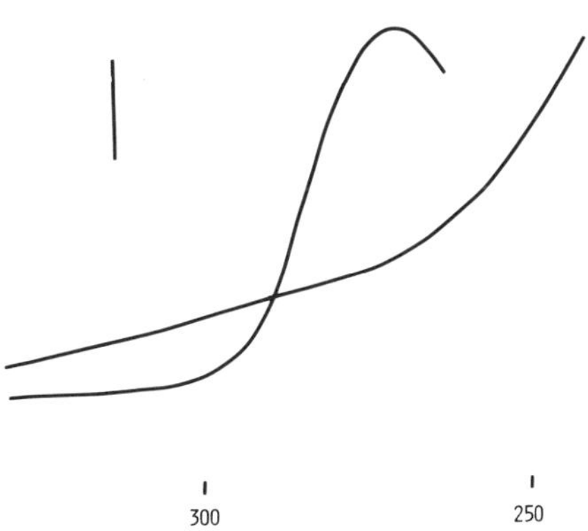

FIG. 6. Electrically induced degradation of theophylline. Conditions were the same as for Fig. 5; the concentration of theophylline was 70 μM.

was a change in absorbance spectra and therefore the molecular alterations would have been in the conjugated double-bond system. Chemical alterations in other parts of the molecule such as the ribose would not have been detected by UV spectrophotometry. The drug concentrations were selected for reliable detection in the spectrophotometer and in many cases were higher than those used in biological experiments.

The above effects were not caused by electrophoresis of the drugs into the gel bridge separating the half-cells as evidenced by experiments with radioactively labeled theophylline. In this case practically all the [^{14}C] label was recovered from the half-cell whereas a complete loss of the absorbance peak was observed. It is also important to note that the changes were not reversible by reversing polarity of the electrodes, which is consistent with the observation that organic electrochemical reactions are generally irreversible (5). For those purposes, then, it may make little difference if one uses biphasic stimulation in biological studies.

The results of these experiments were similar whether compounds were diluted in phosphate or Tyrode's solution. However, there was a major difference when they were added to previously stimulated solutions. In this case, phosphate solutions had no effect on the absorbance profiles of theophylline and adenine derivatives. On the other hand Tyrode's solution itself, after anodic electrical stimulation, altered the absorbance spectra of adenosine, its nucleotides, and theophylline. Figure 6 shows the effect of stimulated Tyrode's solution on theophylline. It now appears that the substance responsible for the destruction of these compounds was derived from glucose. Substitution of glucose in Tyrode's solution with gluconolactone also caused loss of absorbance. It appears, therefore, that there are two mechanisms that can affect purines; one that is directly related to the electrode itself, and a second that is indirect and mediated by a derivative of glucose. At this point it is not clear what the degradative products of either process are, but certainly the possibility of electrochemical effects on drugs used in usual pharmacological and physiological experiments must be entertained. Problems could result from both a reduction of drug concentrations and formation of pharmacologically active degradation products. The magnitude and significance of these electrochemical processes remains to be investigated.

ACKNOWLEDGMENT

This research was supported by the Harry Botterell Foundation.

REFERENCES

1. Ally, A. I., and Nakatsu, K. (1976): *J. Pharmacol. Exp. Ther.*, 199:208–215.
2. Burnstock, G. (1972): *Pharmacol. Rev.*, 24:509–581.
3. Langley, J. N. (1898): *J. Physiol. (Lond.)*, 23:407–414.
4. Nakatsu, K. (1978): *Med. Hypotheses*, 4:472–480.
5. Rifi, M. R., and Covitz, F. H. (1974): *Introduction to Organic Electrochemistry*. Marcel Dekker, New York.
6. Su, C. (1975): *J. Pharmacol. Exp. Ther.*, 195:159–166.

Physiological and Regulatory Functions of Adenosine and Adenine Nucleotides, edited by H. P. Baer and G. I. Drummond. Raven Press, New York © 1979.

Effects of 2-2′-Pyridylisatogen on Responses to Adenosine 5′-Triphosphate in Isolated Smooth Muscle Preparations

M. Hooper, *M. Spedding, **A. J. Sweetman, and D. F. Weetman

School of Pharmacy, Sunderland Polytechnic, Sunderland, Tyne & Wear, England

Considerable evidence has accumulated indicating the presence of nonadrenergic inhibitory nerves in the gastrointestinal tract (5,7–9). Burnstock (8) postulated that the neurotransmitter mediating these effects is a purine nucleotide, the most likely candidate being adenosine 5′-triphosphate (ATP). This hypothesis is attractive in that ATP mimics the effects of inhibitory nerve stimulation in some instances (but see ref. 24), although relatively high concentrations of the nucleotide must be used. However, several criteria must be fulfilled before the hypothesis may be accepted.

Demonstration of the release of [^{3}H]ATP following inhibitory nerve stimulation has been attempted in guinea pig stomach and taenia caeci preparations (10, 27,29,37). However, high-frequency stimulation (10 to 50 Hz) was used in all these studies whereas optimal relaxation of the preparations occurs at 2 to 5 Hz. Furthermore, [^{3}H]ATP is readily released from the smooth muscle cells of these preparations following contraction or relaxation (19,20,23). The proposal of Cook et al. (12) that coenzyme A may be released from cholinergic nerves in sufficient quantities to modulate release of acetylcholine further complicates specific demonstration of [^{3}H]nucleotide release.

Dipyridamole has been used to inhibit adenosine uptake into the cells of guinea pig taenia caeci preparations in an attempt to show equivalent potentiation of relaxations induced by field stimulation and by ATP (30). However, although responses of the taenia to adenosine are potentiated by dipyridamole, the potentiation of responses to ATP and field stimulation is small or insignificant and nonspecific (Fig. 1) (3,17).

Critical evidence for the purinergic nerve hypothesis may be provided by the use of specific ATP receptor antagonists. We were therefore interested in the compound 2-2′-pyridylisatogen tosylate (PIT), synthesized in the Pharmaceutical Chemistry Laboratory at Sunderland Polytechnic, which was found to antagonize ATP-induced relaxations of guinea pig taenia caeci preparations after first causing a direct relaxation of the smooth muscle (Fig. 2). In this experiment, and in subsequent

*Present address: Centre de Recherche Merrell International, 16, rue d'Ankara, 67084 Strasbourg Cedex, France.

**Present address: School of Health and Applied Sciences, Leeds Polytechnic, Leeds, England.

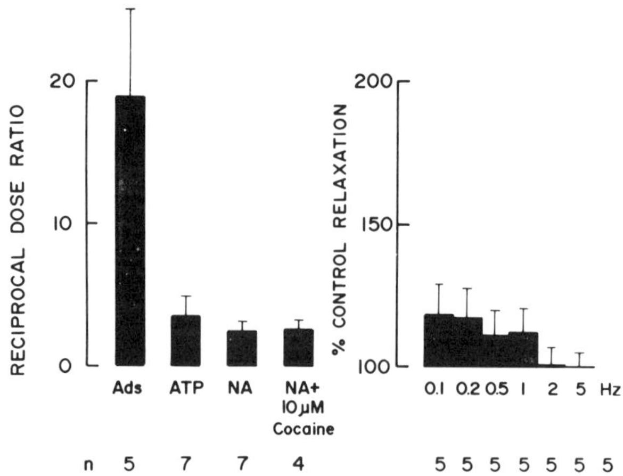

FIG. 1. Effects of dipyridamole (2 μM) on the responses of guinea pig taenia caeci preparations to adenosine (Ads), ATP, noradrenaline (NA), and noradrenaline in the presence of cocaine (10 μM). Dose ratios were calculated as the ratio of EC_{50} values obtained from cumulative concentration-response curves obtained after and before incubation with dipyridamole (2 μM); the results are plotted as reciprocal dose ratio, a measure of potentiation. Note that although the sensitivity of the taenia to all the relaxant drugs was increased ($p > 0.05$) by dipyridamole, the increase in sensitivity to adenosine ($p < 0.01$) was greater than to the other agonists. The effects of dipyridamole (2 μM) on relaxations of the taenia induced by field stimulation in the presence of hyoscine (0.3 μM) are also shown. Relaxations in the presence of dipyridamole (2 μM) were not significantly ($p < 0.1$) greater than control relaxations at any of the frequencies tested. Vertical bars represent SEM.

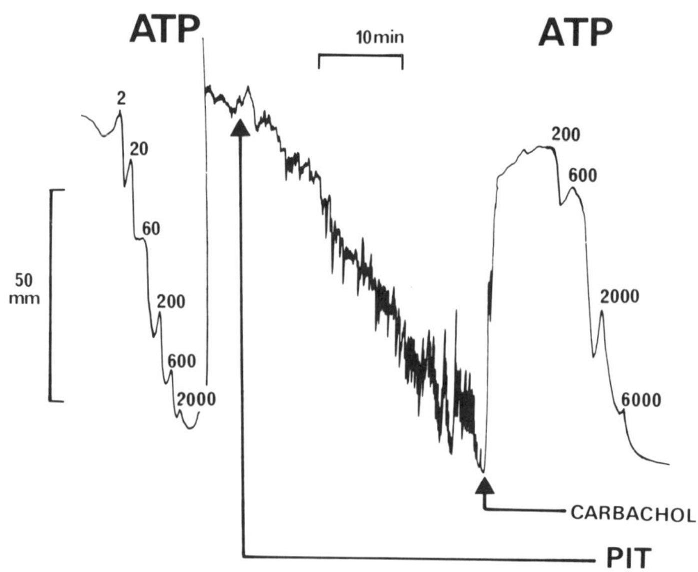

FIG. 2. Effects of PIT (50 μM) on the responses of guinea pig taenia caeci preparation to ATP (μM). The ATP was added cumulatively and the drum speed increased (×8) to show the relaxations. Carbachol (0.2 μM) was used to restore tone.

experiments using isatogen derivatives, the direct relaxant effects of the drugs necessitated the use of a spasmogen to restore the tone of the preparations to within 20% of the original level before the relaxations induced by ATP in the presence of the antagonist could be compared with control responses.

EFFECTS OF ATP

Guinea pig taenia caeci preparations were set up in isolated organ baths filled with McEwen's solution (composition: NaCl, 130 mM; KCl, 5.6 mM; $CaCl_2$, 2.2 mM; $NaHCO_3$, 25 mM; NaH_2PO_4, 1.2 mM; glucose, 11.1 mM; and sucrose, 13.2 mM) (21) at 35 ± 1°C, gassed with 95% and 5% CO_2. Responses were recorded isotonically (magnification 4:1, load 1.5 g). Responses to relaxant drugs were obtained as single relaxations at 5- or 10-min intervals (30- or 60-sec contact) or as cumulative concentration-response curves (33).

Responses to ATP were rapid in onset (1 to 5 sec) and reached maximum effect within 5 to 15 sec. The relaxation was not fully maintained despite the continuing presence of ATP in the bath; the tone of the preparations increased beyond the original level on washout (see control responses in Figs. 2 and 3). Responses to ATP were not antagonized by tetrodotoxin (0.3 μM), guanethidine (17 μM), or by a combination of phentolamine (26 μM) and propranolol (3.4 μM), and can thus be ascribed to a direct inhibition of the smooth muscle cells of the taenia. The responses were not tachyphylactic, whether obtained as cumulative concentration-

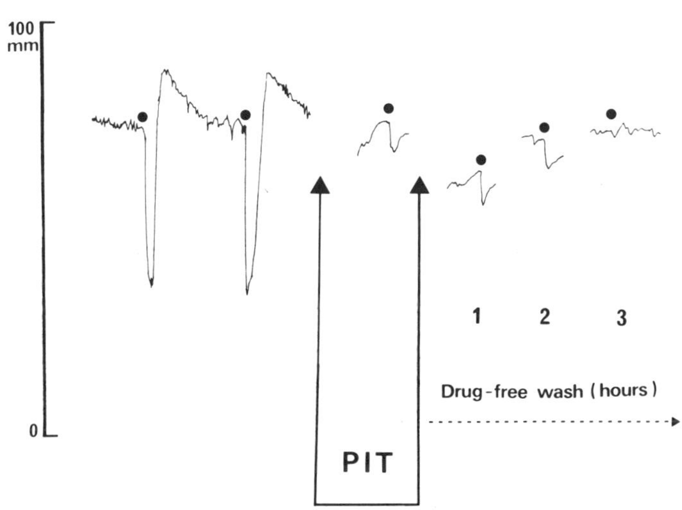

FIG. 3. The effects of washing (1 liter McEwen's solution/hour) on the antagonism of ATP-induced relaxations (• 10 μM) by PIT (50 μM for 30 min) in a guinea pig taenia caeci preparation. Carbachol (0.05 μM) was used to restore tone for the determination in the presence of PIT, but tone returned spontaneously on washing. Note that responses to ATP were still antagonized although both PIT and carbachol were washed from the bath.

response curves (2 to 600 μM) at 20-min intervals or as single responses (2 to 200 μM) at 5-min intervals.

RELAXANT EFFECTS OF 2-2′-PYRIDYLISATOGEN

Incubation with PIT (2.5 to 100 μM) caused a slow relaxation of taenia caeci preparations (see Fig. 2). The relaxation was concentration dependent and was a consequence of a direct action of the drug on the smooth muscle cells (35). In an electrophysiological study, PIT (50 μM for 30 min) abolished the spontaneous spike potentials of the smooth muscle cells but was without effect on the resting membrane potential (control membrane potential, 51.8 ± 0.7 mV; after PIT, 50.2 ± 1.0 mV) (32). The slow relaxation caused by PIT resembled the relaxation caused by gassing with nitrogen or by incubation with papaverine (35). Isatogen derivatives have been shown to inhibit oxidative phosphorylation in mitochondrial preparations (38) and the ability of a series of 2-substituted isatogen derivatives to relax the taenia has been correlated with their ability to inhibit oxidative phosphorylation (13).

ANTAGONISM OF ATP-INDUCED RELAXATIONS BY 2-2′-PYRIDYLISATOGEN

After the tone of the preparations had been restored with carbachol, responses to ATP were antagonized by PIT (50 μM for 30 min) (Fig. 2). The displacement to the right of cumulative concentration-response curves to ATP was dependent on the concentration of PIT (25 μM for 30 min, dose ratio 4.2, range SEM 3.4 to 5.5, $N=8$; 50 μM for 30 min, dose ratio 47, range SEM 40 to 60, $N=26$; 75 μM for 30 min, dose ratio 168, range SEM 135 to 209, $N=8$). The slopes of these curves were not significantly different from the control ($p < 0.1$), except at the highest concentration of PIT (75 μM). High concentrations of PIT (100 μM for 30 min) caused nonspecific antagonism of responses to histamine, carbachol, noradrenaline, and isoprenaline.

The antagonism of ATP-induced relaxations was not competitive, because when the data were analyzed by the method of Arunlakshana and Schild (2), the slope of the plot ($\log_{10}$ dose ratio-1 against molar concentration of antagonist) was 2.6 and not the hypothetical value of 1.0 for a competitive antagonist. However, equilibrium conditions were not established in these experiments, because the antagonism increased progressively with incubation time beyond 120 min (33).

The antagonism was influenced by the individual spasmogen used to restore tone. If acetylcholine or carbachol was used to recontract the tissue, the antagonism was fivefold greater than when histamine was used or in the few experiments when the taenia did not relax following incubation with PIT (50 μM for 30 min) (33).

The antagonism of ATP-induced relaxations was irreversible after prolonged incubation with PIT (25 to 50 μM for 30 min). Figure 3 shows an experiment in which responses to ATP were antagonized by PIT (50 μM for 30 min, tone restored with carbachol, 0.05 μM), after which the taenia caeci preparation was washed with

McEwen's solution (3 liters over 3 hr). The tone of the smooth muscle returned during the washing procedure. Responses to ATP were still antagonized although tone was restored to the original level, i.e., no spasmogen was present.

Pyridylisatogen tosylate has been shown to bind covalently with some amino acids (15) and will react with sulfydryl groups:

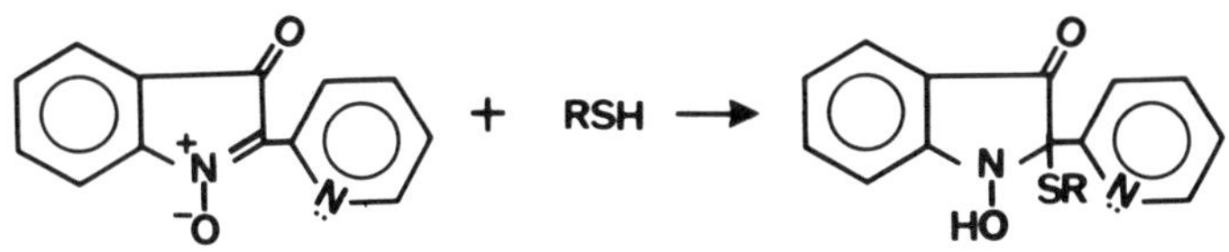

The formation of covalent bonds between PIT and a nucleophilic site at or near the ATP receptor would explain the progressive and irreversible nature of the antagonism of ATP-induced relaxations. Prolonged incubation with high concentrations of PIT would also lead to nonspecific antagonism caused by binding to sites unrelated to the ATP receptor.

SPECIFICITY OF THE ANTAGONISM

By comparison of full concentration-response curves before and after incubation with PIT (50 μM for 30 min), Spedding et al. (33) have shown that, despite some nonspecific antagonism, PIT was 30 times more effective as as antagonist of ATP-induced relaxations than noradrenaline-induced relaxations and 60 times more effective against ATP than against isoprenaline. Pyridylisatogen tosylate (50 μM for 30 min) also causes a slight antagonism of carbachol and histamine-induced contractions (dose ratios 3.6 $\pm$ 1.2 and 4.5 $\pm$ 1.7, respectively, $N=5$).

EFFECTS OF 2-2′-PYRIDYLISATOGEN ON ADENOSINE-INDUCED RELAXATIONS

Adenosine-induced relaxations of taenia caeci preparations differ from ATP-induced relaxations in several respects (34). Responses to adenosine were slow in onset (4 to 15 sec) and slow to reach maximum effect (25 to 60 sec); preparations rarely regained tone in the continuing presence of adenosine. Responses to adenosine 5′-monophosphate (AMP) resembled responses to adenosine, whereas responses to adenosine 5′-diphosphate (ADP) were similar to those to ATP. Taenia caeci preparations were six- to sevenfold more sensitive to ATP and ADP than to AMP and adenosine.

Pyridylisatogen tosylate (50 μM for 30 min) markedly antagonized responses to ATP and ADP but had little effect on the sensitivity of the taenia to AMP and adenosine (Fig. 4). Adenosine-induced relaxations were not antagonized by PIT (50 μM for 30 min) even when the sensitivity of the tissues to adenosine was increased 18-fold by the inclusion of dipyridamole (2 μM) in the McEwen's solution (34).

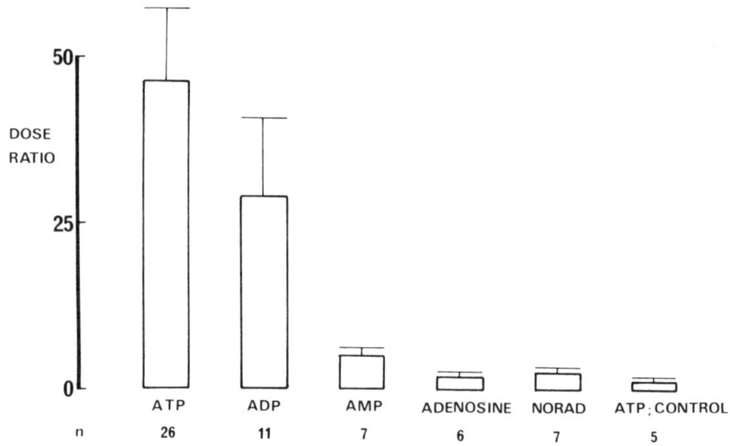

FIG. 4. Dose ratios, obtained from cumulative concentration-response curves to ATP, ADP, AMP, adenosine, and noradrenaline in guinea pig taenia caeci preparations after incubation with PIT (50 μM for 30 min). Carbachol (0.05 to 1.0 μM) was used to recontract the tissue to within 20% of the original level. Control experiments in which cumulative concentration-response curves to ATP were obtained, without PIT, are also shown. Vertical bars represent SEM.

Furthermore, inclusion of adenosine (2 mM) in the McEwen's solution desensitized taenia caeci preparations to subsequent additions of adenosine, yet the sensitivity of the tissues to ATP was increased by the procedure (34). It is therefore likely that ATP and adenosine relax the taenia via separate, distinct mechanisms. This would explain some divergent structure-activity relationships for adenosine and ATP that have been reported for the taenia by Satchell and Maguire (31).

Adenosine and ATP may act via divergent mechanisms in several other tissues. For example, adenosine causes a relaxation of the smooth muscle of guinea pig bladder (11), chick rectum (4), and several dog blood vessel preparations (22,25,28,39), whereas ATP causes a contraction.

EFFECTS OF 2-2′-PYRIDYLISATOGEN ON RESPONSES TO INHIBITORY NERVE STIMULATION

Guinea pig taenia caeci preparations, maintained in McEwen's solution containing hyoscine (0.3 μM), responded to field stimulation (0.5-msec pulse duration, maximal voltage, train duration 10 sec, interval between pulse trains 4 min) with a rapid relaxation (onset < 1 sec, optimal relaxation < 10 sec). The relaxation was optimal following low frequencies of stimulation (2 to 10 Hz), was abolished by tetrodotoxin (0.3 μM), but was not affected by guanethidine (17 μM). These nerve-mediated responses are therefore similar to the relaxations attributed to "purinergic" nerves (8). Theophylline (500 μM) did not antagonize the inhibitory responses to field stimulation (Fig. 5).

Incubation with PIT (50 μM for 30 min) and restoration of tone with histamine

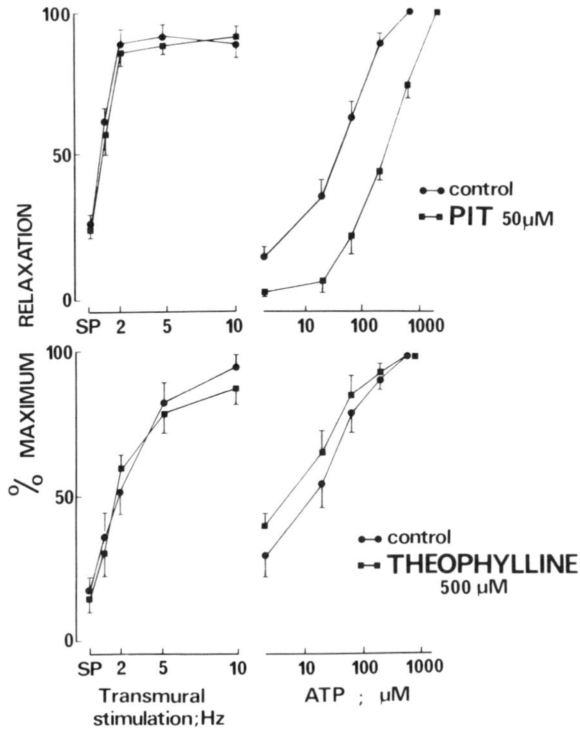

FIG. 5. The effects of PIT (*upper panels:* ■——■, 50 µM for 30 min; ●——●, control) or of theophylline (*lower panels:* ■——■, 500 µM for > 20 min; ●——●, control) on responses of guinea pig taenia caeci preparations to transmural stimulation or to cumulative concentration-response curves to ATP. Hyoscine (0.3 µM) and guanethidine (17 µM) was present in the McEwen's solution. Histamine (0.02 to 0.2 µM) was used to restore tone to within 20% of the original level where preparations were relaxed by PIT.

antagonized ATP-induced relaxations, but did not reduce the responses to field stimulation, even at low frequencies (Fig. 5). Responses to field stimulation were not antagonized following prolonged exposure to PIT (50 µM) (Fig. 6).

The results are difficult to reconcile with the purinergic nerve hypothesis. Prolonged incubation with an irreversible, noncompetitive antagonist, such as PIT, would be expected to antagonize the inhibitory responses to nerve stimulation if ATP mediates the inhibitory response. The gap between nervous structures and the smooth muscle cells is large in the taenia (> 80 nm) (6,41) thereby allowing access of antagonists to the receptors on the smooth muscle cells. The large synaptic cleft would also allow diffusion of the agonist, resulting in relatively low concentrations at the receptors. It is unlikely that the neurotransmitter is adenosine, because of the slow nature of the relaxation caused by the nucleoside. Further experiments must be carried out in order to determine the nature of the neurotransmitter of the nonadrenergic, noncholinergic inhibitory nerves.

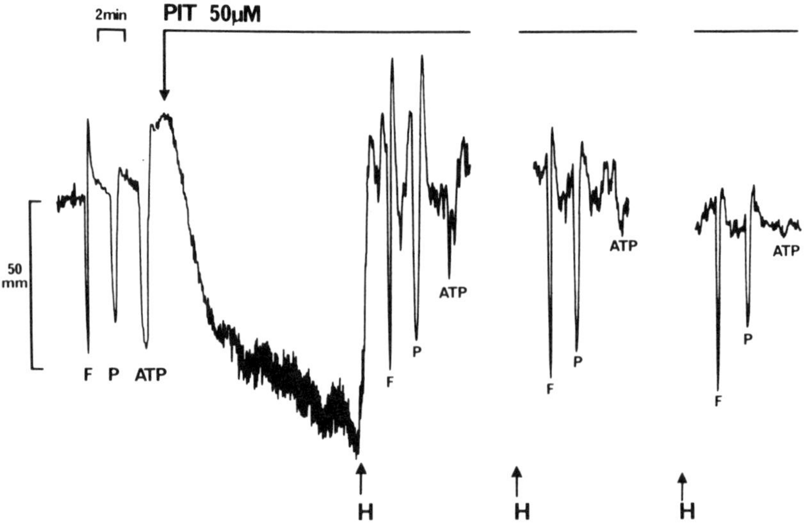

FIG. 6. The effects of PIT (50 μM) on relaxations induced by field stimulation (F: 2 Hz, 10 sec), perivascular stimulation (P: 50 Hz, 15 sec), and ATP (20 μM). Pyridylisatogen tosylate was added to the bathing fluid for the duration of the experiment; the drum speed was halved to show the resulting relaxation. Histamine was added at the lower arrows (0.6, 0.8, and 2.5 μM, respectively) and washed out after 8-min incubation. There is a time interval of 30 min between panels. Hyoscine (0.3 μM) was present throughout.

LIMITATIONS OF 2-2′-PYRIDYLISATOGEN

The experiments described above indicate that PIT exerts two actions on taenia caeci preparations: direct relaxation of the smooth muscle, which may be related to the ability of the compound to inhibit oxidative phosphorylation (13), and an antagonism of ATP-induced relaxations. The antagonism of ATP-induced relaxations appears to be a separate phenomenon because the relaxant effects of a series of isatogen derivatives were not correlated with ATP antagonistic activity (13). Furthermore, antagonism of responses to PIT could be demonstrated when no spasmogen was present (Fig. 3). The relaxant effects of PIT and the noncompetitive nature of the antagonism complicate the use of PIT when defining ATP receptors in other tissues. Thus although PIT antagonizes responses to ATP without concomitant antagonism of the responses to field or vagal stimulation in rat stomach strip (14,16) and in pig retractor penis (1), it does not selectively antagonize responses to ATP in guinea pig fundus (26), guinea pig bladder (11,40), and guinea pig atria preparations (Robertson and Spedding, *unpublished observations*). Pyridylisatogen tosylate strongly antagonized responses to spasmogens in guinea pig ileum preparations (36) making the evaluation of ATP-induced responses difficult in this preparation (see ref. 18). Some of these problems may be answered by the use of theophylline as a competitive antagonist of adenosine. Theophylline (25 to 100 μM) antagonizes responses to ATP and adenosine to an equivalent extent in guinea pig fundus strip (26) and in guinea pig atria (Robertson and Spedding, *unpublished observations*), whereas theophylline is ineffective as an antagonist of ATP-induced relaxations in guinea pig taenia caeci preparations (Fig. 5). The theophylline-sensitive sites may be structurally different from the sites at which ATP acts to cause relaxation of the taenia. The synthesis of isatogen derivatives that are more specific than PIT as antagonists of ATP would help to resolve this point and also the nature of the receptors in the smooth muscle of the guinea pig bladder.

REFERENCES

1. Ambache, N., and Killick, S. W., (1978): *Br. J. Pharmacol.*, 63:25–34.
2. Arunlakshana, O, and Schild, H. O. (1959): *Br. J. Pharmacol. Chemother.*, 14:48–58.
3. Baer, H. P., Frew, R., and Burnstock, G. (1977): *Can. J. Physiol. Pharmacol.*, 55:394–398.
4. Bartlett, A. L. (1974): *Br. J. Pharmacol.*, 51:549–558.
5. Bennett, M. R., Burnstock, G., and Holman, M. E. (1966): *J. Physiol. (Lond.)*, 182:541–558.
6. Bennett, M. R., and Rogers, D. C. (1967): *J. Cell Biol.*, 33:573–596.
7. Burnstock, G. (1969): *Pharmacol. Rev.*, 21:247–324.
8. Burnstock, G. (1972): *Pharmacol. Rev.*, 24:509–581.
9. Burnstock, G., Campbell, G., Bennett, M. R., and Holman, M. E. (1964): *Int. J. Neuropharmacol.*, 3:163–166.
10. Burnstock, G., Campbell, G., Satchell, D., and Smythe, A. (1970): *Br. J. Pharmacol.*, 40:668–688.
11. Burnstock, G., Cocks, T., Crowe, R., and Kasakov, L. (1978): *Br. J. Pharmacol.*, 63:125–138.
12. Cook, M. A., Hamilton, J. T., and Okwuasaba, F. K. (1978): *Nature*, 271:768–771.
13. Foster, H. E., Hooper, M., Spedding, M., Sweetman, A. J., and Weetman, D. F. (1978): *Br. J. Pharmacol.*, 63:378–379P.
14. Heazell, M. A. (1975): *Br. J. Pharmacol.*, 55:285–286P.
15. Hooper, M., and Robertson, J. W. (1971): *Tetrahedron Lett.*, 24: 2139–2140.
16. Hunt, W. B., Parsons, D. G., Wahid, A., and Wilkinson, J. (1978): *Br. J. Pharmacol. (in press).*
17. Jager, L. P. (1976): *Arch. Int. Pharmacodyn.*, 221:40–53.
18. Kazic, T., and Milosavljevic, D. (1977): *J. Pharm. Pharmacol.*, 29:542–545.
19. Kuchii, M., Miyahara, J. T., and Shibata, S. (1973): *Br. J. Pharmacol.*, 49:258–267.
20. Kuchii, M., Miyahara, J. T., and Shibata, S. (1973): *Br. J. Pharmacol.*, 49:642–650.
21. McEwen, L. M. (1956): *J. Physiol. (Lond.)*, 131:678–689.
22. Mentzer, R. M., Jr., Rubio, R., and Berne, R.M. (1975): *Am. J. Physiol.*, 229:1625–1631.
23. Ohga, A., Khoda, T., Saito, K., and Nakazato, Y. (1976): *Jpn. J. Pharmacol.*, 26(Suppl.):618.
24. Ohga, A., and Taneike, T. (1977): *Br. J. Pharmacol.*, 60:221–231.
25. Ohtaki, Y., Obari, A., Nakanishi, H., and Ueda, J. (1977): *Jpn. J. Pharmacol.*, 27 (Suppl.):107P.
26. Okwuasaba, F. K., Hamilton, J. T., and Cook, M. A. (1977): *Eur. J. Pharmacol.*, 46:181–198.
27. Rutherford, A., and Burnstock, G. (1978): *Eur. J. Pharmacol.*, 48:195–202.
28. Sakai, K., Sugano, S., Taira, N., and Hashimoto, K. (1974): *Jpn. J. Pharmacol.*, 24:659–669.
29. Satchell, D. G., and Burnstock, G. (1971): *Biochem. Pharmacol.*, 20:1694–1697.
30. Satchell, D. G., Lynch, A., Bourke, P. M., and Burnstock, G. (1972): *Eur. J. Pharmacol.*, 19:343–350.
31. Satchell, D. G., and Maguire, M. H. (1975): *J. Pharmacol. Exp. Ther.*, 195:540–548.
32. Small, R. C., and Spedding, M. (1978): *Br. J. Pharmacol.*, 63:408–409P.
33. Spedding, M., Sweetman, A. J., and Weetman, D. F. (1975): *Br. J. Pharmacol.*, 53:575–583.
34. Spedding, M., and Weetman, D. F. (1976): *Br. J. Pharmacol.*, 57:305–310.
35. Spedding, M., and Weetman, D. F. (1978): *Br. J. Pharmacol.*, 63:659–664.
36. Spedding, M., and Weetman, D. F. (1978): *J. Pharm. Pharmacol.*, 28:335–336.
37. Su, C., Bevan, J. A., and Burnstock, G. (1971): *Science*, 173:336–338.
38. Sweetman, A. J., Green, A. P., and Hooper, M. (1971): *FEBS Lett.*, 14:306–308.
39. Verhaeghe, R. H. (1977): *Am. J. Physiol.*, 233:H114–121.
40. Weetman, D. F., and Turner, N. (1977): *Arch. Int. Pharmacodyn.*, 228:10–14.
41. Yamauchi, A. (1964): *Acta Anat. Nippon.*, 39:22–38.

Physiological and Regulatory Functions of Adenosine and Adenine Nucleotides, edited by H. P. Baer and G. I. Drummond. Raven Press, New York © 1979.

Action of Adenosine on Isolated Vascular Smooth Muscle

Jeremiah T. Herlihy

Department of Physiology, University of Texas Health Science Center, San Antonio, Texas 78284

Adenosine is a potent vasodilator and has been proposed as a mediator in the local control of blood flow (3). Over the years, much emphasis has been placed on its role in hemodynamic regulation and later sections of these proceedings deal with this aspect of adenosine action. In spite of its importance in cardiovascular homeostasis, only in recent years have the effects of adenosine on vascular smooth muscle (VSM) per se come under close scrutiny. Consequently, the action of adenosine on VSM has not been fully characterized, nor, indeed, is the mechanism known by which adenosine relaxes VSM. The aim of the present discussion is to examine certain characteristics of adenosine relaxation of VSM in order to obtain some insight into the mechanism by which it relaxes this tissue.

DISCUSSION

Adenosine Relaxation of Various VSMs

The use of isolated VSM preparations has proved fruitful in uncovering mechanisms by which vasoactive agents act. Many anatomically different VSMs have been shown to relax in response to adenosine administration. These include the hog carotid artery (10), rat aorta (6), rabbit (8,13) and canine (15) coronary arteries, the rat portal vein (J. T. Herlihy, *unpublished observations*), and the canine saphenous vein (17). No systematic comparison between vascular strips from different sources has been performed, but some differences do exist. For example, Schnaar and Sparks (15) noted that small coronary arteries were much more responsive to adenosine than larger arteries taken from the same bed. In addition, the venous side of the system seems less responsive to adenosine administration than the arterial side. Verhaeghe (17) observed no adenosine effect on the spontaneous activity of the canine saphenous vein, whereas the present author observed only slight increases in contraction frequency and decreases in force generation of the rat portal vein with large (10^{-4}M) concentrations of adenosine (J. T. Herlihy, *unpublished observations*). These differences in adenosine efficacy most likely reflect the well-known differences between the various VSMs. Most of the following studies were carried out with a hog arterial strip preparation composed mainly of carotid media (11).

Effect of Agonist Dose

Because the hog carotid artery exhibits little or no tone *in vitro* (like most isolated arterial preparations) the strips must first be contracted before the effects of a relaxing agent are examined. We have found that the degree of adenosine relaxation depends on the dose of agonist used to contract the strip. Figure 1 shows the relaxant effects of cumulative additions of adenosine on the force generated by arterial strips stimulated with either 1 μM or 0.3 μM norepinephrine (NE). The leftward shift in the dose-inhibition curve, observed with the lower concentration of agonist, demonstrates the importance of agonist dose on the efficacy of adenosine in relaxing this muscle. It can be seen in the same figure that at a dose of 100 μM adenosine completely relaxed the contraction produced by 1 μM NE. This dose of NE caused a contraction greater than 80% of maximal (Fig. 2, control curve). At a dose of 3 μM, adenosine shifted the NE dose-response curve to the right. No significant change in maximal tension was observed (Fig. 2). The importance of agonist dose is not confined to NE; similar effects can be seen with the use of potassium depolarization (K^+) as the contractile agent. As shown in Table 1, the relaxant effects of adenosine on hog arterial strips decreased with increasing K^+ concentrations. No significant relaxation of strips contracted with 60 mM K^+ (Table 1) or 124 mM K^+ (not shown) was observed with 3 μM adenosine. However, high doses of adenosine (1 mM) were able to relax strips subjected to 124 mM K^+ (Fig. 3).

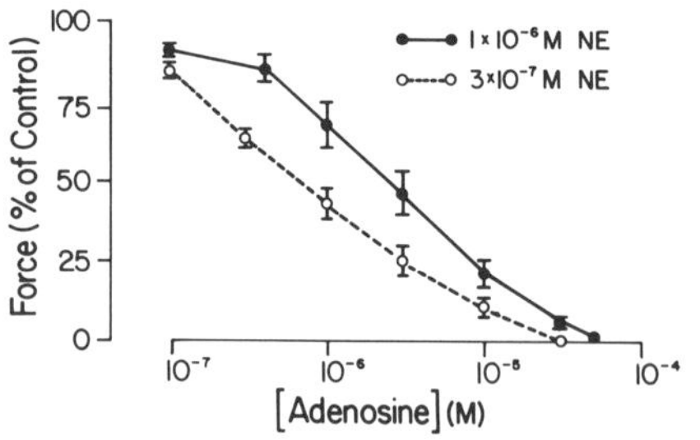

FIG. 1. Effect of increases in adenosine concentration on force developed by hog carotid artery strips contracted with norepinephrine (NE). Force is plotted on ordinate as a percentage of that force developed prior to treatment with adenosine. For 1 μM and 0.3 μM NE, 100% was 7.93 ± 1.50 and 5.2 ± 1.2 g, respectively. Points and bars represent means ± SEM respectively, for six experiments. (From ref. 10, with permission).

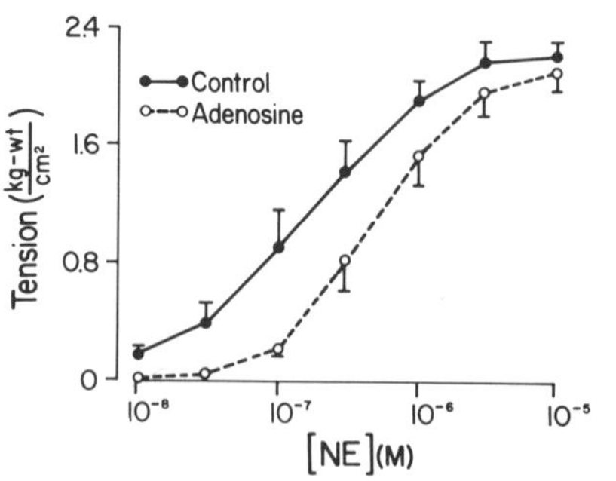

FIG. 2. Effect of 3 μM adenosine on NE dose-response curves for hog carotid artery strips. Points and bars represent means ± SEM, respectively, for four experiments. (From ref. 10, with permission.)

TABLE 1. *Comparison of relaxant effects of adenosine (3 μM) on K^+ and NE contractures at given control tensions*

Stimulation	Control tension (kg wt/cm²)	Decrease in tension (kg wt/cm²)	*N*	Significance
20 mM K^+	0.74 ± 0.05	0.41 ± 0.01	5	$p>0.10$
NE (≃6.8 × 10^{-8}M) [a]	0.74	0.51 ± 0.06	4	
30 mM K^+	1.02 ± 0.09	0.26 ± 0.06	5	$p<0.05$
NE (≃1.3 × 10^{-7}M) [a]	1.02	0.63 ± 0.10	4	
60 mM K^+	1.75 ± 0.20	0.04 ± 0.01	5	$p<0.05$
NE (≃6.7 × 10^{-7}M) [a]	1.75	0.68 ± 0.24	5	

[a] Norepinephrine concentrations represent average NE concentrations at which control tensions were obtained.
From ref. 10, with permission.

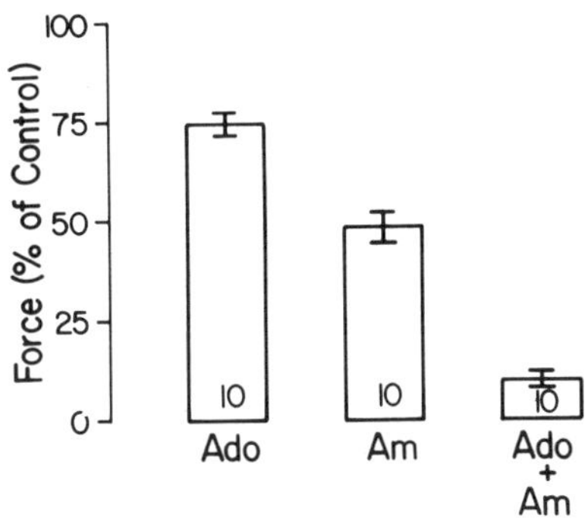

FIG. 3. Effect of 1 mM adenosine (Ado), 1 mM aminophylline (Am), or both (Ado + Am) on force developed by potassium-depolarized hog carotid artery strips. Force is plotted on ordinate as a percentage of the force developed by 124 mM K^+ (100%=1.58 ± 0.16 kg wt/cm²). Heights of columns and vertical bars represent means ± SEM, respectively, for number of experiments indicated within bars. Effects of various treatments were highly significant ($p<0.001$). (From ref. 10, with permission.)

Effect of Agonist Type

In addition to agonist dose, the type of agonist also has an effect on the efficacy of adenosine in relaxing isolated VSM. This can be seen in Table 1 in which the effect of adenosine on the K^+ contractions is compared with its effect on the NE contractions. The basis for comparison rests on the use of the same dose of adenosine (3 μM) on vascular strips contracted to the same tension with either NE or K^+. Tensions of 0.74, 1.02, and 1.75 kg wt/cm² were the mean tensions produced by 20, 30, and 60 mM K^+, respectively. The relaxant effects of 3 μM adenosine were then tested. For NE stimulation, the dose-response curves for the experiments represented in Fig. 2 for both control and adenosine-treated (3 μM) strips were constructed. From these curves the degree of relaxation at initial tensions of 0.74, 1.02, and 1.75 kg wt/cm² was determined. As shown in Table 1, adenosine was more effective in relaxing strips contracted by NE than those contracted by K^+, and the differences between the two were very pronounced at the higher tensions.

Interaction with Other Vasodilators

Synergism and antagonism between adenosine and other vasodilators have been reported. Hypoxia relaxes isolated VSM and has also been proposed as a mediator in the local control of blood flow (5). Recently, hypoxia and adenosine have been shown to relax synergistically isolated coronary arteries (8,13). Gellai et al. (8) suggested that this interaction played a physiological role in the control of the coronary circulation. They proposed that adenosine set the level of sensitivity of the coronary smooth muscle to altered oxygen tension. However, the basis for this interaction may not be physiological but may be a result of using isolated vascular strips. Evidence from the work of Van Harn et al. (16), reproduced here in Fig. 4, supports the concept that, like other tissues, hypoxic VSM produces adenosine. Moreover, they showed that vascular tissue takes up and destroys adenosine. In the case of the isolated coronary arteries, hypoxia may have resulted in the production of adenosine and/or may have inhibited the mechanisms responsible for adenosine destruction. In either case the effective adenosine concentration in strips subjected to lower oxygen tensions would be higher and greater relaxation would result.

Two other vasodilators, aminophylline and dipyridamole, alter the adenosine-induced vasodilation of the coronary circulation. Aminophylline inhibits (1,7) whereas dipyridamole enhances (4) the adenosine action. The mechanisms of the antagonism and synergism are not known, but our work (9) shows that these interactions can occur with an *in vitro* VSM preparation. Dipyridamole (1 μM) caused a significant ($p<0.001$) enhancement (300%) of the adenosine-induced relaxation of hog carotid strips contracted by 1 μM NE (Fig. 5). In contrast, aminophylline (100 μM) caused a significant ($p<0.001$) inhibition of adenosine-induced relaxation (Fig. 6).

Dipyridamole has been shown to be a specific inhibitor of adenosine uptake (4). Therefore, this drug may increase the effective adenosine concentration within the vascular strip by preventing its uptake and subsequent metabolism.

The mechanism responsible for the interaction between adenosine and aminophylline is also not known. A certain parallelism exists between heart and brain with regard to the antagonism between adenosine and aminophylline. In the

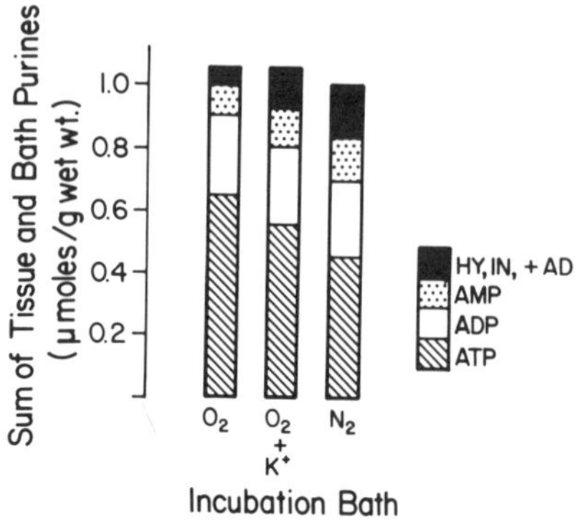

FIG. 4. Changes in adenine nucleotides and purine derivatives during incubation of hog carotid artery strips. Artery strips were incubated in physiological saline bubbled with 95% O_2 and 5% CO_2 (bar labeled O_2), in 124 mM K^+ buffered saline bubbled with 95% O_2 and 5% CO_2 (bar labeled $O_2 + K^+$), and in physiological saline bubbled with 95% N_2 and 5% CO_2 (bar labeled N_2). For each type of experiment, $N = 5$. Hy, hypoxanthine; In, inosine; Ad, adenosine; AMP, adenosine monophosphate; ADP, adenosine diphosphate; ATP, adenosine triphosphate. (From ref. 16, with permission.)

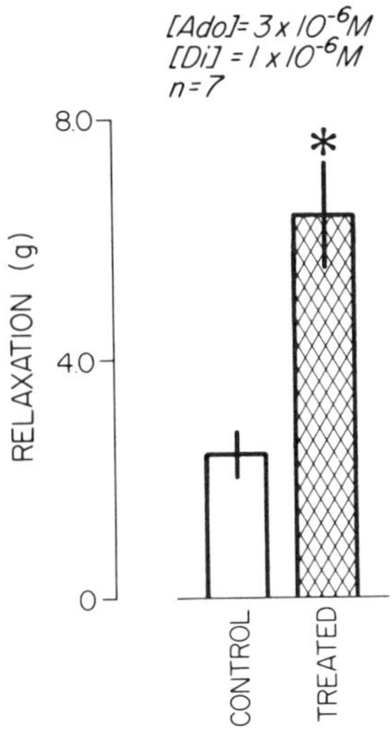

FIG. 5. Effect of dipyridamole (Di) on the relaxation produced by adenosine (Ado) of the force developed by 1 μM NE. Heights of columns and vertical bars represent means $\pm$ SEM, respectively, for the number of experiments indicated within bars. Asterisk represents significant difference ($p<0.001$) from control.

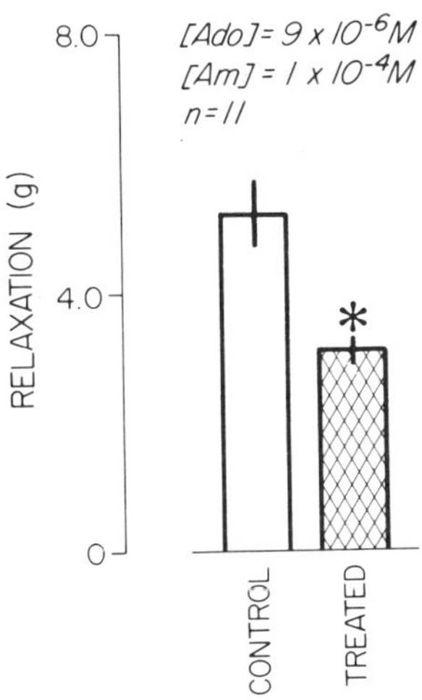

FIG. 6. Effect of aminophylline (Am) on the relaxation produced by adenosine (Ado) of the force developed by 1 μM NE. Heights of columns and vertical bars represent means $\pm$ SEM, respectively, for the number of experiments indicated within bars. Asterisk represents significant difference ($p<0.001$) from control.

coronary circulation both agents are vasodilators, but act antagonistically when present together (1,7). In brain, both adenosine and aminophylline alone cause an increase in the intracellular levels of cyclic adenosine 3′,5′-monophosphate (cAMP), but when present together their combined effect is less than the sum of the individual effects (14). This parallelism suggested to us that (a) adenosine-induced relaxation of VSM could be mediated via changes in the intracellular levels of cAMP, and (b) the antagonism between adenosine and aminophylline observed in the coronary circulation could be explained by an antagonism at the level of cAMP. Unfortunately, our data do not support these hypotheses. As shown in Fig. 7A, no

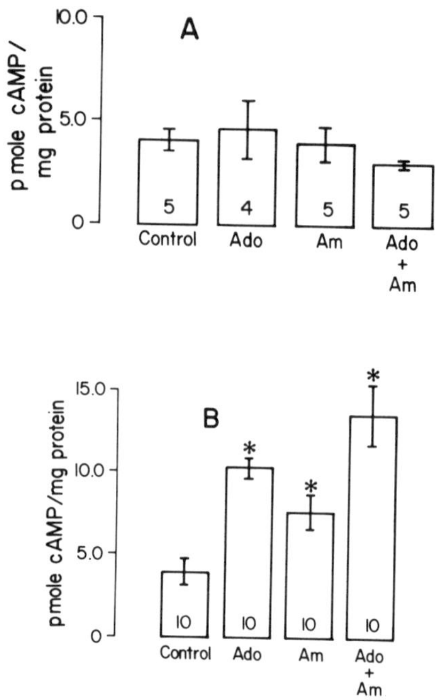

FIG. 7. Effect of adenosine (Ado), aminophylline (Am), and adenosine plus aminophylline (Ado + Am) on cAMP content of the hog carotid artery strips. **A:** 1 μM Adenosine, 10 μM aminophylline. **B:** 1 mM Adenosine, 1 mM aminophylline. Heights of columns and vertical bars represent means ± SEM, respectively, for number of experiments indicated within bars. Asterisk represents significant difference ($p<0.05$) from control. (From ref. 10, with permission).

increase in the cAMP levels of hog arterial strips was observed when adenosine (1 μM) was administered. Furthermore, no interaction between aminophylline and adenosine could be detected statistically. Even at concentrations of 100 μM, adenosine caused no significant increase in cAMP levels (not shown). This concentration of adenosine relaxes a near-maximal contraction of the hog carotid smooth muscle (Fig. 2). Very high doses of adenosine(1 mM) will increase cAMP levels, as will the same high dose of aminophylline (Fig. 7B). However, no antagonistic interaction between these two agents could be detected. Even when increases in cAMP levels do occur with adenosine administration, dissociations between relaxation and changes in these levels are seen. For example, adenosine (1 mM) and aminophylline (1 mM) both cause increases in cAMP in the hog carotid artery (Fig. 7B), with adenosine causing a greater increase than aminophylline. At the same doses both agents relax K^+ contractures of the same muscle (Fig. 3), but aminophylline causes a greater relaxation than adenosine. Thus the degree of relaxation obtained is dissociated from the changes in the cyclic nucleotide levels and these cannot be considered to mediate both adenosine and theophylline action. That adenosine-induced relaxation of VSM does not involve changes in cAMP content is supported by more recent studies. Verhaeghe (17) observed no change in the cAMP content, cyclic guanosine 3′,5′-monophosphate (cGMP) content or their ratio when canine saphenous vein was exposed to adenosine. In rat aorta, Cohen and Berkowitz (6) noted an age-dependent decrease in the cAMP-induced relaxation, but no change in adenosine-induced relaxation.

Other Mechanisms Proposed for Adenosine Relaxation of VSM

Mechanisms of adenosine-induced relaxation of VSM other than increases in cAMP levels have been suggested in the past. Schnaar and Sparks (15) observed adenosine-induced relaxation of dog coronary arteries contracted with 35 mM K^+ and suggested that adenosine acted by decreasing the membrane permeability to Ca^{2+}. This had been suggested earlier by Axelsson and Holmberg (2) who observed a 43% decrease in the contractile force of depolarized taenia coli with 1.5 mM adenosine. Our own work with hog carotid artery supports this concept. As shown in Fig. 3, adenosine at a dose of 1 mM relaxed the contraction developed by potassium depolarization. Moreover the relaxant effects of adenosine have been shown to be dependent on the extracellular Ca^{2+} concentration (10). Since membrane potential changes are unlikely under these conditions, adenosine may act by directly altering the permeability of the membrane to Ca^{2+}. Direct measurements of calcium flux during adenosine administration have never been reported for VSM.

Unfortunately, a direct effect by adenosine on the membrane permeability to Ca^{2+} cannot account for all phenomena. Verhaeghe (17) showed that adenosine inhibited the NE contraction of canine saphenous vein equilibrated in Ca^{2+}-free solution. Presumably, NE is acting in this case via intracellular stores of Ca^{2+} and, therefore, adenosine relaxation would not involve inhibition of Ca^{2+} movement across the membrane. Support for an alternative mechanism is supported by the observation shown in Fig. 2 in which adenosine did not decrease the maximum tension produced by NE. This indicates a competition between adenosine and NE with regard to force generation.

Finally, adenosine could relax VSM by altering its membrane potential. This mechanism was proposed by Imai and Takeda (12) who measured membrane hyperpolarization and relaxation of the taenia coli after adenosine administration. They suggested that adenosine caused this hyperpolarization by decreasing membrane permeability to Na^+. This hypothesis, like that proposed for changes in Ca^{2+} permeability, has never been tested in VSM, nor have possible changes in the membrane potential of VSM caused by adenosine administration been characterized.

SUMMARY

In summary, adenosine has been shown to relax several isolated VSMs *in vitro*, but its efficacy depends on a number of factors. These factors include type of VSM tested, dose and type of agonist used to contract the muscle, and the presence of other vasodilators. The mechanism by which adenosine relaxes VSM is not known.

Changes in intracellular levels of cAMP do not appear to be involved. Changes in membrane permeability to Ca^{2+} or membrane potential may be involved and these possibilities should be studied in the future.

ACKNOWLEDGMENT

This work was supported in part by Public Health Service Grants HL10384 and HL03179 and Grants-in-Aid from the Virginia and Texas Affiliates of the American Heart Association. Dipyridamole (Persantine) was kindly supplied by Boehringer Ingelheim, Ltd.

REFERENCES

1. Afonso, S. (1970): *Circ. Res.,* 26:743–752.
2. Axelsson, J., and Holmberg, B. (1969): *Acta Physiol., Scand.,* 75:149–156.
3. Berne, R. M., Rubio, R., Dobson, J. G., Jr., and Curnish, R. R. (1971): *Circ. Res.,* 28 (Suppl. I):115–119.
4. Bretschneider, H. J., Frank. A., Bernard, U., Kochsiek, K., and Scheler, F. (1959): *Arzneim. Forsch.,* 9:49–59.
5. Carrier, O., Jr., Walker, J. R., and Guyton, A. C. (1964): *Am. J. Physiol.,* 206:951–954.
6. Cohen, M. L., and Berkowitz, B. A. (1974): *J. Pharmacol. Exp. Ther.,* 191:147–155.
7. Curnish, R. R., Berne, R. M., and Rubio, R. (1972): *Proc. Soc. Exp. Med.,* 141:593–598.
8. Gellai, M., Norton, J. M., and Detar, R. (1973): *Circ. Res.,* 32:279–289.
9. Herlihy, J. T. (1977): *Physiologist,* 20:42(abst.).
10. Herlihy, J. T., Bockman, E. L., Berne, R. M., and Rubio, R. (1976): *Am. J. Physiol.,* 230:1239–1243.
11. Herlihy, J. T., and Murphy, R. A. (1971): *Circ. Res.,* 29:375–384.
12. Imai, S., and Takeda, K. (1967): *J. Pharmacol. Exp. Ther.,* 156:557–564.
13. Norton, J. M., Gellai, M., and Detar, R. (1972): *Pfluegers Arch.,* 335:279–286.
14. Sattin, A., and Rall, T. W. (1972): *Mol. Pharmacol.,* 6:13–23.
15. Schnaar, R. L., and Sparks, H. V. (1972): *Am. J. Physiol.,* 223:223–228.
16. Van Harn, G. L., Rubio, R., and Berne, R. M. (1977): *Am. J. Physiol.,* 233:H299–H304.
17. Verhaeghe, R. H. (1977): *Am. J. Physiol.,* 233:H114–H121.

Physiological and Regulatory Functions of Adenosine and Adenine Nucleotides, edited by H. P. Baer and G. I. Drummond. Raven Press, New York © 1979.

Structure-Activity Relationship Studies of Purinergic Agonists: Some Speculations on the Source and Identity of the Purinergic Mediator

M. A. Cook, J. T. Hamilton, and *F. K. Okwuasaba

Department of Pharmacology, University of Western Ontario, London, Ontario, N6A 5C1 Canada

The existence of inhibitory responses at the mammalian gastrointestinal tract that are neither adrenergically nor cholinergically mediated is well established and the evidence that suggests that the mediator is a purine nucleotide, probably adenosine triphosphate (ATP), has been reviewed (7,8). The mimicking of these responses by specific nucleotide receptors is generally accepted although the lack of a specific nucleotide receptor blocking drug has been an important limiting factor in assessing the possible transmitter role of ATP. Recently, however, the methylxanthine theophylline was shown to selectively antagonize the inhibitory effects of adenosine and related nucleotides (1) and Okwuasaba et al. (26) demonstrated the efficacy of theophylline in inhibiting the peristaltic activity of the guinea pig ileum *in vitro,* a preparation in which nucleotide receptors are presumed to be located presynaptically. In addition, the competitive nature of the antagonism between theophylline and purine nucleotides at the guinea pig fundus *in vitro* has also been described (27). Thus postsynaptic receptors with affinity for nucleotides exist, and these have been considered to function in the nonadrenergic nerve-mediated relaxations of the guinea pig colon and taenia caecum (6,9,10).

The existence of "purinergic" nerves in the myenteric plexus that release ATP as the putative transmitter has been suggested (7,10) and the ability to distinguish these nerves on the basis of their axon varicosity profiles from adrenergic and cholinergic nerves has also been claimed (2,7,10). This has led to the suggestion that "purinergic" nerves may represent a third, distinct branch of the autonomic nervous system. Recently however, analysis of the vesicular structures present in neurons of Auerbach's plexus of the rabbit jejunum and opposum esophagus (16) has failed to differentiate between axons in which varicosities contain predominantly large opaque vesicles (putatively purinergic) and those containing predominantly small agranular (cholinergic) vesicles. These studies have questioned the existence of purinergic nerves as a distinct entity.

Our present studies on the effects of various structural analogs of adenosine on

*Present address: Department of Pharmacology, University of Alberta, Edmonton, Alberta, T6G 2H7 Canada.

the guinea pig ileum have led us to suggest that coenzyme A (CoA), a naturally occurring adenine compound, could play the role of a physiological transmitter in "purinergic" nerves. Further, the apparent controversy concerning the existence of distinct "purinergic" nerves prompted a series of studies in which the effects of agents known to interfere with cholinergic transmission were assessed for their ability to interfere with nerve-mediated nonadrenergic inhibition of the guinea pig fundus *in vitro*. These studies have allowed further speculation on the possible source of an endogenous purine nucleotide mediator.

METHODS

Structure-activity relationships were examined using the field-stimulated guinea pig ileum preparation. Longitudinal muscle strips from male guinea pigs (250 to 400 g) were prepared according to the method of Paton and Vizi (28) and set up in 10-ml organ baths containing Krebs' solution. The composition of the Krebs' solution was (mM) KCl 4.75, $CaCl_2$ 2.54, KH_2PO_4 1.19, $MgSO_4$ 1.19, NaCl 118, $NaHCO_3$ 25, and glucose 11. The solution was bubbled with 95% O_2 and 5% CO_2, temperature was maintained at 37°C and the pH was 7.3. Guanethidine (20 μM) was present throughout to eliminate adrenergic responses. Field stimulation was achieved by means of trains of supramaximal rectangular pulses of 1-msec duration applied through platinum electrodes at the top and bottom of the bath. The frequency of stimulation used was 0.2 Hz. Longitudinal contractions were recorded auxotonically with a Harvard apparatus smooth muscle transducer using an applied load of 1 g. Preparations were equilibrated for at least 90 min with washing every 15 min prior to exposure to drugs, and the time between addition of drugs was 15 min with the following exception. Cumulative dose-response curves were obtained and responses were calculated as percentages of maximum twitch height. Log dose-response curves for the inhibitory agonists were constructed using the pooled mean results from at least five preparations.

The postsynaptic efficacy of the inhibitory agonists was assessed using the guinea pig fundic strip preparation. This was prepared as described by Okwuasaba et al. (27). Results were obtained on a minimum of five preparations. Field stimulation of the fundic strips was achieved by placing the preparation between a pair of parallel platinum wires without permitting them to touch the tissue. Periarterial nerve stimulation of a thin mesenteric bundle was achieved by means of a bipolar platinum ring electrode. Guanethidine (20 μM) was present throughout.

RESULTS

A wide range of substituted nucleotides and nucleosides possessed inhibitory activity on the stimulated guinea pig ileum. Figure 1 shows the dose-response curves obtained for some of the compounds tested and the EC_{50} data are listed in Table 1. Among the compounds bearing modifications at the 2′ and 3′ positions of the ribose, only 2′-deoxyadenosine and 2′3′-diacetyladenosine showed striking

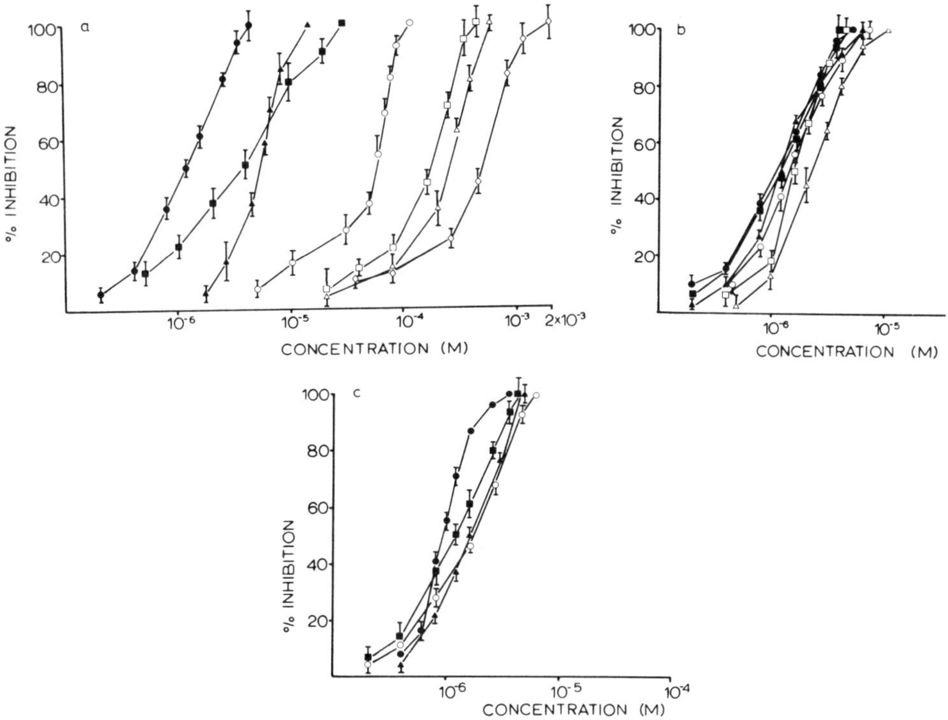

FIG. 1. Comparison of the inhibitory potency of ATP and analogs substituted at the ribose moiety on electrically stimulated guinea pig ileum. **a:** Modifications at the 2′- and 3′-positions. Log dose-response curves are shown to ATP (●), 2′,3′-isopropylidene adenosine (■), 3′-deoxyadenosine (▲), adenine-β-D-arabinofuranoside (○), 2′-deoxyadenosine (□), 2′,3′-diacetyladenosine (Δ), and 2′-deoxyadenosine monophosphate (◇). **b:** Modifications at the 5′-phosphate groups. Log dose-response curves are shown to ATP (●), α,β,-methylene ATP (■), adenosine (□), ADP (▲), AMP (○), and β,γ-methylene ATP (Δ). **c:** Modifications at the 5′ position. Log dose-response curves are shown to ATP (■), CoA (○), adenosine-3′,5′-diphosphate (●), and acetyl CoA (▲). The preparation was stimulated coaxially at 0.2 Hz, 1-msec duration, and with supramaximal voltage. The dose-response curves were obtained cumulatively. Vertical bars indicate ± SEM.

decreases in activity, being 130 and 165 times less potent, respectively, than ATP. Inversion of the 2′-hydroxyl group as in adenine-β-D-arabinofuranoside resulted in a 36-fold reduction in activity compared to ATP whereas 3′-deoxyadenosine was only some four times less active than ATP.

Adenosine triphosphate possessed the same activity as adenosine diphosphate (ADP), adenosine monophosphate (AMP), and adenosine. The methylene isosteres α,β- and β,γ-methylene ATP showed essentially the same potency as ATP; however, the time of offset of these compounds was considerably prolonged. Coenzyme A, an adenine nucleotide with a large substituent group at the 5′ position as well as a 3′-phosphate group, was as potent as ATP in inhibiting the ileum, and several analogs of CoA exhibited similar activity. Figure 2B shows the inhibition produced

TABLE 1. *Relative potency of ATP and analogs substituted at the ribose moiety*

Compound	$EC_{50} \times 10^6$ [a]	pD_2 [b]	αE [c]	Compound	$EC_{50} \times 10^6$ [a]	pD_2 [b]	αE [c]
ATP	1.20	5.92	1.0	2′3′-Isopropylidene adenosine	3.31	5.48	1.0
ADP	1.24	5.91	1.0	α,β-Methylene ATP	1.51	5.82	1.0
AMP	1.30	5.89	1.0	β,γ-Methylene ATP	2.16	5.67	1.0
Adenosine	1.23	5.91	1.0	CoA	1.47	5.83	1.0
2′-Deoxyadenosine	1.52×10^2	3.82	1.0	Acetyl CoA	1.60	5.80	1.0
3′-Deoxyadenosine	5.47	5.26	1.0	Dephospho CoA	1.45	5.84	1.0
Adenine-β-D-arabinofuranoside	44.23	4.35	1.0	Adenosine 3′,5′-diphosphate	0.87	6.06	1.0
2′,3′-Diacetyladenosine	1.97×10^2	3.71	1.0	Oxidized CoA	1.46	5.84	1.0

[a] EC_{50} values from regression lines obtrained by pooling results from five to seven experiments.
[b] Values calculated as the negative logarithm of the EC_{50}.
[c] Intrinsic activity relative to ATP+1.0.

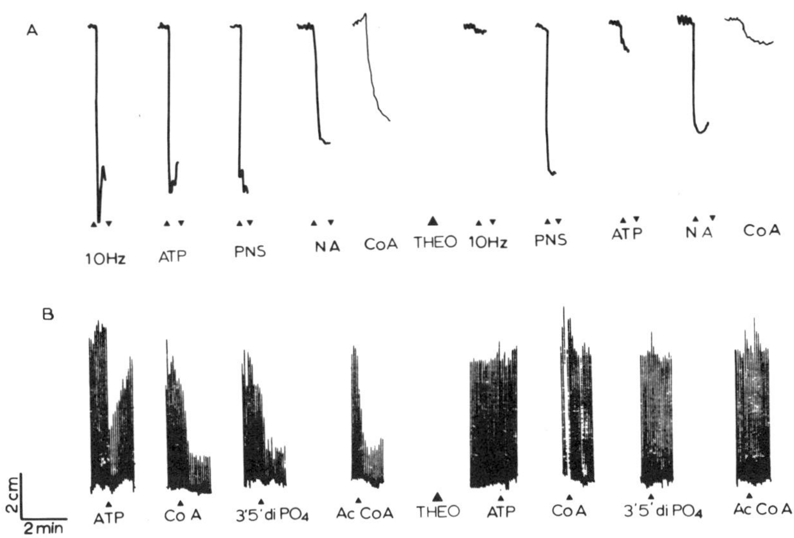

FIG. 2. A: Responses of the guinea pig fundus to inhibitory agonists and nerve stimulation in the absence and presence of theophylline. Responses shown are to field stimulation (10 Hz), ATP (2 μM), periarterial nerve stimulation (PNS) (50 Hz, 1 msec), noradrenaline (NA) (0.1 μM), and CoA (4 μM). Following incubation of the preparation with theophylline (THEO) (50 μM) for 45 min the responses to drugs and stimulation were retested. Atropine (0.5 μM) was present in the bath throughout the experiment. **B:** Responses of the field-stimulated guinea pig ileum to various inhibitory agonists in the absence and presence of theophylline. Responses are shown to ATP (1.7 μM), CoA (3 μM), adenosine-3′,5′-diphosphate (3′5′-diPO$_4$) (0.95 μM), and acetyl CoA (Ac CoA) (3 μM). Following incubation of the preparation with theophylline (THEO) (50 μM) for 45 min, responses to the agonists were retested. Field stimulation was applied at 0.2 Hz and 1-msec duration.

by ATP and CoA as well as its analogs acetyl CoA and adenosine 3′,5′-diphosphate. Although the efficacy of these nucleotides was not significantly different, the rates of offset were. The response to ATP recovered to within 20% of control values during the 1-min observation period, whereas the responses to CoA and its analogs persisted throughout this period. The responses to CoA, like those to ATP, were antagonized by theophylline (50 μM) following incubation of the tissue with the methylxanthine for 30 to 45 min (Fig. 2).

Coenzyme A relaxed the fundic strip preparation in a manner similar to ATP despite being less potent. Figure 2A shows the relaxations of an atropinized fundic strip to ATP, CoA, and noradrenaline, as well as to periarterial nerve and field stimulation. Theophylline (50 μM) antagonized the responses to ATP, CoA, and field stimulation at 10 Hz without affecting the responses to noradrenaline or periarterial nerve stimulation.

Other experiments (25) using the atropinized fundic strip preparation showed that continued exposure of the tissues to ATP, adenosine, adenosine 3′,5′-diphosphate, or CoA in a concentration of 50 μM for 30 to 45 min resulted in tachyphylaxis of the inhibitory responses to any of those agonists as well as to field stimulation. Responses to noradrenaline were not affected by this treatment. Similarly, repetitive field stimulation at 10 Hz every 6 min for 30 min decremented the response both to field stimulation and to the above agonists although not affecting the responses to noradrenaline.

In unatropinized fundic strip preparations (but with guanethidine present) field stimulation resulted in a biphasic response consisting of a primary contraction that was atropine sensitive and an after-relaxation. In the presence of tetrodotoxin (0.5 μM) both phases of the response were abolished. Tissues incubated when phentolamine (5 μM) plus propranolol (1 μM) responded in the same biphasic manner as untreated tissues. However, hemicholinium (HC-3) markedly reduced or abolished both phases of the response to field stimulation without affecting the responses to ATP or acetylcholine.

Figure 3a shows the biphasic responses to field stimulation at 1 and 10 Hz as well as the responses to ATP (1 μM) and acetylcholine (0.5 μM). Following incubation with HC-3 for 2 to 4 hr with intermittent stimulation at 5 to 10 Hz, the biphasic response was abolished whereas the responses to the exogenous agonists were not significantly affected. Identical experiments in the absence of HC-3 showed no change from controls. Following washout of HC-3, incubation of the tissue in choline (10 to 50 μM) for 1 to 2 hr partially antagonized the blockade produced by HC-3, such antagonism appearing to be time dependent.

Responses in the presence of β-bungarotoxin (β-Bu) (5×10^{-5} g/ml) to field stimulation at 1, 2, and 10 Hz (Fig. 3b) were completely abolished whereas the responses to ATP and acetylcholine were unaffected. The blockade produced by β-Bu was usually observed after 3 to 5 hr of incubation, developed more rapidly when the tissue was stimulated at 10 Hz for 1 to 2 min every 20 min, and was not reversible even after prolonged washing.

In the presence of black widow spider venom (BWSV) (2×10^{-5} to 5×10^{-5} g/ml) the responses to field stimulation at 1 and 5 Hz were abolished whereas both phases of the response at 10 Hz were reduced (Fig 3c). The responses to ATP and acetylcholine were not affected. The blockade produced by BWSV was usually observed after 3 to 5 hr of incubation and was more effective and developed more rapidly when the tissue was stimulated at 10 to 20 Hz for 1- to 2-min periods every 20 min. The effects of BWSV were not reversible even after prolonged washing.

Following incubation of the tissues with α-bungarotoxin (α-Bu) (1×10^{-6} g/ml) only the primary contraction in response to field stimulation was abolished, the after-relaxation being unaffected or slightly enhanced (Fig 3d). Responses to ATP were similarly unaffected by the toxin whereas those to acetylcholine were abolished or converted to a relaxation. The blockade produced by α-Bu developed following 15 to 20 min of incubation and was not reversible even after prolonged washing.

DISCUSSION

The activity of various analogs of adenosine bearing modifications at the ribose moiety revealed the importance of an intact 2′-hydroxyl group for inhibitory activity

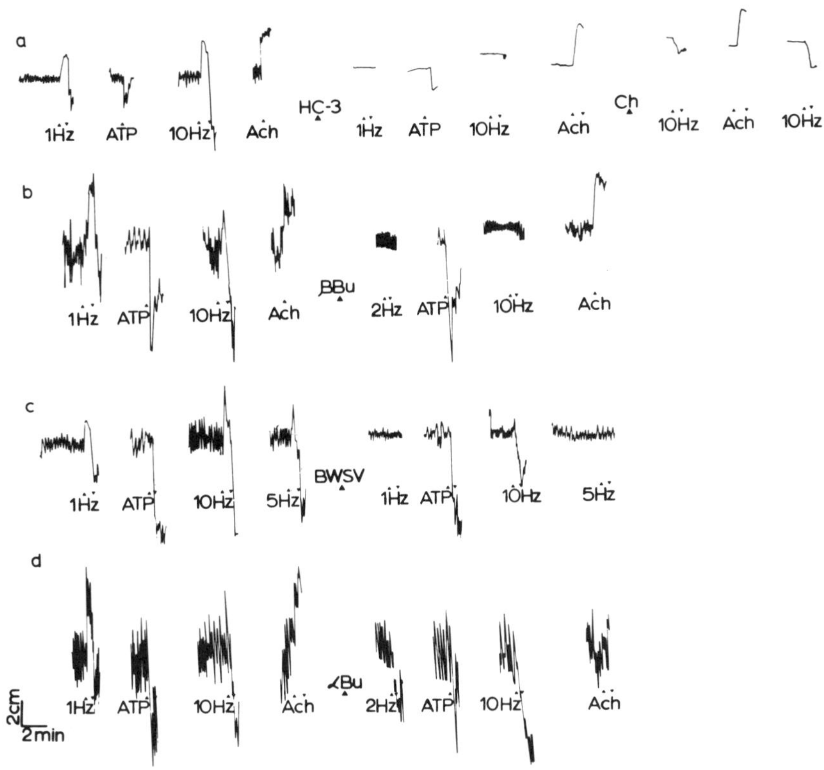

FIG. 3. a: The effects of HC-3 on the biphasic response of the guinea pig fundus induced by field stimulation. Responses are shown to field stimulation (1-msec duration) at 1 and 10 Hz, ATP (1 μM), and acetylcholine (ACh), (0.5 μM). At the mark ▲HC-3 the preparation was incubated with HC-3 (5 μM) for 3 hr with intermittent field stimulation at 10 Hz and the responses retested. After washout of HC-3, the preparation was incubated with choline (▲Ch) (20 μM) for 2 hr and the responses retested. Note that HC-3 abolished both phases of the response to stimulation without affecting those to ACh or ATP and that choline partially restored the response. **b:** The effects of incubation with β-bungarotoxin on the biphasic response of the guinea pig fundus induced by field stimulation. Field stimulation parameters and drug concentrations as in **a**. At the mark (▲β-Bu) the preparation was incubated with β-bungarotoxin (5×10^{-5} g/ml) for 3 hr with intermittent field stimulation at 10 Hz. Note that β-Bu abolished both phases of the response to stimulation without affecting those to ACh or ATP. **c:** The effect of incubation with Black widow spider venom on the biphasic response of the guinea pig fundus induced by field stimulation. Stimulation parameters and drug concentrations as in **a.** At the mark (▲BWSV) the preparation was incubated with black widow spider venom (3×10^{-5} g/ml) for 4 hr with intermittent field stimulation at 10 Hz. Note that BWSV abolished the response at 1 and 5 Hz and reduced that at 10 Hz. **d:** The effect of incubation with α-bungarotoxin on the biphasic response of the guinea pig fundus induced by field stimulation. Stimulation parameters and drug concentrations as in **a**. At the mark (▲α-Bu) the preparation was incubated with α-bungarotoxin for 20 min. Note that α-Bu abolished the primary contraction without affecting the after-relaxation or the response to ATP.

at the stimulated ileum as indicated by the marked decrease in potency of 2′-deoxyadenosine, 2′,3′-diacetyladenosine, and adenine-β-D-arabinofuranoside compared to ATP or adenosine. In contrast, the 3′-hydroxyl group is evidently less important for interaction with the receptor since 3′-deoxyadenosine possessed essentially the same activity as ATP or adenosine.

Substitution at the 5′ position revealed a great deal of tolerance on the part of the receptor since not only ATP, ADP, AMP, and adenosine demonstrated equal potency but CoA and its analogs exhibited essentially similar activity. Although it is possible that all of these compounds are metabolized to the nucleoside and that adenosine is the final mediator, this seems unlikely in view of the activity of the methylene-substituted isosteres. It is not clear whether the introduction of substituent groups changes the affinity of the substituted portions of the molecule for specific sites on the receptor or the presence of substituents alters the conformation of the adenosine moiety so as to alter the affinity of the nucleotide as a whole for the receptor (see D. Miles, *this volume*). The discrepancy between the potency observed for 2′,3′-diacetyladenosine compared to that for 2′,3′-isopropylidene adenosine suggests that the latter possibility may be more important.

The activity of CoA and its analogs was of particular interest since this compound is a naturally occurring purine nucleotide. The efficacy of this compound in relaxing the fundic strip preparation was less than that of ATP by a factor of 5 and less than that of adenosine by a factor of 2. However, otherwise, its effects were identical to those of ATP. The effects of CoA were antagonized competitively by theophylline in a manner similar to that shown for ATP (27), and Okwuasaba (25) has shown that the pA_2 value for such antagonism is also similar to that for ATP. In addition, the ability of dipyridamole to increase the potency of both ATP and CoA as well as the demonstration of cross-tachyphylaxis between ATP, adenosine, CoA, and adenosine 3′,5′-diphosphate as well as between these compounds and the endogenously released mediator (25) has confirmed that the coenzyme interacts with the receptor on fundic muscle in a manner similar to ATP.

At presynaptic receptors, mediating inhibition of acetylcholine output, CoA also acts in the same manner as ATP. Its efficacy on the Trendelenburg preparation of the guinea pig ileum (25), as well as on the stimulated guinea pig ileum preparation (15), has been established and the sensitivity of the responses at the stimulated ileum to competitive antagonism by theophylline and to potentiation by dipyridamole is similar to that shown for ATP. In addition, Cook et al. (15) showed that the acetylcholine output from the stimulated ileum, as measured by bioassay, is inhibited by the nucleotides in a manner similar to the inhibition of the mechanical response.

The demonstration of both the pre- and postsynaptic efficacy of CoA led to the conclusion that since this compound is an endogenous purine nucleotide, if it were released on stimulation it would act in a manner analogous to ATP, the putative purinergic transmitter. As the possible source for CoA release we considered the cholinergic nerve.

The suggestion that the inhibitory mediator is released together with acetylcholine from cholinergic nerves was made by Daniel et al. (16) who used the terms ''simultaneous release'' to explain the findings of both excitatory (EJPs) and inhibitory junction potentials (IJPs) in isolated rabbit jejunum and opposum esophagus. Cook et al. (15), in describing the activity of CoA on the stimulated ileum, suggested that since that compound is present in cholinergic nerve terminals, where it is produced during the synthesis of acetylcholine, it could be released on stimulation together with acetylcholine. Adenosine triphosphate is also known to be located in cholinergic nerve terminals and Meunier et al. (24) have demonstrated the presence of ATP in cholinergic vesicles. The release of ATP from the phrenic nerves has been demonstrated (29,30) and Israel et al. (18) have established that the changes in ATP and acetylcholine levels in *Torpedo* electroplaque synapses are related.

The possible cholinergic source of the endogenous purine nucleotide mediator was tested using cholinergic neuron-blocking drugs. The blockade of both phases of the response to field stimulation by HC-3 constitutes evidence for the involvement of cholinergic nerves in the after-relaxation as well as in the primary contraction. The mechanism of action of HC-3 in causing depletion of acetylcholine in cholinergic nerve terminals through prevention of choline uptake is well known (5,19). In addition, competitive antagonism of the effect of HC-3 by choline has been demonstrated (23,31) and the partial restoration of the response by choline argues for the involvement of cholinergic nerves. It is unlikely that distinct ''purinergic'' nerves would be sensitive to the depleting action of HC-3 since this would imply that they possess a high-affinity uptake mechanism for choline similar to that in cholinergic nerves.

Additional evidence for the cholinergic source of the endogenous inhibitory mediator came from the studies with β-Bu and BWSV. β-Bungarotoxin (an Elapid snake venom) is known to cause marked reduction in the output of acetylcholine by an action at the nerve terminal (12,20,21). It binds at presynaptic nerve terminals and inhibits the release of acetylcholine, for example, from the isolated rat diaphragm–phrenic nerve preparation (11). Black widow spider venom substantially decreases the number of synaptic vesicles present in motor nerve endings (14,22) and changes the structure of such nerve endings consistent with depletion of vesicles of their acetylcholine content (13,14). Although the action of these two toxins is not well understood, the results obtained with them, being similar to those obtained with HC-3, whose mechanism of action has been more clearly delineated, support a cholinergic origin for the endogenous mediator of nonadrenergic inhibition.

The identity of the mediator remains unknown. It is not unreasonable to assume that, if the mediator is released simultaneously with acetylcholine from cholinergic nerves, it should be synthesized and stored in such nerves. The presence of ATP in cholinergic nerves and evidence for its release on stimulation together with the putative transmitter role for ATP (7) not only supports such a role for ATP but also the hypothesis of its cholinergic source.

Equally, the demonstration of the pre- and postsynaptic efficacy of CoA and its presence in cholinergic nerve terminals suggests that it might be considered a candidate for the inhibitory mediator. Whether the moiety responsible for nonadrenergic inhibition is ultimately shown to be ATP or CoA or another purine nucleotide, it seems possible that cholinergic nerves are the source of the mediator and the concept of simultaneous release of purine nucleotide and acetylcholine is offered as being potentially useful in explaining some of the discrepant observations of both EJPs and IJPs in various smooth muscles (3,4,16,17). Were a purine nucleotide to be released with acetylcholine from cholinergic nerve terminals and function as the nonadrenergic inhibitory mediator, the production of either an EJP or an IJP at the muscle could result depending on the balance between the concentrations or activity of the two mediators. Such a balance would be determined by differential rates of breakdown or uptake of the two mediators, the number and distribution of receptors sensitive to the two mediators, as well as the concentrations of the mediators necessary to activate these receptors.

Figure 4 shows a diagrammatic representation of a cholinergic nerve varicosity exhibiting simultaenous release of both acetylcholine and a purine nucleotide modulator. The production of acetylcholine following acetylation, accompanied by the production of CoA and storage of acetylcholine in a vesicle together with ATP, is shown. The release of acetylcholine, its activity at the muscarinic receptor (M) and subsequent hydrolysis by acetylcholinesterase (Ach E) is also shown. The

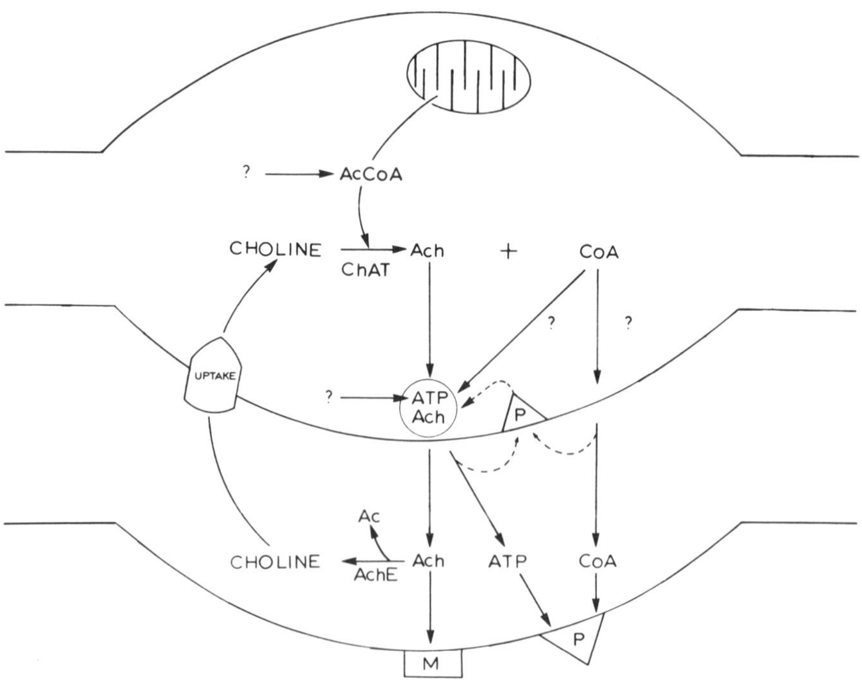

FIG. 4: Diagram of a cholinergic nerve varicosity exhibiting simultaneous release of acetylcholine (the excitatory mediator) and either ATP or CoA (putative inhibitory mediators). The acetylation of choline to ACh and CoA by choline acetyltransferase (ChAT) is shown together with the storage of ACh and ATP in a vesicle. Postsynaptic muscarinic receptors (M) and both pre- and postsynaptic purinergic receptors (P) are present. For other details see text.

occurrence of these events is well accepted. The presence of "purinergic" receptors (P) is also shown and these are placed on both sides of the synaptic cleft in an attempt to synthesize a mechanism from the evidence presented. The release of ATP together with acetylcholine from the same vesicle is tentative. However, if this occurs, activity of the nucleotide at the presynaptic site to inhibit acetylcholine release and at the postsynaptic site to inhibit the muscle directly would be expected. In the absence of any experimental data, CoA is shown possibly being released directly from the terminal membrane or from vesicular stores. The scheme is, of course, speculative and further investigation of cholinergic nerve involvement in "purinergic" functions may be fruitful.

ACKNOWLEDGMENTS

This work was supported by a grant in aid from the Medical Research Council of Canada to M.A.C. F.K.O. is the recipient of a Mid-West Nigerian Government Scholarship. The authors wish to thank Miss M. Ackerman for technical assistance.

REFERENCES

1. Ally, A. I., and Nakatsu, K. (1976): *J. Pharmacol. Exp. Ther.,* 199:208–215.
2. Baumgarten, H. G., Holstein, A. F., and Owman, C. H. (1970): *Z. Zellforsch. Mikrosk. Anat.,* 106:376–397.
3. Beani, L., Bianchi, C., and Crema, A. (1971): *J. Physiol. (Lond.),* 217:259–279.
4. Bennett, M. R. (1966): *J. Physiol. (Lond.),* 185:132–147.
5. Bertolini, A., Greggia, A., and Ferrari, W. (1967): *Life Sci.,* 6:537–543.
6. Bianchi, C., Beani, L., Frigo, G. M., and Crema, A. (1968): *Eur. J. Pharmacol.,* 4:51–61.
7. Burnstock, G. (1972): *Pharmacol. Rev.,* 24:509–581.
8. Burnstock, G. (1976): *J. Theor. Biol.,* 62:491–503.
9. Burnstock, G., Campbell, G., and Rand, M. J. (1966): *J. Physiol. (Lond.),* 155:115–133.
10. Burnstock, G., Campbell, G., Satchell, D. G., and Smythe, A. (1970): *Br. J. Pharmacol.,* 40:668–688.
11. Chang, C. C., Chen, T. J., and Lee, C. Y. (1973): *J. Pharmacol. Exp. Ther.,* 184:339–345.
12. Chang, C. C., and Lee, C. Y. (1963): *Arch. Int. Pharmacodyn. Ther.,* 144:241–257.
13. Clark, A. W., Hurlbut, W. P., and Mauro, A. (1972): *J. Cell Biol.,* 52:1–14.
14. Clark, A. W., Mauro, A., Longenecker, H. E., and Hurlbut, W. P. (1970): *Nature,* 225:703.
15. Cook, M. A., Hamilton, J. T., and Okwuasaba, J. K. (1978): *Nature,* 271:768–771.
16. Daniel, E. E., Taylor, G. S., Daniel, V. P., and Holman, M. E. (1977): *Can. J. Physiol. Pharmacol.,* 55:243–250.
17. Furness, J. B. (1969): *J. Physiol. (Lond.),* 205:549–567.
18. Israel, M., Lesbats, B., Manaranche, R., Marsal, J., Mastour-Frachon, P, and Meunier, F. M. (1977): *J. Neurochem.,* 28:1259–1267.
19. Jones, S. F., and Kwanbunbumpen, S. (1970): *J. Physiol. (Lond.),* 207:31–50.
20. Lee, C. Y. (1970): *Clin. Toxicol.,* 3:457–472.
21. Lee, C. Y. (1972): *Annu. Rev. Pharmacol.,* 12:265–286.
22. Longenecker, H. E., Hurlbut, W. P., Mauro, A., and Clark, A. W. (1970): *Nature,* 225:701.
23. MacIntosh, F. C., Birks, R. I., and Sastry, P. B. (1956): *Nature,* 178:1181.
24. Meunier, F. M., Israel, M., and Lesbats, B. (1975): *Nature,* 257:407–408.
25. Okwuasaba, F. K. (1978): Ph.D. Thesis, University of Western Ontario, London.
26. Okwuasaba, F. K., Hamilton, J. T., and Cook, M. A. (1977): *Eur. J. Pharmacol.,* 43:181–194.
27. Okwuasaba, F. K., Hamilton, J. T., and Cook, M. A. (1977): *Eur. J. Pharmacol.,* 46:181–198.
28. Paton, W. D. M., and Vizi, E. S. (1969): *Br. J. Pharmacol. Chemother.,* 41:381–400.
29. Silinsky, E. M. (1975): *J. Physiol. (Lond.),* 247:145–162.
30. Silinsky, E. M., and Hubbard, J. I. (1973): *Nature,* 243:404–405.
31. Wong, K. C., and Long, J. P. (1961): *J. Pharmacol. Exp. Ther.,* 133:211–215.

Cardiovascular Actions

Physiological and Regulatory Functions of Adenosine and Adenine Nucleotides, edited by H. P. Baer and G. I. Drummond. Raven Press, New York © 1979.

Role of Adenosine as a Mediator of Metabolic Vasodilation in the Heart and Brain: A Brief Overview and Update

Robert M. Berne, Duane H. Foley, William P. Watkinson, Wayne L. Miller, H. Richard Winn, and Rafael Rubio

Department of Physiology, University of Virginia, School of Medicine, Charlottesville, Virginia 22901

Adenosine has been implicated as a regulator of blood flow in heart, brain, skeletal muscle, lung, kidney, and intestine. The main interest of this laboratory has been with respect to the function of the nucleoside in the circulation of the heart and brain. This discussion is restricted to the main historical developments of the adenosine hypothesis of blood flow regulation in these systems.

HEART

That a close connection exists between myocardial metabolic activity and coronary blood flow (CBF) has been known for many years (7) and a large number of factors have been suggested to control CBF in relation to oxygen supply and demand. Of these, adenosine appears to most closely meet the criteria for such an agent. It is a potent vasodilator, readily crosses cell membranes, is effectively metabolized, and can be supplied readily from intracellular adenosine triphosphate (ATP). Early studies revealed that inosine and hypoxanthine, i.e., degradation products of adenosine, were released from isolated perfused or *in situ* dog hearts subjected to anoxia or ischemia (1). Subsequent experiments in the presence of an adenosine deaminase inhibitor (10) or an uncoupler of oxidative phosphorylation (15) resulted in the demonstration of adenosine release with anoxia. These studies were followed by experiments in the open-chest dog in which adenosine was observed in coronary sinus blood during the period of reactive hyperemia following 30- to 60-sec coronary occlusions (19) and was found to increase in the myocardium during brief periods of coronary occlusion (13) and decrease during the period of reactive hyperemia (14). If adenosine should serve as a regulator of coronary resistance it should also be present in normal, well-oxygenated heart muscle as well as under conditions of impaired oxygen supply. In attempts to investigate this requirement, indirect estimates of myocardial adenosine levels by measurements of adenosine concentrations in pericardial fluid (16) and direct measurements of adenosine levels in the myocardium (13) demonstrated the presence of adenosine in the normal heart.

Since it appeared that an increase in adenosine production and release was associated with an inadequate myocardial oxygen supply, time studies were conducted on rat hearts subjected to different periods of ischemia (3). With cross-clamping at the atrioventricular groove in the rat heart, cardiac adenosine levels increased several-fold, reaching three times normal levels in 5 sec. There was also an excellent correlation between CBF and myocardial adenosine production and release in the isolated perfused guinea pig heart with progressive hypoxia (21). The source of this adenosine was the myocardial cells, and not vascular or other nonmyocardial tissue in the heart, since cultures of pure embryonic chick heart cells also released adenosine into the medium under conditions of hypoxia (11).

The enzyme 5′-nucleotidase, presumably responsible for adenosine formation from adenosine monophosphate (AMP), is located at the myocardial cell margins and hence adenosine produced at this site can immediately enter the interstitial space and induce dilation of the arterioles (18). Adenosine entering the myocardial cell is either phosphorylated by adenosine kinase to AMP and incorporated into adenine nucleotides or deaminated to inosine (26). Adenosine that crosses the capillary wall is, to a large degree, deaminated to inosine which in turn is further catabolized to hypoxanthine and ribose-1-PO_4 by nucleoside phosphorylase present in endothelial cells, pericytes, and erythrocytes (20). Nucleoside phosphorylase was not found in myocardial cells or vascular smooth muscle cells. Studies on red cell ghosts (22) and cultured chick embryo heart cells (12) indicated that adenosine is undetectable as such within cells even when external concentrations greatly exceed physiological levels.

In the studies summarized above, enhanced adenosine production was always accomplished by limiting the oxygen supply to the myocardium. In fact, recent observations reveal that when a flow gradient is produced across the wall of the canine left ventricle by left coronary artery constriction, the adenosine concentration in the endocardium increases proportionately more than in the epicardium. This is depicted in Fig. 1 in which the ratio of endocardial to epicardial adenosine concentrations is plotted against the quotient of the diastolic pressure time index and the time tension index (4). Under control conditions, when blood flow to the endocardium and epicardium is equal, adenosine concentrations are also equal. One might expect higher levels in the endocardium to counteract the extravascular compression of the endocardial vessels and thereby maintain an endocardial-to-epicardial flow ratio of 1.0. However, current tissue-sampling techniques are probably not fast enough to permit detection of such a difference, if it indeed exists.

Since the balance between oxygen supply and demand appears to be the critical factor in determining adjustments of CBF, attention has now been focused on changes in oxygen demand without restriction of oxygen supply. In one series of experiments the aortas of rats were constricted just beyond the origin of the innominate artery and the increase in cardiac pressure work was compared to increases in myocardial adenosine, inosine, and hypoxanthine levels (8) (Fig. 2). Not

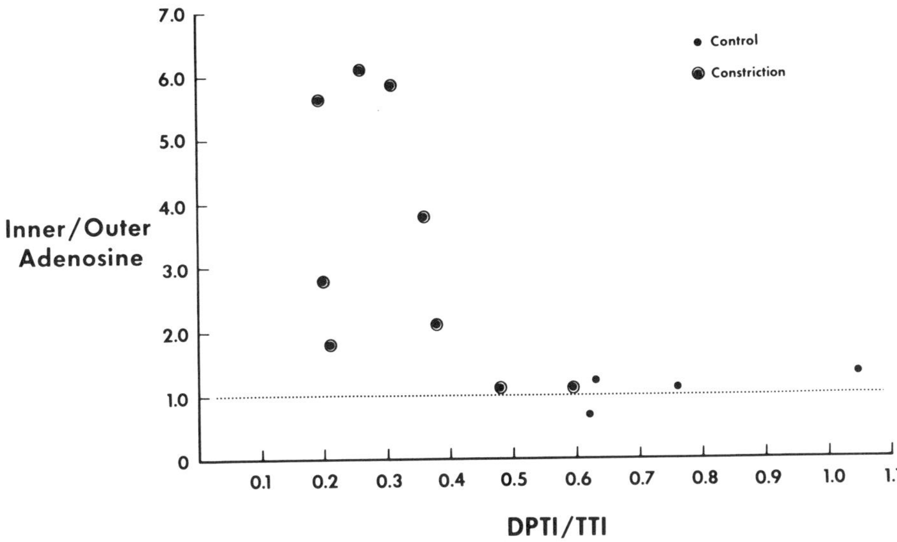

FIG. 1. Relationship between the ratio of the diastolic pressure time index (DPTI): tension time index (TTI), and the ratio of the inner:outer left ventricular wall adenosine concentration in control experiments and during partial constriction of the left coronary artery. The DPTI is the area under the diastolic aortic pressure curve minus left atrial pressure, and is an index of diastolic perfusion of the endocardium. The TTI is the area under the systolic portion of the left venticular pressure curve and is an index of myocardial oxygen requirements.

only is there no restriction of CBF in these experiments, but flow is actually increased, which may be due in part to the greater coronary perfusion pressure produced by the aortic constriction.

In another series of experiments, myocardial metabolism was increased by different procedures and the release of adenosine from the epicardial surface of the dog heart was determined in Krebs-Hensleit solution introduced into the pericardial sac. This fluid was withdrawn after 4.5 min and analyzed for its adenosine content. Stimulation of the stellate ganglia (right, left, or both simultaneously) produced increases in heart rate, blood pressure, and adenosine levels in the pericardial fluid (24), and the increments in adenosine levels were proportional to the intensity of stimulation (9). The CBF correlated well with myocardial oxygen consumption (Fig. 3) which itself showed good correlation with the adenosine release (Fig. 4) during graded ganglion stimulation. Simulated exercise, produced by stimulating the motor nerves to the four extremities, produced similar increments of adenosine levels in pericardial fluid. Finally, as shown in Fig. 5, the unanesthetized dog with a special catheter implanted in the pericardial sac also showed a striking increase in the release of adenosine with treadmill exercise (25).

These studies on adenosine release from hearts subjected to increased metabolic activity without restriction of oxygen supply in both the anesthetized and unanesthetized animal lend strong support to the concept that adenosine plays a physiological role in the regulation of CBF.

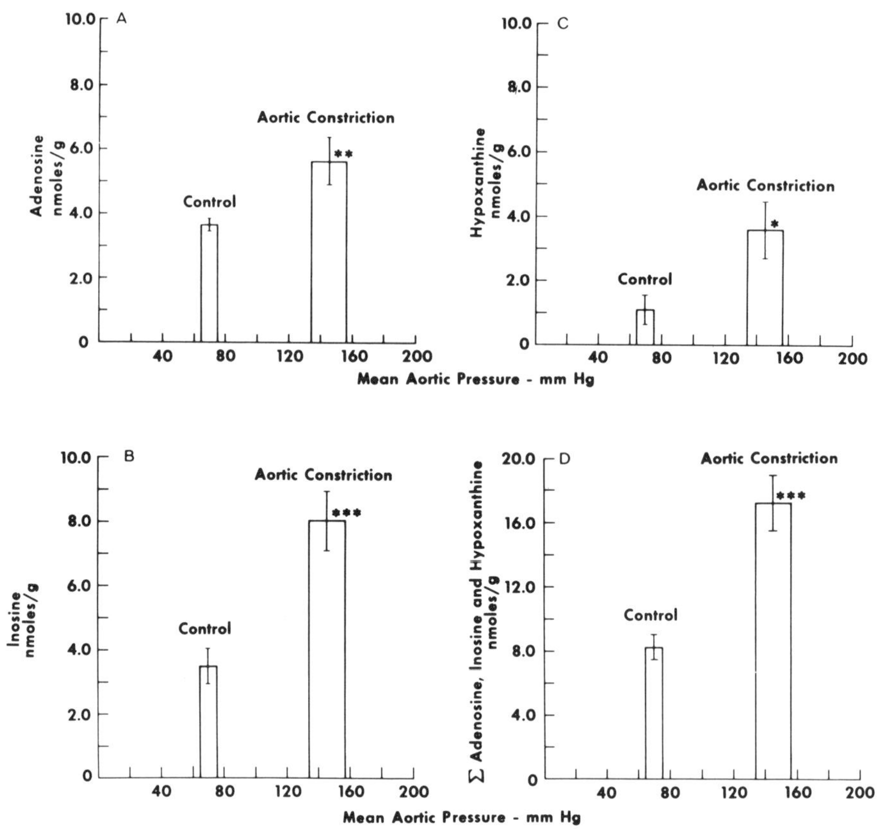

FIG. 2. Adenosine **(A)**, inosine **(B)**, hypoxanthine **(C)**, and sum of adenosine, inosine, and hypoxanthine **(D)** contents of rat myocardium as a function of mean aortic pressure. The values represent mean±SEM. The widths of the bars represent ±SEM for mean aortic pressure. Controls, $N = 16$; aortic constriction, $N = 7$. Asterisks indicate mean values of purine contents in the experimental group significantly greater than corresponding values in the control group: $^{*}p<0.02$; $^{**}p<0.005$; $^{***}p<0.001$. (From ref. 8, with permission.)

BRAIN

As in the case of the heart, the preferential pathway for the breakdown of cellular adenine nucleotides with ischemia appeared to occur via dephosphorylation of AMP to adenosine (6). Thus it seemed logical to suspect that adenosine might serve as a metabolic mediator of cerebral blood flow in a manner similar to that proposed for the heart. However, when adenosine was infused into a carotid artery, it was essentially ineffective in altering cerebrovascular resistance (5). We considered the possibility that adenosine may not cross the blood-brain barrier and that vasodilation might be elicited by direct application of the nucleoside to cerebral vessels.

Studies on dog and rat brain in our laboratory confirmed the formation of adenosine in ischemic brain and also showed that the nucleoside was released into

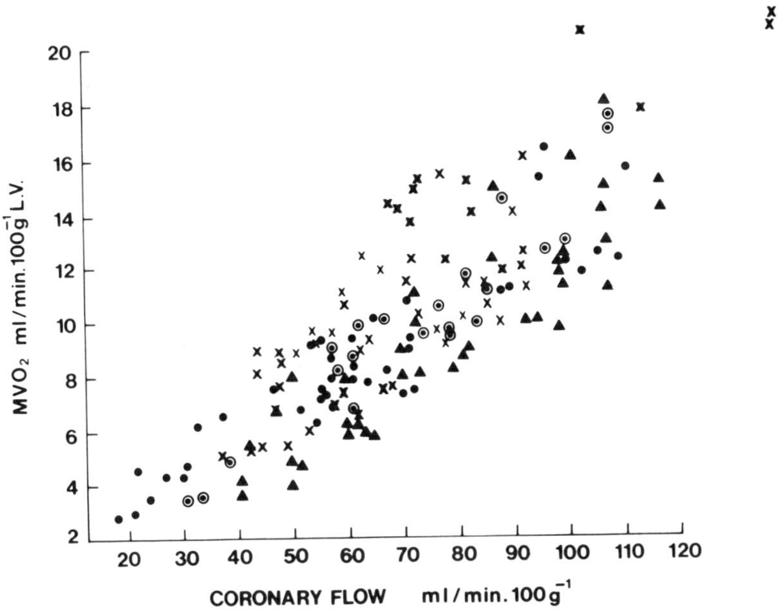

FIG. 3. Relationship between CBF and myocardial oxygen consumption in the dog heart. ● = Controls, ◉ = measurements obtained when oxygen consumption was changed either by increasing blood volume by hemorrhage or by i.v. administration of epinephrine, ▲ = the effect of intracoronary administration of 2,4-dinitrophenol, X = effect of bilateral stellate ganglion stimulation in dogs with fluid in the pericardial sac. (The data on 2,4-dinitrophenol and stellate ganglion stimulation have been added to the figure of Eckenhoff et al., ref. 7, with permission.)

the cerebrospinal fluid (CSF) (2). We also confirmed the ineffectiveness of intracarotid adenosine on cerebral blood flow and demonstrated a striking increment in blood flow to the noncerebral cephalic tissue (2). However, when the pial vessels of the cat were exposed to adenosine dissolved in mock CSF, vasodilation occurred and the response was dose dependent (2). In some experiments on cats, adenosine was administered intravascularly and topically to the same arteriole and vasodilation was only observed with topical application.

When [^{14}C]-labeled adenosine was introduced into the arterial circulation of the dog brain, practically no label was found in the CSF, and when the adenosine was injected into the CSF only insignificant amounts of label were found in the jugular (venous) blood (2). Intravenous infusion of labeled adenosine in the rat revealed that incorporation of the label into cerebral adenine nucleotides was only 5% of the amount of label incorporation into cardiac nucleotides. These observations indicated that adenosine could serve as a dilator of pial arterioles but could not reach the vascular smooth muscle from the luminal side of the vessel in concentrations sufficient to elicit vasodilation because of the endothelial cell barrier.

Several procedures known to influence cerebral blood flow were undertaken in the rat to determine if they also altered the adenosine concentration of the brain. Indeed, electrical stimulation of the brain, a sustained reduction in arterial pressure,

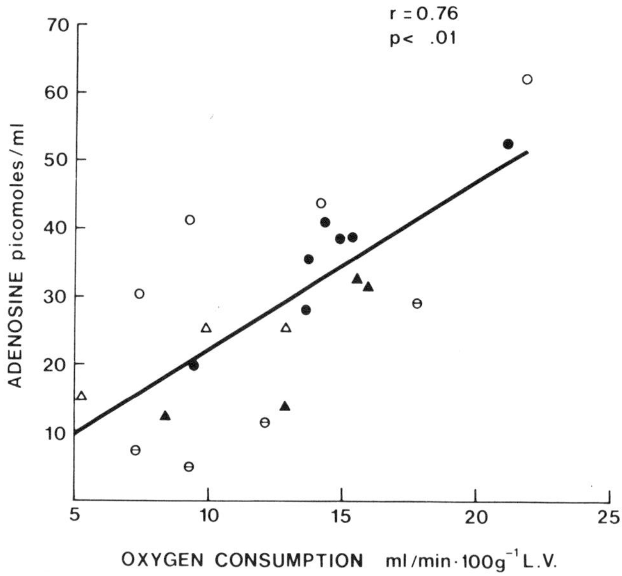

FIG. 4. Effect of different intensities of bilateral stellate ganglion stimulation on myocardial oxygen consumption and the adenosine release from the epicardial surface of the heart in pmoles/ml of pericardial injectate (Krebs-Henseleit solution). Each symbol represents an individual dog heart.

and hypoxemia all resulted in a significant increase in the adenosine levels (17). Hypocapnia produced by hyperventilation in the rat caused a decrease in cerebral adenosine levels, whereas hypercapnia caused a decrease to below control values (17).

Adenosine is quite stable in CSF since incubation for 10 min at 37°C did not diminish the initial concentration whereas a similar incubation of adenosine in cell and platelet-free plasma showed a significant decrease in the adenosine concentration (2).

Since adenosine is practically absent in CSF under normal conditions and does not escape into the blood, it must be reincorporated into cerebral adenine nucleotides following its release from brain subjected to brief periods of stimulation, hypotension, hypoxia, or ischemia. To study the uptake of adenosine by brain, convexity-cisternal superfusion was carried out in dogs with mock CSF containing [U-^{14}C]adenosine (27). After 20 min of superfusion and a 40-min wash with unlabeled mock CSF, brain samples were frozen *in situ* with liquid nitrogen and the adenine nucleotides of the tissue subjected to enzymatic degradation to nucleosides (inosine) and base (hypoxanthine). The ratio of the specific activity of the nucleoside:base was determined and compared to the ratio of nucleoside:base of the adenosine-containing superfusion fluid. If the ratios of the specific activity of nucleoside:base are the same for degraded brain tissue and superfusion fluid, it would indicate that adenosine was directly phosphorylated by adenosine kinase to AMP and subsequently incorporated into ATP and adenosine diphosphate (ADP)

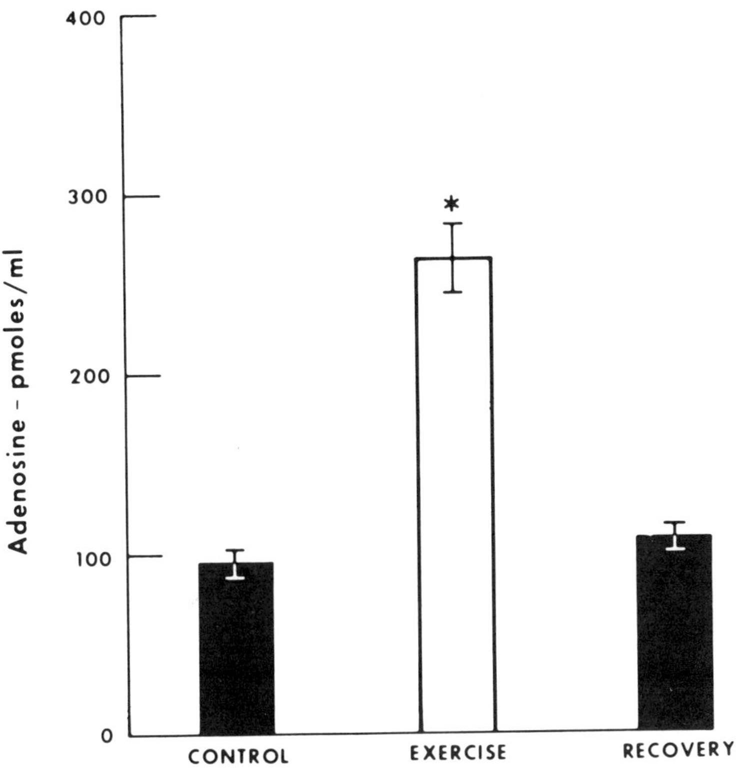

FIG. 5. Concentration of adenosine in pericardial infusates during control, exercise, and recovery conditions in the unanesthetized dog. (*), $p<0.02$.

without prior degradation and dilution of the labeled substrate by unlabeled substrates. This proved to be the case and the results are shown in Table 1.

When labeled inosine was used as substrate, no radioactivity was detected in the cerebral adenine nucleotides. This result is probably due to the fact that inosine cannot be directly phosphorylated to inosinic acid nor can it be degraded by brain tissue to hypoxanthine and ribose-1-PO_4 by nucleoside phosphorylase. Histochemical studies have revealed that in the brain the nucleoside phosphorylase is located primarily in the endothelial cells and it is postulated that the inosine cannot reach these cells because of the blood-brain barrier. Studies currently under way using labeled hypoxanthine as substrate should shed some light on this hypothesis.

Since changes in cerebral blood flow occur promptly to such stimuli as hypoxia, it is important to determine the rapidity with which adenosine is formed when the brain is subjected to a reduced oxygen supply. Experiments using the brain-blowing technique of Veech et al. (23) were performed on rats subjected to different periods of cerebral ischemia. Blood flow to the brain was rapidly reduced to near-zero levels by acute severance of the aorta at the level of the diaphragm. It can be seen from Fig. 6 that cerebral adenosine levels increased over twofold within 5 sec of the onset

TABLE 1. *Specific activity of nucleoside:base moieties from hydrolyzed adenine nucleotides*

Substance hydrolyzed	Nucleoside moiety (inosine) $SA = \frac{CPM}{nmoles}$	Base moiety (hypoxanthine) $SA = \frac{CPM}{nmoles}$	$\frac{\text{Sa nucleoside moiety}}{\text{SA base moiety}}$	Ratio of SA $\frac{\text{perfusion fluid}}{\text{Brain}}$
Adenosine perfused into CSF	374,612	211,700	1.76	
Adenine nucleotides from brain				
1	1.57	0.99	1.59	
2	1.13	0.73	1.55	
3	4.28	2.57	1.67	$\frac{1.76}{1.70} = 1.03$
4	4.29	2.48	1.81	
5	2.78	1.46	1.90	
			Avg. 1.70 ± 0.07(SEM)	

SA, specific activity; CPM, counts per minute.

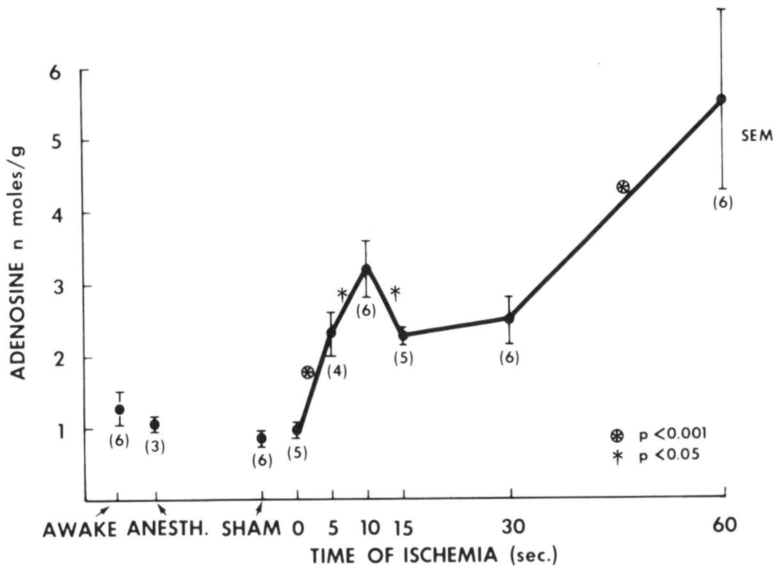

FIG. 6. Adenosine concentrations of rat brain after different periods of ischemia. Adenosine levels in conscious, anesthetized, and sham-operated (loose ligature placed around aorta) rats were not significantly different from each other.

of cerebral ischemia. Adenosine levels in the brain were five times greater at 60 sec than they were at zero time. The reason for the reproducible decrease at 15 and 30 sec is unknown at present. Similar rapid increases in brain adenosine levels (15 ×) were observed within 10 sec of the onset of bicuculline-induced seizures (28).

In summary, adenosine appears to be an important mediator of blood flow in the myocardium and may play a similar role in the brain. However, much additional information must be obtained about the relationship between cerebral blood flow and brain adenosine release before a definitive statement can be made regarding the importance and physiological role of adenosine in the regulation of cerebral blood flow.

ACKNOWLEDGMENTS

The research of the authors in this paper was supported by USPHS Grant HL10384.

REFERENCES

1. Berne, R. M. (1963): *Am. J. Physiol.,* 204:317–322.
2. Berne, R. M., Rubio, R., and Curnish, R. R. (1974): *Circ. Res.,* 35:262–271.
3. Berne, R. M., Rubio, R., and Duling, B. R. (1971): In: *Myocardial Ischemia,* edited by R. S. Ross and F. Hoffmann, pp. 28–43. Excerpta Medica, Amsterdam.
4. Buckberg, G. D., Fixler, D. E., Archie, J. P., and Hoffman, I. E. (1972): *Circ. Res.,* 30:67–81.
5. Buyniski, J. P., and Rapela, C. E. (1969): *Am. J. Physiol.,* 217:1660–1664.

6. Deuticke, B., and Gerlach, E. (1966): *Arch. Ges. Physiol.,* 292:239–254.
7. Eckenhoff, J. E., Hafkenschiel, J. H., Landmesser, C. M., and Harmel, M. (1947): *Am. J. Physiol.,* 149:634–649.
8. Foley, D. H., Herlihy, J. T., Thompson, C. I., Rubio, R., and Berne, R. M. (1978): *J. Mol. Cell. Cardiol.,* 10:292–300.
9. Foley, D. H., Watkinson, W. P., Rubio, R., and Berne, R. M. (1978): *Fed. Proc.,* 37:565.
10. Katori, M., and Berne, R. M. (1966): *Circ. Res.,* 19:420–425.
11. Mustafa, S. J., Berne, R. M., and Rubio, R. (1975): *Am. J. Physiol.,* 228:1474–1478.
12. Mustafa, S. J., Rubio, R., and Berne, R. M. (1975): *Am. J. Physiol.,* 228:62–67.
13. Olsson, R. A. (1970): *Circ. Res.,* 26:301–306.
14. Olsson, R. A., Snow, J. A., and Gentry, M. K. (1978): *Circ. Res.,* 42:358–362.
15. Richman, H. G., and Wyborny, L. (1964): *Am. J. Physiol.,* 207:1139–1145.
16. Rubio, R., and Berne, R. M. (1969): *Circ. Res.,* 25:407–415.
17. Rubio, R., Berne, R. M., Bockman, E. L., and Curnish, R. R. (1975): *Am. J. Physiol.,* 228:1896–1902.
18. Rubio, R., Berne, R. M., and Dobson, J. G., Jr. (1973): *Am. J. Physiol.,* 225:938–953.
19. Rubio, R., Berne, R. M., and Katori, M. (1969): *Am. J. Physiol.,* 216:56–62.
20. Rubio, R., Wiedmeier, V. T., and Berne, R. M. (1972): *Am. J. Physiol.,* 222:550–555.
21. Rubio, R., Wiedmeier, V. T., and Berne, R. M. (1974): *J. Mol. Cell. Cardiol.,* 6:561–566.
22. Schrader, J., Berne, R. M., and Rubio, R. (1972): *Am. J. Physiol.,* 223:159–166.
23. Veech, R. L., Harris, R. L., Veloso, D., and Veech, E. H. (1973): *J. Neurochem.,* 20:183–188.
24. Watkinson, W. P., Foley, D. H., Rubio, R., and Berne, R. M. (1976): *Circulation,* 54:11–107.
25. Watkinson, W. P., Foley, D. H., Rubio, R., and Berne, R. M. (1977): *Fed. Proc.,* 36:598.
26. Wiedmeier, V. T., Rubio, R., and Berne, R. M. (1972): *Am. J. Physiol.,* 223:51–54.
27. Winn, H. R., Curnish, R., Rubio, R., and Berne, R. M. (1978): *Fed. Proc.,* 37:653.
28. Winn, H. R., Welsh, J. E., Rubio, R., and Berne, R. M. (1978): *Physiologist,* 21:129.

Physiological and Regulatory Functions of Adenosine and Adenine Nucleotides, edited by H. P. Baer and G. I. Drummond. Raven Press, New York © 1979.

Sites and Mode of Action of Adenosine in the Heart. I. Coronary System

E. Gerlach, J. Schrader, and S. Nees

Department of Physiology, University of Munich, D-8000 Munich, Germany

Adenosine is known to have different sites of action in the heart: (a) it causes vasodilation (1,2,4); (b) it affects the atrioventricular node resulting in a decrease of heart rate or even blockade of conduction (4); and (c) it exerts a negative inotropic effect on the atrial myocardium (4,23) and attenuates the catecholamine-induced increase in ventricular force development (19). Despite the different sites of action of adenosine, the basic mechanism of its effects may be similar in all cases, namely interference with transmembranal Ca^{2+} movement (23) and/or Ca^{2+} availability.

In recent years we have performed extensive studies concerning adenosine metabolism in the heart and the possible involvement of this nucleoside in the regulation of coronary blood flow under various conditions (19–22). The main results obtained are summarized in this chapter.

All our investigations were carried out on an isolated perfused guinea pig heart preparation. The hearts, perfused with an oxygenated, pyruvate containing Krebs-Henseleit solution, had been previously shown to be metabolically and functionally stable for several hours (3). Analyses of the adenine nucleotides [adenosine triphosphate (ATP), adenosine diphosphate (ADP), and adenosine monophosphate (AMP)] and of their breakdown products (adenosine, inosine, and hypoxanthine) present in myocardial tissue and/or coronary effluent perfusate were performed by use of chromatographic methods for separation (6,20) and by application of microspectrophotometric techniques and enzyme assay procedures for quantitation (10,11,13). Cyclic 3′,5′-AMP (cAMP) was determined according to the method of Gilman (7). Measurements of specific activities of [^{14}C]-labeled purine compounds in cardiac tissue and perfusate after prelabeling the hearts with [^{14}C]adenine or [^{14}C]adenosine were performed as described recently (20).

RELEASE OF ADENOSINE FROM HEARTS DURING HYPOXIA, AUTOREGULATION, AND β-ADRENERGIC STIMULATION

Our experiments, designed to evaluate the significance of adenosine for coronary flow regulation, first required definition of the dose-response relationship between exogenously applied adenosine and the degree of coronary vasodilation. The threshold concentration of adenosine for inducing coronary dilation was in the range of about 10 nM, whereas maximal dilation was obtained at 5 μM.

Hypoxia, which was induced by lowering the oxygen content in the perfusate (equilibration with 30 and 20% oxygen, respectively), resulted in an increase of coronary flow as well as in an augmented release of adenosine, inosine, and hypoxanthine. From Fig. 1 it is evident that the concentrations of adenosine in the coronary effluent perfusate from the hypoxic hearts (30 nM and 100 nM, respectively) were within the coronary effective range determined with exogenously applied adenosine. The concentration changes of inosine, always released from hearts as the major nucleotide degradative, as well as of hypoxanthine, paralleled those of adenosine. It is interesting that the concentration of adenosine in the perfusate from normoxic hearts proved to be slightly below the threshold concentration at which adenosine initiates vasodilation.

Autoregulation of coronary flow following reduction of perfusion pressure is known to be caused by an active vasodilation (9). This phenomenon could also be elicited in the isolated perfused heart. From the data in Fig. 2 it appears that a gradual reduction of perfusion pressure below 60 cm H_2O was associated with a

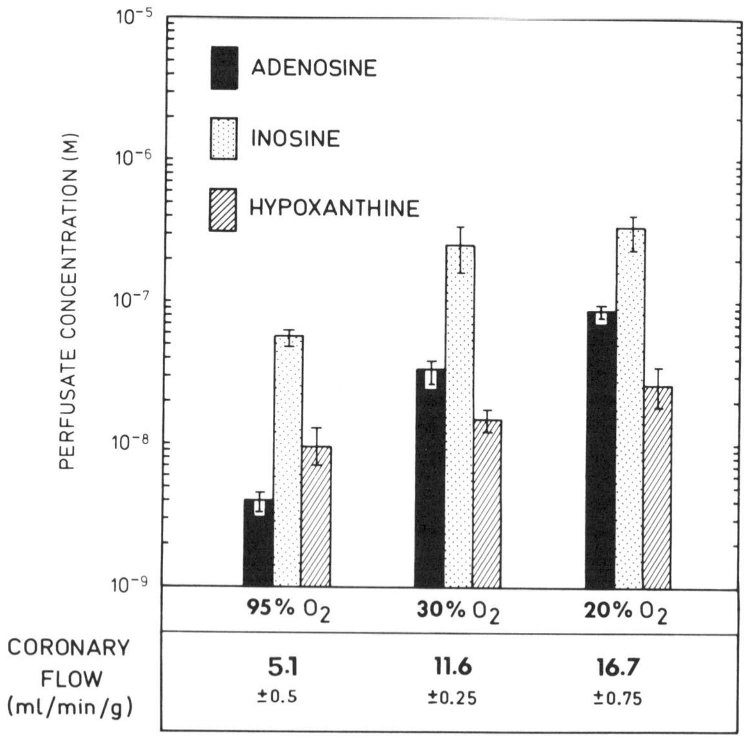

FIG. 1. Hypoxia-induced changes of coronary flow and of the concentrations of adenosine, inosine, and hypoxanthine released into the coronary effluent perfusate from isolated perfused guinea pig hearts. Oxygen content of the coronary perfusate was altered by equilibrating the perfusion medium with a gas mixture containing 95, 30, or 20% O_2, and 5% CO_2, the balance being N_2. $\overline{X} \pm$ SEM; $N=5$.

progressive increase in the formation and release of adenosine as well as of inosine. Again, the concentration of adenosine reached the vasodilating range of exogenously applied adenosine.

β-Adrenergic stimulation of the heart leads to an increase of ventricular force

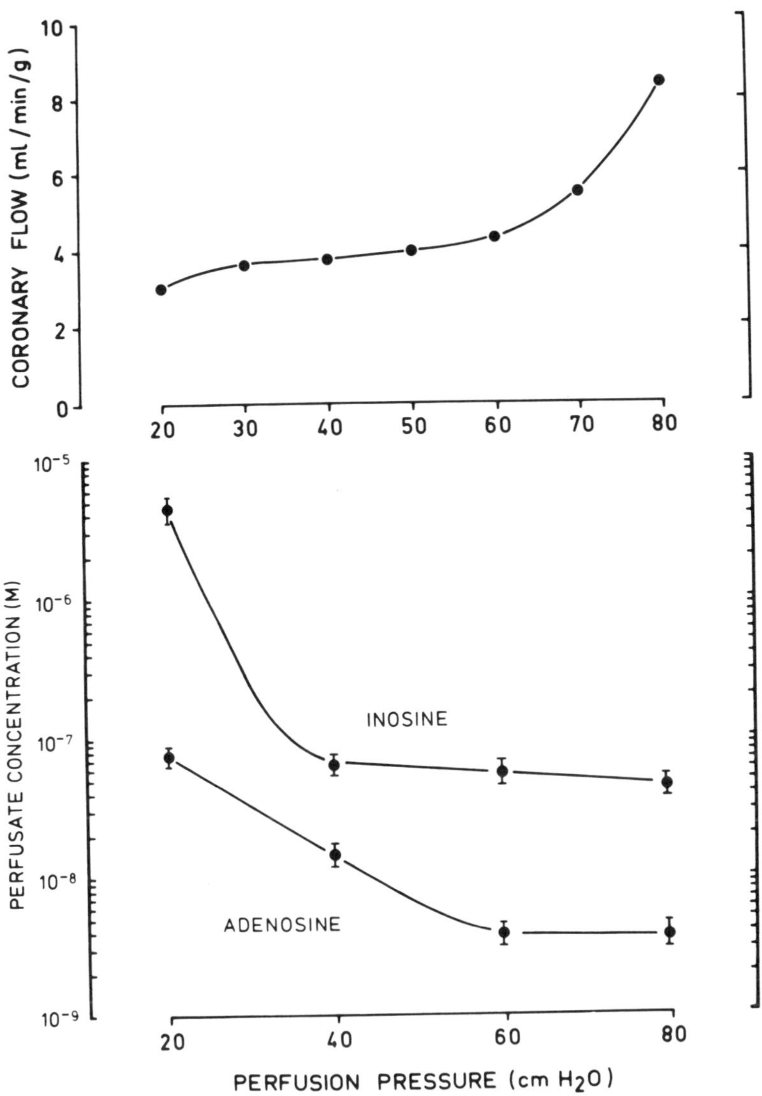

FIG. 2. Relationship between coronary flow *(upper panel)* and concentration changes of adenosine and inosine in the coronary effluent perfusate *(lower panel)* during autoregulation in the isolated perfused guinea pig heart. Pressure changes were initiated from a control pressure of 60 cm H_2O. Data for coronary flow are mean values from 10 experiments each, those for adenosine and inosine from five experiments. $\overline{X} \pm$ SEM.

development followed by an augmentation of cardiac oxygen consumption as well as of coronary flow. Under these conditions the utilization and rate of ATP breakdown are known to be accelerated (24). Furthermore, β-adrenergic stimulation by isoproterenol (Fig. 3) resulted in a pronounced increase in the formation and rate of adenosine release, which reached a concentration of about 0.2 μM in the effluent perfusate. The β-blocking agent propranolol completely inhibited all the isoproterenol-induced hemodynamic and metabolic alterations but had no effect when applied alone.

The kinetics of hemodynamic changes and of the release from prelabeled hearts (20,21) of [^{14}C]adenosine, [^{14}C]inosine, and [^{14}C]hypoxanthine on a bolus injection of isoproterenol are shown in Fig. 4. Obviously there was a close parallelism between the changes of coronary flow and of the rate of release of [^{14}C]adenosine.

COMPARTMENTATION OF CARDIAC ADENINE NUCLEOTIDES AND FORMATION OF ADENOSINE

It is well recognized that myocardial adenine nucleotides must be compartmentalized (8,12,25). On the other hand, it has not been investigated to what extent adenosine, as well as inosine and hypoxanthine, originate from the different adenine nucleotide compartments. Accordingly, studies were performed on isolated perfused guinea pig hearts, in which the adenine nucleotides had been labeled with [^{14}C]adenine or [^{14}C]adenosine in a preperfusion period of 35 min (20). Under steady state conditions, specific radioactivities of ATP, ADP, AMP, cAMP, and adenosine, inosine, and hypoxanthine were determined, both in myocardial tissue and/or in the coronary perfusate. The rationale for this approach was that in a

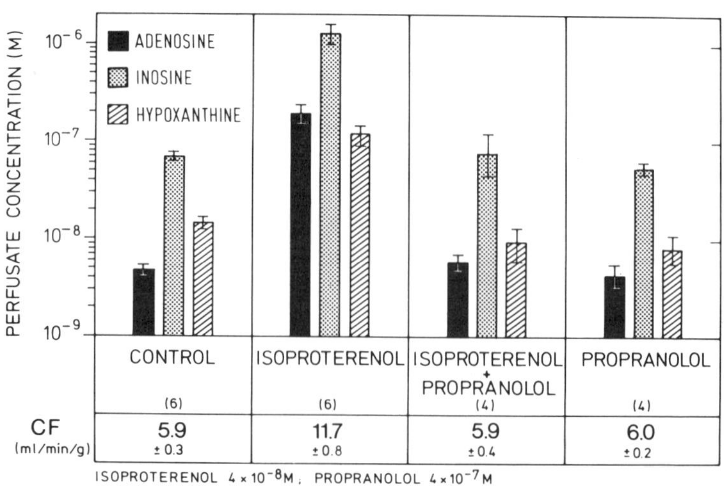

FIG. 3. Changes of coronary flow as well as of the concentration of adenosine, inosine, and hypoxanthine in the coronary effluent perfusate of isolated perfused guinea pig hearts under the influence of isoproterenol (40 nM) and propranolol (400 nM). Analyses were performed 5 min after the onset of the infusions of the respective substances. $\overline{X} \pm$ SEM.

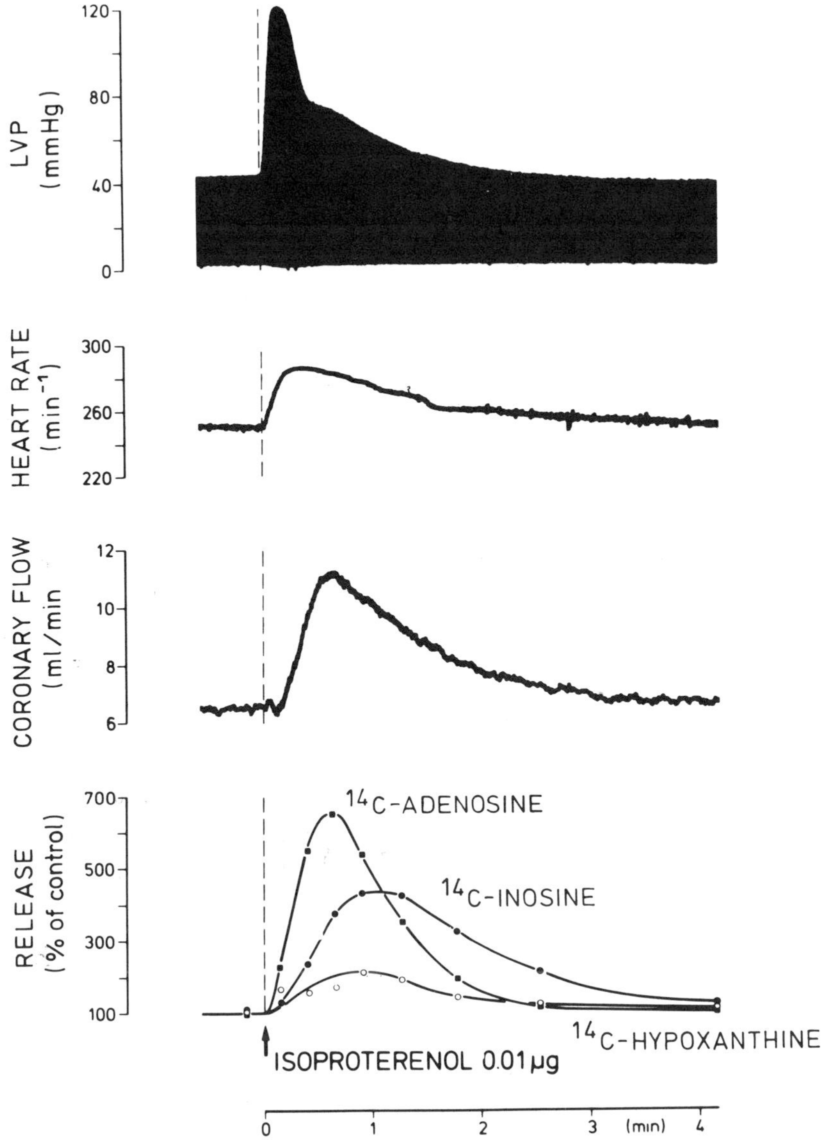

FIG. 4. Time course of hemodynamic changes and of the release of [^{14}C]-labeled purine compounds into the coronary effluent perfusate of guinea pig hearts after a bolus injection of isoproterenol (0.01 μg). The heart was labeled with [8-^{14}C]adenine (0.4 μM) during a preperfusion period of 35 min (20).

homogenous biochemical system the specific activities of a precursor substance and of its immediate metabolic products should be identical (26). Consequently, any differences observed in the specific activities of these compounds can only result from compartmentation of the precursor substances.

Our major findings concerning relative specific activity values of the different purine compounds are summarized in Table 1. It is obvious that in the myocardium labeled with [^{14}C]adenine the activities of cAMP, adenosine, inosine, and hypoxanthine were markedly higher than the respective activities of their precursors ATP, ADP, and AMP. Furthermore, in the coronary perfusate the relative specific activities of cAMP and adenosine proved to be as much as one order of magnitude above the respective values of these compounds in myocardial tissue. A similar pattern of relative specific activity values was also found in hearts after prelabeling with [^{14}C]adenosine. Although there were some differences, mainly with respect to the distribution of [^{14}C] label among the adenine nucleotides, in these experiments too, the specific activities of cAMP and adenosine in the perfusate exceeded those of their precursors in the heart manyfold.

In interpreting these findings it must be considered that cAMP and adenosine of very high specific activities released from well-oxygenated hearts must be derived from precursor nucleotide fractions of equal high activities. Since the relative specific activities of ATP, ADP, and AMP in the myocardium proved to be rather low, these values can only represent mean specific activities of at least two differently labeled compartmentalized nucleotide fractions. This interpretation implies that a rather large nucleotide pool of low specific activity must be present in addition to a small fraction of very high activity that become mixed during tissue extraction. In this context it should be mentioned that the observed differences between myocardial adenine nucleotides and adenosine and cAMP released from the heart were not caused by diffusion gradients effective during the period of pulse labeling. This conclusion is derived from experiments in which prelabeling of hearts at different adenine concentrations (0.4 to 10 μM) did not significantly alter the

TABLE 1. *Relative specific activities (RSA × 10^3) of different purine compounds in the myocardium and perfusate after 35 min prelabeling isolated guinea pig hearts with [8-^{14}C]adenine (adenine concentration 0.4 μM) or with [U-^{14}C]adenosine (adenosine concentration 0.4 μM)*

	[^{14}C]Adenine		[^{14}C]Adenosine	
	Myocardium (N=4–7)	Perfusate (N=9)	Myocardium (N=5)	Perfusate (N=4)
ATP	1.46 ± 0.09	N.D.	3.91 ± 0.20	N.D.
ADP	0.98 ± 0.05	N.D.	3.99 ± 0.12	N.D.
AMP	0.31 ± 0.06	N.D.	3.92 ± 0.44	N.D.
cAMP	3.20 ± 0.28	49.7 ± 3.8	12.40 ± 0.98	147.65 ± 17.4
Adenosine	3.28 ± 0.32	54.6 ± 7.3	9.66 ± 1.10	95.26 ± 17.7
Inosine	5.41 ± 0.63	3.3 ± 0.6	12.12 ± 1.06	17.62 ± 5.3
Hypoxanthine	2.23 ± 0.47	4.1 ± 1.0	16.33 ± 3.74	17.95 ± 3.0

Myocardial tissue was analyzed at the end of the prelabeling period after a brief washout. Specific activities in the perfusate were determined after collection for 30 min following the prelabeling period. Relative specific activity values were calculated by relating the measured specific activities of the individual purine compounds to the specific activities of [8-^{14}C]adenine and [U-^{14}C]adenosine, respectively. $\overline{X}$ ± SEM. N.D., not detectable.

described pattern of specific activities of the purine compounds in tissue and perfusate.

The validity of the outlined concept could be further substantiated by studies in which prelabeled hearts were subjected to a 6-min period of ischemia followed by determination of relative specific activities of adenosine. Ischemia caused a slight decrease in the specific activity of adenosine in the tissue as well as a very pronounced reduction of the activity of adenosine in the perfusate collected immediately after the release of ischemia (time of collection, 5 min). Although the relative specific activity of adenosine in the effluent perfusate of normoxic hearts was found to be about 0.055, the respective value after ischemia was only 0.003. Furthermore, the adenosine concentration in the perfusate from ischemic hearts was shown to be elevated more than 12-fold. All these findings can best be explained by assuming that the ischemia-induced breakdown of adenine nucleotides also involves the large nucleotide pool of low specific activity which then results in a dilution of the specific activity of adenosine in the perfusate.

As far as the localization of the two differently labeled nucleotide compartments is concerned, our data do not permit one to decide whether this compartmentation occurs at the cellular or subcellular level. However, in view of the well-known localization of 5′-nucleotidase (EC 3.1.3.5) and adenylate cyclase (EC 4.6.1.1) at the sarcolemmal membrane (17), it seems reasonable to assume that the small nucleotide compartment characterized by its high specific activity is preferentially localized in the close vicinity of cell membranes.

EVIDENCE FOR ADENOSINE RECEPTOR SITES ON THE SURFACE OF CORONARY SMOOTH MUSCLE CELLS

Two possibilities were discussed in the past regarding the site of action of adenosine in the coronary vessels, namely an interaction of adenosine with binding sites at the membrane of the smooth muscle cells or uptake of the nucleoside into the smooth muscle cells followed by some intracellular mechanisms. In an attempt to decide between these possibilities, studies were performed with a newly synthesized adenosine derivative of high molecular weight which was shown to be confined to the extracellular space (22).

The adenosine derivative was prepared by coupling 5′-AMP via phosphoamide linkages to a protein (carbonic anhydrase, MW 30,000). The schematic structure of the compound, named protein-AMP-conjugate, is depicted in Fig. 5. It is evident that in the protein-AMP-conjugate the structural features of the adenosine molecule, which are essential for inducing vasodilation, are unaltered (amino group in the 6 position of the purine ring and the hydroxyl groups in the 2′ and 3′ position of the ribose moiety). The compound, in which about 2 moles of AMP were bound to 1 mole of protein, proved to be stable under physiological conditions. Hydrolytic cleavage of the protein-AMP-conjugate resulting in the formation of AMP and adenosine, respectively, amounted to less than 1% within 1 hr (37°C, pH 7.0).

In Fig. 6 results are summarized concerning the coronary dilating effects of the

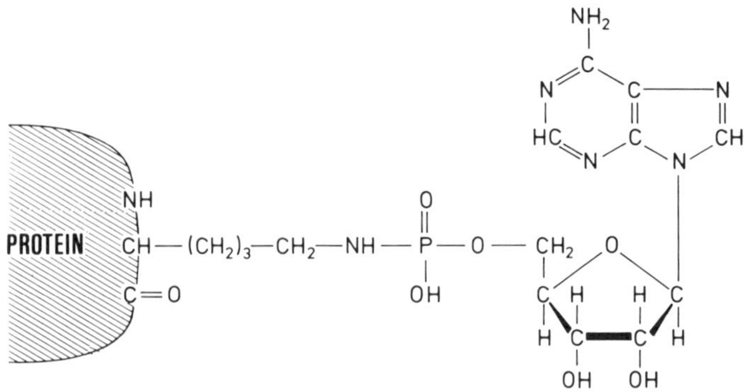

FIG. 5. Schematic structure of the protein-AMP-conjugate. Synthesis was accomplished by coupling 5′-AMP via phosphoamide linkage to lysine residues of a protein (carbonic anhydrase). For methodological details see ref. 22.

protein-AMP-conjugate in the isolated perfused heart. Obviously the conjugate proved to be as potent as adenosine and AMP. Although application of adenosine deaminase (EC 3.5.4.4) and 5′-nucleotidase (EC 3.1.3.5) plus adenosine deaminase abolished the dilatory actions of adenosine and AMP, respectively, these enzymes did not affect the coronary response to the protein-AMP-conjugate.

From all studies with the protein-AMP-conjugate it appears that the dilating effect

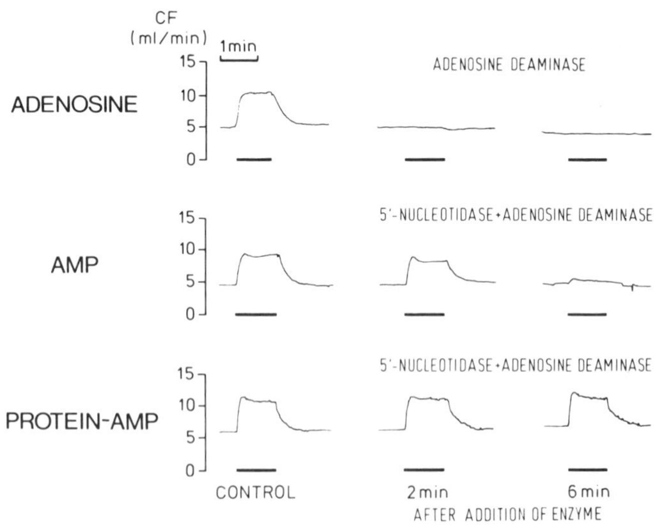

FIG. 6. Changes of coronary flow induced by adenosine, AMP, and protein-AMP-conjugate and influence of adenosine deaminase and 5′-nucleotidase. Adenosine (10 μg/ml), AMP (10 μg/ml), and protein-AMP-conjugate (38 mg protein/ml) were infused at a rate yielding similar vascular responses. Enzymes were added to the solutions of the substances before their infusion to yield a final concentration of 2 mU/ml for adenosine deaminase and 0.5 mU/ml for 5′-nucleotidase. (From ref. 22, with permission.)

of this compound must be attributed to an interaction of its adenosine moiety with a receptor site that is accessible from the interstitial space. Such an adenosine receptor is most likely localized on the surface of the coronary smooth muscle cells. This interpretation is in agreement with conclusions drawn by Olsson et al. (14,15) who also studied coronary effects of high molecular weight adenosine derivatives, namely of a stacchyose-linked adenosine and of a polylysine adenosine compound.

CONCLUSIONS

Under the experimental conditions tested (hypoxic perfusion, autoregulation, β-adrenergic stimulation), formation of adenosine by the heart and its release into the coronary effluent perfusate was shown to be considerably enhanced. The concentrations of adenosine always proved to be in the coronary effective range of exogenously applied adenosine. Moreover, a good parellelism was found between the degree of coronary dilation and the amount of adenosine released. Thus our findings lend further support to the hypothesis that adenosine may be an important factor in the metabolic regulation of coronary blood flow (1,5,16,18).

As was revealed by experiments in which cardiac adenine nucleotides were prelabeled with [^{14}C]adenine or [^{14}C]adenosine, at least two different nucleotide pools must contribute to the formation of adenosine. In the normoxic heart a rather small nucleotide compartment characterized by a high rate of renewal was mainly responsible for the production of adenosine; in the hypoxic heart a second adenine nucleotide compartment of greater size and much lower rate of renewal became the major precursor pool. Therefore, it is evident that adenosine formed by the heart cannot be regarded to be of uniform origin.

Concerning the site of action of adenosine in the coronary vessels, it was demonstrated by experiments with a protein-AMP-conjugate, an adenosine derivative of high molecular weight, that adenosine must act via an adenosine receptor localized at the membrane of the smooth muscle cell. Neither transfer of adenosine across the cell membrane nor a metabolic conversion within the smooth muscle cell appears to be a prerequisite for the vasodilatory action of the nucleoside. It is suggested that ATP, ADP, and AMP, which do not readily penetrate across cellular membranes and which are as potent in causing vasodilation as adenosine, likewise exert their effects by an interaction of their adenosine moiety with the proposed adenosine receptor at the smooth muscle cell membrane.

ACKNOWLEDGMENTS

The skillful technical assistance of Mrs. Christa Schäefer-Grunitz is gratefully acknowledged.

REFERENCES

1. Berne, R. M. (1963): *Am. J. Physiol.,* 204:317–322.
2. Bretschneider, H. J. (1964): In: *Kreislaufmessungen. 4. Freiburger Colloquium,* edited by A. Fleckenstein, Werk-Verlag Dr. E. Banaschewski, München-Gräfelfingo.

3. Bünger, R., Haddy, F. J., Querengasser, A., and Gerlach, E. (1975): *Pfluegers Arch.*, 353:317–326.
4. Drury, A., and Szent-Györgyi, A. (1929): *J. Physiol. (Lond.)*, 68:213–226.
5. Gerlach, E., Deuticke, B., and Dreisbach, R. H. (1963): *Naturwissenschaften*, 50:228–229.
6. Gerlach, E., Deuticke, B., Dreisbach, R. H., and Rosarius, C. W. (1963): *Pfluegers Arch.*, 278:296–315.
7. Gilman, A. G. (1970): *Proc. Natl. Acad. Sci. U.S.A.*, 67:305–312.
8. Gudbjarnason, S., Mathes, P., and Ravens, K. G. (1970): *J. Mol. Cell. Cardiol.*, 1:325–339.
9. Guyton, A. C., Ross, J. M., Carrier, O., and Walker, J. R. (1964): *Circ. Res.*, 14 (Suppl. I):60–80.
10. Kalckar, H. M. (1947): *J. Biol. Chem.*, 167:429–443.
11. Kalckar, H. M. (1947): *J. Biol. Chem.*, 167:445–459.
12. Liu, M. S., and Feinberg, H. (1971): *Am. J. Physiol.*, 220:1242–1248.
13. Olsson, R. A. (1970): *Circ. Res.*, 26:301–306.
14. Olsson, R. A., Davis, C. J., and Khouri, E. M. (1978): *Life Sci.*, 21:1343–1350.
15. Olsson, R. A., Davis, C. J., Khouri, E. M., and Patterson, R. E. (1976): *Circ. Res.*, 39:93–98.
16. Rubio, R., and Berne, R. M. (1969): *Circ. Res.*, 25:407–415.
17. Rubio, R., Berne, R. M., and Dobson, J. R. (1973): *Am. J. Physiol.*, 225:938–953.
18. Rubio, R., Berne, R. M., and Katori, M. (1969): *Am. J. Physiol.*, 216:56–62.
19. Schrader, J., Baumann, G., and Gerlach, E. (1977): *Pfluegers Arch.*, 372:29–35.
20. Schrader, J., and Gerlach, E. (1976): *Pfluegers Arch.*, 367:129–135.
21. Schrader, J., Haddy, F. J., and Gerlach, E. (1977): *Pfluegers Arch.*, 369:1–6.
22. Schrader, J., Nees, S., and Gerlach, E. (1977): *Pfluegers Arch.*, 369:251–257.
23. Schrader, J., Rubio, R., and Berne, R. M. (1975): *J. Mol. Cell. Cardiol.*, 7:427–433.
24. Takenaka, F., and Higuchi, M. (1974): *J. Mol. Cell. Cardiol.*, 6:123–135.
25. Yagi, K., and Mase, R. (1965): In: *Molecular Biology of Muscular Contraction*, edited by S. Ebashi, F. Oosawa, T. Seine, and Y. Tonomura, pp. 109–123. Elsevier, Amsterdam.
26. Zilversmit, D. B., Entman, C., and Fishler, M. C. (1943): *J. Gen. Physiol.*, 26:325–331.

Physiological and Regulatory Functions of Adenosine and Adenine Nucleotides, edited by H. P. Baer and G. I. Drummond. Raven Press, New York © 1979.

Sites and Mode of Action of Adenosine in the Heart. II. Ventricular Myocardium

J. Schrader, E. Gerlach, and G. Baumann

Department of Physiology, University of Munich, D-8000 Munich, Germany

It has been known for many years that adenosine, in addition to its effects on the coronary system, reduces contractile force development in the atrial myocardium (7,8). Only recently was evidence obtained from electrophysiological studies that adenosine also inhibits transmembrane Ca^{2+} movement into catecholamine-stimulated atrial muscle (20). The latter action of adenosine is most likely caused by an interaction of the nucleoside with specific receptor sites localized on the surface of the atrial cell (19). This conclusion was derived from experiments in which a high molecular weight protein-AMP-conjugate, restricted to the extracellular space, was shown to be as effective as adenosine in inhibiting Ca^{2+} influx into the cells.

Contradictory results concerning possible effects of adenosine on contractile force development in ventricular myocardium have been reported in the past (2,10,22). Thus it was of interest to evaluate under what conditions adenosine affects cardiac contractility and to define its mechanism of action. Investigations concerning this problem were carried out using the isolated perfused guinea pig heart preparation (4). Details of the method as well as of the analytical procedures for the determination of adenosine, inosine, and hypoxanthine are summarized in the preceding chapter. The activity of adenylate cyclase was determined as described recently (17).

ATTENUATION BY ADENOSINE OF THE HEMODYNAMIC EFFECTS OF CATECHOLAMINES AND HISTAMINE

Figure 1 shows an original recording of an experiment from which the first evidence for a specific action of adenosine on the ventricular myocardium was obtained. The hearts, perfused at constant flow, were stimulated by bolus injections of isoproterenol in the absence and presence of adenosine and papaverine, respectively. As can be seen, the typical positive inotropic responses to isoproterenol (transient increases in left venticular pressure and dP/dt) were almost completely abolished in the presence of adenosine. In contrast, adenosine did not alter either contractile parameter when applied to the unstimulated heart. The isoproterenol-antagonistic action of adenosine was not caused by the reduction of perfusion

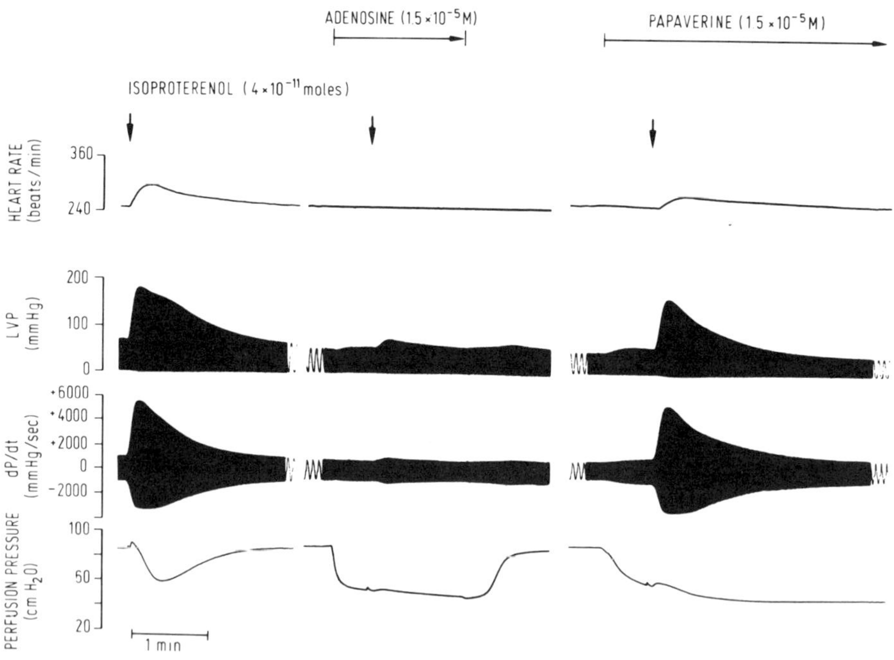

FIG. 1. Effects of adenosine and papaverine on changes of some hemodynamic parameters caused by bolus injection of isoproterenol in the isolated guinea pig heart. Since adenosine at the given concentration induces an atrioventricular block, the ventricles were electrically paced (8 V, 3 msec) during application of the nucleoside. Perfusion at constant flow (12 ml/min). (From ref. 17, with permission.)

pressure induced by adenosine, since the inotropic response to isoproterenol remained almost unchanged in the presence of papaverine which lowered perfusion pressure as did adenosine.

Similar experiments were carried out with other β-adrenergic stimulants (epinephrine, norepinephrine, and dopamine) and with histamine. In all these cases, adenosine was found to attenuate strongly the inotropic effects of these agents. On the other hand, adenosine did not inhibit inotropic responses elicited by elevating the extracellular Ca^{2+} concentration. It is interesting to note that the adenosine-induced attenuation of inotropic effects of catecholamines and histamine could be prevented almost completely by theophylline (50 μM), a compound known to antagonize the coronary dilating action of adenosine (1,3).

The dose-response relationship obtained for the inhibitory effects of adenosine on dP/dt_{max} in isoproterenol-, dopamine-, and histamine-stimulated hearts are summarized in Fig. 2. Obviously, the dose-response curves for the different agents proved to be very similar. The adenosine-dependent inhibition of the increase in dP/dt_{max} started at concentrations of about 0.1 μM and reached maximal values (80 to 90% inhibition) at concentrations of about 20 μM. It is of particular interest that these dose-response curves are shifted to the right by one order of magnitude

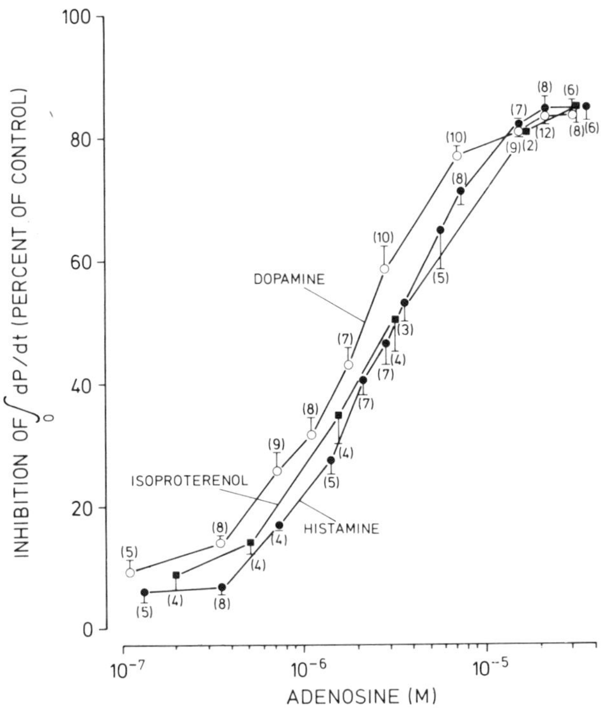

FIG. 2. Dose-response curves of the inhibition by adenosine of the isoproterenol- (40 pmoles), histamine- (7.5 nmoles), and dopamine- (15 nmoles) induced increases in contractile force development. The contractile responses to the inotropic agents were quantitated by measuring the total area of the hormone-induced rise of dP/dt_{max}. $\overline{X} \pm$ SEM.

compared with the dose-response curve of adenosine for inducing coronary dilation (18). Thus the coronary vessels can be considered to be about 10 times more sensitive to adenosine than the ventricular myocardium.

In an attempt to elucidate if inotropic effects of endogenously liberated catecholamines can also be antagonized by adenosine, experiments with tyramine were performed. This compound exerts its cardiac action by causing a release of catecholamines stored in the heart. As can be seen from Fig. 3, infusion of tyramine induced an increase of left ventricular pressure and of dP/dt. These changes were paralleled by a greatly augmented release of norepinephrine into the coronary effluent perfusate which was determined by measuring the efflux of [^{3}H]norepinephrine from prelabeled hearts (26). In the presence of adenosine, tyramine still elicited an almost identical release of norepinephrine. However, the contractility effects proved to be remarkably attenuated under these circumstances. From all these findings it is evident that there is no principal difference in the inhibitory effects of adenosine on endogenously liberated or exogenously applied catecholamines.

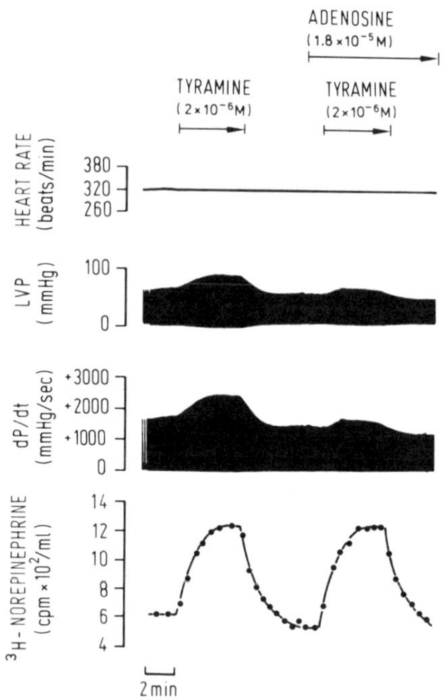

FIG. 3. Influence of adenosine on the increase in contractile force development and release of [^{3}H]noradrenaline in the isolated perfused guinea pig heart. The hearts were labeled with [^{3}H]noradrenaline (6 nM) in a preperfusion period of 20 min.

INTERACTION OF ADENOSINE WITH THE CARDIAC ADENYLATE CYCLASE SYSTEM

It has been well established that catecholamine- and histamine-induced increases in ventricular contractile force are associated with an activation of adenylate cyclase leading to an enhanced formation of cyclic 3′,5′-adenosine monophosphate (cAMP), the second messenger of the hormones (23,24). Based on these facts and taking into account that adenosine was already shown in various tissues to either inhibit (9,13,15,21,25) or to stimulate (16) adenylate cyclase, it was of interest to analyze whether or not the adenosine-mediated attenuation of contractile force was paralleled by respective alterations in cAMP levels and/or changes in the activity of the adenylate cyclase.

It is evident from the data in Fig. 4 that adenosine (10 μM) inhibited the isoproterenol-induced elevation of cAMP levels in the ventricular myocardium. Apparently the cAMP-dependent enhancement of glycogenolysis was strongly attenuated as evidenced by the tissue levels of glucose-1-phosphate and glucose-6-phosphate which increased only slightly when isoproterenol was applied in the presence of adenosine.

Since it could be excluded that the effects of adenosine on cAMP levels resulted from an activation of the membrane-bound phosphodiesterase (17), an inhibitory action of adenosine on adenylate cyclase appeared most likely. In order to define the

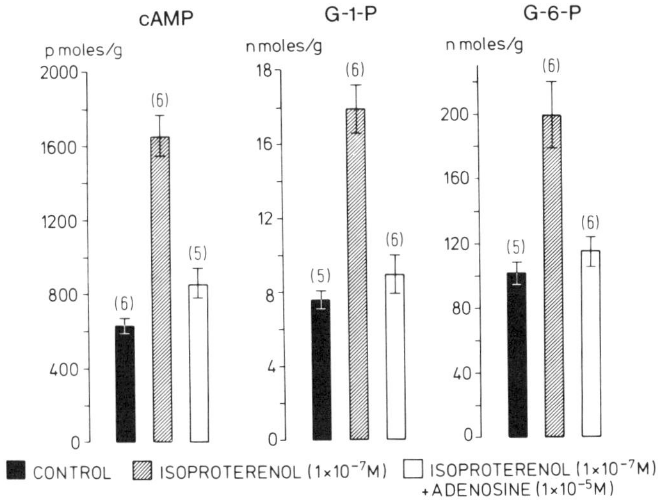

FIG. 4. Effect of adenosine on the isoproterenol-induced changes in myocardial tissue levels of cAMP, glucose-1-phosphate (G-1-P), and glucose-6-phosphate (G-6-P). The hearts were freeze-clamped 20 sec after the addition of isoproterenol and/or adenosine. $\overline{X} \pm$ SEM.

influence of adenosine on the adenylate cyclase the activity of this enzyme was studied in a crude membrane preparation of guinea pig hearts.

In Fig. 5 our data are summarized concerning the influence of adenosine on the isoproterenol-, dopamine-, and histamine-stimulated cyclase as well as on the basal enzyme activity. Obviously, adenosine in a concentration above 10 μM strongly inhibited the stimulated adenylate cyclase. In contrast, basal activity of the enzyme was affected only in concentrations above 100 μM.

These findings on a crude membrane preparation did not permit us to decide whether the adenosine-induced inhibition of the stimulated adenylate cyclase resulted from an interaction with the binding of the hormones to their respective receptor sites or from an interference with the catalytic unit of the enzyme. In order to clarify this problem, studies were carried out using a Lubrol-PX solubilized membrane preparation of guinea pig hearts, in which the adenylate cyclase is no longer sensitive to hormones but can still be activated by F^- and guanosine triphosphate (GTP) (11,12). Results obtained with this membrane preparation revealed that F^- and GppNHp-stimulated cyclase was inhibited by about 40% at an adenosine concentration of 10 μM, whereas at 100 μM inhibition was almost complete. In additional experiments with the solubilized membrane preparation it could be demonstrated that adenosine did not compete with adenosine triphosphate (ATP) as substrate for the cyclase.

From all these studies it appears reasonable to conclude that adenosine exerts its inhibitory effect by interfering with the catalytic unit of the adenylate cyclase. In view of the proposed role of cAMP in the hormone-induced augmentation of cardiac contractile force (24) it seems very likely that it is the inhibition of the adenylate

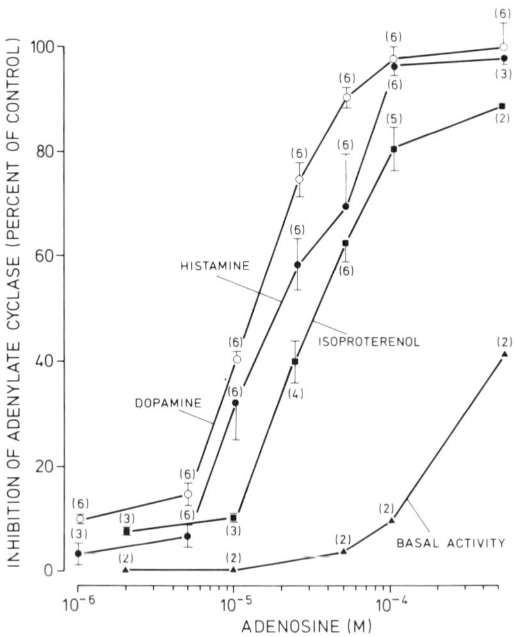

FIG. 5. Inhibition by adenosine of basal activity as well as of isoproterenol- (10 μM), histamine- (10 μM), and dopamine- (100 μM) stimulated activity of adenylate cyclase in a crude membrane preparation of guinea pig ventricles. Cyclase activity was measured in the presence of papaverine (1 mM) as inhibitor of the phosphodiesterase. For further methodological details see ref. 17. $\overline{X} \pm$ SEM.

cyclase by adenosine that is responsible for the inhibitory effects of the nucleoside on the catecholamine- and histamine-stimulated heart.

POSSIBLE PHYSIOLOGICAL IMPLICATIONS OF THE CATECHOLAMINE-ANTAGONISTIC EFFECTS OF ADENOSINE

After evaluation of the basic action of adenosine on the hormone-stimulated ventricular myocardium further investigations were undertaken to determine if the observed effects of adenosine could be of physiological significance. Therefore, it was of particular interest to clarify if the living heart is capable of forming adenosine in amounts sufficient to influence the hemodynamic and metabolic effects of β-adrenergic stimulation.

Experiments concerning this problem were performed using isoproterenol applied to isolated perfused hearts under normoxic and hypoxic conditions. As demonstrated by the data in Fig. 6, left ventricular pressure and dP/dt rapidly increased immediately on onset of isoproterenol infusion; then both parameters decreased to half-maximal values. Under these conditions the release of adenosine and of its breakdown product inosine into the coronary effluent perfusate was

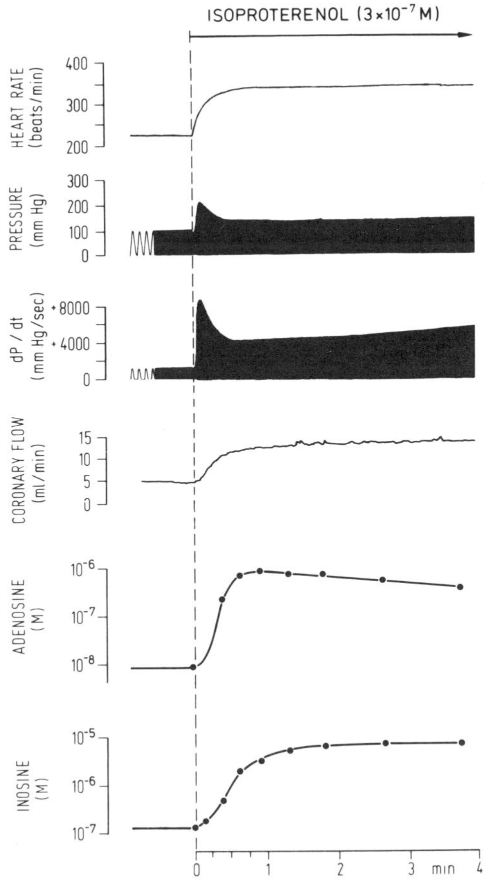

FIG. 6. Effects of an infusion of isoproterenol on hemodynamic parameters of the isolated perfused guinea pig heart and on concentrations of adenosine and inosine in the coronary effluent perfusate. Perfusion at constant pressure (60 cm H_2O). (From ref. 17, with permission.)

strongly augmented. Most interesting, the changes of the adenosine concentration closely paralleled not only the increase in coronary flow, but also the secondary decrease in contractile force. Of particular importance is the observation that the concentration of adenosine rose within only 45 sec by two orders of magnitude and reached a value of 1 μM. Such a concentration of adenosine was demonstrated to effectively inhibit the inotropic response to bolus injections of isoproterenol (Fig. 2). It thus appears likely that endogenously formed adenosine may well participate in causing the secondary decrease of the inotropic responses during the infusion of isoproterenol (Fig. 6).

A further piece of evidence that endogenously formed adenosine may actually limit the response of the heart to β-adrenergic stimulation was derived from experiments on hypoxic hearts. Also under these conditions the reduced responsiveness of the hypoxic heart to catecholamines, which is a well-recognized phenomenon (5,6,14), was associated with a greatly enhanced release of endogenously formed adenosine.

CONCLUSIONS

According to these results adenosine attenuated the positive inotropic responses of the heart resulting from stimulation with different catecholamines and histamine as well as from liberation of endogenously stored sympathetic transmitter substances. Furthermore, adenosine was shown to greatly inhibit the formation of cAMP in the catecholamine- and histamine-stimulated heart. Evidence was obtained that all these phenomena are related to an inhibition of myocardial adenylate cyclase resulting from an interference of the adenosine molecule with the catalytic unit of this enzyme. It is suggested that such an adenosine-mediated mechanism could also become effective to prevent sympathetic overstimulation of the heart *in vivo*.

ACKNOWLEDGMENTS

We are indebted to Mrs. Christa Schäfer-Grunitz for her excellent technical assistance.

REFERENCES

1. Afonso, S. (1970): *Circ. Res.,* 26:743–752.
2. Buckley, N. M. (1970): *Am. J. Physiol.,* 218:1399–1405.
3. Bünger, R., Haddy, F. J., and Gerlach, E. (1975): *Pfluegers Arch.,* 358:213–224.
4. Bünger, R., Haddy, F. J., Querengässer, A., and Gerlach, E. (1975): *Pfluegers Arch.,* 353:317–326.
5. Daniell, H. B., Bagwell, E. E., and Walton, R. P. (1967): *Circ. Res.,* 21:85–98.
6. Davidson, S., Maroko, P. R., and Braunwald, E. (1974): *Am. J. Physiol.,* 227:439–443.
7. De Gubareff, T., and Sleator, W. (1965): *J. Pharmacol. Exp. Ther.,* 148:202–214.
8. Drury, A., and Szent-Györgyi, A. (1929): *J. Physiol. (Lond.),* 68:213–226.
9. Fain, J. N., Pointer, R. H., and Ward, W. F. (1972): *J. Biol. Chem.,* 247:6866–6872.
10. Lammerant, J., and Becsei, I. (1973): *J. Physiol. (Lond.),* 229:41–49.
11. Lefkowitz, R. J. (1975): *J. Biol. Chem.,* 250:1006–1011.
12. Levey, G. S. (1970): *Biochem. Biophys. Res. Commun.,* 38:86–92.
13. Londos, C., and Preston, M. S. (1977): *J. Biol. Chem.,* 252:5951–5956.
14. Mahler, J. T., Manchanda, S. C., Cymeran, A., Wolfe, D. L., and Hartley, L. H. (1975): *Am. J. Physiol.,* 228:477–481.
15. McKenzie, S. G., and Bär, H. P. (1973): *Can. J. Physiol. Pharmacol.,* 51:190–196.
16. Sattin, A., and Rall, T. W. (1970): *Mol. Pharmacol.,* 6:13–23.
17. Schrader, J., Baumann, G., and Gerlach, E. (1977): *Pfluegers Arch.,* 372:29–35.
18. Schrader, J., Haddy, F. J., and Gerlach, E. (1977): *Pfluegers Arch.,* 369:1–6.
19. Schrader, J., Nees, S., and Gerlach, E. (1977): *Pfluegers Arch.,* 369:251–257.
20. Schrader, J., Rubio, R., and Berne, R. M. (1975): *J. Mol. Cell. Cardiol.,* 7:427–433.
21. Schwabe, U., Schönhofer, P. S., and Ebert, R. (1974): *Eur. J. Biochem.,* 46:537–545.
22. Shah, A., Kechejian, S. J., Kavaler, F., and Fisher, V. J. (1974): *Am. Heart J.,* 87:740–749.
23. Sutherland, E. W., and Robison, G. A. (1966): *Pharmacol. Rev.,* 18:145–161.
24. Tsien, R. W. (1977): In: *Advances in Cyclic Nucleotide Research,* vol. 8, edited by P. Greengard and G. A. Robison, pp. 363–411. Raven Press, New York.
25. Weinryb, J., and Michel, J. (1974): *Biochim. Biophys. Acta,* 334:218–225.
26. Westfall, T. C., and Hunter, P. E. (1974): *J. Pharm. Pharmacol.,* 26:458–460.

Physiological and Regulatory Functions of Adenosine and Adenine Nucleotides, edited by H. P. Baer and G. I. Drummond. Raven Press, New York © 1979.

Adenosine Production in Skeletal Muscles of Different Fiber Types

Emma L. Bockman and Jack E. McKenzie

Department of Physiology, Louisiana State University, School of Medicine, New Orleans, Louisiana 70112

Adenosine has been proposed as a mediator of local control of the circulation in a number of organs. Present evidence suggests a role for the nucleoside in the regulation of coronary circulation; however, a similar role in the local control of the circulation to skeletal muscle remains to be defined. Clearly, the ability to accumulate adenosine in response to ischemia is less in skeletal muscle than in cardiac muscle. Imai et al. (13) and Deuticke and Gerlach (6) could detect no adenosine in skeletal muscle after 30 to 60 min of ischemia whereas the adenosine content of cardiac muscle increased several-fold after 5 to 10 min of ischemia. The higher adenosine monophosphate (AMP) deaminase activity in skeletal muscle compared with that in cardiac muscle was thought to result in adenine nucleotide degradation to inosine monophosphate (IMP) rather than to adenosine (Fig. 1). Although adenosine was not detectable in skeletal muscle in these studies, bioassay studies provided indirect evidence that adenosine might have a role in the local control of skeletal muscle vasculature. Scott et al. (18) perfused bioassay organs with venous blood from contracting skeletal muscle. The venous blood caused vasodilation when the bioassay organ was the forelimb, but caused vasoconstriction when the bioassay organ was the kidney. Adenosine and AMP mimicked the differential response of the two bioassay organs.

With the development of a more sensitive method for measuring adenosine (15), it is now possible to detect adenosine in resting skeletal muscle as well as increases in the adenosine content of skeletal muscle under a number of conditions. Dobson et

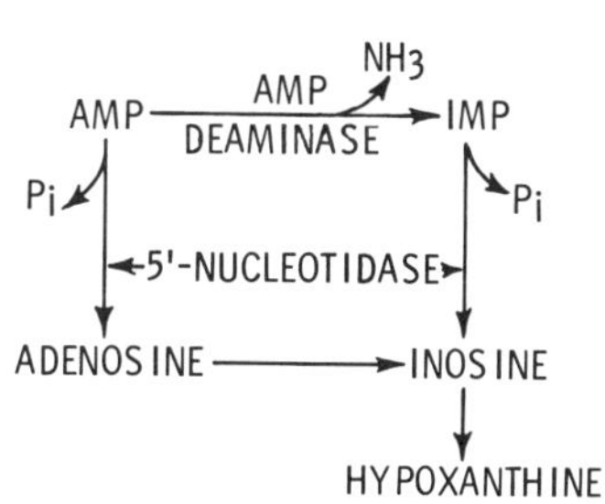

FIG. 1. Pathways of AMP degradation.

al. (7) found that the concentration of adenosine in venous blood from muscle was increased about fivefold after 5 min of ischemic, tetanic contraction. Bockman et al. (4) have demonstrated that adenosine is released from an isolated, artificially perfused rat hindlimb preparation and that the rate of adenosine release is enhanced by muscle contraction.

In the studies discussed so far, the production of adenosine was elicited by severe conditions such as ischemia or hypoxia. We wanted to study the production of adenosine in skeletal muscle under more physiological conditions. We also considered the possibility that the fiber type of which the muscle was composed, and thus its inherent metabolic pattern, might have an influence on the local control of the circulation. This possibility was particularly intriguing with respect to the role of adenosine, since Rubio et al. (17) have reported that the pathway for AMP degradation in skeletal muscle differs in dog compared with rat. In dog skeletal muscle, which is composed predominantly of highly oxidative, "red" fibers (Fig. 2), AMP appeared to be degraded primarily via 5′-nucleotidase to adenosine. In rat muscle, composed predominantly of low oxidative, "white" fibers, AMP degradation appeared to occur primarily via AMP deaminase to IMP.

ADENOSINE PRODUCTION IN SITU

To study if the fiber type influences the local control of circulation, we have developed soleus and gracilis muscle preparations in the cat. In the cat, the soleus muscle is composed entirely of oxidative, slow-twitch fibers ("intermediate") (Fig. 2). About 75% of the fibers in gracilis muscle are low-oxidative, fast-twitch fibers ("white") (3). The circulation to the muscle was isolated by ligating all branches of the femoral artery and vein except those directly supplying the muscle. A venous cannula diverted blood from the muscle to a drop counter for measurement of blood flow and then returned the blood to the animal. Venous blood samples could also be obtained from the cannula through a T-connector. The muscles were denervated and stimulated, via the nerve, to contract isometrically. The oxygen contents of arterial and venous blood samples were measured for calculations of oxygen consumption. Blood flow was allowed to vary during the period of stimulation (free flow). Muscle samples were frozen *in situ* with aluminum tongs precooled in liquid nitrogen. The adenosine and lactate contents of the muscle were measured in neutralized, perchloric acid extracts of the muscle (11,15). The contralateral (unisolated) muscle served as control.

RED:	INTERMEDIATE:	WHITE:
Fast twitch	Slow twitch	Fast twitch
High oxidative	Oxidative	Low oxidative
Glycolytic	Low glycolytic	High glycolytic

FIG. 2. Characteristics of three major fiber types of skeletal muscle.

In contrast to previous reports (10,12), we have found resting blood flow and oxygen consumption of soleus muscle to be similar to that of gracilis muscle. Resting blood flow is in the range of 5 ml/min/100 g for both muscles. Resting oxygen consumption is in the range of 0.2 to 0.3 ml O_2/min/100 g. Since Hilton et al. (10) and Hudlická (12) have reported resting blood flow in the range of 50 ml/min/100 g and resting oxygen consumption in the range of 3 ml O_2/min/100 g for soleus muscle, we were concerned that our isolation procedure limited blood flow to the soleus muscle preparations. Therefore, we used radiolabeled microspheres to measure blood flow to soleus muscle before and after isolation of its circulation. In 4 animals, blood flow to soleus muscle, as measured by the microspheres, was 2.0 ± 0.3 ml/min/100 g before and 4.6 ± 2.0 ml/min/100 g after isolation of its circulation. After isolation of the circulation to the soleus muscle, blood flow measured with microspheres was linearly related to blood flow measured with the drop counter over a range of flows from 3.5 to 28 ml/min/100 g. From these data, we concluded that our isolation procedure did not limit blood flow to the soleus muscle.

Both soleus muscle and gracilis muscle respond to contraction with increased blood flow and oxygen consumption. Although blood flows ranging from a resting value of 5 ml/min/100 g to 30 ml/min/100 g for gracilis muscle and to 40 ml/min/100 g for soleus muscle have been elicited by varying the frequency of stimulation, we have been unable to find a correlation between muscle adenosine content and blood flow in either of the muscles. The adenosine content of gracilis muscle did not change even in muscles in which force development decreased to less than 40% of the initial value or in those in which lactate content increased more than fivefold. However, there was a positive correlation between muscle adenosine content and muscle lactate content in soleus muscle. This may indicate that some degree of hypoxia is necessary to stimulate adenosine production and that adenosine may then play a role as a mediator of reactive hyperemia in soleus muscle.

Our data do not support a role for adenosine as a mediator of active hyperemia under free flow conditions in soleus and gracilis muscles of cats. However, we have previously found the muscle content of adenosine to be increased in response to muscle contraction in a dog hindlimb preparation perfused at constant flow (5). The muscles were stimulated to contract at a frequency that resulted in a reduction in perfusion pressure of approximately 50%. Although isolated muscle adenosine content tended to be higher than that of the contralateral control muscle after 5 min of stimulation, the difference was not statistically significant (Fig. 3). After 10 and

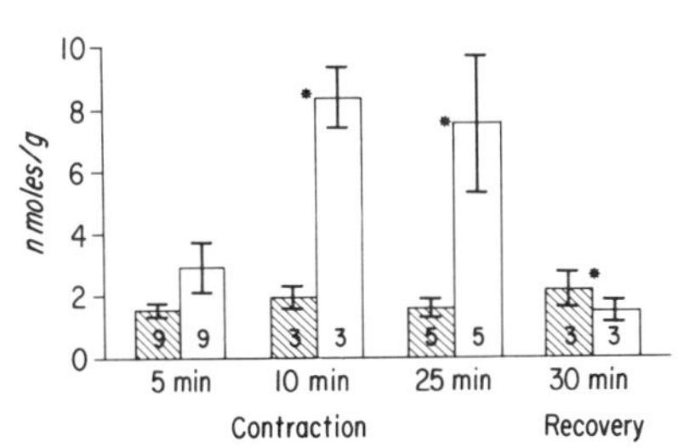

FIG. 3. Adenosine content of control *(hatched bars)* and experimental *(open bars)* muscles. Vertical lines represent ± SEM, and numbers within bars indicate the number of animals in each group. Asterisk indicates significant difference (p<0.05) between experimental value and value in contralateral control muscle. (From Bockman et al., ref. 5, with permission.)

25 min of stimulation, the adenosine content of the contracting muscles was about fourfold greater than the control values. After a 30-min recovery period, muscle adenosine content was slightly less than the control value. During muscle contraction, there was an inverse relationship between vascular resistance and muscle adenosine content when each was expressed as a percentage of the control value (Fig. 4). These data support a role for adenosine in the control of vascular resistance in dog muscle perfused at constant flow.

However, Phair and Sparks (16) have reported recently that they could find no correlation between muscle adenosine content and blood flow in a dog skeletal muscle preparation perfused under free flow conditions. Adenosine production increased markedly when blood flow was restricted to the point at which oxygen consumption equaled oxygen delivery.

Experiments in which skeletal muscle is perfused under free flow conditions fail to support a role for adenosine as a mediator of active hyperemia. Moreover, we did not find a significant increase in adenosine content of skeletal muscle perfused at constant flow within the first 5 min of contraction. However, Rubio et al. (17) have shown that 5′-nucleotidase is localized primarily around blood vessels in skeletal muscle. Thus it is possible that the concentration of adenosine around blood vessels may increase sufficiently to cause vasodilation without significantly increasing the total tissue pool of adenosine. With longer periods of stimulation or with greater increases in the concentration of adenosine around blood vessels, the adenosine content of the whole tissue would be significantly increased. Belardinelli et al. (2) have attempted to answer the question about the localized production of adenosine by infusing [^{14}C]adenosine intraarterially into a dog hindlimb preparation perfused at constant flow. As the concentration of adenosine around blood vessels increases, the [^{14}C]adenosine will be diluted in a larger pool, resulting in a decrease in its specific activity in venous blood. When the muscle was stimulated to contract, perfusion pressure fell to about 40% of the control value. A decrease in the specific activity of adenosine in venous blood paralleled the decrease in perfusion pressure. This indicated that the intraarterially infused [^{14}C]adenosine was being diluted in a

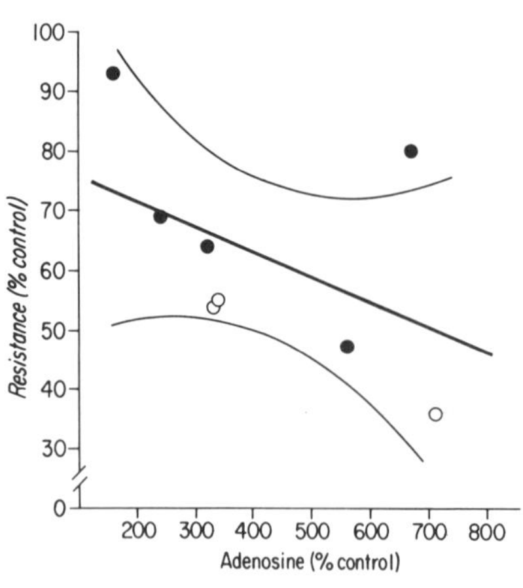

FIG. 4. Relationship between vascular resistance and adenosine content of dog skeletal muscle perfused at constant flow. Open circles represent values after 10 min of contraction. Closed circles represent values after 25 min of contraction. Lines are calculated regression and 95% confidence limits.

larger pool with a time course similar to that for the change in perfusion pressure. However, these experiments do not resolve the problem of the apparent differences in adenosine production under free flow conditions compared with constant flow perfusion.

Many questions remain concerning the role of adenosine in the local control of the circulation of skeletal muscle. However, it seems clear that there are differences in the ability of various muscles to produce adenosine. At one extreme is the low-oxidative gracilis muscle of cat that accumulated little adenosine even under conditions that caused the muscle to fatigue. At the other extreme is the highly oxidative cardiac muscle that accumulated adenosine in response to very brief periods of ischemia or to increased work load. The ability of cat soleus muscle and dog skeletal muscle to accumulate adenosine lies between these two extremes.

ADENOSINE MONOPHOSPHATE DEAMINASE AND 5′-NUCLEOTIDASE ACTIVITIES

As indicated above, the ability of muscle to produce adenosine seems to vary with the inherent metabolic pattern of the muscle. The inherent metabolic pattern is characterized by differences in the activities of various enzymes, such as those associated with glycolysis and oxidative phosphorylation. Therefore, we considered the possibility that the different abilities to produce adenosine among the muscles might result from differences in the activities of the enzymes associated with AMP degradation (Fig. 1). Obviously, the maximum rate of adenosine production is determined by the activity of 5′-nucleotidase. However, as AMP deaminase activity increases relative to the activity of 5′-nucleotidase, adenosine production will be decreased in two ways. First, for a given rate of production of AMP, more of the AMP will be converted to IMP. Second, IMP, which is also a substrate for 5′-nucleotidase, will compete with AMP, thereby decreasing the rate of adenosine production. The hypothesis that the activities of the enzymes associated with AMP degradation vary with the metabolic pattern of the muscle is supported by data obtained from the literature. As discussed previously, the difference in the rate of adenosine production between skeletal and cardiac muscle has been attributed to the higher AMP deaminase activity in skeletal muscle (6,13). Moreover, Winder et al. (20) have shown that AMP deaminase activity correlates with fiber type in rat muscle, being highest in low-oxidative "white" fibers and lowest in oxidative "intermediate" fibers. Rubio et al. (17) have reported that 5′-nucleotidase activity is greatest around blood vessels in skeletal muscle. Since capillarity has been reported to vary as a function of fiber type (8), it is possible that 5′-nucleotidase activity also varies as a function of fiber type. To test our hypothesis, we measured the AMP deaminase and 5′-nucleotidase activities in gracilis and soleus muscles of cats ("white" and "intermediate" fibers, respectively) and gracilis muscle of dogs ("red" fibers). In view of the difference in the rate of adenosine production between skeletal muscle and cardiac muscle, we also measured the activities of these two enzymes in dog cardiac muscle (3).

To measure AMP deaminase activity (19), muscle samples were homogenized in a solution containing KCl (180 mM), phosphate buffer (88 mM, pH 6.5), and mercaptoethanol (1 mM). The homogenate was centrifuged (14,000 × *g*, 15 min, 4°C) and the supernatant fraction assayed for AMP deaminase activity. Enzymatic activity was measured by recording the change in absorbance at 285 nm as AMP was converted to IMP after an aliquot of the supernatant fraction (2.5 mg tissue) was added to the assay solution (0.1 M sodium succinate and 4 mM AMP at pH 7.0, 30°C).

To measure 5′-nucleotidase activity (1), an acetone powder of muscle was homogenized in phosphate buffer (pH 7.5). The homogenate was dialyzed overnight against the phosphate buffer. Sodium deoxycholate (1% final concentration) was added to the dialyzed homogenate. After standing for 4 hr at 0°C, the homogenate was centrifuged (30,000 × *g*, 15 min, 4°C) and the supernatant fraction was assayed for 5′-nucleotidase activity. The assay contained 1 mM AMP, 1 mM p-nitrophenylphosphate, 50 mM phosphate buffer (pH 7.5, 30°C), and appropriate amounts of muscle enzyme preparation. After 0, 30, and 60 min, aliquots of the assay mixture were added to perchloric acid. The concentrations of adenosine, inosine, and hypoxanthine were measured in the neutralized perchloric acid extracts (15).

The activity of AMP deaminase in the muscles varies from a high of 92 ± 8.2 (mean ± SEM) μmoles/min/g in cat gracilis muscle to a low of 1.0 ± 0.2 μmoles/min/g in dog cardiac muscle. That is, there is an inverse relationship between AMP deaminase activity and the oxidative capacity of the muscle. These data confirm the difference in AMP deaminase activity between skeletal and cardiac muscles. The activity in dog cardiac muscle is only 4% of the activity in dog gracilis muscle. However, the data also indicate that there are large differences in AMP deaminase activity among different skeletal muscles. The activities in cat soleus and dog gracilis muscles are about 35% of the activity of cat gracilis muscle.

The activity of 5′-nucleotidase shows a direct relationship with the oxidative capacity of the muscle (3). Dog gracilis muscle (80 ± 18 nmoles/min/g) has a much higher 5′-nucleotidase activity than does either of the cat muscles, nearly 300% of the activity of cat soleus muscle (27 ± 4 nmoles/min/g) and nearly 700% of the activity of cat gracilis muscle (12 ± 3 nmoles/min/g). The 5′-nucleotidase activity of dog cardiac muscle (190 ± 42 nmoles/min/g) is about 250% of the activity in dog gracilis muscle and nearly 1,600% of the activity in cat gracilis muscle.

These data support our hypothesis that differences in the ability of various muscles to produce adenosine result, at least in part, from differences in the activities of the enzymes associated with AMP degradation. On the basis of these data, one would predict that the ability of a muscle to produce adenosine would increase as oxidative capacity increased, both as a result of decreased AMP deaminase activity and increased 5′-nucleotidase activity. To get an index of the relative potential of a muscle to produce adenosine, we have plotted the ratio of the activity of 5′-nucleotidase to AMP deaminase as a function of the estimated maximum oxygen consumption of the muscle (Fig. 5). The estimated maximum

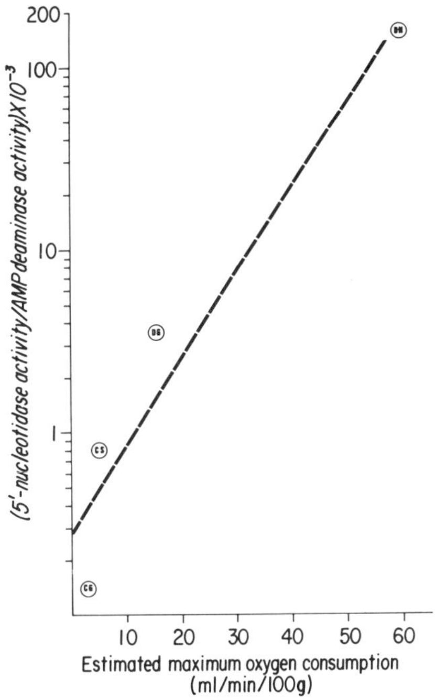

FIG. 5. Relationship between potential for adenosine production, as measured by ratio of activity of 5′-nucleotidase to AMP deaminase, and estimated maximum oxygen consumption. See text for explanation of estimated maximum oxygen consumption. CG = cat gracilis; CS = cat soleus; DG = dog gracilis; DH = dog heart.

oxygen consumptions are from our work with cat gracilis and soleus muscles (E. L. Bockman and J. E. McKenzie, *unpublished observations),* from Maxwell et al. (14) for dog skeletal muscle, and Gregg and Fischer (9) for cardiac muscle. As illustrated in Fig. 5, there is a direct relationship between the log of the ratio of the activities of the two enzymes and the estimated maximum oxygen consumption. Although the use of the ratio of the activity of 5′-nucleotidase to AMP deaminase is a simplistic approach to predicting the potential for adenosine production, the ratios correlate well with the relative rates of adenosine production *in situ* among the various muscles. For example, cat gracilis muscle, which has the lowest relative potential for adenosine production as predicted by the ratio of the two enzymes, did not produce adenosine in response to muscle contraction. Dog cardiac muscle, which has the highest relative potential for adenosine production, produced adenosine in response to brief periods of occlusion and to increased cardiac work. The ratio of the activity of 5′-nucleotidase to AMP deaminase may be looked on as providing a relative index of the maximum rate of adenosine production as might occur in response to prolonged ischemia. Certainly, the rate of adenosine production under less severe conditions would depend on a number of factors, including the tissue concentrations of AMP and IMP as well as the K_ms, the presence of activators or inhibitors, and perhaps the localization of the enzymes. Since 5′-nucleotidase seems to be localized toward the periphery of skeletal muscle cells (16), it may have access to a pool of AMP not available to AMP deaminase. Thus the rate of adenosine production might be higher than would be predicted from the ratio of the two enzymes.

SUMMARY

The role of adenosine in the local control of the circulation of skeletal muscle is still uncertain. The adenosine content of dog skeletal muscle perfused at constant flow increases as vascular resistance decreases in response to muscle contraction. However, it has not been possible to demonstrate an increase in the adenosine content of dog or cat muscle perfused under free-flow conditions even though blood flow increased several-fold in response to muscle contraction. It is possible that the localized production of adenosine around blood vessels, sufficient to cause vasodilation but not to increase muscle adenosine content, may explain the differences between experiments performed under the conditions of free flow compared with constant flow. However, it is also possible that muscles perfused at constant flow are slightly hypoxic during muscle contraction and that it is the hypoxia that stimulates adenosine production. This would provide evidence for a role of adenosine as a mediator of reactive hyperemia but not active hyperemia.

There appears to be a difference in the ability of various muscles to produce adenosine as evidenced both by *in situ* measurements of muscle adenosine content and by the ratio of the activity of 5′-nucleotidase to AMP deaminase. In general, the production of adenosine *in situ* and the relative potential for adenosine production predicted from the ratio of the two enzymes increase as the oxidative capacity of the muscle increases.

ACKNOWLEDGMENTS

This work was supported by Public Health Service Grant HL 17932. Travel funds were provided by the LSU Foundation. The authors gratefully acknowledge the excellent assistance of Mrs. Jane Anderson and Mr. Philip Karam.

REFERENCES

1. Baer, H. P., Drummond, G. I., and Duncan, E. L. (1966): *Mol. Pharmacol.,* 2:67–76.
2. Belardinelli, L., Tominaga, S., Curnish, R., Rubio, R., and Berne, R. M. (1978): *Fed. Proc.,* 37:734(abst.).
3. Bockman, E. L., and McKenzie, J. E. (1979): *Am. J. Physiol. (Submitted for publication).*
4. Bockman, E. L., Berne, R. M., and Rubio, R. (1975): *Pfluegers Arch.,* 355:229–241.
5. Bockman, E. L., Berne, R. M., and Rubio, R. (1976): *Am. J. Physiol.,* 230:1531–1537.
6. Deuticke, B., and Gerlach, E. (1966): *Pfluegers Arch.,* 292:239–254.
7. Dobson, J. G., Jr., Rubio, R., and Berne, R. M. (1971): *Circ. Res.,* 29:375–384.
8. Gray, S. D., and Renkin, E. M. (1977): *Fed. Proc.,* 36:522.
9. Gregg, D. E., and Fischer, L. C. (1963): In: *Handbook of Physiology,* Vol. II, pp. 1517–1584. American Physiological Society, Washington, D.C.
10. Hilton, S. M., Jeffries, M. G., and Vrbová, G. (1970): *J. Physiol. (Lond.),* 206:543–562.
11. Hohorst, H.-J. (1965): In: *Methods of Enzymatic Analysis,* edited by H. U. Bergmeyer, pp. 266–270. Academic Press, New York.
12. Hudlická, O. (1975): *Microvasc. Res.,* 10:17–28.

13. Imai, S., Riley, A. L., and Berne, R. M. (1964): *Circ. Res.*, 15:443–450.
14. Maxwell, L. C., Barclay, J. K., Mohrman, D. E., and Faulkner, J. A. (1977): *Am. J. Physiol.*, 233:C14–C18.
15. Olsson, R. A. (1970): *Circ. Res.*, 26:301–306.
16. Phair, R. D., and Sparks, H. V. (1977): *Physiologist*, 20:74(abst.).
17. Rubio, R., Berne, R. M., and Dobson, J. G., Jr. (1973): *Am. J. Physiol.*, 225:938–953.
18. Scott, J. B., Daugherty, R. M., Jr., Dabney, J. M., and Haddy, F. J. (1965): *Am. J. Physiol.*, 208:813–824.
19. Smiley, K. L., Jr., Berry, A. J., and Suelter, C. H. (1967): *J. Biol. Chem.*, 242:2502–2506.
20. Winder, W. W., Terjung, R. L., Baldwin, K. M., and Holloszy, J. O. (1974): *Am. J. Physiol.*, 227:1411–1414.

Physiological and Regulatory Functions of Adenosine and Adenine Nucleotides, edited by H. P. Baer and G. I. Drummond. Raven Press, New York © 1979.

Cardiovascular and Metabolic Actions of Adenosine and Adenosine Analogs

Gerhard Raberger

Pharmakologisches Institut der Universität Wien, 1090 Wien, Austria

Although adenosine has been known to be a potent vasodilator for a long time (3), real interest in its possible physiological role was aroused much more recently by the adenosine hypothesis of coronary blood flow regulation (1). Enhancement of the coronary activity of adenosine by long-acting coronary vasodilators like dipryridamole and hexobendine (2,6) further stimulated interest in this nucleoside. In the following, results are summarized from this laboratory, dealing with studies on the possible mechanism of adenosine action as well as with the effects of a series of adenosine derivatives modified in particular at the 5′-position.

PREVIOUS WORK

We began with investigations on the effects of intracoronary infusion of adenosine (10 μg/kg/min) on myocardial metabolism as measured by arteriovenous differences of oxygen, carbon dioxide, H^+ ions, glucose, free fatty acids, and lactate (11), using anesthetized dogs. The results revealed a decrease in oxygen consumption together with an increase in myocardial glucose uptake and lactate and H^+ ion release (Fig. 1). These results were interpreted as evidence of a metabolic action of adenosine leading to increased glycolysis. Results of further investigations in isolated cat hearts perfused under constant pressure or volume did not confirm this assumption (26). An increase in glucose uptake and lactate release was found only in the hearts perfused with constant pressure. In an additional set of experiments, glucose uptake by isolated cat hearts in response to adenosine was shown to be solely flow dependent (24).

The above-mentioned results obtained in dogs led to the conclusion that the coronary action of adenosine is likely to be influenced by changes in acid-base equilibria. Investigations carried out in anesthetized dogs (16) did indeed show a clear dependence of the coronary dilatory action of adenosine on arterial pH and pCO_2 (Fig. 2). At the same time the coronary vascular tone was shown to be mainly pCO_2 dependent, i.e., dependent on intracellular pH.

Since adenosine was proposed to be a mediator of hypoxic and reactive hyperemic coronary vasodilation, it was of interest to investigate if the coronary vasodilation induced by the nucleoside during hypoxia or occlusion could be

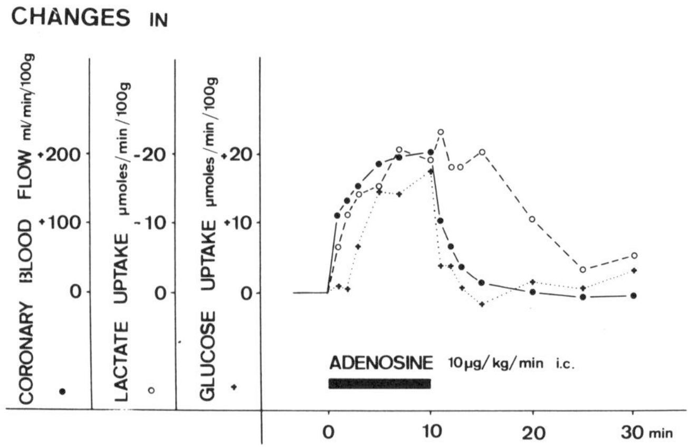

FIG. 1. Experiments on seven dogs (urethan/chloralose/N_2O). Comparison of the changes in coronary blood flow, lactate uptake, and glucose uptake during and following an intracoronary infusion of adenosine (10 μg/kg). Statistical evaluation with *t*-test for paired data. (+):$0.05 > p > 0.01$; (++):$0.01 > p > 0.001$.

similarly influenced. Hexobendine which greatly enhances the coronary activity of adenosine (10) led to a marked increase in hypoxic coronary vasodilation during 100% N_2O respiration. It was of particular interest that the increase in heart rate caused by hypoxia which certainly is sympathomimetic in origin was not influenced by pretreatment with hexobendine (17). These results indicate that adenosine participates to some extent in hypoxic coronary vasodilation.

ADENOSINE AND REACTIVE HYPEREMIA

Rubio and Berne (18) and Olsson (8) showed an increased release of adenosine and its degradation products in the coronary effluent during reactive hyperemia after coronary occlusion. Since we had shown that the coronary activity of adenosine was dependent on the arterial pH and pCO_2, we decided to investigate the influence of changes in acid-base equilibria on coronary reactive hyperemia after occlusion periods of 2 to 100 beats. Acid-base changes were induced either by CO_2 respiration (1 to 8%) or by hypoventilation (12). The dependence of the coronary flow on arterial pH and pCO_2 was proved once again under these conditions. Occlusions of the descending branch of the left coronary artery were carried out in thoracotomized dogs. The reactive hyperemia was positively correlated to the duration of occlusion. A similar correlation between occlusion time and reactive hyperemia was also obtained with hypercapnic acidosis. The repayment, which was independent of the duration of occlusion during normal respiration, exhibited a dependence on the duration of occlusion during hypercapnic acidosis. This arises from an enhancement of coronary vasodilation after long occlusion periods (above 10 to 12 beats) and a diminution of vasodilation after short occlusion periods (up to 10 beats) during hypercapnic acidosis (Fig. 3). These results led to the conclusion that the mediator

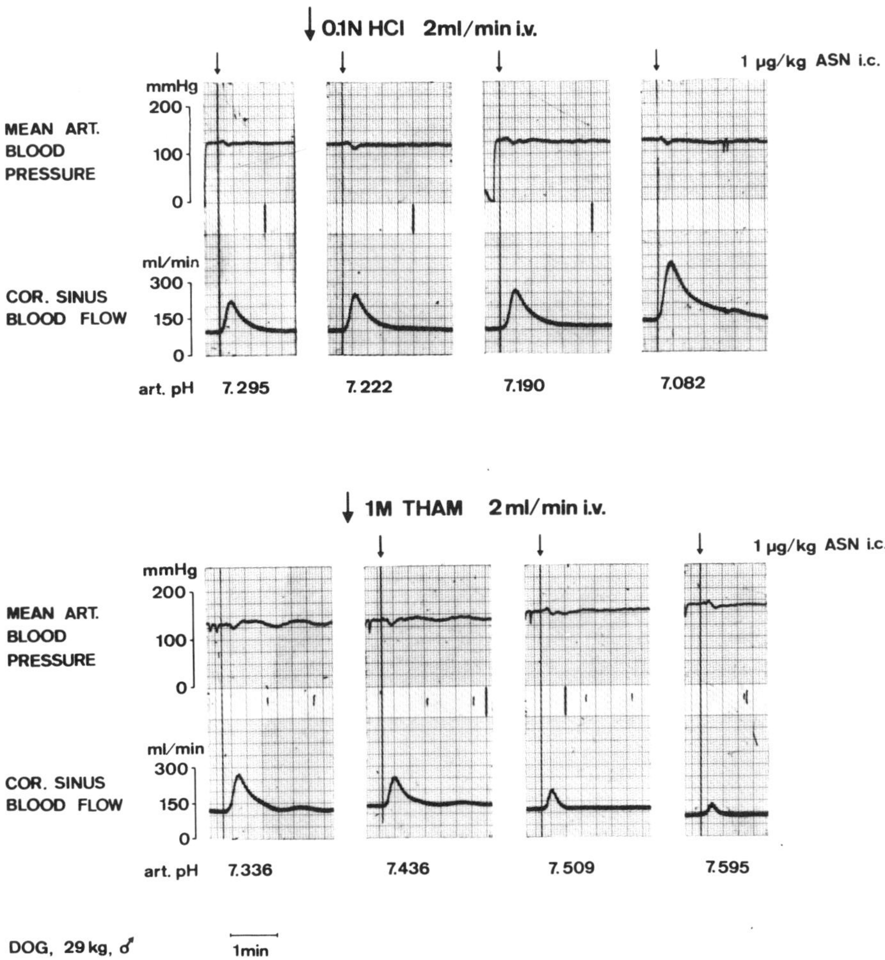

FIG. 2. Experiments on two dogs (urethan/chloralose/N_2O). Effects of intracoronary administration of adenosine (1 μg/kg) on mean arterial blood pressure and coronary sinus outflow at different pH values.

of reactive hyperemia after long and short occlusion periods cannot be the same, or alternatively that reactive hyperemia is the consequence of a number of events that eventually develop during coronary occlusion. Since both reactive hyperemia after long occlusion periods and vasodilation caused by intracoronarily administered adenosine are enhanced by acidosis, it seems likely that adenosine is to some extent involved in reactive hyperemia after long occlusion periods (as illustrated particularly well by the response to occlusion over 100 beats, Fig. 4).

This rather unexpected discrepancy between the reactive hyperemic response after short and long occlusion periods was further supported by data obtained in experiments carried out with intracoronary infusion of aminophylline (25 μg/cm^3

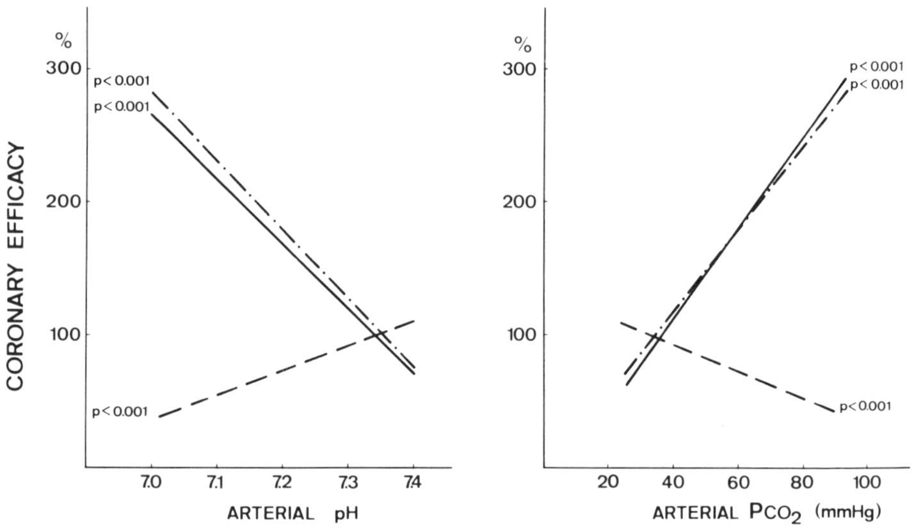

FIG. 3. Experiments on dogs (urethan/chloralose/N_2O). Dependence of the dilator effect of adenosine (-·-·-) and reactive hyperemic response after long (—) and short (----) coronary artery occlusions on the arterial pH and pco_2 during experiments with increasing hypercapnic acidosis. Control values during normal respiration are taken as 100% for reference. Statistical evaluation with correlation analysis and linear regression lines.

coronary blood) (22), angiotensin II (3 ng/cm^3 coronary blood) (Raberger et al., *unpublished data),* or on intravenous injection of dihydroergotamine (10 μg/kg) (15). The effect of intracoronary administration of adenosine was significantly reduced in the presence of each of these substances. The repayment after long coronary occlusion periods was also diminished, although to a lesser extent. The repayment after short occlusion periods was not significantly altered in any of the experiments. Thus the conclusion must be drawn that adenosine seems to be responsible only for reactive hyperemia after coronary occlusion lasting longer than 10 to 12 beats (heart rate 70 to 100) and that reactive hyperemia after shorter occlusion periods is caused either by a Bayliss effect or a mediator other than adenosine, for instance H^+ ions. These assumptions are further supported by recently published work by Olsson et al. (9) and Merrill et al. (7), who also propose the increased formation of H^+ ions during reactive hyperemia as an important factor for reactive hyperemic flow increase.

HEMODYNAMIC AND METABOLIC ACTIONS OF ADENOSINE ANALOGS

Adenosine has a very short-lived action on coronary blood flow since it is rapidly taken up into red blood cells, platelets, and lungs and converted to inosine and adenosine 5′-monophosphate (5′-AMP). It was, therefore, desirable to have analogs

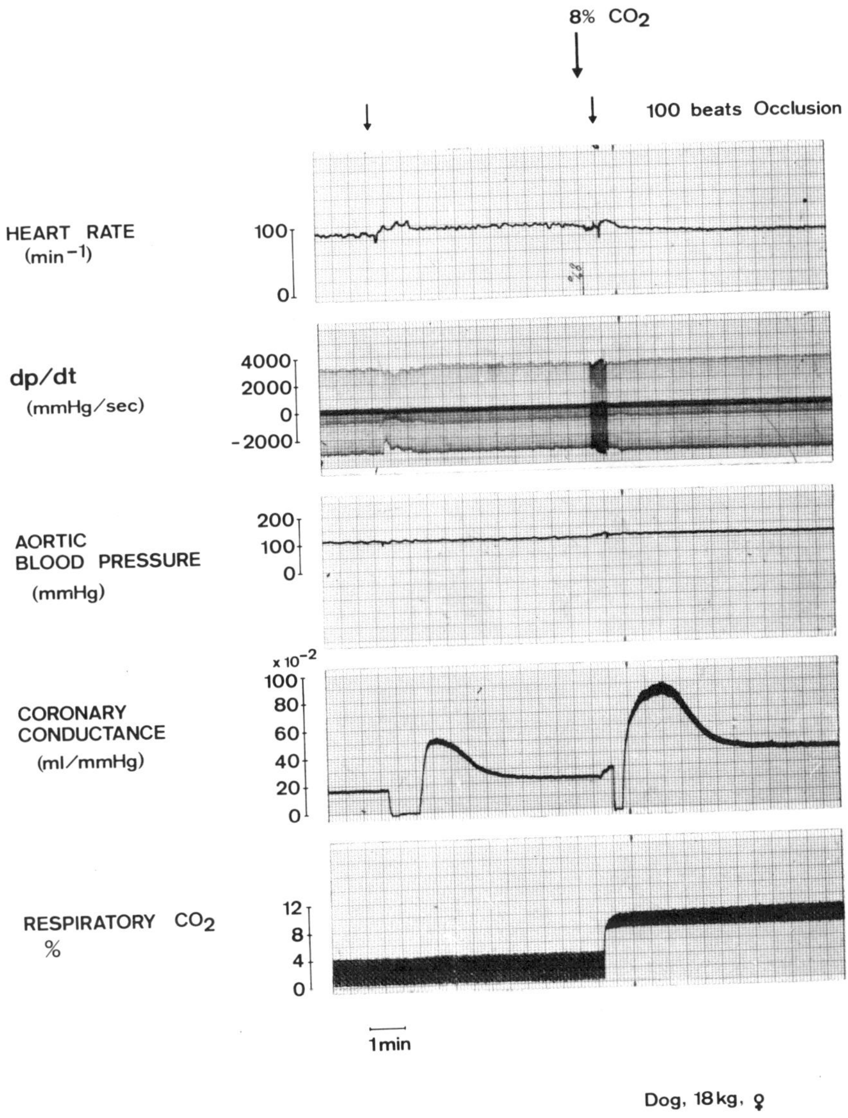

FIG. 4. Experiment on one dog (urethan/chloralose/N_2O). Influence of 10 min of breathing 8% CO_2 on heart rate, left ventricular dP/dt, aortic blood pressure, coronary conductance, reactive hyperemia after 100 beats occlusion, and on respiratory CO_2.

with a prolonged duration of action to establish other possible physiological roles of adenosine. Analogs of adenosine that have structural changes in only one or two positions enable further information to be gained on the structural requirements for interaction with the adenosine receptor.

Over 60 adenosine analogs were synthesized in the Byk-Gulden Laboratories and at the University of Konstanz. I report on the coronary activity of 23 of these substances and focus attention on hemodynamic and metabolic activities of three of the most potent analogs.

In an initial set of experiments, the coronary activity (increase in blood flow integrated over time of action) of these analogs was tested in anesthetized dogs (13). Of the analogs substituted only at the 5′ position the activity of adenosine-5′-esters was generally less pronounced than that of the respective 5′-carboxamides (Table 1), a finding already reported by Stein and Somani (23). The most active compound was adenosine-5′-ethylcarboxamide (744–96) which was already active at a dose of 0.01 μg/kg i.v. Adenosine carboxamides that were interlinked by a polymethylene chain exhibited coronary activity only if the interposed chain had a length of more than eight methylene groups. Adenosine analogs with sterically altered substitution in the 2′- and 3′-positions were generally inactive. A rather interesting group of

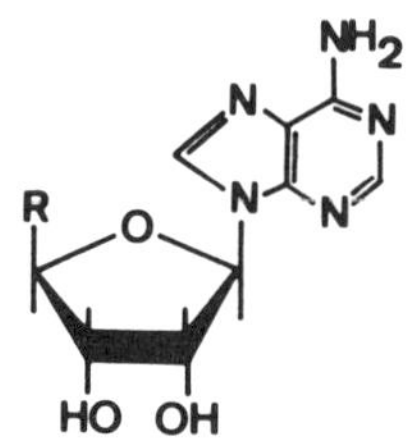

	R	DOSE RANGE μg/kg i.v.	RELATIVE EFFICACY	ADENOSINE ENHANCEMENT
Adenosine	$-CH_2OH$	100 - 800	1	-
7199-19	$-COOH$	1 - 10	53	-
7199-20	$-COOCH_3$	1 - 20	28	-
744-84	$-COOC_2H_5$	2 - 10	112	-
7199-21	$-CONH_2$	0.1 - 1	550	-
7299-58	$-CONHCH_3$	0.2 - 1	934	-
7199-22	$-CON(CH_3)_2$	1 - 5	229	-
744-96	$-CONHC_2H_5$	0.01 - 0.1	22 900	±
7499-28	$-CONH(CH_2)_2NH_2$	5 - 20	60	-

TABLE 1. Formulas and relative coronary efficacies of 5′-substituted adenosine analogs as well as their influence on adenosine efficacy.

analogs was the one with 2′-, 3′-, and 5′-substitution. Adenosine-5′-ethylcarboxamide was taken as the parent compound and substitution of the 2′- and/or 3′-hydroxyl group was carried out with NO_2, $COCH_3$, methoxymethylidene, or methoxyethylidene. These analogs were characterized by a markedly delayed onset and prolonged duration of action. The above results can be summarized as follows: 5′-substituted adenosine analogs exhibit a coronary activity superior to that of adenosine, but the time course is similar. Additional substitution at the 2′- and 3′-positions leads to an altered kinetic behavior after i.v. administration. The 2′-, 3′-, and 5′-substituted adenosine analogs seem to display coronary activity only after metabolic degradation to the 5′-substituted analog.

All further investigations were carried out with only three of the above-mentioned adenosine analogs (Table 2). The complete hemodynamic profile was established with 744-98 (2′3′-*O*- methoxyethylidene-adenosine-5′-ethylcarboxamide) and 744-99 (2′,3′-dinitroadenosine-5′-ethylcarboxamide) in anesthetized dogs (19). Coronary vasodilation was accompanied by an increase in cardiac output and a decrease in total peripheral resistance. At the same time an increase in left ventricular dP/dt and heart rate and a decrease in left ventricular end diastolic pressure was observed. Pulmonary hemodynamics were not markedly influenced. Blood flow measurements in the superior mesenteric and hepatic artery revealed a marked increase in flow with 744-96 and 744-98 especially in the mesenteric artery. Blood flow in the renal artery was decreased as might have been expected from an analog with an adenosine-like action. The analog 744-98 markedly increased arterial glucose level and decreased arterial free fatty acid level. Pretreatment with bupranolol (50 μg/kg i.v.) (Fig. 5) or propranolol (1.5 mg/kg i.v.) resulted in only partial inhibition of the increase in heart rate and left ventricular dP/dt. Thus apart from the increase in blood flow, these analogs showed two distinct actions in anesthetized dogs that merit comment. One is the marked metabolic action and the other the only partial reversibility of the positive chronotropic and inotropic action by adrenergic β-blockade.

In order to exclude any possible interactions between the anesthetic and these analogs, another set of experiments was carried out in conscious dogs. The dogs were trained to either stand or lie quietly, depending on the intended investigations. In some dogs flow probes were implanted on the descending branch of the left coronary artery and on one renal artery. The administration of 744-98 led to a dose-dependent increase in coronary blood flow and, in contrast to the results obtained in anesthetized dogs, to no change or even an increase in renal blood flow, although a rapid and marked decrease in renal blood flow could be obtained with adenosine. Approximately twice the dosage was required in conscious dogs to achieve similar coronary effects as in anesthetized dogs. A comparison of the i.v. to the oral activity in conscious dogs revealed a ratio of 3:1 (14).

The metabolic actions of the adenosine analogs were further investigated in conscious dogs with a femoral catheter (Intranule® 112.12, Vygon). Arterial blood pressure and heart rate were monitored continuously and arterial blood samples were drawn at certain intervals before and after the administration of the adenosine

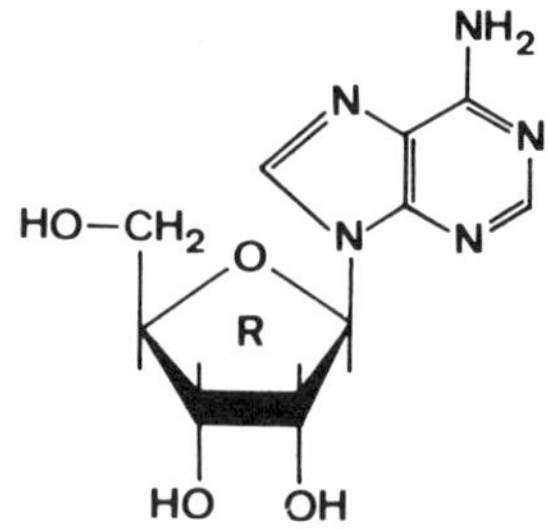

	R	DOSE RANGE µg/kg i.v.	RELATIVE EFFICACY
Adenosine	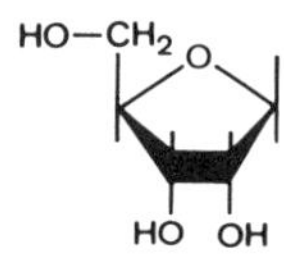	100-800	1
744-96	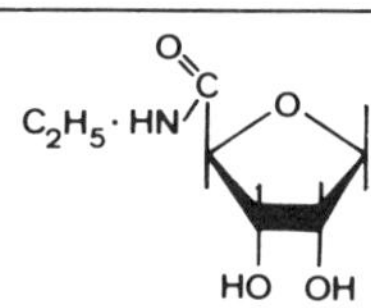	0.01-0.1	22 900
744-98	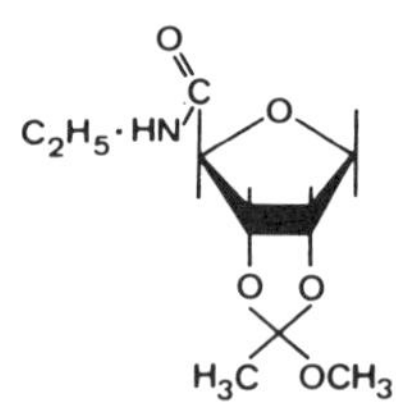	0.1-2.0	31 600 (0.5µg) 4 000 000 (0.6 µg)
744-99	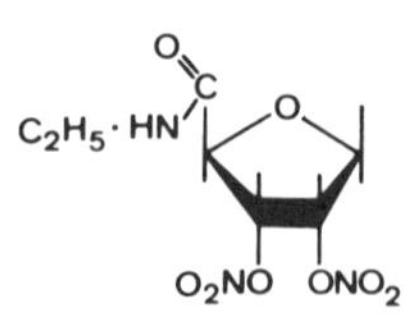	0.5-5.0	6 000 (1µg) 170 000 (2µg)

TABLE 2. Formulas and relative coronary efficacies of adenosine-5′-ethylcarboxamide and the 2′- and 3′-substituted analogs as well as their influence on adenosine efficacy.

analog 744-98. These experiments showed that the adenosine analog increases heart rate at low i.v. dosage but leads to a decrease in heart rate at relatively high dosage, despite a marked fall in blood pressure. The increase in arterial glucose and decrease in arterial free fatty acid levels, already observed in anesthetized dogs, was shown to be also dose dependent (14). Plasma insulin and glucagon levels were investigated in order to further characterize the metabolic effects obtained with the three adenosine analogs 744-96, 744-98, and 744-99. The increase in glucose levels was preceded by a two- to threefold increase in arterial glucagon levels. This increase persisted for longer than 2 hr, indicating a long-acting glucagon-releasing action, since glucagon has a very short half-life. It would have been anticipated that the

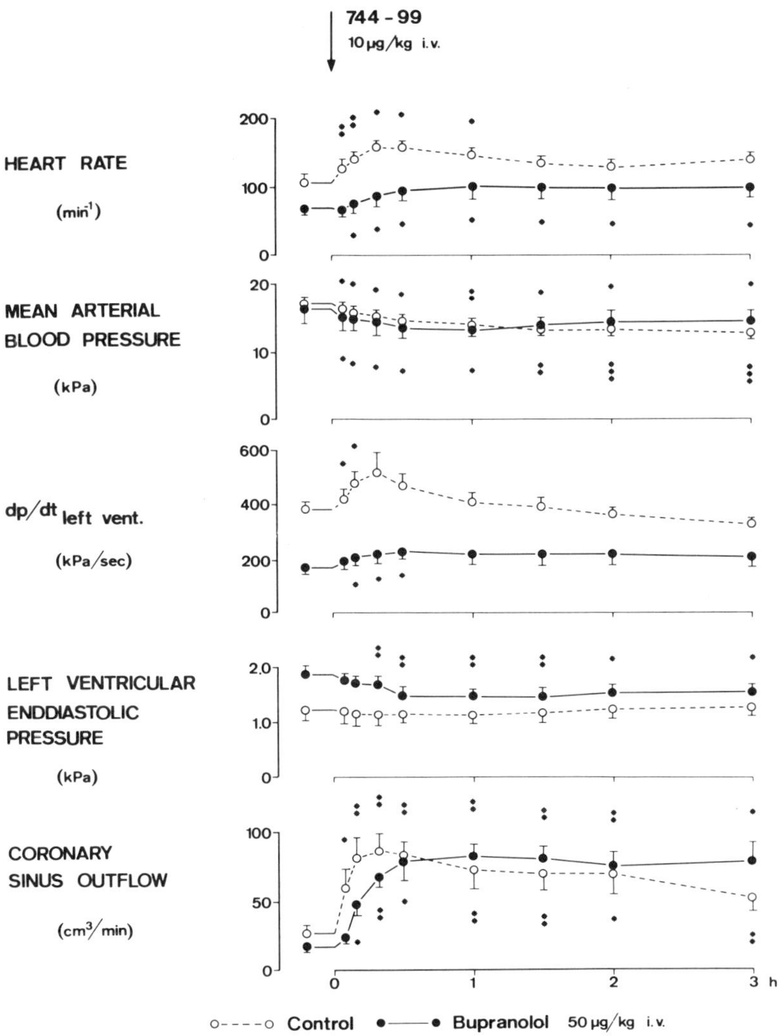

FIG. 5. Experiments on 10 dogs (chloralose/N_2O). The effect of 10 μg/kg 744-99 i.v. on heart rate, mean arterial blood pressure, left ventricular dP/dt and end diastolic pressure, and on coronary sinus outflow with and without pretreatment of bupranolol. Statistical evaluation with *t*-test for paired data. For symbols see Fig. 1.

marked increase in arterial glucose level observed with the adenosine analogs should lead to a stimulation of insulin secretion, but, in fact, the arterial insulin level remained unchanged (20). Thus the decrease in arterial free fatty acid level, which was constantly observed in these experiments, seems to be a direct adenosine-like effect (4). These conclusions were further substantiated in experiments on alloxan-diabetic dogs in which 744-98 led to a marked increase in glucose and decrease in

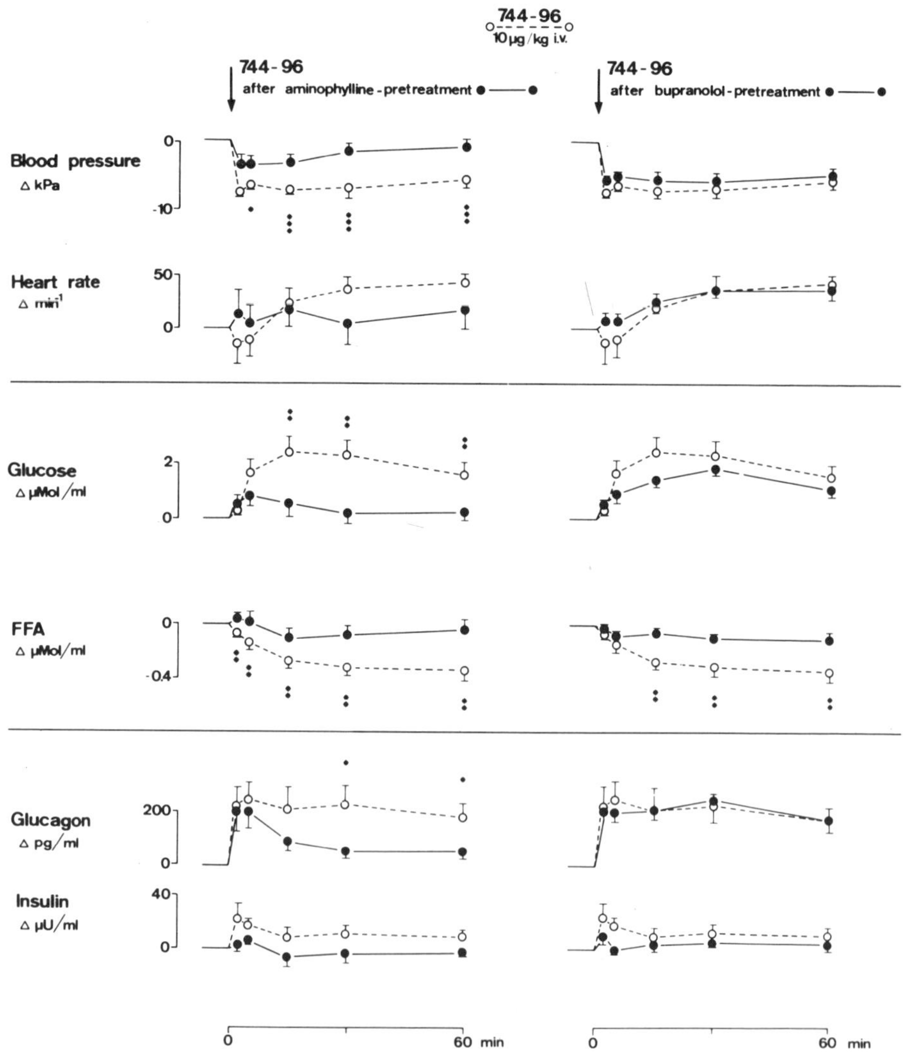

FIG. 6. Influence of pretreatment with aminophylline or bupranolol on hemodynamic and metabolic changes induced by 744-96. Changes are drawn as difference values. Statistical evaluation with *t*-test. For symbols see Fig. 5.

free fatty acid level. To assess a reflex sympathomimetic release of glucagon secondary to the decrease in blood pressure, dogs were infused with sodium nitroprusside. Sodium nitroprusside did not show any marked increase in arterial glucagon, although blood pressure and heart rate were affected similarly as seen in experiments with adenosine analogs. Thus the glucagon release caused by the adenosine analogs appears to be a direct adenosine-like action and not reflexly mediated. Indeed, a glucagon-releasing action of adenosine was found in the isolated perfused rat pancreas by Weir et al. (25).

Investigations in conscious dogs carried out with 744-96, the parent compound with rapid onset of action, gave further evidence of the direct adenosine-like, glucagon-releasing action of this analog. Experiments were carried out either without pretreatment or following bupranolol (50 μg/kg i.v.) or aminophylline pretreatment (10 mg/kg i.v.) (21). The hemodynamic and metabolic effects of 744-96 were almost completely abolished by aminophylline at the chosen dosage, but remained almost unchanged by bupranolol pretreatment (Fig. 6). These results indicate that the adenosine analog causes glucagon release via a receptor that can be blocked by aminophylline-like adenosine receptors in the vascular bed. Moreover adenosine analogs must inhibit insulin release since a large increase in arterial glucose level (3 to 5 μM/cm^3) led to no increase in plasma insulin level. Experiments in isolated rat pancreatic islets (5), which showed a decrease in glucose-stimulated insulin release after addition of adenosine, provide support for this theory. Preliminary results from our laboratory on isolated rat pancreatic islets demonstrate this inhibitory action on glucose-induced insulin release for the adenosine analog 744-96 also.

SUMMARY

Adenosine has so far been considered to play a role in hypoxic coronary blood flow regulation. It certainly participates in the reactive hyperemic increase in blood flow after long occlusion periods, but other mediators, such as H ions, for instance, must also be considered to take part in this adaptive process. An additional role of adenosine in the release of glucagon and insulin seems very likely. These actions were investigated using long-acting adenosine analogs. The analogs exhibit adenosine-like actions, but also have effects that differ from those of adenosine, including an increase in heart rate and myocardial contractility at moderate dosage which can be blocked only partially by adrenergic β-blockade. These unexpected actions might result from altered intracellular and extracellular distribution of the analogs and thus might reflect adenosine actions at intracellular sites.

ACKNOWLEDGMENTS

This work was supported by the Herbert Quandt-Stiftung (Varta AG) and Fonds zur Förderung der wissenschaftlichen Forschung in Osterreich. The adenosine analogs were synthesized by Prof. Schmidt and Dr. Mengel, University of

Konstanz, and kindly supplied by Byk Gulden, Lomberg Chemische Fabrik, Konstanz, BRD.

REFERENCES

1. Berne, R. M. (1963): *Am. J. Physiol.*, 204:317–322.
2. Bretschneider, H. J., Frank, A., Bernard, U., Kochsiek, K., and Scheler, F. (1959): *Arzneim. Forsch.*, 9:49–59.
3. Drury, A. N., and Szent-Györgyi, A. (1929): *J. Physiol.*, 68:213–237.
4. Ebert, R., and Schwabe, U. (1973): *Naunyn Schmiedebergs Arch. Pharmacol.*, 278:247–259.
5. Ismail, N. A., El Denshary, E. S. M., and Montague, W. (1977): *Biochem. J.*, 164:409–413.
6. Kraupp, O., Wolner, E., Adler-Kastner, L., Ploszczanski, B., and Tuisl, E. (1965): *Arzneim. Forsch.*, 15:1187–1196.
7. Merrill, G. F., Haddy, F. J., and Dabney, J. M. (1978): *Circ. Res.*, 42:225–229.
8. Olsson, R. A. (1969): *Fed. Proc.*, 28:779.
9. Olsson, R. A., Snow, J. A., and Gentry, M. K. (1978): *Circ. Res.*, 42:358–362.
10. Raberger, G., and Kraupp, O. (1971): *Eur. J. Pharmacol.*, 13:312–319.
11. Raberger, G., Kraupp, O., Stühlinger, W., Nell, G., and Chirikdjian, J. J. (1970): *Pfluegers Arch.*, 317:20–34.
12. Raberger, G., Schütz, W., and Kraupp, O. (1975): *Clin. Exp. Pharmacol. Physiol.*, 2:373–382.
13. Raberger, G., Schütz, W., and Kraupp, O. (1977): *Arch. Int. Pharmacodyn.*, 230:140–149.
14. Raberger, G., Schütz, W., and Kraupp, O. (1978): *Arch. Int. Pharmacodyn.*, 232:150–157.
15. Raberger, G., Schütz, W., Zimpfer, M., and Kraupp, O. (1976): *Basic Res. Cardiol.*, 71:645–651.
16. Raberger, G., Weissel, M., and Kraupp, O. (1971): *Naunyn Schmiedebergs Arch. Pharmacol.*, 271:301–310.
17. Raberger, G., Weissel, M., and Kraupp, O. (1973): *Arzneim. Forsch.*, 23:764–768.
18. Rubio, R., and Berne, R. M. (1969): *Circ. Res.*, 25:407–415.
19. Schütz, W., Raberger, G., and Kraupp, O. (1978): *Arzneim. Forsch.*, 11:2079–2082.
20. Schütz, W., Raberger, G., and Kraupp, O. (1978): *Naunyn Schmiedebergs Arch. Pharmacol.*, 304:249–254.
21. Schütz, W., Raberger, G., and Kraupp, O. (1978): *Arch. Int. Pharmacodyn.*, 234:214–220.
22. Schütz, W., Zimpfer, M., and Raberger, G. (1977): *Cardiovasc. Res.*, 11:507–511.
23. Stein, H. H., and Somani, P. (1975): *Ann. N.Y. Acad. Sci.*, 255:380–389.
24. Turnheim, K., Donath, R., Weissel, M., and Kolassa, N. (1976): *Pfluegers Arch.*, 365:197–202.
25. Weir, G. C., Knowlton, St. D., and Martin, D. B. (1975): *Endocrinology*, 97:932–936.
26. Weissel, M., Brugger, G., Raberger, G., and Kraupp, O. (1974): *Pharmacology*, 12:120–128.

Physiological and Regulatory Functions of Adenosine and Adenine Nucleotides, edited by H. P. Baer and G. I. Drummond. Raven Press, New York © 1979.

Cardiac Adenosine: Electrophysiological Effects, Possible Significance in Cell Function, and Mechanisms Controlling Its Release

Rafael Rubio, Luiz Belardinelli, Carl I. Thompson, and Robert M. Berne

Department of Physiology, School of Medicine, University of Virginia, Charlottesville, Virginia 22903

Adenosine is a compound found in many tissues and its production can be greatly enhanced by various procedures that either retard energy regeneration or accelerate its utilization (12,32,34,35,43; and Berne et al., *this volume)*. The possible role of adenosine in the metabolic control of blood flow in heart, brain, and other tissues has been the subject of study in our laboratory for several years. In addition, adenosine has been found to be a compound that affects various functions in different tissues at physiological concentrations (16,17,22,27,36,40). Hence, it is important to determine by what mechanism this nucleoside acts and if its various effects on cell function have physiological meaning.

ELECTROPHYSIOLOGICAL EFFECTS OF ADENOSINE IN CARDIAC TISSUES

Atrial and ventricular mammalian myocardium is capable of releasing adenosine under different physiological and pathophysiological conditions (32,43). Adenosine has been shown to exert a negative inotropic effect on atrial muscle that is associated with a shortening of the normal action potential (8), and to produce atrioventricular block (41) and a decrease in heart rate when injected directly into the artery supplying the sinoatrial node (18). The factor common to all these processes is their dependence on the concentration of extracellular calcium (20,45,49), which suggests that these effects of adenosine are caused by a depression of calcium influx.

Effects of Adenosine on the Slow Atrial Action Potential

To investigate the action of adenosine on slow atrial action potentials, we selected potassium-depolarized (K_o = 25 mM) guinea pig or rat atria treated with catecholamines because they generate slow action potentials whose amplitudes increase 30 mV for a 10-fold increase in extracellular calcium concentration

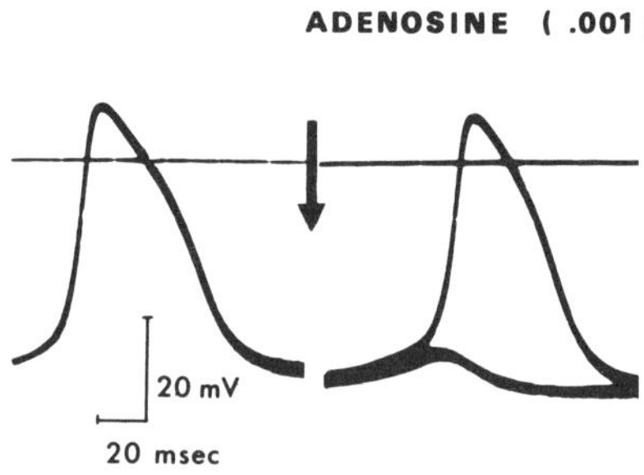

FIG. 1. Depressant effects of adenosine on the slow Ca^{2+}-dependent action potentials of the guinea pig atrium. **Left:** Control. **Right:** Successive tracings following addition of adenosine. Frequency of stimulation 0.5 Hz.

($[Ca^{2+}]_o$) (3,4,26,45), and are not dependent on the concentration of extracellular Na^+ (3,4). These observations suggest that Ca^{2+} is the main carrier of the inward current. Previous observations with this preparation showed that addition of adenosine to the superfusion solution in a final concentration of 1 μM (Fig. 1) rapidly depressed the slow action potentials (39).

Further studies with the same preparation demonstrated that the overshoot of the

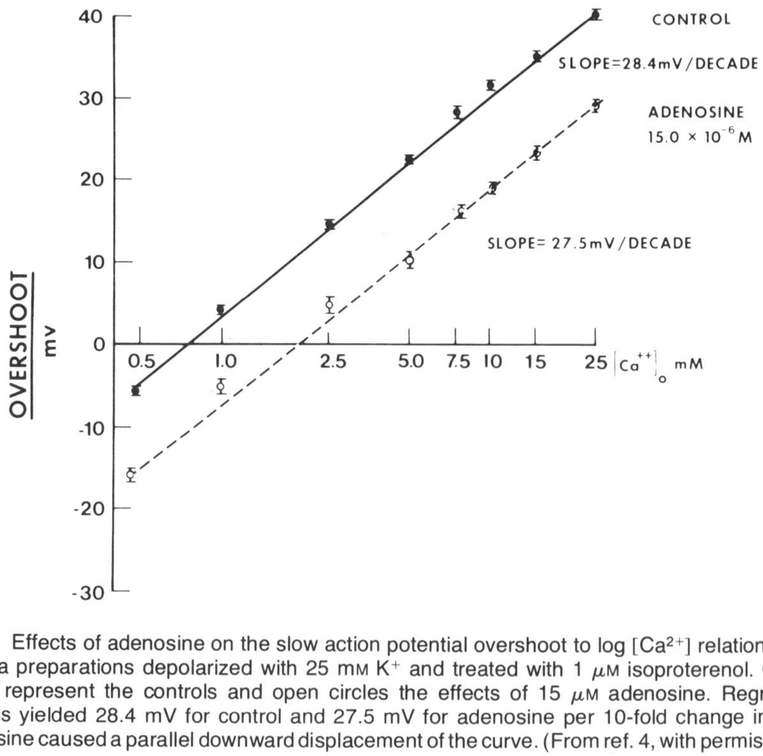

FIG. 2. Effects of adenosine on the slow action potential overshoot to log $[Ca^{2+}]$ relationship in rat atria preparations depolarized with 25 mM K^+ and treated with 1 μM isoproterenol. Closed circles represent the controls and open circles the effects of 15 μM adenosine. Regression analysis yielded 28.4 mV for control and 27.5 mV for adenosine per 10-fold change in Ca^{2+}. Adenosine caused a parallel downward displacement of the curve. (From ref. 4, with permission).

slow action potential changed linearly with the logarithm of $[Ca^{2+}]_o$ within a range of 0.5 to 25 mM with a slope of 30 mV for a 10-fold change in $[CA^{2+}]_o$ (Fig. 2). This linear relationship was found to be largely independent of extracellular Na^+ (3,4). The slope of 30 mV/decade is usually interpreted to be in accordance with the Nernst equation for membranes selectively permeable to Ca^{2+}. Adenosine caused a parallel downward displacement of this curve in proportion to its concentration, and the effect of 15 μM adenosine is depicted in the lower line of Fig. 2 (3,4). The fact that the downward displacement of the overshoot to log $[Ca^{2+}]_o$ caused by adenosine is parallel to the control curve suggests that this endogenous metabolite decreases the density of the slow Ca^{2+} channels. The selective ion channels may be either in a "resting" or an "inactive" state. These two states are membrane potential dependent. Only those channels in a "resting" state can be made available for activation when the tissue is electrically stimulated. We have provided evidence (4) that adenosine does not exert its depressant effects on the slow Ca^{2+}-dependent action potential by increasing the number of channels in the "inactive" state, as occurs with cocaine on the fast Na^+-dependent channels (48).

Adenosine may act by reducing the maximal number of slow Ca^{2+} channels available for activation, or by reducing the conductance of the individual channels. It may block the channels physically or prevent their availability for activation. These possibilities were studied in vascular smooth muscle.

Using intracellular recordings we have observed action potentials of the slow type in smooth muscle cells of tetraethylammonium-treated dog coronary arteries. The amplitude of the action potential increased as a function of log $[Ca^{2+}]_o$ with a slope of 30 mV per decade, indicating that Ca^{2+} is the main carrier of the electrical current. Adenosine at 10^{-5}M completely blocked these action potentials in arteries of less than 0.5 mm diameter, but was without effect on larger vessels (17). These observations suggest that adenosine may exert a negative inotropic effect on the smooth muscle of the resistance vessels, as it does in atria, by interfering with Ca^{2+} influx. A similar line of reasoning has been invoked in the case of inhibition of the action potential–induced calcium influx into the presynaptic endings to explain the depression of the evoked postsynaptic potentials induced by adenosine in olfactory cortex slices (20).

Effects of Other Adenosine Derivatives on the Slow Atrial Action Potential

The action of adenosine on atrial slow Ca^{2+}-dependent action potentials is shared by the adenine nucleotides 5′-adenosine monophosphate (AMP), adenosine diphosphate (ADP), adenosine triphosphate (ATP), nicotinamide-adenine dinucleotide (NAD), and cyclic AMP (cAMP). All of these compounds with the exception of cAMP are equipotent and act with equal effectiveness in the presence of a high concentration of adenosine deaminase (Fig. 3). Furthermore, similar depression of these responses has been demonstrated using a protein-AMP conjugate that is stable in the presence of 5′-nucleotidase and adenosine deaminase (38). This indicates that these compounds act directly and that prior hydrolysis to adenosine is not necessary

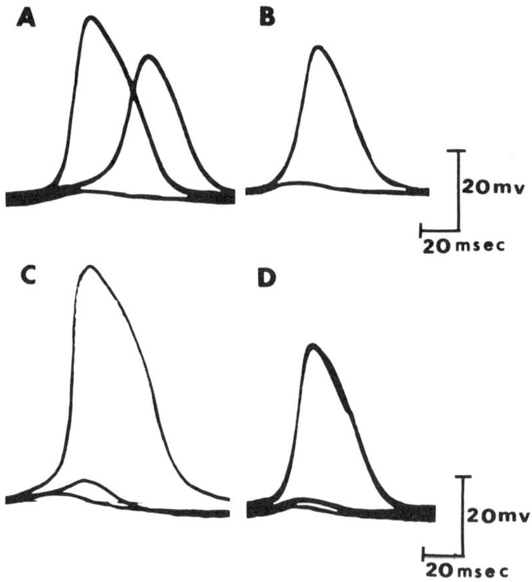

FIG. 3. Depressant effects of various adenine nucleotides on the atrial slow Ca^{2+}-dependent action potentials in the presence of an excess of adenosine deaminase that prevented the action of 1 mM adenosine. In all cases successive traces are superimposed. The nucleotides were added just before the largest action potential was photographed. **A:** Effects of 4 μM 5′-AMP. **B:** Effect of 4 μM ATP. **C:** Effect of 80 μM cAMP. **D:** Effect of 4 μM nicotinamide adenine dinucleotide. The magnification used in **C** was greater than that used in the other recordings.

for their blockade of the Ca^{2+} current. This is in contrast to what is observed in nervous tissues, in which adenosine and its derivatives stimulate adenylate cyclase and nucleotides must be hydrolyzed to adenosine in order to be effective (27,28). The specificity of the neural adenylate cyclase "receptor" to adenosine undoubtedly differs from that of the atrial slow Ca^{2+} current "receptor."

Effects of Adenosine on the Slow Ca^{2+}-Dependent Action Potential of Other Cardiac Tissues

Auriculoventricular (AV) transmission is particularly sensitive to the depressant effects of adenosine (38), and AV nodal cells normally generate slow action potentials that are Ca^{2+} dependent (49). Adenosine, at physiological concentrations, blocked the slow action potentials of nodal cells of the rabbit heart (Fig. 4A, B, and C). The blockade of the impulse-generating capacity of these cells by adenosine may account for the AV blockade induced by this agent.

Conflicting results have been reported concerning the effects of adenosine on the contractility of ventricular myocardium, and it has been proposed that the reason for this controversy resides in the fact that adenosine exerts a depression of the catecholamine-induced positive inotropism as well as a stimulation of cardiac adenylate cyclase (37). Furthermore, there is suggestive evidence consistent with

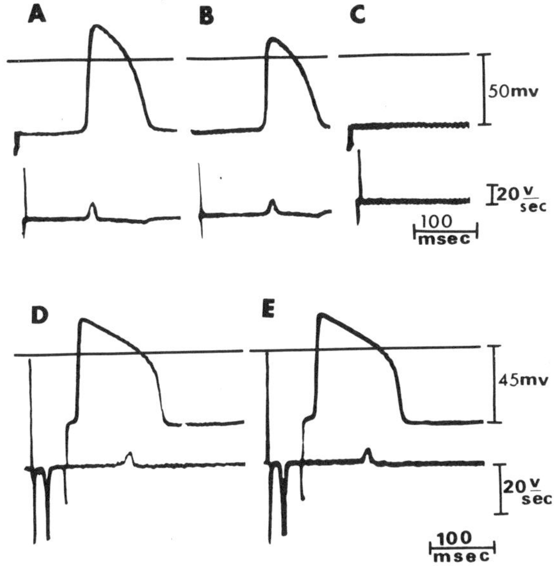

FIG. 4. A, B, and **C:** Depressant effect of adenosine on the slow Ca^{2+} action potential of the rabbit AV nodal cells. These cells normally generate this type of action potential. Upper traces are the action potential and the lower traces are the first derivative. **A:** Control, **B** and **C:** 1 and 2 min, respectively, after adding adenosine to the bathing solution to a final concentration of 20 μM. **D** and **E:** Lack of effect of adenosine on the slow action potentials of guinea pig ventricular cells depolarized with 24 mM K^+ and treated with 1 μM isoproterenol. **D:** Control. **E:** After treatment with 50 μM adenosine.

the idea that cAMP acts as a mediator of catecholamine-induced increases in Ca^{2+} currents and it has been proposed that altered phosphorylation of membrane channels is the mechanism that links cAMP to increased inward slow current (46). Hence, it would be expected that adenosine would depress catecholamine-induced slow action potentials in ventricular myocardium. However, such an effect was not observed in the guinea pig papillary muscle (Fig. 4D and E). This finding indicates that the negative inotropism exerted by adenosine in catecholamine-treated muscle is not mediated by affecting the influx of Ca^{2+} associated with the action potential, and that the mechanism must reside either at the level of the sarcoplasmic reticulum (19), or at the level of the contractile proteins (31). Furthermore, since catecholamine-stimulated cAMP levels are depressed by adenosine without depression of the slow action potential, these observations are not in accordance with the concept that elevated cAMP is a requirement for the induction of the slow action potentials, at least in ventricular muscle.

In summary, adenosine depresses the slow Ca^{2+}-mediated action potentials of some cardiac tissues whether or not these action potentials are catecholamine induced. The implication is that this action of adenosine may not be closely coupled to the effects of catecholamines.

POSSIBLE SIGNIFICANCE OF ADENOSINE ACTION AND RELEASE ON CARDIAC FUNCTION

Understanding the significance of adenosine release on tissue function depends on several factors: (a) whether or not endogenously produced adenosine can elicit the same responses as exogenous adenosine; (b) knowledge of the mechanism of action of adenosine; (c) elucidation of the mechanisms that control the release of adenosine; and (d) demonstration that the proposed mechanism for the control of adenosine release operates *in vivo*.

An endogenous function of adenosine exists most likely in fat cells (40). Recently Dobson (9) showed that hypoxic slices of rat ventricular muscle (a tissue with high adenosine levels) failed to increase cAMP levels in response to catecholamine. However, previous incubation of the slices in a medium containing adenosine deaminase partially restored the response, indicating that endogenous adenosine prevented the tissue from responding to the catecholamine.

Possible Mechanisms Controlling Adenosine Release

5′Nucleotidase hydrolyzes 5′AMP *in vivo* at about 1/100 the rate observed *in vitro;* accordingly the question arises as to the control of adenosine production and its coupling to oxygen supply and use (13,25,33,34; and Berne et al., *this volume)*.

Role of the Balance of Adenine Nucleotides and Mg^{2+} as Regulators of 5′-Nucleotidase Activity

A possible role of the adenine nucleotides (1) in the regulation of 5′-nucleotidase *in vivo* was suggested by Baer et al. (2), since ATP strongly inhibited this enzyme. Other investigators (32,33) have also proposed that reduction in ATP associated with a myocardial oxygen imbalance leads to enhancement of adenosine release. This proposal has been criticized (25) on the basis that changes in the levels of ATP in the heart are not detected under conditions in which adenosine levels are high (24,25). Furthermore, a decrease in ATP leads to an increase in ADP which has been shown to be a more powerful inhibitor of 5′-nucleotidase than ATP (7). Consequently, ATP hydrolysis should result in greater inhibition of 5′-nucleotidase. However, the increase in AMP, plus the decrease in ATP, could act synergistically to overcome the increased inhibition by ADP (32).

Atkinson (1) proposed the energy charge of the adenylate pool, which he defines as a parameter intended to describe quantitatively the energy status of the cell:

$$\text{Energy charge (E)} = \frac{\text{ATP} + \frac{1}{2}\text{ ADP}}{\text{ATP} + \text{ADP} + \text{AMP}}$$

In the cell, the sum of ATP + ADP + AMP tends to remain relatively constant, but the relative concentrations of each nucleotide would be determined by the energy status of the system and the equilibrium constant of the myokinase reaction:

$$2\ \text{ADP} \rightleftarrows \text{ATP} + \text{AMP};\ K = \frac{[\text{ATP}][\text{AMP}]}{[\text{ADP}]^2}$$

Therefore, a more meaningful approach to the study of the regulation of 5′-nucleotidase would be to change the concentrations of ATP, ADP, and AMP simultaneously according to the myokinase equilibrium reaction and use the expression of energy charge as an independent variable.

We have performed such studies using 5′-nucleotidase, isolated from rat hearts, as described by Baer et al. (2) and with commercially available 5′-nucleotidase from snake venom (Sigma type III). The results were similar with both enzymes and are illustrated in Fig. 5. As the energy charge decreased, the rate of formation of adenosine increased rapidly at first and then more gradually. Accordingly, adenosine was formed at a faster rate despite an increase in ADP. The energy charge *in vivo* can be estimated to be between 0.95 and 0.8, where a small change in E results in a large change in adenosine production. However, this relationship can be modulated by the concentration of Mg^{2+} (Fig. 5B). Low concentrations of Mg^{2+} depress the production of adenosine whereas high concentrations accelerate it until a maximum activation is reached. Sullivan and Alpers (42) have suggested that Mg^{2+} reduces inhibition of 5′-nucleotidase by complexing with ATP or ADP, as well as by a direct action. However, as shown in Fig. 5B, modulation of inhibition requires Mg^{2+} concentrations in excess of ATP and ADP and thus may be of minor importance *in vivo* where its concentration has been estimated to be below 0.5 to 1 μM (30). Evaluation of 5′-nucleotidase regulation *in vivo* is difficult because only small changes in ATP levels or the energy charge are associated with adenosine release (33).

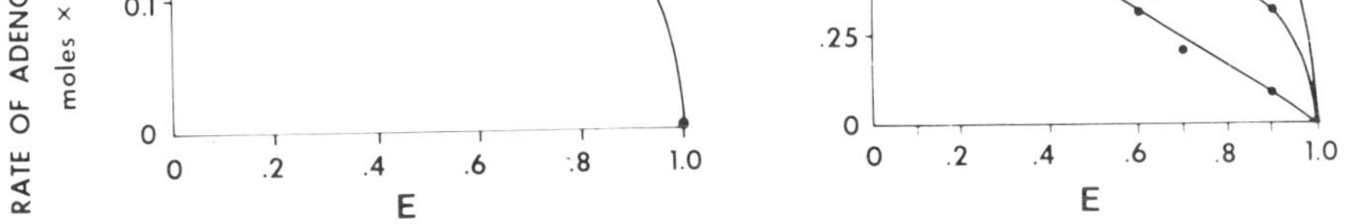

FIG. 5. Increase in 5′-nucleotidase activity *(ordinate)* associated with a decrease in energy charge [E = (ATP + ½ ADP)/(ATP + ADP + AMP) *(abscissa)*]. **A:** Rat heart 5′-nucleotidase. **B:** Snake venom 5′-nucleotidase. In both cases the total amount of nucleotides was 2.5 mM. For the cardiac enzyme Mg^{2+} concentration was 7.5 mM. For the snake venom enzyme various concentrations of Mg^{2+} were used.

Role of Creatine Phosphate in the Control of 5′-nucleotidase

Dr. Michael Mahler suggested to us that perhaps 5′-nucleotidase was also importantly regulated by creatine phosphate, which is known to change significantly with myocardial oxygen imbalance (10). Indeed, we found that creatine phosphate strongly inhibited 5′-nucleotidases and that this action could be abolished almost fully by Mg^{2+} (Figs. 6–8). In these experiments creatine phosphate concentrations

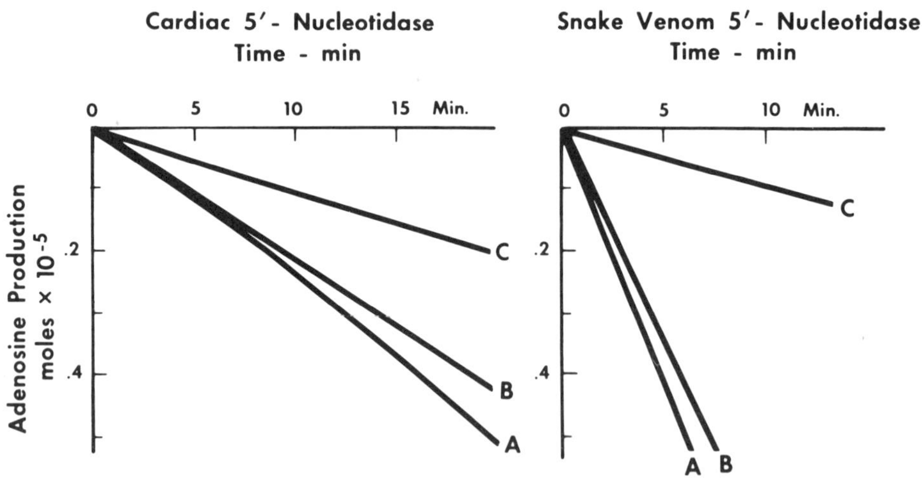

FIG. 6. Inhibition of cardiac *(left panel)* and snake venom *(right panel)* 5′-nucleotidase by creatine phosphate (CrP) at low Mg^{2+}, and removal of this inhibition by addition of Mg^{2+}. Enzymes were incubated with an excess of adenosine deaminase. Lines correspond to superimposed recordings of the decrease in absorbance taking place at 265 nm when 5′-AMP was hydrolyzed to adenosine under various conditions. Reaction time was started by addition of 5′-nucleotidase (zero time). **A:** Control traces at 0.02 mM Mg^{2+}. **B:** Traces in 10 mM CrP and 1 mM Mg^{2+}. **C:** Traces in the presence of 10 mM CrP and 0.02 mM Mg^{2+}.

are within the range estimated for brain (5 mM), heart (14 mM), and skeletal muscle (25 mM) (6,10,14,21,29,47). In view of this one can speculate that skeletal muscle would require prolonged periods of contraction in order to deplete its creatine phosphate stores sufficiently to remove the inhibition of the enzyme, and this may explain why elevation of adenosine levels in skeletal muscles requires at least 5 min sustained contraction (5).

Fine Modulation of the Inhibitory Effect of Creatine Phosphate by Free Mg^{2+}

The removal of the creatine phosphate–induced inhibition by Mg^{2+} (Fig. 6) indicates that control of adenosine release *in vivo* could be based on a creatine phosphate–Mg^{2+} interaction at the level of the enzyme. Therefore, studies of the effects of Mg^{2+} on the inhibition induced by various concentrations of creatine phosphate were performed and are illustrated in Fig. 8. The results of two separate

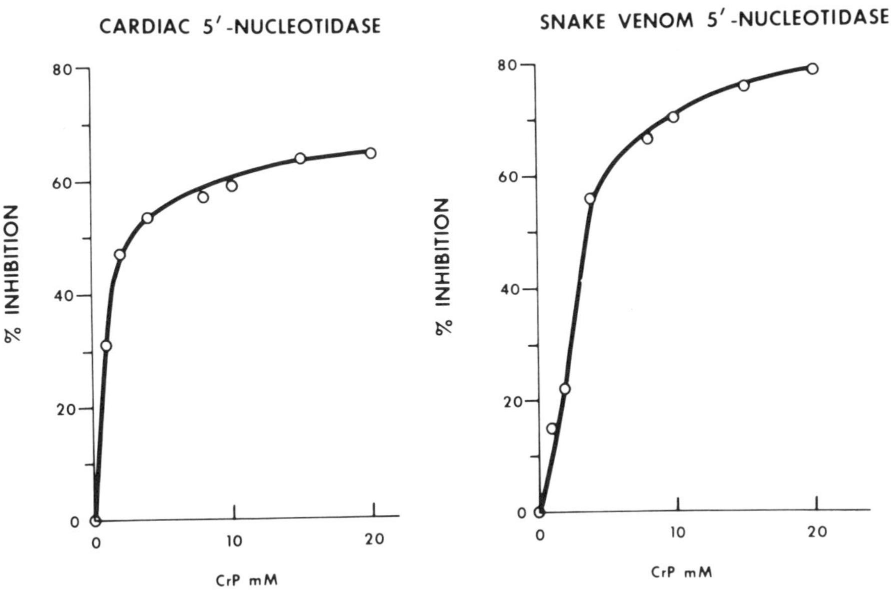

FIG. 7. Inhibitory effects of different concentrations of creatine phosphate on the cardiac and snake venom 5'-nucleotidases in the presence of 0.02 μM Mg^{2+}. One hundred percent represents zero enzyme activity. CrP = creatine phosphate.

experiments with snake venom enzyme are shown in panels A and B. For each curve, the concentration of creatine phosphate remained constant. Enzyme activity expressed as percent inhibition (ordinates) was varied by changing the Mg^{2+} concentration. The results are plotted on a semilogarithmic scale. For each creatine

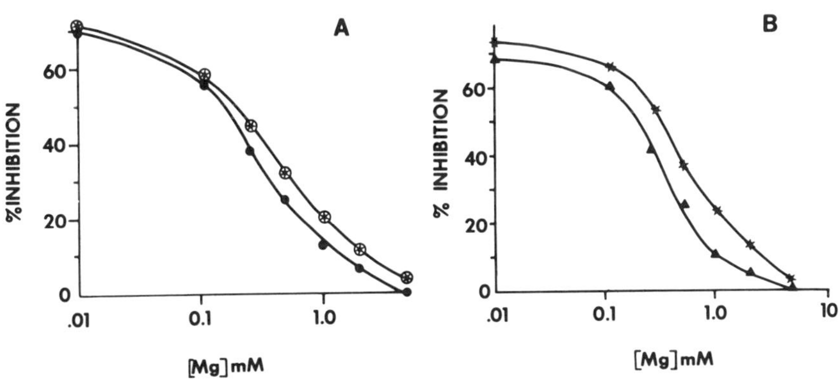

FIG. 8. Removal by Mg^{2+} of 5'-nucleotidase inhibition caused by creatine phosphate. The various curves were obtained at four different creatine phosphate concentrations. Panels **A** and **B** correspond to two separate experiments. Panel **A** upper and lower curves, 8 and 4 mM CrP, respectively. Panel **B** upper and lower curves, 16 and 8 mM CrP, respectively. The largest removal of the inhibition in all cases occurred with 0.1 to 1.0 mM Mg^{2+}.

phosphate concentration an increase in Mg^{2+} was associated with a decrease in the enzyme inhibition which followed a sigmoidal curve with the inflection point at about 0.4 mM Mg^{2+} which is within the range at which the largest degree of modulation occurs. It is of interest that the estimated concentrations of intracellular free Mg^{2+} in cardiac muscle fall within this range (30). As shown in Fig. 8, a decrease in creatine phosphate displaces the curve downward and to the left whereas an increase causes a shift upward and to the right.

How a creatine phosphate–Mg^{2+} interaction could act as a regulatory mechanism to control adenosine release is illustrated in Fig. 9 which shows a family of hypothetical curves (solid lines) for various creatine phosphate concentrations. If it is assumed that as creatine phosphate is hydrolyzed a small but undetectable amount of ATP is also hydrolyzed, this will result in the release of ATP-chelated Mg^{2+}. The result would be a decreased inhibition by three mechanisms: (a) decrease in creatine phosphate; (b) a decrease in ATP; and (c) an elevated free Mg^{2+}. As creatine phosphate decreases and free Mg^{2+} increases, a greater sensitivity to changes in free Mg^{2+} would result. A decrease of 0.01 in energy charge within the 0.9 level estimated to exist *in vivo* (1,33) with an estimated sum of adenylates of 5mM would result in an increase in free Mg^{2+} of approximately 0.1 mM. This represents the

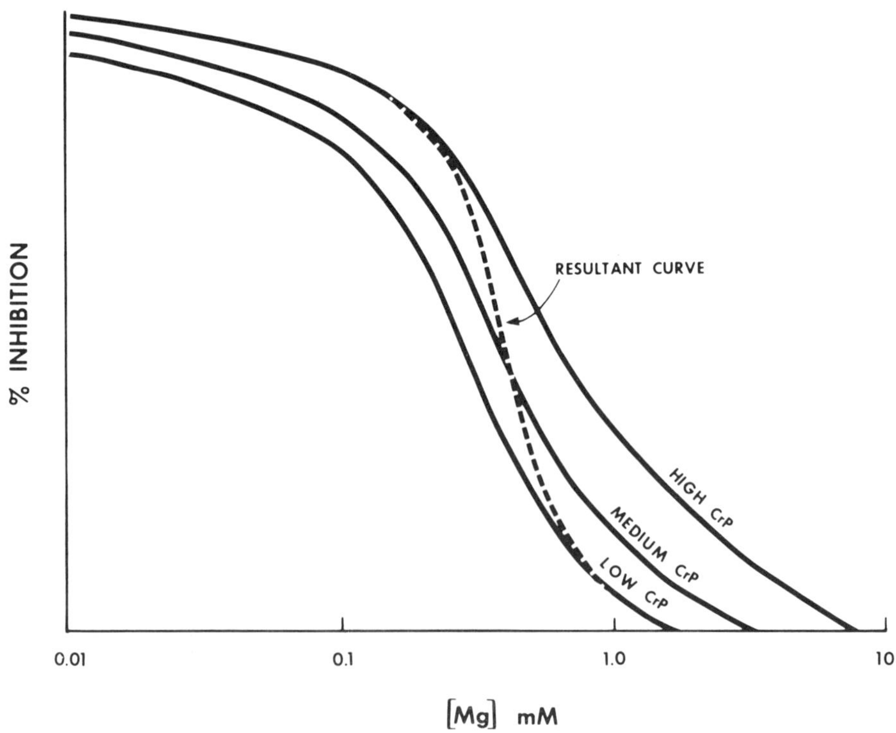

FIG. 9. Hypothetical curves of the removal by Mg^{2+} of inhibition of 5′-nucleotidase caused by various concentration of creatine phosphate (see text for discussion).

hydrolysis of about 0.1 μmole/g of ATP, a change that cannot be detected by available procedures. Thus the decrease in energy charge of 0.035 that we estimated to occur in association with an eightfold increase in adenosine levels could be attributed to an important disinhibition of the 5′-nucleotidase enzyme by the mechanism described above.

As noted above, a modulatory role for Mg^{2+} similar to that proposed here has been suggested by Sullivan and Alpers (42) for the removal of the ADP- and ATP-induced inhibition of cardiac 5′-nucleotidase. However, the removal of the inhibition imposed by the ADP-Mg and ATP-Mg complexes is not a fine control mechanism because in order to detect significant disinhibition, concentrations of Mg^{2+} of about 100 times that of inhibitory nucleotide are required. If one estimates a cellular concentration of ADP and ATP of 0.5 mM and 5 mM, respectively, it means that concentrations of 50 mM and 500 mM Mg^{2+} would be needed which do not exist in the cell (30). Thus, *in vivo,* the modulation of the adenylate-induced 5′-nucleotidase inhibition by Mg^{2+} may be restricted to the formation of the nucleotide-Mg complexes which are less powerful inhibitors than the free nucleotide.

An Integrated Scheme of the Modulatory Role of Adenine Nucleotides, Creatine Phosphate, and Free Mg^{2+}

The scheme shown in Fig. 10 illustrates our views on how adenosine release may be controlled and how it affects various functions. This diagram is a modification of one published previously (32). In this scheme the rate of adenosine release is controlled by a balance of the cellular energy stores and its modulation by free Mg^{2+}. Our *in vitro* studies with 5′-nucleotidase demonstrate that the energy charge and the levels of creatine phosphate represent inhibitory influences at the input of the integrator as indicated by the minus sign. These two input signals are in turn modulated by the concentration of free Mg^{2+} which is represented as a variable resistor whose value is proportional to the Mg^{2+} concentration. The modulatory role of Mg^{2+} at the creatine phosphate input is one of very fine control, in contrast to that at the level of energy charge input. The Mg^{2+} concentration is in turn determined by the energy charge, mostly because of chelation by ATP. In this scheme adenosine would tend to repress certain energy-consuming cell functions such as atrial contraction, atrial AV node electrical activity, ventricular hormonal responses, and vascular smooth muscle tone.

Steady State Versus Transient Release of Adenosine

Considerable evidence is available to indicate that adenosine release in cardiac muscle maintains a balance with those mechanical parameters (dP/dt, heart rate × aortic pressure) that determine oxygen consumption (12; and Berne et al., *this volume*). However, in these studies adenosine production was measured under steady state conditions and they carry the assumption that per contraction a certain amount of adenosine is released proportional to the amount of energy consumed. If the amount of energy consumed per contraction increases, this would have to be paralleled by adenosine release. This would result in the proportionality observed between adenosine release and the different degrees of cardiac performance in steady state conditions. In addition, it is known that throughout the period of contraction energy is consumed only during the period of time between the onset of systole to peak tension (23). Furthermore, all evidence suggests that in cardiac muscle the activation of the energy-generating mechanisms occurs almost simultaneously with the development of contraction (15). Since during the single muscle contraction adenosine production may result from a slight imbalance that occurs between energy-utilizing and -regenerating processes, then a transient enhancement of adenosine release would be expected.

We have attempted to attack the difficult question of whether or not there is enhancement of adenosine release during contraction. Since the period of enzyme inactivation during sampling (freezing) cannot be reduced below 50 msec we have increased the duration of contraction by lowering the temperature from 37° C to 33° C and by replacing Ca^{2+} with Sr^{2+} in isolated perfused guinea pig hearts, resulting in a contraction four to six times slower than normal. To be able to sample the heart at various points during the cardiac cycle an automated system was designed (Fig.

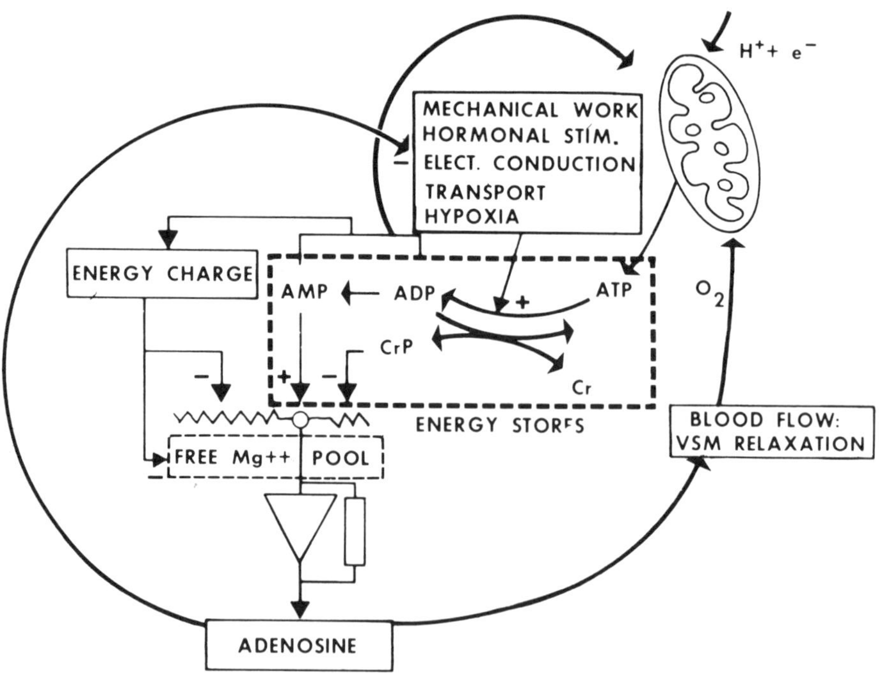

FIG. 10. Schematic representation of the relationships between cellular energy stores, free Mg^{2+} pool, and adenosine formation in the regulation of certain functions (see text for discussion).

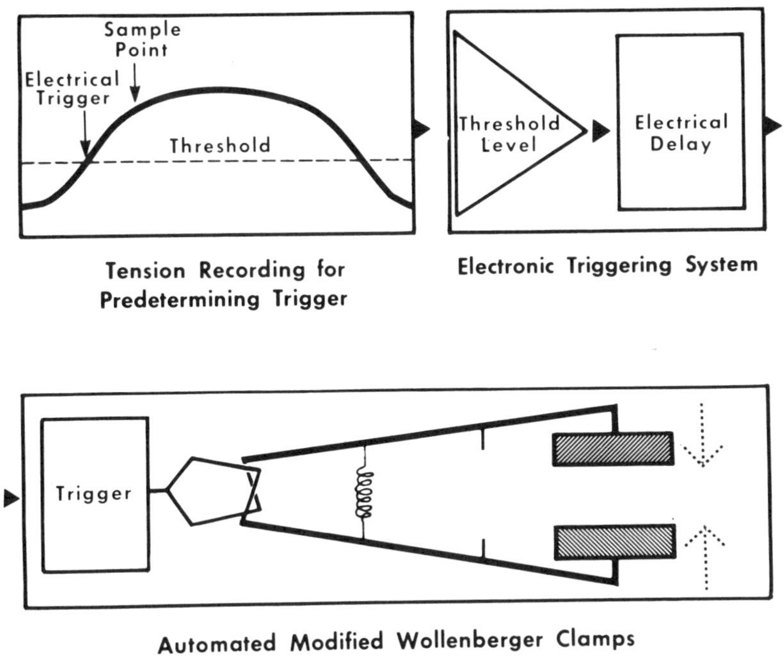

FIG. 11. Diagram of the automated freezing apparatus designed to arrest enzymatic activity in the heart at various points during the cardiac cycle.

11) with electronically triggered, modified Wollenberger clamps. The electronic triggering device can be activated by the mechanical event by means of a variable triggering level and the use of a variable time delay and is capable of sampling tissue at predetermined points in the contractile cycle (44). Illustration of actual tracings showing the moments at which tissue freezing occurred is shown in Fig. 12. Tissue analysis indicated that the sum of the tissue levels of adenosine, inosine, and hypoxanthine paralleled tension development (Fig. 13) and regression analysis shows a significant correlation between these two parameters. However, adenosine tissue levels failed to show a statistically significant parallelism with tension development. Failure to show such a pattern is possibly the result of the large variability we have observed with the adenosine determinations. Nevertheless, since the formation of adenosine as well as inosine and hypoxanthine requires, as a first step, activation of 5′-nucleotidase, the results shown in Fig. 13 are indicative that the enzyme can respond rapidly and that it may be capable of following the transient changes in energy stores that are assumed to occur during a single contraction.

In summary, adenosine is normally released by cardiac tissues and may act as a repressor of functions that trigger its release. The release of adenosine appears to be controlled by the balance of the energy stores and this process may be finely modulated by the concentration of free Mg^{2+}.

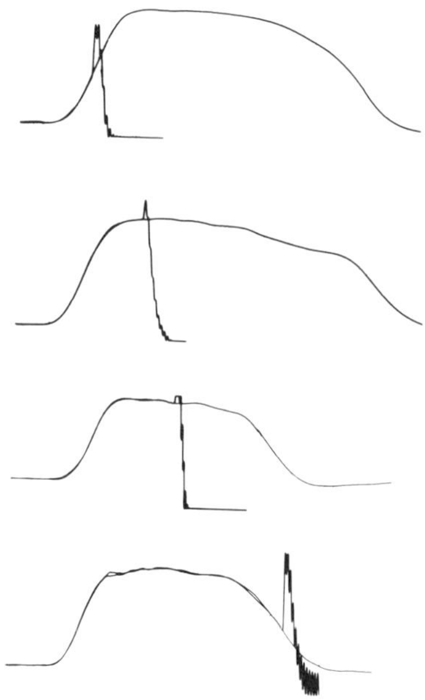

FIG. 12. Recording of contractions of guinea pig ventricles perfused with Sr^{2+}-Krebs-Henseleit buffer at 33° C. Under these conditions contraction lasted approximately 1 sec. The moment of freezing is indicated by the disturbance of the smooth tracings. The freezing instrument was programmed to freeze the hearts at these specific time points shown on the recordings.

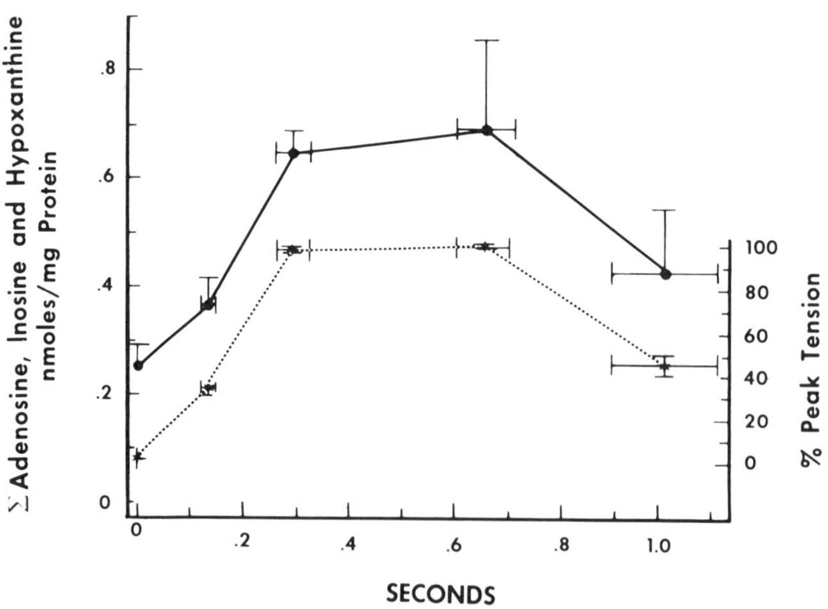

FIG. 13. Parallelism between cardiac levels of the sum of adenosine, inosine, and hypoxanthine *(solid line)* and the tension developed by guinea pig ventricles as percent of peak tension *(broken line)*. Hearts were frozen at five different points during ventricular contraction. Zero time indicates initiation of contraction. The increase in the sum of purines associated with the tension development suggests that there is activation of 5′-nucleotidase during contraction.

ACKNOWLEDGMENTS

This research was supported by USPHS Grant no. HL10384. Dr. Luiz Belardinelli is the recipient of fellowship no. 1112.1273/76 from Conselho Nacional de Desenvolvimento Cientifico e Tecnologico (CNPq) and Fundacao Universitaria de Cardiologia do Rio Grande do Sul, Brasil.

REFERENCES

1. Atkinson, D. E. (1968): *Biochemistry,* 7:4030–4034.
2. Baer, H. P., Drummond, G. I., and Duncan, E. L. (1966): *Mol. Pharmacol.,* 2:67–76.
3. Belardinelli, L. Rubio, R., and Berne, R. M. (1977): *Fed. Proc.,* 36:435.
4. Belardinelli, L., Rubio, R., and Berne, R. M. (1979): *Pfluegers Arch.* (In press.)
5. Bockman, E. L., Berne, R. M., and Rubio, R. (1976): *Am. J. Physiol.,* 230:1531–1537.
6. Boerth, R. C., Covell, J. W., Seagren, S. C., and Pool, P. E. (1969): *Am. J. Physiol.,* 216:1103–1106.
7. Burger, R. M., and Lowenstein, J. M. (1970): *J. Biol. Chem.,* 245:6274–6280.
8. DeGubareff, T., and Sleator, W., Jr. (1965): *J. Pharmacol. Exp. Ther.,* 148:202–214.
9. Dobson, J. G., Jr. (1978): *Fed. Proc.,* 37:782.
10. Dunn, R. B., and Griggs, D. M. (1977): *Cir. Res.,* 37:438–445.
11. Dunn, R. B., Hickey, K. M., and Griggs, D. M. (1977): *Fed. Proc.,* 36:598.
12. Foley, D. H., Herlihy, J. T., Thompson, C. I., Rubio, R., and Berne, R. M. (1978): *J. Mol. Cell. Cardiol.,* 10:293–300.
13. Frick, P. G., and Lowenstein, J. M. (1976): *J. Biol. Chem.,* 251:6372–6378.
14. Gerlach, E., and Janke, J. (1958): *Biochem. Z.,* 330:565–570.
15. Gibbs, C. L. (1974): In: *The Mammalian Myocardium,* edited by G. A. Langer and A. Brady, p. 105. Wiley & Sons, New York.
16. Ginsborg, B. L., and Hirst, G. D. S. (1972): *J. Physiol.,* 224:629–645.
17. Harder, D., Belardinelli, L., Sperelakis, N., Rubio, R., and Berne, R. M. (1977): *Circ. Res. (in press).*
18. James, T. N. (1965): *J. Pharmacol. Exp. Ther.,* 149:233–247.
19. Katz, A. M., Tada, M., and Kirchberger, M. A. (1975): *Adv. Cyclic Nucleotide Res.,* 5:453–472.
20. Kuroda, Y., Saito, M., and Kobayashi, K. (1976): *Proc. J. Acad.,* 52:86–89.
21. Medina, M. A., Jones, D. J., Stavinoha, W. B., and Ross, D. H. (1975): *J. Neurochem.,* 24:223–227.
22. Miller, W. L., Thomas, R. A., Berne, R. M., and Rubio, R. (1978): *Circ. Res.,* 43:390–397.
23. Monroe, R. G. (1964): *Circ. Res.,* 14:294–299.
24. Olsson, R. A. (1970): *Circ. Res.,* 26:301–306.
25. Olsson, R. A., and Patterson, R. E. (1976): *Prog. Mol. Subcell. Biol.,* 4:222–248.
26. Pappano, A. J. (1970): *Circ. Res.,* 27:379–391.
27. Phillis, J. W. (1977): *J. Can. Sci. Neurol.,* 4:151–197.
28. Phillis, J. W., and Kostopoulos, G. K. (1975): *Life Sci.,* 17:1085–1094.
29. Piiper, J., and Spiller, P. (1970): *J. Appl. Physiol.,* 28:657–666.
30. Polimeni, P. I., and Page, E. (1973): *Circ. Res.,* 23:367–374.
31. Rubio, R., Bailey, C., and Villar-Palasi, C. (1975): *J. Cyclic Nucleotide Res.,* 1:143–150.
32. Rubio, R., and Berne, R. M. (1969): *Circ. Res.,* 25:407–415.
33. Rubio, R., and Berne, R. M. (1975): *Prog. Cardiovasc. Dis.,* 18:105–122.
34. Rubio, R., Berne, R. M., Bockman, E. L., and Curnish, R. R. (1975): *Am. J. Physiol.,* 228:1896–1902.
35. Rubio, R., Berne, R. M., and Katori, M. (1969): *Am. J. Physiol.,* 216:56–62.
36. Sattin, A. (1971): *J. Neurochem.,* 18:1087–1096.
37. Schrader, J., Baumann, G., and Gerlach, E. (1977): *Pfluegers Arch.,* 372:29–35.
38. Schrader, J. Nees, S., and Gerlach, E. (1977): *Pfluegers Arch.,* 369:251–257.
39. Schrader, J., Rubio, R., and Berne, R. M. (1975): *J. Mol. Cell. Cardiol.,* 7:427–433.
40. Schwabe, U., Ebert, R., and Erbler, H. C. (1975): *Adv. Cyclic Nucleotide Res.,* 5:569–584.
41. Stafford, A. (1966): *Br. J. Pharmacol. Chemother.,* 28:218–227.

42. Sullivan, J. M., and Alpers, J. B. (1971): *J. Biol. Chem.*, 246:3057–3063.
43. Thomas, R. A., Rubio, R., and Berne, R. M. (1975): *J. Mol. Cell. Cardiol.*, 7:115–123.
44. Thompson, C. I., Rubio, R., and Berne, R. M. (1976): *Physiologist*, 19:389.
45. Thyrum, P. T. (1974): *J. Pharmacol. Exp. Ther.*, 188:166–179.
46. Tsien, R. W. (1977): *Adv. Cyclic Nucleotide Res.*, 8:363–420.
47. Veech, R. L., Harris, R. L., Veloso, D., and Veech, E. H. (1973): *J. Neurochem.*, 20:183–188.
48. Weidmann, S. (1955): *J. Physiol. (Lond.)*, 129:568–582.
49. Zipes, D. P., and Mendez, C. (1973): *Circ. Res.*, 32:477–454.

Physiological and Regulatory Functions of Adenosine and Adenine Nucleotides, edited by H. P. Baer and G. I. Drummond. Raven Press, New York © 1979.

Cardiovascular Effects of Adenosine 5′-Carboxylates and Carboxamides

Herman H. Stein and *Raj N. Prasad

*Abbott Laboratories, North Chicago, Illinois 60064; and *Abbott Laboratories, Limited, Montreal, Quebec, H4P 1A5, Canada*

The dilator activity of adenosine and adenine nucleotides on the coronary vasculature was reported as early as 1929 by Drury and Szent-Györgi (5), and other investigators (6,17,18) have confirmed the observation. Berne and co-workers (2,4) have postulated that adenosine may be an important physiological regulator of blood flow. The effects of exogenously administered compounds are of short duration, however, because their concentrations are controlled by preexisting enzymes and regulatory systems. Adenosine, for example, is taken up rapidly by red blood cells and other tissues (7,11,12) and converted to inosine (1) or adenosine 5′-monophosphate (5′-AMP).

In an attempt to obtain analogs with a longer duration of action, esters and amides of adenosine 5′-carboxylic acid were synthesized and tested for their cardiovascular effects (13,15,16).

METHODS

Chemistry

Adenosine 5′-carboxylic acid was prepared according to the procedure described by Harmon et al. (8), and esters and amides were obtained from alcohols and amines by standard chemical procedures; a detailed description of the synthesis of the carboxylates has been published (13). Representative structures for the two series of compounds are shown in Fig. 1.

Pharmacology

The compounds were evaluated initially in a cardiovascular test in which blood pressure, heart rate, and coronary sinus PO_2 were monitored continuously on a Grass Model 7 Polygraph (14). Dogs of either sex were anesthetized with 250

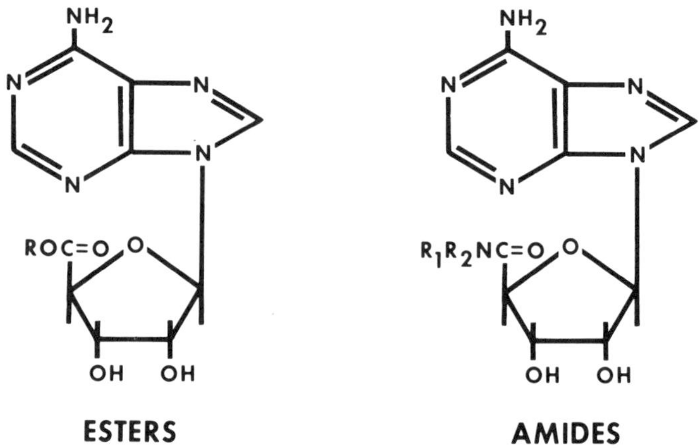

FIG. 1. Chemical structures of experimental compounds.

mg/kg i.v. of barbital sodium after sedation with 3 mg/kg, subcutaneously, of morphine sulfate. Under artificial respiration, an incision was made in the fourth right intercostal space, and the pericardium was incised. A Beckman polarographic PO_2 microelectrode (in a Riley needle) was placed in the coronary sinus via the superior vena cava. Compound, dissolved in isotonic saline, was administered i.v. (jugular); each animal was given only one compound.

Coronary artery blood flow experiments were carried out in a different series of anesthetized dogs in which an incision was made in the left fifth intercostal space, and, under positive pressure artificial ventilation, the pericardial sac was cut and sutured to the chest wall to form a cradle. The left circumflex coronary artery was isolated and a calibrated flow probe (Carolina Medical Electronics) was placed around the vessel near its origin. A suture was placed just distal to the flow probe for producing temporary occlusion of the circumflex flow for the purpose of calibrating the flow probe and inducing a reactive hyperemic response (3) in the dog. Compounds were administered via the femoral vein in isotonic saline.

Blood pressure studies were carried out also in unanesthetized, spontaneously hypertensive rats (SHR) (10); the animals were placed in a warming box at 33° C for 30 min and the systolic blood pressure was measured indirectly, utilizing an inflatable tail cuff sphygmomanometer and audio detection of the systolic pulse. Test compound was administered p.o. in isotonic saline.

RESULTS AND DISCUSSION

The effects of the esters and amides on coronary sinus PO_2, blood pressure, and heart rate are listed in Tables 1 and 2, respectively. In the simple aliphatic ester series, the ethyl derivative was the most potent, being more than 10 times as active in the PO_2 test as the methyl, propyl, or butyl esters. Coronary sinus PO_2 is directly related to coronary flow when oxygen extraction by the myocardium does not alter

TABLE 1. *Effect of esters of adenosine 5'-carboxylic acid on cardiac parameters in anesthetized dogs*[a]

		PO_2[b] (mm Hg)		Mean aortic pressure (mm Hg)		Heart rate (beats/min)	
Ester (R)[c]	Dose (mg/kg, i.v.)	Change (%)	Duration (min)	Change (%)	Duration (min)	Change (%)	Duration (min)
Methyl	1	13	30	−8	30	0	0
Ethyl	0.1	145	50	−5	26	17	50
i-Propyl	1	29	60	−6	60	2	60
n-Propyl	1	47	60	0	0	3	5
Propargyl	1	−5	14	−7	14	9	15
n-Butyl	1	8	20	0	0	0	0
Phenethyl	10	9	18	−8	9	19	11

[a] Peak changes are shown; the duration is the total time elapsed from this initiation of the change until return to pretest value. The range of initial basal values of the preparations was: PO_2, 15 to 25 mm Hg; blood pressure, 90 to 125 mm Hg; heart rate, 95 to 140 beats/min.
[b] Partial pressure of oxygen.
[c] See Fig. 1.

markedly, and increases in this parameter were utilized as the initial indicator for activity as a coronary vasodilator. Adenosine 5'-carboxylic acid, the major metabolite of the ethyl ester in the dog, rat, and mouse (9) was inactive at 1.0 mg/kg, 10 times the active dose of the ethyl ester. The PO_2 increases occurred within 5 sec after i.v. administration. In general, only small decreases in blood pressure were

TABLE 2. *Effect of amides of adenosine 5'-carboxylic acid on cardiac parameters in anesthetized dogs*[a]

		PO_2[b] (mm Hg)		Mean aortic pressure (mm Hg)		Heart rate (beats/min)	
Amide (R_1)[c]	Dose (mg/kg, i.v.)	Change (%)	Duration (min)	Change (%)	Duration (min)	Change (%)	Duration (min)
Unsubstituted	0.01	79	25	−8	20	38	25
Methyl	0.01	105	14	−18	60	27	60
Dimethyl	1.0	0	0	0	0	0	0
Ethyl	0.01	250	300	−16	120	74	300
Cyclopropyl	0.01	136	150	−39	150	66	150
i-Propyl	0.01	43	240	−5	70	29	240
n-Propyl	0.1	61	120	−14	37	47	120
Allyl	0.1	53	120	−12	45	29	120
Cyclobutyl	0.01	100	120	−45	90	4	120
n-Butyl	0.1	30	40	−5	40	16	40
Cyclopropylmethyl	0.01	27	60	0	0	4	60
Piperidino	1.0	0	0	0	0	0	0
Phenyl	0.1	4	15	−5	4	0	0

[a,b] See Table 1 for explanations.
[c] See Fig. 1; R_2 = H, except for dimethyl and piperidino. $R_1 = R_2$ = H for unsubstituted amide.

observed, and the respective slight increases in heart rate were probably compensatory to the blood pressure change. Unsaturation caused a loss of activity (compare *n*-propyl and propargyl esters), whereas the aromatic phenethyl ester was less than 1/100 as potent as the ethyl ester.

The amides (Table 2) were generally more potent than the esters; analogous to the latter series, the ethyl derivative was the most active in the PO_2 test, although the structure-activity profile was somewhat broader with the amides. In addition to large increases in coronary sinus PO_2, markedly prolonged hypotension and tachycardia were observed. The inactivity of the dimethyl amide is noteworthy. It appears that a hydrogen atom is required at the amide nitrogen atom for binding to the receptor, or for orientation of the purine moiety into a more favorable conformation. This effect was confirmed with the piperidino carboxamide which also lacks a proton on the amide N and which was also inactive in the test. Aromatic amides, represented by the phenyl derivative, were much less active than the aliphatic amides.

The hypotensive properties of the amides were also observed in SHR. These data are shown in Table 3. The cyclopropyl carboxamide was the most potent on a mg/kg basis, but marked decreases in blood pressure were obtained with higher doses of the less potent compounds such as the allyl carboxamide. The degree and duration of the hypotension was dose dependent and correlated generally with activity in the PO_2 test.

The coronary vasodilator properties of three compounds, deduced from the increases in coronary sinus PO_2, were verified by direct measurement in dogs, and these data, the averages of two to four animals per compound, are shown in Table 4; the durations of the effects are also given. Results obtained with adenosine are included for purposes of comparison. In every test, the increase in flow at the cited dose was equal to 65 to 90% of the reactive hyperemic response (3) of the particular animal, indicative of near-maximal dilation. The greater potency of the amides compared with the ethyl ester is apparent, as is the much shorter duration of activity of adenosine.

Previous work has demonstrated that aminophylline exerts a marked inhibitory effect, as it does with adenosine, on the increases in coronary blood flow caused by the ethyl ester of adenosine 5′-carboxylic acid, and that neither the ethyl ester nor the ethyl amide are substrates for, or are inhibitory to, adenosine or adenylate

TABLE 3. *Hypotensive effects of carboxamides in SHR*

Amide[a]	Dose (*mg/kg, p.o.*)	Decrease in systolic blood pressure[b] (%)
Unsubstituted	5	6 ± 3
Ethyl	0.5	14 ± 2
Cyclopropyl	0.5	28 ± 6
Allyl	3	23 ± 3

[a] See Table 2 and Fig. 1 for explanation.
[b] Mean of four rats ± SEM, 2 hr after dose.

TABLE 4. *Effect of selected derivatives on circumflex coronary artery blood flow in dogs*

Derivative and compound type	Dose (mg/kg, i.v.)	Maximum[a] increase in flow (%)	Duration of increase (min.)
Ethyl ester	0.050	285	30
Cyclopropyl amide	0.0004	180	15
Cyclobutyl amide	0.0016	290	25
Adenosine	0.3	160	2

[a] Range of control values; 12 to 27 ml/min.

deaminase *in vitro* (16). In addition, the adenosine 5′-carboxylates and -carboxamides have no 5′-hydroxyl group available for phosphorylation. All of these factors lend support to the hypothesis that the adenosine 5′-carboxylates and -carboxamides are acting directly on an adenosine receptor, and that their great potency and prolonged duration of action stem at least in part from an increased biological half-life resulting from a lesser degree of metabolism compared with adenosine.

ACKNOWLEDGMENTS

The authors thank D. S. Bariana, A. Fung, and K. Tietje for synthesizing some of the compounds described, H. Brondyk and J. Stansbury for carrying out the experiments in dogs, and J. Kyncl for providing the data in rats.

REFERENCES

1. Baer, H. P., Drummond, G. I., and Duncan, E. L. (1966): *Mol. Pharmacol.*, 2:67–76.
2. Berne, R. M. (1963): *Am. J. Physiol.*, 204:317–322.
3. Berne, R. M., and Levy, M. N. (1972): *Cardiovascular Physiology*. C. V. Mosby, St. Louis.
4. Berne, R. M., Rubio, R., Dobson, J. G., Jr., and Curnish, R. R. (1971): *Circ. Res.*, 29 (Suppl. 1): I-115–I-119.
5. Drury, A. N., and Szent-Györgi, A. (1929): *J. Physiol. (Lond.)*, 68:213–237.
6. Folkow, B. (1949): *Acta Physiol. Scand.*, 17:311–316.
7. Hansing, C. E., Folts, J. D., Afonso, S., and Rowe, G. G. (1972): *J. Pharmacol. Exp. Ther.*, 181:498–511.
8. Harmon, R., Zenarosa, C., and Gupta, S. (1969): *Chem. Ind. (Lond.)*: 1141.
9. Merits, I., and Anderson, D. J. (1973): *Xenobiotica*, 3:381–388.
10. Okamoto, K. (editor) (1972): *Spontaneous Hypertension*. Igaku Shoin, Tokyo.
11. Olsson, R. A., Snow, J. A., Gentry, M. K., and Frick, G. P. (1972): *Circ. Res.*, 31:767–778.
12. Pfleger, K. E., Seifen, E., and Schondorf, H. (1969): *Biochem. Pharmacol.*, 18:43–51.
13. Prasad, R. N., Fung, A., Tietje, K., Stein, H. H., and Brondyk, H. D. (1976): *J. Med. Chem.*, 19:1180–1186.
14. Schoepke, H. G., Darby, T. D., and Brondyk, H. D. (1966): *Pharmacologist*, 8:179.
15. Stein, H. H. (1973): *J. Med. Chem.*, 16:1306–1308.
16. Stein, H. H., Somani, P., and Prasad, R. N. (1975): *Ann. N.Y. Acad. Sci.*, 255:380–389.
17. Wedd, A. M. (1931): *J. Pharmacol. Exp. Ther.*, 41:355–366.
18. Winbury, M. M., Papierski, D. H., Hemmer, M. L., and Hambourger, W. E. (1953): *J. Pharmacol. Exp. Ther.*, 109:255–260.

Physiological and Regulatory Functions of Adenosine and Adenine Nucleotides, edited by H. P. Baer and G. I. Drummond. Raven Press, New York © 1979.

Adenosine Receptors of the Blood Platelet: Interactions with Adenylate Cyclase

*Richard J. Haslam, *Monica M. L. Davidson, *Byron W. G. Lemmex, **Janet V. Desjardins, and ***Brian E. McCarry

*Departments of *Pathology, **Biochemistry, and ***Chemistry, McMaster University, Hamilton, Ontario, Canada, L8S 4J9*

The discovery that adenosine diphosphate (ADP) causes blood platelets to aggregate (14) led to a search for compounds related to ADP that might block this effect. Adenosine was found to be such a compound and was naturally first thought of as an antagonist of a putative ADP receptor (4). However, the inhibitory effects of adenosine and its analogs, such as 2-chloroadenosine, were found to increase after preincubation of these compounds with platelets (3,4). Moreover, adenosine inhibited the aggregating actions of a wide variety of agents including epinephrine, 5-hydroxytryptamine (5-HT), and thrombin (16,43). Initially, the latter effects were thought to be caused by the inhibition by adenosine of the action of ADP released from the platelets, but eventually it was recognized that, in addition to causing the secretion of ADP from platelet granules, most aggregating agents have primary direct aggregating actions that are also inhibited by adenosine (26). Of particular interest was the finding that adenosine prevented aggregation of nonmammalian thrombocytes by thrombin, since these cells are insensitive to ADP (32). Doubts about the hypothesis that adenosine interacted with the ADP receptor led to the suggestion that its action might depend in some way on its uptake and metabolism by the platelets, for example, that competition for metabolic adenosine triphosphate (ATP) between adenosine uptake and some aspect of the aggregation mechanism might inhibit the latter (46). However, subsequent studies showed that inhibitors of adenosine transport do not prevent the inhibition of platelet aggregation by adenosine (17,27,42).

Little progress was made in understanding the action of adenosine on platelets until it was found that prostaglandin (PG) E_1, the most potent inhibitor of platelet aggregation known at that time, activated platelet adenylate cyclase (54) and that adenosine increased adenosine 3′,5′-cyclic monophosphate (cyclic AMP) levels in brain slices (49). These observations triggered the initial studies of the effects of adenosine on platelet cyclic AMP (41) and adenylate cyclase (23), elaboration of which has provided evidence for our present concept of adenosine as an inhibitory agonist in its own right that interacts with a specific platelet membrane receptor, causing activation of platelet adenylate cyclase (17,27). Many lines of evidence indicate that the consequent elevation of platelet cyclic AMP concentrations results in the inhibition of platelet aggregation (18–20). In addition, studies of the actions of adenosine and its analogs on platelet adenylate cyclase have shown that certain deoxyadenosines inhibit the enzyme at an intracellular site. This effect has been used experimentally to establish the role of cyclic AMP in platelet function with greater certainty than has previously been possible (19–21).

EFFECTS OF ADENOSINE AND ITS ANALOGS ON THE ADENYLATE CYCLASE ACTIVITY OF A PARTICULATE FRACTION FROM PLATELETS

Biphasic Effect of Adenosine

Activation of isolated platelet adenylate cyclase by adenosine (23) was observed several years before comparable results were obtained with adenylate cyclase from other biological systems (for example, see refs. 2 and 8). Important factors in obtaining this effect were the use of a mild freezing and thawing procedure to

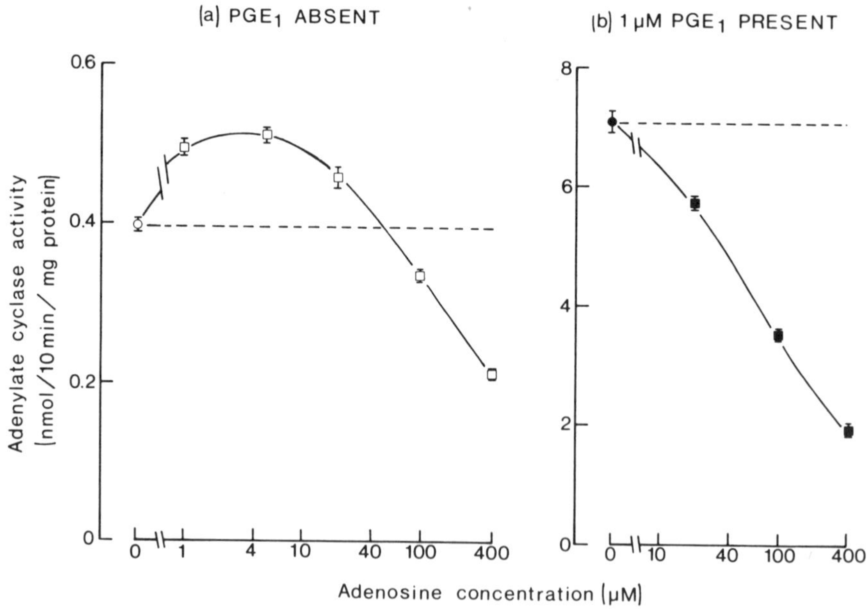

FIG. 1. Effects of different adenosine concentrations on the adenylate cyclase activities of a human platelet particulate fraction assayed in the absence **(a)** and presence **(b)** of 1 μM PGE_1. A 48,000 *g* × 20 min particulate fraction from platelets frozen and thawed only once was used. Assays (10-min incubations at 30° C) were carried out in a medium containing 75 mM Tris-HCl, pH 7.4, 5 mM $MgCl_2$, 1 mM cyclic AMP, 1 mM papaverine, 0.4 mM [^{3}H]ATP, 4 mM phosphocreatine, 8 units of creatine phosphokinase/ml, 1 mg of bovine serum albumin/ml, 0.4 mM dithiothreitol, and 0.4 mM Mg ethylene glycol-*bis*-(β-aminoethyl ether)N, N′-tetraacetate. Values are means ± SEM from three determinations on the same particulate fraction. Further details of the methods used are given elsewhere (21).

disrupt the platelets, omission from the assay medium of methylxanthines, which had been shown to block the increases in cyclic AMP caused by addition of adenosine to brain slices (49), and use of a range of adenosine concentrations. Whereas 1 to 40 μM adenosine caused activation of the enzyme, concentrations above 100 μM reproducibly caused inhibition (Fig. 1a). A maximum increase in enzyme activity of 25 to 50% was obtained with approximately 5 μM adenosine. This activation was very weak compared with the 15- to 20-fold stimulation of platelet adenylate cyclase observed on addition of 1 μM PGE_1. Moreover, in the presence of PGE_1 (Fig. 1b), no additional stimulatory effect of adenosine could be demonstrated and adenosine appeared to have enhanced inhibitory activity. These observations suggested that activation and inhibition of adenylate cyclase by adenosine were two independent and opposing effects with different concentration dependencies, and that, whereas in the presence of PGE_1 only the inhibitory action of adenosine was expressed, in the absence of PGE_1 the adenylate cyclase activity observed represented the sum of both effects (23).

Structure-Activity Relationships

In the initial study on the effects of adenosine on platelet adenylate cyclase, 2-chloroadenosine was found to be at least as potent an activator of the enzyme as adenosine but a weaker inhibitor (23). This suggested that the two actions of adenosine might have different structural determinants. The effects of analogs of adenosine on platelet adenylate cyclase were therefore tested at two concentrations, one that gave optimal activation with adenosine (5 μM) and one that gave approximately 50% inhibition with adenosine (400 μM) (Table 1).

Substitution of adenosine in the 2-position was found to decrease (2-chloroadenosine) or abolish (2-hydrazinoadenosine and 2-azidoadenosine) the inhibitory activity observed with 400 μM adenosine, whereas some or all of the stimulatory activity observed with 5 μM adenosine was retained. 2-Azidoadenosine was synthesized from 2-chloroadenosine via 2-hydrazinoadenosine (50) and was the only compound found that stimulated adenylate cyclase as effectively as adenosine at 5 μM and was also stimulatory at 400 μM (21). Monosubstitution in the N^6 position also gave compounds that lacked inhibitory activity but were less effective than adenosine. As both the 2-substituted and N^6-monosubstituted adenosine derivatives studied have been reported to be effective inhibitors of platelet aggregation (3,5,10,33), these results support the view that cyclic AMP mediates the inhibitory effects on platelets of adenosine and its derivatives. The actions of 2-azidoadenosine are of particular interest since this compound is a potential photoaffinity label (9,10).

In contrast, adenosine derivatives in which the ribosyl moiety was modified by removal of one or more oxygen atoms had no stimulatory activity but were inhibitors of adenylate cyclase more potent than adenosine. Compounds in this group have been found to be relatively ineffective as inhibitors of platelet aggregation (5,21). The two compounds with the greatest inhibitory activity against adenylate cyclase were 2′,5′-dideoxyadenosine (DDA), which has previously been shown

TABLE 1. *Effects of adenosine and its analogs on the adenylate cyclase activity of a human platelet particulate fraction*

Compounds tested	Changes in adenylate cyclase activity (%) 5 μM	400 μM
Adenosine	+30 ± 2[b]	−44 ± 3[b]
2-Chloroadenosine	+25 ± 2[b]	−14 ± 5[a]
2-Hydrazinoadenosine	+15 ± 2[b]	+29 ± 2[b]
2-Azidoadenosine	+34 ± 4[b]	+41 ± 5[b]
N^6-γ,γ-Dimethylallyladenosine	+12 ± 2[b]	+29 ± 2[b]
N^6-Cyclohexyladenosine	+19 ± 2[b]	+37 ± 4[b]
N^6-Phenyladenosine	+24 ± 2[b]	+44 ± 4[b]
2′-Deoxyadenosine	−13 ± 5	−71 ± 3[b]
5′-Deoxyadenosine	− 5 ± 6	−72 ± 2[b]
3′-Deoxyadenosine	−11 ± 4[a]	−79 ± 2[b]
2′,5′-Dideoxyadenosine	−46 ± 2[b]	−86 ± 2[b]
2′-Deoxyadenosine 3′-monophosphate	−55 ± 3[b]	−92 ± 2[b]

Assays were carried out in triplicate or quintuplicate, as described in the legend to Fig. 1. Percent changes in adenylate cyclase activity with respect to controls containing no added nucleoside are given as means ± SE of the differences; significant changes (Student's *t*-test) are indicated. The control adenylate cyclase activity was 0.59 ± 0.01 nmoles/10 min/mg protein.

[a] $2p > 0.05$; [b] $2p > 0.001$.

to inhibit adenylate cyclase from fat, liver, and epithelial cells (11,12,36,55), and 2′-deoxyadenosine-3′-monophosphate, a naturally occurring compound that has also been reported to inhibit adenylate cyclase (47). DDA was slightly more potent in inhibiting platelet adenylate cyclase in the presence than in the absence of PGE_1. Kinetic studies in the presence of PGE_1 indicated that this compound was a noncompetitive inhibitor with respect to MgATP, showed negative cooperativity, and had an IC_{50} of 4 μM under the assay conditions used, which included 5 mM $MgCl_2$ (21). Studies in other laboratories have shown that the inhibitory activities of adenosine and its analogs are greatly potentiated by Mg^{2+} ions (36,53).

These structure-activity studies have confirmed the earlier indications (23,27) that the stimulatory and inhibitory actions of adenosine on platelet adenylate cyclase are two independent effects and suggest that whereas optimal activation of platelet adenylate cyclase requires an intact ribose residue, optimal inhibition of the enzyme requires an unmodified adenine moiety. The structure-activity relationships for activation of platelet adenylate cyclase by adenosine analogs are similar to those reported for elevation of cyclic AMP levels in slices of brain (30) and myocardium (29), whereas the structural requirements for inhibition of the platelet enzyme resemble those observed with liver preparations (36). Our results are also in accord with a recent study by Londos and Wolff (37) of the effects of a somewhat different range of adenosine derivatives on platelet adenylate cyclase. These authors have also suggested the existence of distinct stimulatory and inhibitory adenosine-reactive sites associated with adenylate cyclase.

Effects of Theophylline and Adenosine Deaminase

Addition of theophylline blocked the stimulation of platelet adenylate cyclase by adenosine or its derivatives; 0.5 mM theophylline decreased and 4 mM theophylline abolished the effects of 5 μM adenosine or 2-azidoadenosine (Fig. 2). In addition, 4 mM theophylline inhibited the adenylate cyclase activity of platelet particulate fraction by up to about 40% in the absence of added adenosine (Fig. 2; Table 2). This suggested that the apparent basal enzyme activity was in fact partially stimulated by adenosine present in the enzyme preparation or formed by breakdown of ATP during the assay. Indeed, addition of adenosine deaminase inhibited platelet adenylate cyclase activity by 20 to 30% in the absence of theophylline, but had no further effect when theophylline was present (Table 2). However, 4 mM theophylline consistently inhibited the enzyme to a significantly greater extent than adenosine deaminase, suggesting that theophylline not only blocked activation by endogenous adenosine but had an additional direct inhibitory effect, as has also been reported by Premont et al. (45). In support of this view, the fully PGE_1-activated enzyme adenylate cyclase was not affected by adenosine deaminase but was inhibited by 4 mM theophylline by about 20% (Table 2). When the PGE_1-stimulated adenylate

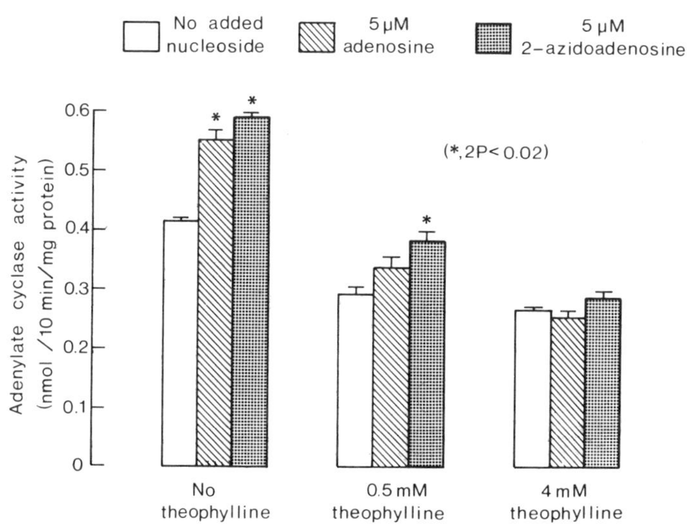

FIG. 2. Effects of theophylline on the basal and adenosine- or 2-azidoadenosine–stimulated adenylate cyclase activities of a human platelet particulate fraction. Assays were carried out in triplicate as described in the legend to Fig. 1; values given are means ± SEM.

TABLE 2. *Effects of theophylline and adenosine deaminase on the basal and PGE_1-stimulated adenylate cyclase activities of a human platelet particulate fraction*

Conditions of assay	Other additions	Changes in adenylate cyclase activity (%)
Without PGE_1	Theophylline (4 mM)	-41 ± 6^a
	Adenosine deaminase (0.1 unit/ml)	-22 ± 4^a
	Adenosine deaminase (2 units/ml)	-18 ± 3^a
	Adenosine deaminase (2 units/ml) + theophylline (4 mM)	-41 ± 4^a
With PGE_1	Theophylline (4 mM)	-23 ± 3^a
	Adenosine deaminase (2 units/ml)	$+3 \pm 3$
	Adenosine deaminase (2 units/ml) + theophylline (4 mM)	-21 ± 3^a
	Adenosine (0.1 mM)	-55 ± 2^a
	Adenosine (0.1 mM) + theophylline (4 mM)	-53 ± 2^a

Assays were carried out in triplicate as described in the legend to Fig. 1. Percent changes in adenylate cyclase activity with respect to controls without PGE_1 (1.74 ± 0.05 nmoles/10 min/mg protein) or with 1 μM PGE_1 (15.2 ± 0.3 nmoles/10 min/mg protein) are given as means ± SE of the differences; significant changes (Student's *t*-test) are indicated.

[a] $2p > 0.01$.

cyclase of platelet particulate fraction was inhibited by over 50% by 0.1 mM adenosine, no further inhibition was seen on addition of 4 mM theophylline. This suggests that theophylline may exert an adenosine-like inhibitory effect on platelet adenylate cyclase in addition to blocking the stimulatory action of adenosine.

Because of the presence of endogenous adenosine, the activation of adenylate cyclase by added adenosine or its derivatives observed in these and previous experiments represents an underestimate of the potential effects of these compounds. A similar situation has been reported in rat striatal homogenates in which the endogenous adenosine concentration was sufficient to maximally stimulate adenylate cyclase activity (45). We have studied the potential of the platelet enzyme for stimulation by adenosine analogs by addition of 2-azidoadenosine in the presence of a large excess of adenosine deaminase. 2-Azidoadenosine is a relatively weak inhibitor rather than a substrate for this enzyme (9). It was found that the maximum stimulation of platelet adenylate cyclase by 2-azidoadenosine (observed with 10 μM) was increased either by addition of adenosine deaminase or by lowering the $MgCl_2$ content from 5 mM to 0.8 mM (Table 3). When both conditions were met, 2-azidoadenosine increased human platelet adenylate cyclase activity up to about 100%. Even greater stimulations of up to 200% were observed with particulate fractions from rabbit platelets. Although enhancement of the stimulation of platelet adenylate cyclase by adenosine has previously been observed on decreasing the Mg^{2+} concentration in the assay (37), this was attributed to a decrease in the inhibitory activity of adenosine and would therefore not be expected to occur with 2-azidoadenosine which has no inhibitory action at 10 μM. Thus the results imply that stimulation of adenylate cyclase by adenosine and its derivatives is also directly favored by the lower Mg^{2+} concentration.

TABLE 3. *Effects of adenosine deaminase and of* $MgCl_2$ *concentration on the activation of adenylate cyclase in a human platelet particulate fraction by 2-azidoadenosine*

Additions				
Adenosine deaminase (units/ml)	$MgCl_2$ (mM)	2-Azidoadenosine (μM)	Adenylate cyclase activity (nmoles/10 min/mg protein)	Changes in adenylate cyclase activity (%)
0	5.0	0	0.110 ± 0.005	—
		0.5	0.132 ± 0.002	+20 ± 5 [a]
		10.0	0.146 ± 0.002	+33 ± 5 [b]
	0.8	0	0.047 ± 0.001	—
		0.5	0.059 ± 0.001	+26 ± 3 [b]
		10.0	0.071 ± 0.001	+51 ± 3 [b]
12	5.0	0	0.078 ± 0.003	—
		0.5	0.113 ± 0.004	+45 ± 6 [b]
		10.0	0.135 ± 0.004	+73 ± 6 [b]
	0.8	0	0.035 ± 0.001	—
		0.5	0.053 ± 0.001	+51 ± 4 [b]
		10.0	0.070 ± 0.002	+100 ± 6 [b]

Assays were carried out in triplicate as described in the legend to Fig. 1, except that 0.04 mM [^{3}H]ATP was used. Adenylate cyclase activities are means ± SEM and changes in activity caused by 2-azidoadenosine are given as means ± SE of the differences; significant changes (Student's *t*-test) are indicated.

[a] $2p > 0.02$
[b] $2p > 0.005$.

EFFECTS OF ADENOSINE AND ANALOGS ON CYCLIC AMP CONCENTRATIONS IN INTACT PLATELETS

Methodology

Intact platelets can be most easily studied in a functionally responsive state when suspended in plasma containing citrate or heparin as the anticoagulant. However, because of the considerable amount of cyclic AMP in plasma, the sensitivity with which changes in platelet cyclic AMP can be detected in platelet-rich plasma is relatively low when the total cyclic AMP present is measured by conventional methods (19). For this reason, several workers including ourselves have found it advantageous to measure changes in labeled cyclic AMP in platelets by labeling the metabolic adenine nucleotide pool by preincubation of the platelets with radioactive adenine (for example, see refs. 27 and 41). This is justified by the finding that the large increases in labeled cyclic AMP and in total cyclic AMP in platelet-rich plasma caused by addition of PGE_1 are linearly related (19). Moreover, calculation of the average specific activity of the newly formed labeled cyclic AMP suggests that this material is derived from the metabolic pool of ATP as a whole, rather than from some nonrepresentative subcompartment of this ATP. Although adenosine is also readily incorporated into the platelet adenine nucleotides, the maximum rate of uptake could only dilute the metabolic pool of ATP by about 1%/min (27). Thus the prelabeling technique can be applied to studies of the effects of adenosine on platelet cyclic AMP concentrations, provided incubation times are relatively short (27). The results of these studies have been confirmed in general terms by less accurate measurements of changes in the total cyclic AMP in platelet-rich plasma (27,39,40).

Studies in the Absence of Inhibitors of Cyclic AMP Phosphodiesterase

Addition of adenosine to platelet-rich plasma that had been preincubated with labeled adenine increased the labeled cyclic AMP present in the platelets two- to fourfold within 0.5 min, after which the concentration of cyclic AMP slowly declined (27). The largest increases were observed with 10 to 40 μM adenosine; higher concentrations were less effective, although the net inhibition of cyclic AMP formation obtained with particulate fraction adenylate cyclase was not observed (Fig. 3). Nevertheless, the shape of this dose-response curve suggests a weak inhibitory component to the effect of high adenosine concentrations in the intact platelet. The role of adenosine uptake in the cyclic AMP response of the platelet was investigated using *p*-nitrobenzylthioguanosine (NBTGR) to inhibit adenosine transport (44). In contrast to inhibitors of adenosine transport such as papaverine and dipyridamole, this compound did not enhance the action of PGE_1 in increasing platelet cyclic AMP and is therefore not an inhibition of cyclic AMP phosphodiesterase

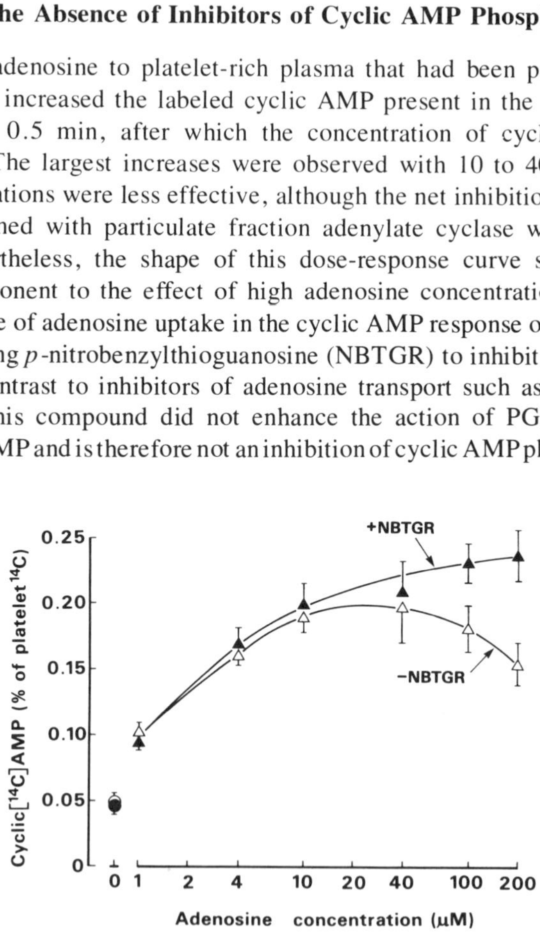

FIG. 3. Effects of different concentrations of adenosine added with or without NBTGR on platelet cyclic AMP levels. Human heparinized platelet-rich plasma that had been preincubated with [^{14}C]adenine to label platelet adenine nucleotides was used. Samples were incubated for 0.5 min at 37° C with the indicated concentrations of adenosine with and without 10 μM NBTGR. Platelet cyclic [^{14}C]AMP was then isolated and counted and was expressed as a percentage of the total ^{14}C incorporated into the platelets. Details of the experimental procedures are given elsewhere (27). Values for cyclic [^{14}C]AMP are means ± SEM from three identical incubation mixtures. Additions are indicated as follows: ○, none; ●, NBTGR; Δ, adenosine; ▲, adenosine + NBTGR.

(27). As shown in Fig. 3, NBTGR had no effect on the increases in platelet cyclic AMP caused by adenosine concentrations up to 10 μM, although it prevented the decrease in the cyclic AMP response observed with high adenosine concentrations. As NBTGR inhibited the uptake of adenosine present at concentrations up to 10 μM by at least 95%, these results indicated that the action of adenosine in increasing platelet cyclic AMP is exerted at the platelet surface, presumably at externally oriented membrane receptors (17,27). Comparable results indicating that increases in cellular cyclic AMP caused by adenosine were not dependent on adenosine uptake have subsequently been obtained in a variety of other biological systems (7,28,29). Although NBTGR was less effective in preventing adenosine uptake when the concentration of the latter compound was high, an apparent blockade of the weak inhibitory component of the action of high adenosine concentrations by NBTGR suggests that adenosine may inhibit platelet adenylate cyclase at intracellular sites. A more marked inhibitory effect of adenosine on cyclic AMP accumulation in intact platelets was obtained in experiments carried out in the presence of PGE_1; this effect was also blocked by NBTGR (27). Since NBTGR did not affect the inhibition of adenylate cyclase by adenosine when platelet particulate fractions were used, it probably acts only by blocking the uptake of adenosine and thus its access to intracellular inhibitory sites.

Studies in the Presence of Papaverine

Papaverine is a potent inhibitor of both platelet cyclic AMP phosphodiesterase activity (1,41) and adenosine uptake by platelets (27,42). Addition of this compound alone increased platelet cyclic AMP and it also greatly enhanced the increases in platelet cyclic AMP caused by adenosine (Fig. 4) (17,27). As the rates of increase in the concentration of platelet cyclic AMP were linear for up to 1 min after addition of papaverine with or without adenosine, it has proved possible to measure the activity of adenylate cyclase in intact platelets by using short incubation times. In longer incubations, rates of cyclic AMP accumulation decreased rapidly, possibly because papaverine prevents the replenishment of platelet ATP by inhibiting oxidative phosphorylation (6). As shown in Fig. 4, the stimulation of adenylate cyclase by adenosine or 2-azidoadenosine in intact platelets reached a maximum of around 10-fold, far greater than could be demonstrated with broken cell preparations. Half-maximal activation was observed with 1 μM adenosine or 2 to 2.5 μM 2-azidoadenosine. This result contrasts with the relative activities of the two compounds on particulate fraction adenylate cyclase, when 2-azidoadenosine appeared to be the more potent, perhaps because adenosine breaks down at low concentrations and exerts inhibitory effects at higher concentrations. Double reciprocal plots of the stimulation of adenylate cyclase in intact platelets by adenosine were linear, indicating that adenosine interacts with a single population of saturable sites (27). Theophylline caused roughly equal and parallel displacements of the log dose-response curves for activation of adenylate cyclase in intact platelets by adenosine and 2-azidoadenosine (Fig. 4), suggesting that all three compounds interact with the same

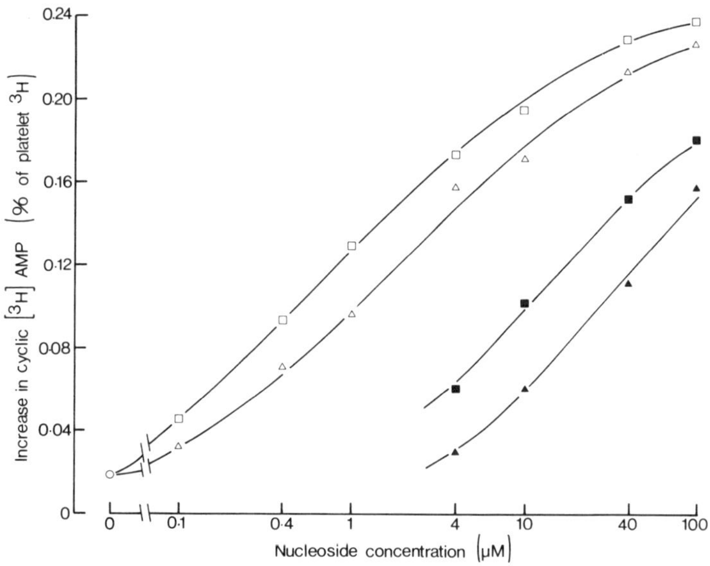

FIG. 4. Activation of adenylate cyclase by different concentrations of adenosine or of 2-azidoadenosine in intact platelets; antagonism by theophylline. Human heparinized platelet-rich plasma that had been preincubated with [^{3}H]adenine to label platelet adenine nucleotides was used. Samples were incubated for 0.5 min at 37° C with the cyclic AMP phosphodiesterase inhibitor, papaverine (2 mM), and the indicated concentrations of adenosine or 2-azidoadenosine in both the presence and absence of theophylline (0.4 mM). Platelet cyclic [^{3}H]AMP was then extracted and counted (21) and was expressed as a percentage of the total ^{3}H incorporated into the platelets. The cyclic [^{3}H]AMP found in additional control samples containing no added compounds was subtracted from each experimental value to give the increment occurring in the presence of papaverine (a measure of the intracellular adenylate cyclase activity). Additions are indicated as follows: ○, none; □, adenosine; Δ, 2-azidoadenosine; ■, adenosine + theophylline; ▲, 2-azidoadenosine + theophylline.

receptors. The apparent K_I for the inhibitory effect of theophylline was about 25 μM, which is similar to the values obtained in studies of other adenylate cyclases that are activated by adenosine (2,45).

Adenosine diphosphate has been shown to inhibit the activation of adenylate cyclase in intact platelets by PGE_1 noncompetitively, with an apparent K_I of about 1 μM (18). Identical results were obtained when the inhibitory effect of ADP on the activation of the enzyme by adenosine was investigated (27). These results show that the ADP and adenosine receptors linked to platelet adenylate cyclase are functionally distinct entities.

Relationship Between Platelet Cyclic AMP and the Inhibition of Platelet Aggregation

A wide range of correlative studies have supported the concept that the inhibition of platelet aggregation by PGE_1 is mediated by cyclic AMP (for reviews see refs. 18

through 20). Similar though less extensive studies have been carried out with adenosine. Firstly, phosphodiesterase inhibitors other than the methylxanthines synergistically enhanced both the increases in platelet cyclic AMP and the inhibitions of aggregation caused by adenosine, whereas appropriate concentrations of methylxanthines antagonized both effects (17,27,41). Second, adenosine inhibited vasopressin-induced platelet aggregation much more potently than that caused by ADP (26), a finding that correlates with the ability of ADP but not vasopressin to inhibit the activation of adenylate cyclase by adenosine (27). Third, differences between the effects of adenosine and its derivatives on platelet adenylate cyclase and cyclic AMP levels correlate, insofar as they have been investigated, with the effects of these compounds on platelet aggregation (21,27). Finally, the relationship between the actual increases in platelet cyclic AMP during inhibition of ADP-induced aggregation by adenosine with or without papaverine was very similar or identical to that obtained when PGE_1 with or without papaverine was used (17).

Inhibition of Adenylate Cyclase in Intact Platelets

Since the action of high concentrations of adenosine on adenylate cyclase in intact platelets appeared to include an inhibitory component, we investigated the effects on platelets of adenosine derivatives that inhibited particulate fraction adenylate cyclase (21). In this study, DDA was found to be a potent inhibitor of the increases in platelet cyclic AMP caused by PGE_1 as well as of its inhibitory effects on ADP-induced aggregation. Virtually identical observations were made when DDA was tested together with the more potent activator of platelet adenylate cyclase, prostacyclin (PGI_2) (Fig. 5). Studies with a somewhat less potent inhibitor of adenylate cyclase, 9-(tetrahydro-2-furyl)adenine, have given very similar results (15,21). These investigations have provided cogent evidence that the inhibition of platelet aggregation by PGE_1 or PGI_2 is mediated by increases in platelet cyclic AMP content. Dideoxyadenosine also inhibited the increases in platelet cyclic AMP induced by papaverine or by PGE_1 plus papaverine, though in these instances it was less effective than with PGE_1 alone. This could result from a partial inhibition of the entry of DDA into the platelets by papaverine, but since 10 μM NBTGR did not affect the actions of DDA, it appears that the latter compound is not taken up by the same mechanism as adenosine. In contrast with DDA, 2′-deoxyadenosine 3′-monophosphate, i.e., a more potent inhibitor of particulate fraction adenylate cyclase, did not block the inhibition of platelet aggregation by PGE_1. This finding supports previous evidence that the inhibitory site on adenylate cyclase is intracellular, as phosphorylated compounds would not be expected to penetrate the platelet plasma membrane.

In contrast with its marked effects on the inhibition of platelet aggregation by prostaglandins, DDA neither caused aggregation when added alone nor potentiated the effects of most aggregating agents, when tested in the absence of prostaglandins (21). Since DDA did inhibit the basal adenylate cyclase activity of intact platelets, as indicated by its effect on the increase in cyclic AMP caused by papaverine, these findings suggest that the steady state concentration of cyclic AMP in resting platelets

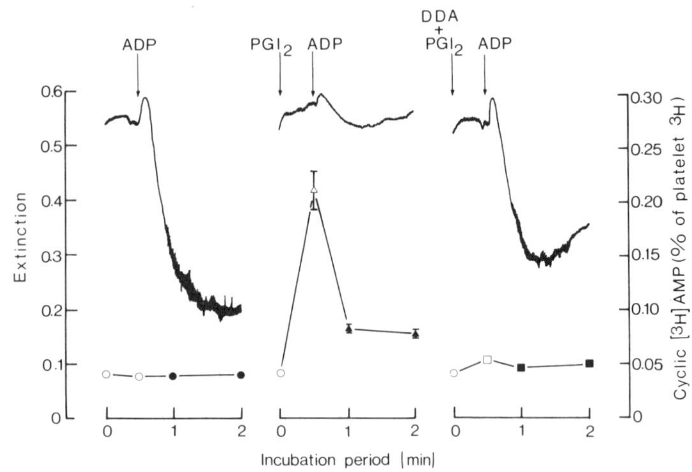

FIG. 5. Effects of DDA on the inhibition by PGI_2 of ADP-induced aggregation of human platelets and on the associated increase in platelet cyclic AMP. Heparinized platelet-rich plasma that had been preincubated with [^{3}H]adenine to label platelet adenine nucleotides was used. Samples were stirred at 37° C with additions made at the times indicated; the final concentrations of additions were: ADP, 1 μM; PGI_2, 4 nM; DDA, 100 μM. Aggregation was recorded turbidometrically as the decrease in extinction of the stirred platelet suspension; recordings from 2-min incubations are reproduced. Platelet cyclic [^{3}H]AMP was measured as described elsewhere (21) in sample incubations terminated after the periods indicated; mean values ± SEM from four determinations on the same platelet-rich plasma are shown. Cyclic [^{3}H]AMP values corresponding to particular additions are indicated as follows: ○ no additions; ●, ADP; Δ, PGI_2, ▲, PGI_2 + ADP; □, DDA + PGI_2; ■, DDA + PGI_2 + ADP.

is too low to affect the responses of platelets to aggregating agents. No decrease in the resting level of platelet cyclic AMP could be detected on addition of inhibitors of adenylate cyclase such as DDA, even using the prelabeling technique, so it is probable that most of this cyclic AMP is protein bound and functionally inert. We have obtained identical results to those described above with the weaker inhibitor of platelet adenylate cyclase, 9-(tetrahydro-2-furyl) adenine (21), although one group has reported that this compound can potentiate the aggregating action of PGG_2 (48). The latter effect may have been caused by suppression by the adenosine analog of increases in platelet cyclic AMP caused by PGD_2 formed from the PGG_2. Thus experiments with both DDA and a related compound have indicated that cyclic AMP is a unidirectional regulator of platelet function, in the sense that cyclic AMP is apparently not required to maintain platelets in a disaggregated state and is only of importance in mediating the effects of inhibitory agonists.

MECHANISM OF ACTION OF CYCLIC AMP WITHIN PLATELETS

Cyclic AMP is believed to exert its effects in eukaryotic cells by activating the phosphorylation of regulatory proteins. Although platelets are known to contain both soluble and membrane-bound protein kinases and several potential substrates for these enzymes (38,51), the physiologically important substrates have not been clearly identified. In an attempt to resolve this problem, we have preincubated platelets with [^{32}P]P$_i$, exposed them to agents that increase platelet cyclic AMP and then resolved platelet [^{32}P]phosphopolypeptides by SDS-polyacrylamide gel electrophoresis (22,24,25). In the earliest of these studies (24), PGE$_1$ was found to increase the labeling of 25,000- to 19,000 MW polypeptides by 30 to 40%. Subsequently, by using discontinuous SDS-polyacrylamide gel electrophoresis (35), it proved possible to resolve several phosphopolypeptides in this gels region. Labeling of both a major phosphopolypeptide with an apparent MW of 24,000 (P24) and of a minor one with an apparent MW of 22,000 (P22) was roughly doubled on exposure of the labeled platelets to PGE$_1$ (25). Further improvements in methodology, including use of slab gel electrophoresis followed by detection of [^{32}P]-labeled phosphopolypeptides by autoradiography, has now revealed two additional minor phosphopolypeptides (P50 and P36) showing increased labeling in the presence of PGE$_1$ (Fig. 6) (22). Enhanced phosphorylation of the identical polypeptides was observed after incubation of the labeled platelets with adenosine in the presence of a low concentration of papaverine (Fig. 6), suggesting that the increased platelet cyclic AMP mediates these effects. After rapid subcellular fractionation of [^{32}P]-labeled platelets that had been exposed to PGE$_1$, labeled P24 was found exclusively in the particulate fractions and was present at highest concentration in a 19,000- to 90,000-g_{av} fraction containing vesicles derived from both plasma and intracellular membranes (13). Käser-Glanzmann et al. (31) have reported that addition of cyclic AMP and platelet protein kinase to a similar membrane fraction potentiated the ATP-dependent uptake of Ca^{2+} ions by the membranes. We therefore suggest that phosphorylation of P24 may mediate this effect of cyclic AMP and that cyclic AMP may inhibit platelet aggregation primarily by decreasing the concentration of Ca^{2+} ions in the platelet cytosol. This hypothesis closely parallels that proposed for the action of cyclic AMP on cardiac sarcoplasmic reticulum, in which phosphorylation of a 22,000 MW regulatory protein appears to stimulate the ATP-dependent uptake of Ca^{2+} ions (34,52).

SUMMARY

Adenosine exerts independent stimulatory and inhibitory effects on platelet particulate fraction adenylate cyclase; the former predominates with low adenosine concentrations (1 to 40 μM) and the latter with high concentrations (>100 μM). The structural requirements of these two effects are different; for example, 2-azidoadenosine has only stimulatory activity, whereas DDA has only inhibitory activity. Theophylline prevents the stimulation of particulate fraction adenylate cyclase by adenosine and its derivatives and also decreases basal activity, partly by blocking the action of endogenous adenosine. Addition of adenosine deaminase also decreases the basal enzyme activity and enhances the stimulatory effects of the deaminase-resistant 2-azidoadenosine.

Adenosine increases the cyclic AMP concentration in intact platelets. Maximum

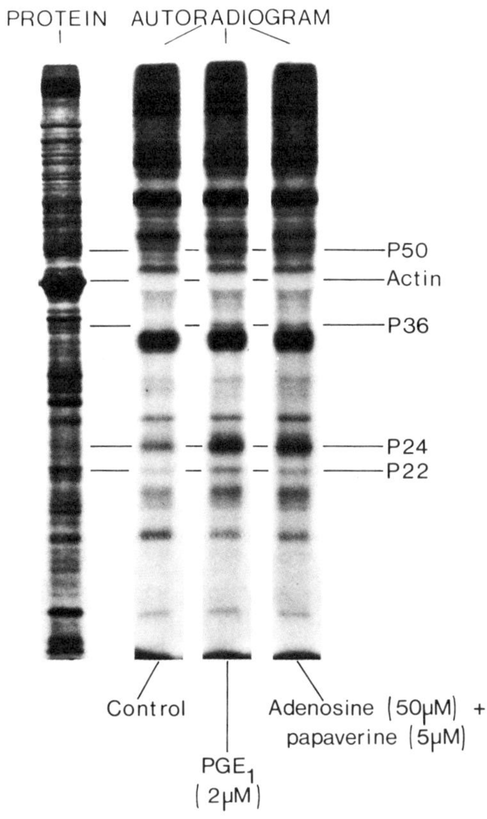

FIG. 6. Effects of agents that increase the concentration of platelet cyclic AMP on the phosphorylation of proteins in intact platelets. Human platelets suspended in a modified Tyrode's solution were labeled by preincubation with [^{32}P]P$_i$ (25). After resuspension of the platelets in fresh medium, aliquots were incubated for 3 min at 37° C with no further additions (control), with 2 μM PGE$_1$, or with 50 μM adenosine + 5 μM papaverine (final concentrations). Incubations were stopped by addition of perchloric acid and the precipitated material was dissolved in 62.5 mM Tris-HCl, pH 6.8, containing 3% (wt/vol) SDS, 6% (wt/vol) glycerol and 5% (vol/vol) mercaptoethanol. Platelet polypeptides were then resolved by discontinuous SDS-polyacrylamide slab gel electrophoresis using 13% (wt/vol) polyacrylamide (35). The gels were stained with Coomassie Brilliant Blue R, and [^{32}P]-labeled phosphopolypeptides were detected by autoradiography of the dried gels. The stained polypeptide pattern observed with whole platelets is shown on the left and the autoradiograms obtained after electrophoresis of protein from platelets incubated with the indicated additions are shown on the right. [^{32}P]Phosphopolypeptides of particular interest are designated P50, P36, P24, and P22, the numerical value indicating the molecular weight ($\times 10^{-3}$) determined from protein standards electrophoresed in parallel.

increases of two- to four-fold are obtained with 10 to 40 μM adenosine; higher concentrations are somewhat less effective. Inhibition of adenosine uptake by NBTGR has no effect on the increases in platelet cyclic AMP with 10 μM adenosine, but enhances the effects of adenosine concentrations above 40 μM. This indicates that adenosine stimulates adenylate cyclase by interaction with an externally orientated membrane receptor and inhibits the enzyme at an intracellular site.

In the presence of papaverine, added to inhibit cyclic AMP phosphodiesterase and adenosine transport, adenosine increases the rate of cyclic AMP accumulation in intact platelets up to 10-fold; a half-maximal effect requires 1 μM adenosine. Theophylline competitively inhibits this activation of adenylate cyclase by adenosine. The increases in platelet cyclic AMP and inhibitions of platelet aggregation caused by adenosine correlate under a wide variety of conditions.

DDA inhibits the adenylate cyclase activity of intact platelets and decreases markedly the inhibition of platelet aggregation by PGE_1 or PGI_2, providing new evidence that cyclic AMP mediates the effects of these agents. However, DDA does not potentiate aggregation in the absence of compounds that increase platelet cyclic AMP, indicating that the concentration of cyclic AMP in resting platelets is too low to affect platelet function.

Addition of adenosine with papaverine increases the phosphorylation of the same platelet polypeptides as does PGE_1, presumably via activation of a cyclic AMP–dependent protein kinase. One of these phosphopolypeptides (P24) is membrane bound and it is suggested that the increased phosphorylation of P24 may mediate the effects of inhibitory agonists, such as adenosine, by stimulating the active transport of Ca^{2+} ions out of the platelet cytosol.

ACKNOWLEDGMENTS

This work was supported by grants from the Medical Research Council of Canada (MT 5626) and the Ontario Heart Foundation (T.15–6). J.V.D. was in receipt of an M.R.C. Studentship.

REFERENCES

1. Asano, T., Ochiai, Y., and Hidaka, H. (1977): *Mol. Pharmacol.,* 13:400–406.
2. Blume, A. J., and Foster, C. J. (1975): *J. Biol. Chem.,* 250:5003–5008.
3. Born, G. V. R. (1964): *Nature,* 202:95–96.
4. Born, G. V. R., and Cross, M. J. (1963): *J. Physiol. (Lond.),* 168:178–195.
5. Born, G. V. R., Haslam, R. J., Goldman, M., and Lowe, R. D. (1965): *Nature,* 205:678–680.
6. Browning, E. T., Groppi, V. E., and Kon, C. (1974): *Mol. Pharmacol.,* 10:175–181.
7. Clark, R. B., Gross, R., Su, Y.-F., and Perkins, J. P. (1974): *J. Biol. Chem.,* 249:5296–5303.
8. Clark, R. B., and Seney, M. N. (1976): *J. Biol. Chem.,* 251:4239–4246.
9. Cusack, N. J., and Born, G. V. R. (1976): *Proc. R. Soc. Lond. [Biol.],* 193:307–311.
10. Cusack, N. J., and Born, G. V. R. (1977): *Proc. R. Soc. Lond. [Biol.],* 197:515–520.
11. Fain, J. N., Pointer, R. H., and Ward, W. F. (1972): *J. Biol. Chem.,* 247:6866–6872.
12. Fain, J. N., and Shepherd, R. E. (1977): *J. Biol. Chem.,* 252:8066–8070.
13. Fox, J. E. B., Lynham, J. A., and Haslam, R. J. (1978): *Fed. Proc.,* 37:799.
14. Gaarder, A., Jonsen, J., Laland, S., Hellem, A., and Owren, P. A. (1961): *Nature,* 192:531–532.
15. Harris, D. N., Phillips, M. B., and Goldenberg, H. J. (1975): *Fed. Proc.,* 34:617.
16. Haslam, R. J. (1964): *Nature,* 202:765–768.
17. Haslam, R. J. (1973): *Ser. Haematol.,* 6:333–350.
18. Haslam, R. J. (1975): In: *Biochemistry and Pharmacology of Platelets,* Ciba Foundation Symposium 35 (new series), pp. 121–151. Elsevier/Excerpta Medica/North-Holland, Amsterdam.
19. Haslam, R. J. (1978): In: *Platelet Function Testing,* edited by H. J. Day, H. Holmsen, and M. B. Zucker, pp. 487–503. U.S. Department of Health, Education, and Welfare, Publication No. (NIH) 78–1087, Washington, D.C.

20. Haslam, R. J., Davidson, M. M. L., Davies, T., Lynham, J. A., and McClenaghan, M. D. (1978): *Adv. Cyclic Nucleotide Res.,* 9:533–552.
21. Haslam, R. J., Davidson, M. M. L., and Desjardins, J. V. (1978): *Biochem. J.,* 176:83–95.
22. Haslam, R. J., Davidson, M. M. L., Desjardins, J. V., Fox, J. E. B., and Lynham, J. A. (1979): *Proc. 7th Int. Congr. Pharmacol. (in press).*
23. Haslam, R. J., and Lynham, J. A. (1972): *Life Sci.,* 11 (Part II):1143–1154.
24. Haslam, R. J., and Lynham, J. A. (1978): *Thromb. Res.,* 12:619–628.
25. Haslam, R. J., Lynham, J. A., and Fox, J. E. B. (1978): *Biochem. J.,* 178:397–406.
26. Haslam, R. J., and Rosson, G. M. (1972): *Am. J. Physiol.,* 223:958–967.
27. Haslam, R. J., and Rosson, G. M. (1975): *Mol. Pharmacol.,* 11:528–544.
28. Huang, M., and Daly, J. W. (1974): *Life Sci.,* 14:489–503.
29. Huang, M., and Drummond, G. I. (1976): *Biochem. Pharmacol.,* 25:2713–2719.
30. Huang, M., Shimizu, H., and Daly, J. W. (1972): *J. Med. Chem.,* 15:462–466.
31. Käser-Glanzmann, R., Jakábová, M., George, J. N., and Lüscher, E. F. (1977): *Biochim. Biophys. Acta,* 466:429–440.
32. Kien, M., Belamarich, F. A., and Shepro, D. (1971): *Am. J. Physiol.,* 220:604–608.
33. Kikugawa, K., Iizuka, K., and Ichino, M. (1973): *J. Med. Chem.,* 16:358–364.
34. Kirchberger, M. A., Tada, M., and Katz, A. M. (1974): *J. Biol. Chem.,* 249:6166–6173.
35. Laemmli, U. K. (1970): *Nature,* 227:680–685.
36. Londos, C., and Preston, M. S. (1977): *J. Biol. Chem.,* 252:5951–5956.
37. Londos, C., and Wolff, J. (1977): *Proc. Natl. Acad. Sci. U.S.A.,* 74:5482–5486.
38. Lyons, R. M., Stanford, N., and Majerus, P. W. (1975): *J. Clin. Invest.,* 56:924–936.
39. McElroy, F. A., and Philp, R. B. (1975): *Life Sci.,* 17:1479–1494.
40. Michel, H., Caen, J. P., Born, G. V. R., Miller, R., d'Auriac, G. A., and Meyer, P. (1976): *Br. J. Haematol.,* 33:27–38.
41. Mills, D. C. B., and Smith, J. B. (1971): *Biochem. J.,* 121:185–196.
42. Mills, D. C. B., Smith, J. B., and Born, G. V. R. (1970): *Thromb. Diath. Haemorrh.,* Suppl. 42:175–184.
43. O'Brien, J. R. (1964): *J. Clin. Pathol.,* 17:275–281.
44. Paterson, A. R. P., and Oliver, J. M. (1971): *Can. J. Biochem.,* 42:271–274.
45. Prémont, J., Perez, M., and Bockaert, J. (1977): *Mol. Pharmacol.,* 13:662–670.
46. Rozenberg, M. C., and Holmsen, H. (1968): *Biochim. Biophys. Acta,* 155:342–352.
47. Sahyoun, N., Schmitges, C. J., Siegel, M. I., and Cuatrecasas, P. (1976): *Life Sci.,* 19:1961–1970.
48. Salzman, E. W. (1977): *Biochim. Biophys. Acta,* 499:48–60.
49. Sattin, A., and Rall, T. W. (1970): *Mol. Pharmacol.,* 6:13–23.
50. Schaeffer, H. J., and Thomas, H. J. (1958): *J. Am. Chem. Soc.,* 80:3738–3742.
51. Steiner, M. (1975): *Arch. Biochem. Biophys.,* 171:245–254.
52. Tada, M., Kirchberger, M. A., and Katz, A. M. (1975): *J. Biol. Chem.,* 250:2640–2647.
53. Weinryb, I., and Michel, I. M. (1974): *Biochim. Biophys. Acta,* 334:218–225.
54. Wolfe, S. M., and Shulman, N. R. (1969): *Biochem. Biophys. Res. Commun.,* 35:265–272.
55. Zenser, T. V. (1976): *Proc. Soc. Exp. Biol. Med.,* 152:126–129.

Physiological and Regulatory Functions of Adenosine and Adenine Nucleotides, edited by H. P. Baer and G. I. Drummond. Raven Press, New York © 1979.

Evidence for an Adenylate Cyclase–Linked Adenosine Receptor Mediating Coronary Relaxation

Walter R. Kukovetz, Annelies Wurm, Sigrid Holzmann, Gerald Pöch, and Ingo Rinner

Institut für Pharmakodynamik und Toxikologie, Universitätsplatz 2, A–8010 Graz, Austria

The mechanism by which adenosine and adenosine phosphates—including externally applied non-substituted cyclic adenosine monophosphate (cAMP)—relax smooth muscle is not clear (see ref. 1). There is evidence suggesting that adenosine binds to a surface receptor of the membrane in coronary smooth muscle (13,16) and that theophylline acts as a weak but competitive antagonist. In various tissues, e.g., in blood platelets (4), the theophylline-sensitive binding site of adenosine is linked to adenylate cyclase. In vascular smooth muscle tissue, however, no significant increases in cAMP were found by Herlihy et al. (5) using 100 μM adenosine in dog coronary and hog carotid arteries, and by Verhaeghe (18) with 10 μM adenosine in dog saphenous vein. Furthermore, McKenzie et al. (11) found no significant increases in cAMP with 1 μM adenosine in rabbit colon longitudinal muscle and inhibitory effects on isolated adenylate cyclase from this tissue were obtained.

This chapter deals with a study of the actions of adenosine and isoprenaline on contractility and cAMP content of circular strips of coronary arteries from beef and pig hearts (8,9). The actions of adenosine on particulate preparations of adenylate cyclase as well as on the detergent-solubilized enzyme are also discussed (14,19).

METHODS

Circular strips of beef coronary arteries, carefully freed from adherent myocardial tissue, were suspended in modified high-K^+ (10-fold) Tyrode's solution (NaCl, 112.8; KCl, 26.8; $CaCl_2$, 1.26; $MgSO_4$, 1.16; $NaHCO_3$, 11.9; glucose, 10.1 mM), gassed with 95% O_2 and 5% CO_2 at a temperature of 37° C. Length changes were recorded with mechanoelectronic transducers simultaneously in 16 strips from the same heart and expressed as percent of the maximum relaxation obtained at the end of the experiment with a large dose (100 μg/ml) of papaverine.

Cyclic AMP was measured by the radioimmunoassay method of Steiner et al. (17); cyclic guanosine monophosphate was similarly measured according to Harper and Brooker (3).

Adenylate cyclase was assayed using a particulate preparation (1,000 × *g* precipitate) of coronary vascular tissue, homogenized for 30 sec with a Microdismembrator (Braun). Incubation was carried out with 75 μM ATP and 10 mM Mg in Tris-HCl buffer at pH 7.5 in the presence of an ATP-regenerating system and 0.8 mM papaverine; cAMP formed was determined by radioimmunoassay. In later experiments, [α ^{32}P]ATP (1 mM) was used as substrate and [^{32}P]cAMP was measured according to the procedure of Salomon et al. (15). Mean control activities were 14 and 71 pmoles cAMP/mg protein/min by radioimmunoassay and ^{32}P, respectively.

RESULTS AND DISCUSSION

Effect of Adenosine and Isoprenaline on Coronary Strips

Dose-Response Experiments

Adenosine produced dose-dependent increases in cAMP in coronary strips from cattle and pig hearts that were closely associated with the relaxant effects (Fig. 1) but did not significantly increase cGMP (8).

Pig coronary strips were more sensitive than those of cattle. The threshold concentration of adenosine that significantly increased cAMP and length in pig hearts was 0.37 μM. Bovine strips were less sensitive and concentrations of 18.7 μM adenosine caused about 25% increases in cAMP and relaxation. Isoprenaline produced increases in cAMP and relaxant effects similar to adenosine in concentrations of 1 nM to 1 μM (Fig. 1).

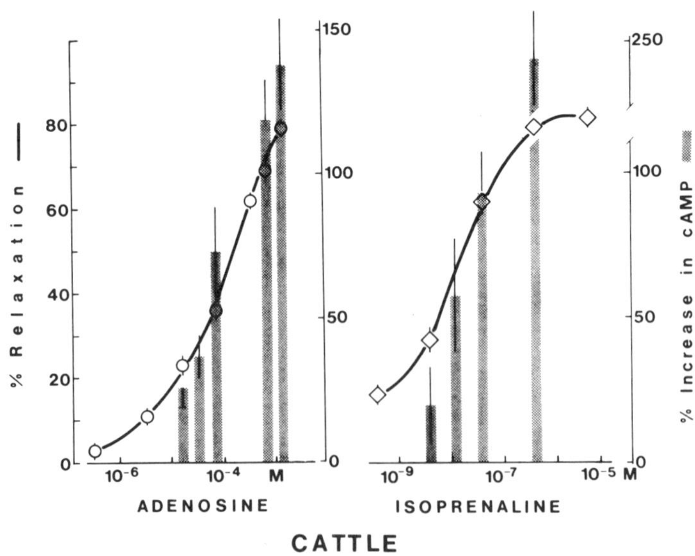

FIG. 1. Dose-response relation of percent relaxation *(curves)* and percent increases in cAMP *(columns)* produced by adenosine *(left)* and isoprenaline *(right)*. Mean values ± SEM *(N* = 6 to 91) of experiments with coronary artery strips from cattle. All increases in cAMP are significant *(p* < 0.05) except those produced by the lowest dose of isoprenaline *(p* < 0.1).

Time Courses

In Fig. 2 the time course of the effect of adenosine on cAMP and tone is shown. Thirty seconds after addition of a submaximal concentration of adenosine (750 μM), the increase in cAMP was considerably larger than the increase in length. It is quite obvious from Fig. 2 that the rise in cAMP preceded the relaxant effect. The time course obtained with 0.4 μM isoprenaline was quite similar (Fig. 2) to that of adenosine.

Adrenergic Beta-Receptor Blockade

The relaxant and cAMP-increasing effect of adenosine was not caused by adrenergic beta-receptor stimulation since it could not be blocked by sotalol in concentrations of 160 μM which completely inhibited similar effects of isoprenaline (9).

Influence of Theophylline

There are several reports on the adenosine antagonistic effect of theophylline. Haslam and Rosson (4) demonstrated that this antagonism on platelet adenylate cyclase was competitive. Thus we used theophylline in our studies, although it was possible that the phosphodiesterase inhibitory property of this compound could potentiate the actions of adenosine in smooth muscle and might obscure its antagonistic effect.

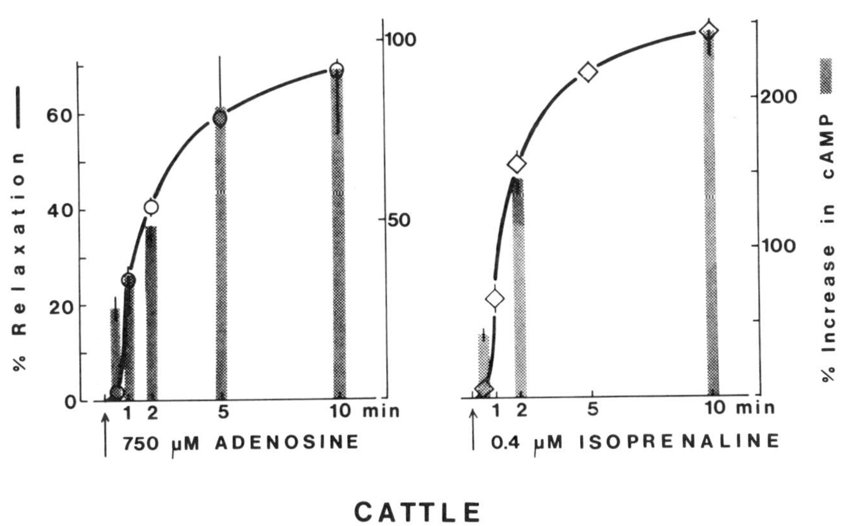

FIG. 2. Time course of the relaxant *(curves)* and cAMP-increasing effect *(columns)* of 750 μM adenosine *(left)* and 0.4 μM isoprenaline *(right)* in experiments with coronary strips from cattle. Mean values ± SEM *(N = 8 to 28)*. All increases in cAMP are significant *($p < 0.05$)*.

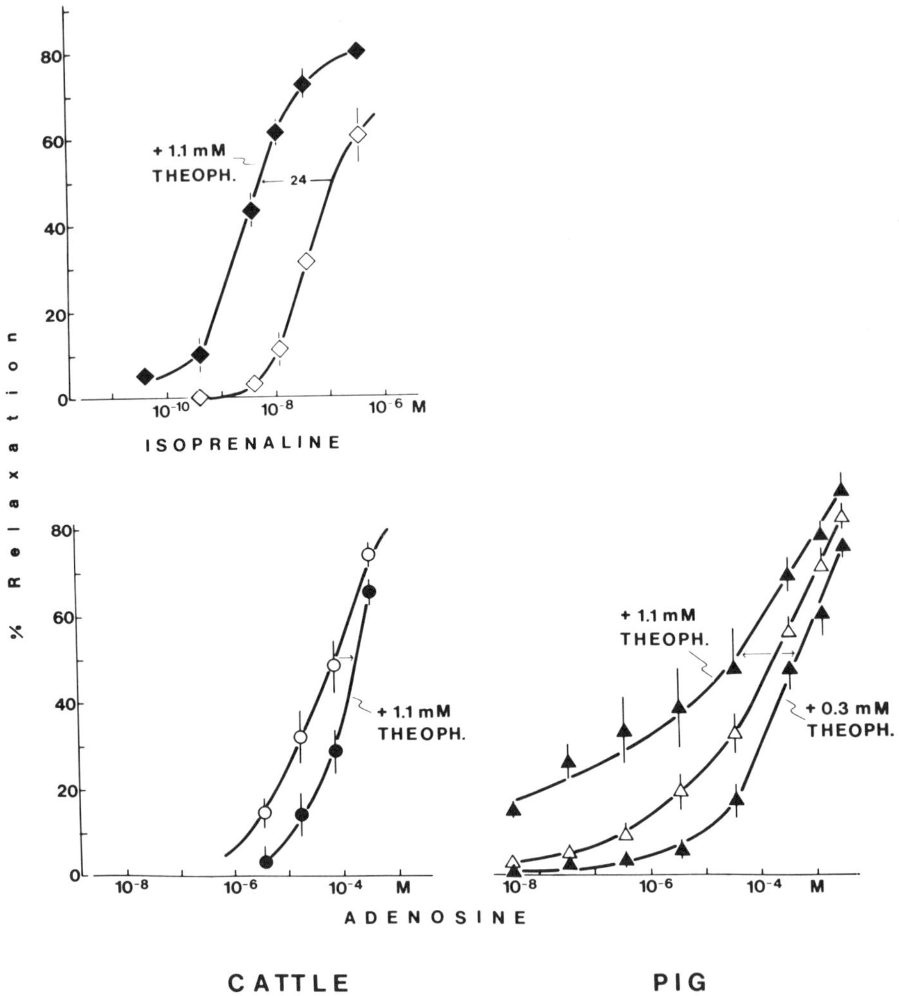

FIG. 3. Influence of theophylline on relaxant effects of adenosine *(lower panel)* and of isoprenaline *(upper panel)*, calculated and expressed as percent of the maximally possible relaxation remaining in the presence of theophylline. Mean values ± SEM of 7 to 14 experiments with isolated coronary artery strips from cattle *(left)* and pig heart *(right)*.

Theophylline (Fig. 3) potentiated dose-dependently isoprenaline relaxation of cattle strips with dose ratios of 15 and 24 at 0.6 and 1.1 mM, respectively. Concomitantly, a two- to threefold rise in cAMP levels was seen with 1.1 mM theophylline (7).

At the same concentration (1.1 mM) theophylline significantly reduced the relaxant effects of adenosine in this tissue. The increases in cAMP produced by low concentrations (19 and 75 μM) of adenosine were not clearly diminished by theophylline but the net rise produced by the high concentration of 750 μM was significantly reduced. Thus it appears that the antagonistic action of theophylline under the conditions employed is not fully shown, and this is likewise evident from the results obtained in pig strips (Fig. 3). At this concentration theophylline inhibited phosphodiesterase activity *in vitro* by 70% (7). In pig coronary strips about the same small but significant inhibition of adenosine relaxation as in cattle strips was found in the presence of 0.3 mM theophylline. At higher concentrations of 1.1 mM theophylline, however, the effect of adenosine was enhanced significantly.

Effects of Adenosine and Isoprenaline on Adenylate Cyclase

Adenosine was found to stimulate dose-dependently adenylate cyclase activity at low concentrations of 1 to 100 nM whereas at higher concentrations this effect was reversed and inhibition occurred.

Influence of Theophylline

Theophylline (1 μM) abolished the stimulatory component of the action of adenosine, whereas the inhibitory effect at 0.1 mM adenosine was not significantly altered (Fig. 4).

In contrast to adenosine, the actions of isoprenaline were purely stimulatory over the entire tested dose range of 1 nM to 0.1 mM. Isoprenaline stimulation of adenylate

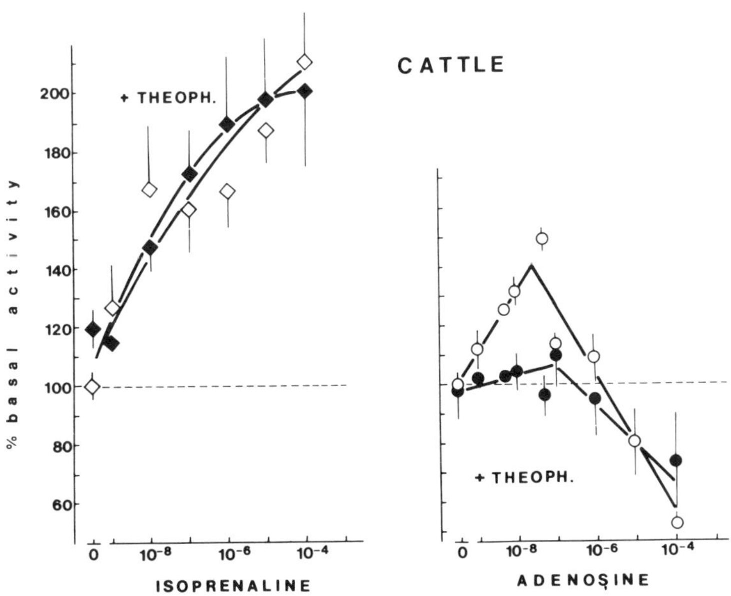

FIG. 4. Adenylate cyclase from coronary arteries of cattle (particulate preparation). Effect of isoprenaline *(left)* and adenosine *(right)* in the absence *(open symbols)* and presence *(filled symbols)* of 1 μM theophylline. Mean values ± SEM *(N* = 3 to 36).

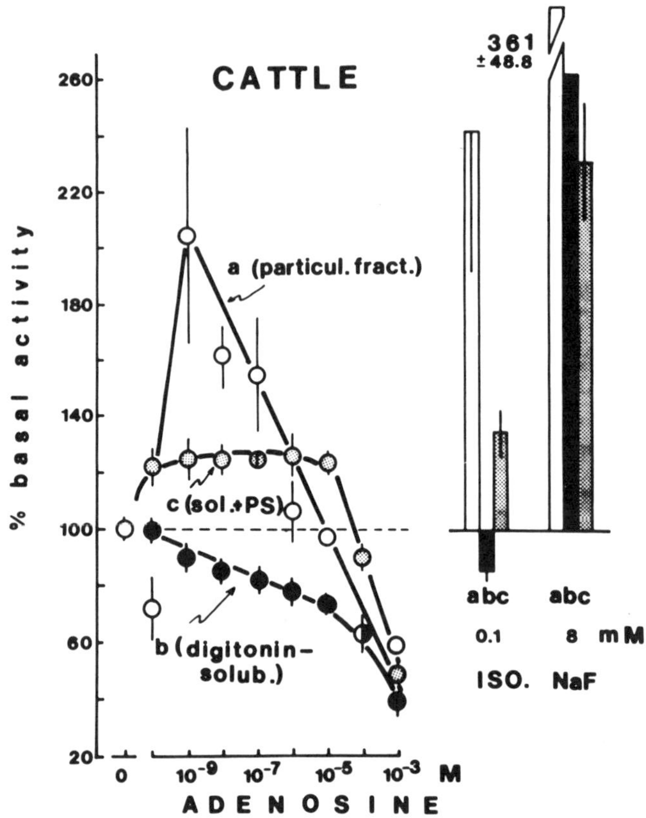

FIG. 5. Adenylate cyclase from coronary arteries of cattle heart. Influence of digitonin solubilization and phosphatidylserine on dualistic effect of adenosine *(curves)* and on the cAMP-increasing effect of 0.1 mM isoprenaline and 8 mM NaF *(columns)*. **a:** Particulate preparation. **b:** Solubilized preparation (2% digitonin). **c:** Solubilized preparation plus 0.1 mM phosphatidylserine.

cyclase was unchanged in the presence of theophylline. The adrenergic beta-blocker sotalol inhibited the isoprenaline-induced stimulation but did not affect the action of adenosine (8). These results support a mediator function of cAMP in the relaxant effects of adenosine and isoprenaline.

Effect of Solubilization of Adenylate Cyclase

It was observed that adenylate cyclase lost its sensitivity to adenosine or isoprenaline when the homogenization procedure was extended from 30 to 50 sec (8). This prompted us to look for similar effects of detergents known to abolish hormonal sensitivity of adenylate cyclases.

For this purpose, the tissue was homogenized for exactly 30 sec in the presence of

2% digitonin, centrifuged at 12,000 × *g* for 15 min and the digitonin removed on a DEAE-cellulose column.

As shown in Fig. 5, digitonin solubilization completely abolished the stimulatory, but not inhibitory effects of adenosine. The stimulation seen in the presence of isoprenaline was also inhibited, but fluoride sensitivity remained. This indicated that the enzyme had lost its receptor sensitivity but retained its catalytic function.

Addition of phosphatidylserine (100 μg/ml) partially restored the responsiveness of the solubilized enzyme to adenosine and to isoprenaline. Other membrane phospholipids, e.g., phosphatidylinositol or phosphatidylethanolamine, were also tested but found to be less effective.

Similar results were obtained when polyoxyethylene ether was used as detergent. Again, phosphatidylserine was the most potent phospholipid partially restoring the responsiveness of the solubilized enzyme to the receptor stimulants adenosine and isoprenaline (data not shown).

The effects obtained on particulate and solubilized adenylate cyclase indicate the existence of two different sites of action of adenosine: A theophylline-sensitive stimulatory site at the surface of the receptor which has also been described in blood platelets (4) and ventricular myocardium (6), and an inhibitory site that is not sensitive to theophylline antagonism is located at the catalytic moiety and are not accessible to adenosine in the intact cell. These sites are symbolized on the left side of a hypothetical diagram (Fig. 6) of the various sites of action of adenosine, which is based on the mobile receptor hypothesis (2). There is evidence that such inhibitory sites are also uncovered in other tissues on homogenization as already shown by McKenzie and Bär (10).

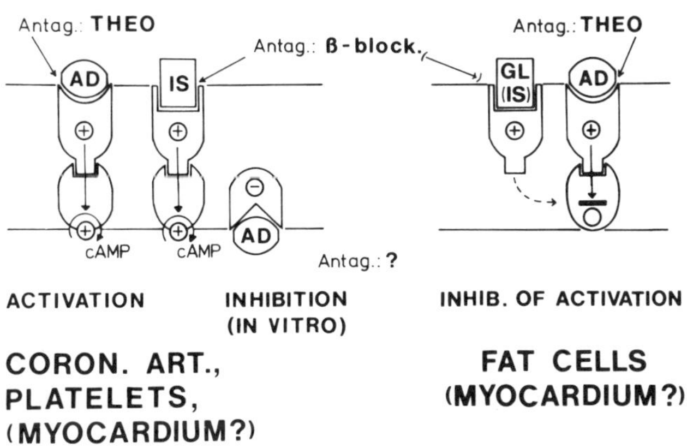

FIG. 6. Hypothetical scheme of stimulatory and inhibitory effects of adenosine (AD) on membrane receptors and on adenylate cyclase of coronary arteries, platelets *(left)* and fat cells *(right)*. Stimulatory effect of adenosine appears similar to that of isoprenaline (IS). Inhibitory effects may be caused by direct inactivation of cyclase or indirectly by inefficient coupling of the adenosine receptor to the cyclase protein. The latter effect could result in the inability of the glucagon- (GL) or isoprenaline-stimulated receptor to activate the cyclase in fat cells and myocardium, respectively.

There is probably still another site of adenosine binding. Malbon and Fain (12) reported that adenosine in rat adipocytes inhibited *glucagon*-stimulated accumulation of cAMP. Kinetic analysis revealed two classes of binding sites at the plasma membrane and a 15% inhibition of adenosine binding to these sites by theophylline.

Similarly, Schrader et al. (16) reported that adenosine in the guinea pig heart inhibited isoprenaline-stimulated accumulation of cAMP (as well as its inotropic effect) and *in vitro* stimulation of adenylate cyclase. Theophylline antagonized these inhibitory actions of adenosine.

SUMMARY AND CONCLUSIONS

Pharmacological evidence is presented suggesting the existence of a specific adenosine receptor, located at the surface of the membranes of coronary arterial smooth muscle, which mediates smooth muscle relaxation by stimulation of adenylate cyclase. Isoprenaline was shown to act by the same type of mechanism but at a clearly different receptor.

Receptor-mediated stimulation of adenylate cyclase by adenosine as well as isoprenaline could also be demonstrated in the particulate enzyme preparation *in vitro*. Receptor sensitivity was abolished by detergent solubilization of adenylate cyclase and was partially restored by phospholipids. A direct inhibitory site of adenosine action on the enzyme becomes predominant on separation of the receptor from the enzyme.

ACKNOWLEDGMENTS

This work was supported by the Austrian Research Council (Österreichischer Forschungsrat: Nr. 1453,2857). The technical assistance of Mrs. Betty Oberer and Ms. Christa Hollegger during the course of the experiments and of Ms. Annemarie Buchmann in the preparation of the manuscript is gratefully acknowledged by the authors.

REFERENCES

1. Bär, H. P. (1974): *Adv. Cyclic Nucleotide Res.,* 4:195–237.
2. Cuatrecasas, P. (1974): *Annu. Rev. Biochem.,* 43:169–214.
3. Harper, J. F., and Brooker, G. (1975): *J. Cyclic Nucleotide Res.,* 1:207–218.
4. Haslam, R. J., and Rosson, G. M. (1975): *Mol. Pharmacol.,* 11:528–544.
5. Herlihy, J. T., Bockmann, E. L., Berne, R. M., and Rubio, R. (1976): *Am. J. Physiol.,* 230:1239–1243.
6. Huang, M., and Drummond, G. I. (1976): *Biochem. Pharmacol.,* 25:2713–2719.
7. Kukovetz, W. R., Holzmann, S., Wurm, A., Pöch, G., and Rinner, I. (1978): *Naunyn Schmiedebergs Arch. Pharmacol (in preparation).*
8. Kukovetz, W. R., Pöch, G., Holzmann, S., Wurm, A., and Rinner, I. (1978): *Adv. Cyclic Nucleotide Res.,* 9:397–409.
9. Kukovetz, W. R., Wurm, A. Rinner, I., Holzmann, S., and Pöch, G. (1977): In: *Excitation-Contraction Coupling in Smooth Muscle,* edited by R. Casteels, T. Godfraind, and J. C. Rüegg, pp. 399–406. Elsevier/North Holland, Amsterdam–New York–Oxford.
10. McKenzie, S. G., and Bär, H. P. (1973): *Can. J. Physiol. Pharmacol.,* 51:190–196.

11. McKenzie, S. G., Frew, R., and Baer, H. P. (1977): *Eur. J. Pharmacol.,* 41:193–203.
12. Malbon, C. C., and Fain, J. N. (1977): *Fed. Proc.,* 36:710.
13. Olsson, R. A., Davis, C. J., Khouri, E. M., and Patterson, R. E. (1976): *Circ. Res.,* 39:93–98.
14. Rinner, I., and Wurm, A. (1978): *Naunyn Schiedebergs Arch. Pharmacol.* 302 (Suppl.): R 27.
15. Salomon, Y., Londos, C., and Rodbell, M. (1974): *Anal. Biochem.,* 58:541–548.
16. Schrader, J., Baumann, G., and Gerlach, E. (1977): *Pfluegers Arch.,* 372:29–35.
17. Steiner, A. L., Kipnis, D. M., Utiger, R., and Parker, C. (1969): *Proc. Natl. Acad. Sci. U.S.A.,* 64:367–373.
18. Verhaeghe, R. H. (1977): *Am. J. Physiol.,* 233:H114–H121.
19. Wurm, A., Holzmann, S., Pöch, G., and Kukovetz, W. R. (1976): *Naunyn Schmiedebergs Arch. Pharmacol.* 294 (Suppl.):R13.

Cyclic AMP and Adenylate Cyclase Involvement: Structure-Activity Studies

Physiological and Regulatory Functions of Adenosine and Adenine Nucleotides, edited by H. P. Baer and G. I. Drummond. Raven Press, New York © 1979.

Regulation of Cyclic Adenosine Monophosphate Accumulation in Brain Tissue: Interactions of Adenosine with Other Agonists

Theodore W. Rall

Department of Pharmacology, University of Virginia, Charlottesville, Virginia 22901

Historically, the first demonstration of adenosine receptors in neural tissue emanated from the study of the regulation of adenosine 3′,5′-monophosphate (cyclic AMP) accumulation in brain slices. Kakiuchi et al. (13) observed that brief periods of electrical pulses applied to slices of guinea pig cerebral cortex increased cyclic AMP accumulation about 10-fold. Paradoxically, the inclusion of theophylline, one of the earliest known inhibitors of 3′,5′-cyclic ribonucleotide phosphodiesterases (PDE), was found to inhibit the effect of electrical pulses about 85%. The effect of electrical pulses was greater than that of maximal amounts of either norepinephrine (NE) or histamine (HIST) and was not diminished in slices prepared from animals treated with reserpine. Further, the effect of electrical pulses was increased two- or threefold in the presence of maximal amounts of either amine.

These and other unexplained observations led to the finding that extracts of brain and other tissues were capable of increasing cyclic AMP accumulation in slices of guinea pig cerebrum up to 25-fold. Since ion-exchange chromatography of extracts yielded four active fractions in the regions occupied by adenosine and the adenine nucleotides, these substances were tested and found to increase cyclic AMP accumulation in brain slices as much as 30-fold after 5 min of exposure (23). Adenosine and adenosine 5′-monophosphate (5′-AMP) were the most potent substances with an EC_{50} of about 30 μM. The effect of adenosine was prevented completely by the presence of a 10- to 20-fold molar excess of theophylline. Most important for the purposes of this discussion, it was also observed that in the presence of maximal amounts of either NE or HIST, the addition of either maximal or submaximal amounts of adenosine produced accumulations of cyclic AMP in slices of guinea pig cerebral cortex that were between two- and threefold the sum of the values produced by the agents acting individually (23). This synergistic action of adenosine could also be obliterated by theophylline (24). The main thrust of this chapter is a discussion of the information concerning the underlying mechanisms and the cellular locale of potentiative interactions between adenosine and other agonists.

LOCALE OF ADENOSINE-STIMULATED CYCLIC AMP ACCUMULATION

Studies with Neuroma Cells

Only infrequently has adenosine been observed to increase cyclic AMP accumulation appreciably in cultured tumor cells of glial origin. Perhaps the only example is provided by 1321N1 cells, derived from a human astrocytoma, in which adenosine increased cyclic AMP levels as much as 30-fold within 10 min (6); this response was almost completely blocked by a 10-fold molar excess of theophylline. Norepinephrine, mediated by a β-adrenergic receptor, produced even larger effects, but combinations of adenosine and NE were no more effective than NE alone.

Beginning in 1973, adenosine and certain analogs were found to increase cyclic AMP accumulation in some, but not all, neuroblastoma cell lines (1,26). In these and later studies, theophylline was an effective antagonist of adenosine responses, and combinations of adenosine with other active agents [e.g., prostaglandin (PG)E_1] were never observed to produce greater than additive effects. Even though these studies were performed with tumor cells of neural origin, it is not possible to decide how the properties expressed in culture relate to those expressed *in vivo* or in brain slices. Thus no conclusions can be drawn as to the neuronal versus glial locale of responses to adenosine acting alone or in concert with other agents.

Studies with Cultured Perinatal Brain Cells

There have been relatively few reports on the adenosine responsiveness of primary cultures derived from mammalian brain tissue. Gilman and Schrier (8) observed that 100 μM adenosine increased cyclic AMP accumulation more than 10-fold in cultures of late fetal rat brain cells grown for several weeks. Although these cultures displayed choline acetyl transferase activity, the ratio of glial to neuronal cell mass probably increased considerably during culture (25). Recently, McCarthy and de Vellis (16) prepared cultures from newborn rat brains that consisted of a mixture of astroglia and oligodendroglia with few, if any, neurons. These cultures responded to adenosine with a more than 10-fold increase in cyclic AMP accumulation. The effect of a combination of NE and adenosine was less than that of adenosine alone owing to an inhibitory α-adrenergic action of NE that was apparent when NE was presented along with other agonists (e.g., PGE_1 or adenosine) or when phenylephrine was combined with isoproterenol. The cyclic AMP content of slices of newborn rat cerebral cortex is not increased by adenosine or adrenergic agents (18). A response to adenosine first becomes evident 4 to 5 days postpartum, at least 6 days before an effect of β-adrenergic agonists is apparent; beginning at the fourth day after birth, NE in the presence of adenosine increases cyclic AMP levels about twofold above that with adenosine alone and has no effect by itself. Thus the glial cell culture developed prominent responses to adenosine and β-adrenergic agonists similar to those displayed by the neonatal cortex, but did not demonstrate a positive synergistic interaction between NE and adenosine. One possible conclusion is that a major portion of the response to adenosine in brain slices is contributed by glial elements and that the synergistic interaction between NE and adenosine requires the presence of neurons. However, it is really not justified to view the response characteristics of the cultured cells as representing the simple subtraction of neuronal elements from the response characteristics of brain slices since there has been no evidence generated yet that rat brain slices possess an α-adrenergic receptor reducing cyclic AMP accumulation. Thus the response characteristics developed during the culture of glial cells could reflect the influence of culture conditions, including the absence of neurons, rather than the inherent properties of the cells that can be expected to be expressed *in situ*.

Ontogenetic Studies

In a study completed recently, the effects of NE, HIST, and adenosine, singly or in combinations, on the accumulation of cyclic AMP in slices of cerebral cortex, were examined using tissue from guinea pigs at 40 to 68 days of gestation (27 days before to 1 day after birth) (29). A response to adenosine was not detectable until 44 days of gestation; after 46 days of gestation, this response developed rapidly, reaching a more than 40-fold stimulation 9 days later. A synergistic response to the combination of adenosine with either NE or HIST first became apparent at 42 days of gestation; the degree of synergism increased for 4 to 5 days when the effect of a combination of adenosine and either NE or HIST reached three- to fivefold the sum of the individual responses. Thereafter, as the response to adenosine by itself increased, the calculated degree of synergism declined toward the adult value. Between days 41 and 46 of gestation, the brain weight of the guinea pig increases about 80% without an appreciable increase in total DNA content (7); there is also histologic evidence for a rapid increase in the number and size of nerve processes and for "maturation" of neuronal nucleoli (20). It is also during this period that the synergistic responses to adenosine and the biogenic amines begin a rapid development; this correlation may indicate that the growing and maturing neuronal elements are the principal sites of these synergistic responses. Between days 46 and 56 of gestation, the total brain DNA content approximately doubles, reaching nearly 75% of the adult value (7); this is followed by a burst of cholesterol accumulation and further DNA accumulation to the adult value by birth. This spurt of DNA and cholesterol accumulation has been attributed to a major wave of glial proliferation and "maturation" (7), and corresponds in time to the rapid development of the response to adenosine by itself and to a decline in the degree of synergism between adenosine and the amines (29). This correlation may indicate that glial elements are the major site of response to adenosine by itself. However, these suggestions assume that the response characteristics of the cells (neurons) appearing and differentiating earlier are not markedly altered during subsequent developmental periods, which may or may not be true.

Regional and Species Variation in Responses to Adenosine

The literature and the investigators dealing with the regulation of cyclic AMP accumulation in brain slices have been plagued by variations in the pattern and characteristics of responses to various agents or depolarizing maneuvers as a function of the brain region, species, and even strain utilized. There are really only two responses that have been consistently observed. First, cerebellar tissue from all mammalian species tested is prominently responsive to adrenergic agonists. Second, and more important for this discussion, adenosine by itself produces large increases in cyclic AMP accumulation in all regions of the central nervous system of all mammalian species examined so far, from rodents to primates. In nearly all cases, adenosine is more effective than any other individual agent, especially in cases in which the participation of endogenous adenosine has been excluded in the evaluation of the other agents. By contrast, synergistic responses to adenosine in combination with other agents or with depolarizing stimuli vary dramatically in pattern and intensity as a function of region and/or species examined. Since there is ample precedent for neuronal differentiation expressed by selective responsiveness to various neurotransmitter substances, it seems logical to ascribe at least regional variation in synergistic responses involving adenosine to differences in the response characteristics of neuronal elements. Species and strain variations might be rationalized in a similar fashion. The complimentary conclusion would be that since differences in properties of glial elements in various regions of the central nervous system have not been noted, the large response to adenosine by itself resides predominantly in these cell types. Both conclusions rely heavily on the virtually untested assumption that the hormonal responsivity of the various glial cell types will not be different regardless of the differences in humoral substances elaborated by their neuronal neighbors.

MECHANISMS OF ADENOSINE ACTIONS

Evidence for Adenosine Receptors

Since adenosine presented to brain slices can be incorporated into intracellular adenosine triphosphate (ATP) and cyclic AMP (23,28), one possible mechanism by which adenosine might increase cyclic AMP accumulation could involve augmenting the availability of substrate in the immediate vicinity of adenylate cyclase. However, several lines of evidence argue against this possibility. These include: (a) theophylline, although competitively inhibiting adenosine stimulation of cyclic AMP accumulation, does *not* reduce the incorporation of adenosine into the ATP of slices (11,23); (b) compounds that do inhibit adenosine uptake by brain slices and incorporation into ATP also potentiate the accumulation of cyclic AMP elicited by low concentrations of adenosine (11); and (c) the marked stimulatory effects of radioactive 2-chloroadenosine on cyclic AMP accumulation in cultured Va-2 cells are not accompanied by significant formation of 2-chloro ATP, 2-chloro cyclic

AMP, or radioactive cyclic AMP (30). The dissociation of adenosine penetration and cyclic AMP accumulation coupled with the feeble potency of adenosine as a PDE inhibitor and with theophylline antagonism of adenosine action (23) would argue against the possibility that adenosine exerted its effects by directly interacting with enzymes hydrolyzing cyclic AMP.

The arguments against these two possible mechanisms can thus be used in favor of the existence of an "external" adenosine receptor stimulating adenylate cyclase or inhibiting PDE. This view is further supported by the observation that a restricted number of purine analogs of adenosine are effective stimulators of cyclic AMP accumulation (12). Theophylline and certain other methylxanthines therefore appear to be competitive receptor antagonists. Although 5′-AMP and other adenine nucleotides were potent stimulators of cyclic AMP accumulation in brain slices (23), it seems more appropriate to use the term "adenosine receptor" since the effects of these substances, but not of certain adenosine analogs, are severely reduced when adenosine deaminase is included in incubation media (15).

Evidence for Adenosine Stimulation of Adenylate Cyclase

In the initial study, no evidence was found for adenosine stimulation of adenylate cyclase prepared from homogenates of guinea pig cerebral cortex (23). However, recently Premont et al. (21) showed that the activity of adenylate cyclase preparations from rat caudate nucleus was reduced by adenosine deaminase; in the presence of this enzyme, the rate of accumulation of cyclic AMP was increased up to twofold by deaminase-insensitive analogs of adenosine at concentrations as low as 1 μM. The dose-dependent stimulation of this enzyme was competitively antagonized by theophylline. This demonstration of an adenosine-stimulated adenylate cyclase from brain tissue had been preceded by the observations of Haslam and Lynham (10) on the platelet enzyme of Blume and Foster (2) on neuroblastoma adenylate cyclase, and of others. It should be pointed out that McKenzie and Bär (17) observed inhibitory effects of adenosine on brain adenylate cyclase. However, Londos and Wolff (14) have recently provided evidence for two adenosine-sensitive sites on the adenylate cyclase of brain and other tissues. One, called the "P" site, interacts with a limited number of purine analogs of adenosine, is blocked by methylxanthines and, with a few exceptions including adipose tissue, results in increased adenylate cyclase activity. The other, called the "R" site, interacts with adenosine at higher concentrations and with a limited number of ribose analogs of adenosine, leading invariably to reduced adenylate cyclase activity. These two actions of adenosine may account for the hypersteep, bell-shaped dose-response curves observed in brain slices (23).

The demonstration of adenosine-sensitive adenylate cyclase activity in brain was preceded by the observation in slices of guinea pig cerebral cortex exposed to short pulses of radioactive inorganic phosphate that low concentrations of adenosine increased the specific activity of cyclic AMP relative to that of the phosphate of ATP (22). The radioactive phosphate was not added until after cyclic AMP had been

TABLE 1. *Effects of adenosine or PDE inhibitors on incorporation of [³H]adenine into cyclic AMP in the presence of HIST*

Additions	Cyclic AMP Accumulation (pmole/mg) protein)	Specific activity (cpm/nmole)		
		Cyclic AMP	ATP	Cyclic AMP/ATP
HIST (100 μM) + adenosine (10 μM) + theophylline (200 μM)	74	5,040	1,820	2.76
HIST (100 μM) + adenosine (10 μM) + theophylline (500 μM) + isobutylmethylxanthine (500 μM)	386	2,250	2,230	1.01
HIST (100 μM) + adenosine (10 μM)	347	5,250	1,840	2.85

Individual 240- to 300-mg portions of slices of adult Hartley guinea pigs were preincubated for 50 min before being transferred to fresh medium containing 100 μM HIST and the indicated additions. After 5.5 min of incubation, 85 μCi of [^{3}H]adenine was added to each sample and the incubation continued for 1.5 min. The slices were frozen in liquid N_2 and extracted with methanolic HCl (24), and a small portion of the extract was assayed for total cyclic AMP (9). The extracts were purified by column chromatography using both ion-exchange resins and neutral alumina. The ATP content of appropriate isolated fractions was estimated by ultraviolet spectrometry, whereas cyclic AMP was determined by radioimmunoassay (9).

allowed to accumulate to a steady state value of about sixfold above control. This result indicated that a major portion of the action of adenosine was accomplished by increasing the rate of formation of cyclic AMP rather than by decreasing the rate of its degradation and/or extrusion into the incubation medium.

Since no adenylate cyclase preparation has yet displayed synergistic responses to adenosine combined with another agent, similar pulse-labeling experiments with brain slices have been attempted recently. The results of a typical experiment are presented in Table 1. In this example, slices of guinea pig cerebral cortex were exposed to maximal amounts of HIST and a low concentration of adenosine (10 μM) in the presence and absence of methylxanthines for a period of time sufficient to allow nearly maximal accumulation of cyclic AMP. This was followed by a brief exposure to radioactive adenine. It can be seen that the combination of HIST and adenosine alone produced a level of cyclic AMP nearly fivefold that achieved in the presence of 200 μM theophylline, an amount sufficient to completely block the effects of adenosine. However, the relative specific activity of the larger pool of cyclic AMP was at least as great as the smaller pool. In the presence of the potent PDE inhibitor, isobutylmethylxanthine, and 500 μM theophylline, cyclic AMP accumulation was also raised about fivefold, but the relative specific activity of cyclic AMP was reduced more than 2.5-fold.

Taken together, these data indicate that adenosine receptor activation resulted in an increased rate of formation of cyclic AMP to maintain the specific activity of an enlarged pool of product. If adenosine had acted predominantly to slow degradation and/or extrusion of cyclic AMP, then a result resembling the effect of isobutylmethylxanthine would have been expected.

Potential Mechanisms of Synergistic Stimulation of Cyclic AMP Formation

If we accept for the time being the proposition that the synergistic effects of adenosine result from an augmentation of the rate of formation of cyclic AMP, the possible mechanisms involved fall into two broad categories: (a) effector-effector interactions, in which the consequences of one agonist-receptor interaction would serve to augment the relative stimulation of adenylate cyclase activity produced by another agonist-receptor interaction; and (b) receptor-receptor interactions, in which either the agonist-receptor binding or the maximal receptor–adenylate cyclase ''coupling'' associated with one agonist required or was promoted by another ''cooperating'' agonist-receptor interaction. In both of these categories, the agonists ultimately interacting need not be the same as those added to incubation media and the complication of mediation by unknown humoral agents must be considered. In this event one special category distinguishable from those listed above might exist in which one agent would elicit the release of mediators with both positive and negative effects on adenylate cyclase, and the other would selectively suppress the release of the inhibitory mediators while also exerting its independent stimulatory action on adenylate cyclase.

Since synergistic effects of low concentrations of adenosine on cyclic AMP accumulation have so far been observed only in brain slice systems, it is no surprise that little evidence exists that would allow a choice among these categories. However, there are some recent interesting observations dealing with manipulation of Ca^{2+} ion concentration in incubation media that may be related to the first category of mechanisms. Schwabe and Daly (27) have reported that the addition of EGTA in amounts exceeding Ca^{2+} ion concentration transforms the interaction between NE and adenosine in slices of rat cerebral cortex from twofold additive to additive. As in guinea pig cerebral slices (24), the interaction appears to involve α-adrenergic receptors (19), and EGTA did not alter the effect of the β-agonist, isoproterenol. These observations can be coupled with those of Bradham et al. (3) that both particulate and detergent-dispersed preparations of adenylate cyclase from various brain regions of a number of species were inhibited more than 50% by EGTA; this inhibition was reversed by Ca^{2+} ions. Subsequently it was found that the same calcium-binding protein that activated species of brain PDE was also required for the Ca^{2+} ion stimulation of solubilized brain adenylate cyclase (4,5). Therefore, there is the possibility that NE, via α-adrenergic reçeptors, would bring about an increased influx of Ca^{2+} ions, which in conjunction with the calcium-binding protein would magnify the adenylate cyclase activity stimulated by adenosine-receptor interaction. This proposition leaves a number of questions unanswered, among them the reason why the increased Ca^{2+} ion influx would not also lead to an increase in PDE activity, opposing and perhaps overwhelming any increase in the

TABLE 2. *Effect of calcium deprivation on response to NE in the presence of adenosine: adult guinea pig cerebral cortical tissue*

Final incubation conditions	Cyclic AMP accumulation: Total (pmoles/mg protein)	Cyclic AMP accumulation: NE increment (pmoles/mg protein)
5 min, 0 Ca^{2+} + EGTA (100 μM)		
Adenosine	311 ± 1.6	—
Adenosine + NE	412 ± 6.4	101
5 min, 0.8 mM Ca^{2+}		
Adenosine	265 ± 7.2	—
Adenosine + NE	377 ± 19	112
10 min, 0 Ca^{2+} + EGTA (100 μM)		
Adenosine	427 ± 29	—
Adenosine + NE	490 ± 49	63
10 min, 0.8 mM Ca^{2+}		
Adenosine	346 ± 20	—
Adenosine + NE	553 ± 41	207
15 min, 0 Ca^{2+} + EGTA (100 μM)		
Adenosine	456 ± 16	—
Adenosine + NE	516 ± 22[a]	60
15 min, 0.8 mM Ca^{2+}		
Adenosine	365 ± 27	—
Adenosine + NE	684 ± 37	319

Slices of adult Hartley guinea pig cerebral cortex (30 to 75 mg portions) were incubated at 37°C for 50 min in Krebs Ringer Bicarbonate Glucose (KRBG) solution containing 0.8 mM $CaCl_2$. The slices were transferred to KRBG containing either no $CaCl_2$ and 100 μM EGTA or 0.8 mM $CaCl_2$ and incubated for an additional 10 min. The slices were then transferred to media containing the indicated additions for an additional 5-, 10-, or 15-min period. The calcium concentration during the second and third incubation period was identical. When present, the concentration of adenosine was 200 μM and that of NE was 50 μM. The values represent the mean ± SEM of three samples treated identically. In the absence of adenosine, NE usually produced an increment of about 10 pmoles/mg protein.

[a] Significantly greater than adenosine only, $p < 0.05$

rate of formation of cyclic AMP. There are a number of alternative interpretations of the data of Schwabe and Daly (27), including the possibility that NE acts by causing the release of some active substances(s) by a Ca^{2+}-dependent mechanism.

The impact of calcium deprivation on the response to NE in the presence of adenosine in slices of guinea pig cerebral cortex has been variable and seems to depend on the duration both of the calcium deprivation and of the exposure to the agonists. An experiment using adult brain tissue is summarized in Table 2. It can be seen that after 10 min of equilibration with 0 Ca^{2+} + 100 μM EGTA the increment resulting from NE is only slightly reduced during the first 5 min of exposure to a combination of NE and adenosine. With longer exposures, the NE increment is severely reduced, almost to extinction; this was a consequence both of an increased

TABLE 3. *Effect of calcium deprivation on response to NE in the presence of adenosine: fetal guinea pig cerebral cortical tissue*

Final incubation conditions	Cyclic AMP accumulation	
	Total	NE increment
	(pmoles/mg protein)	
10 min, 0 Ca^{2+} + EGTA (100 μM)		
Control	13 ± 3.1	—
Adenosine	30 ± 1.4	—
Adenosine + NE	75 ± 7.0	45
10 min, 0.8 mM Ca^{2+}		
Adenosine	30 ± 1.5	—
Adenosine + NE	66 ± 4.1	36
10 min, 1.1 mM Ca^{2+} + EGTA (100 μM)		
Control	10 ± 0.7	—
Adenosine	34 ± 9.2	—
Adenosine + NE	93 ± 3.9	59

Slices of fetal strain 2 guinea pigs (47 days gestation) were incubated for 50 min at 37°C in KRBG solution containing 0.8 mM $CaCl_2$. The slices were transferred to KRBG containing either no $CaCl_2$ and 100 μM EGTA or 0.8 mM $CaCl_2$ and incubated for an additional 10 min. The slices were then transferred to media containing the indicated additions for an additional 10-min period. Samples incubated with 0.8 mM $CaCl_2$ during the final 10 min were exposed to 0.8 mM $CaCl_2$ during the preceding 10 min. Samples incubated with EGTA (± Ca^{2+}) during the final 10 min were exposed to 0 Ca^{2+} + 100 μM EGTA during the preceding 10 min. When present, the concentration of adenosine was 200 μM and that of NE was 50 μM. The values represent the mean ± SEM of three samples treated identically. In the absence of adenosine, NE usually produced an increment of about 10 pmoles/mg protein.

accumulation of cyclic AMP with adenosine alone in the absence of calcium and of a decreased accumulation with both agents present in the absence of calcium. By contrast, calcium deprivation had relatively little impact on the response to a combination of NE and adenosine in slices of fetal guinea pig cerebral cortex (Table 3). The meaning of these data is not at all clear. It may mean that several mechanisms are operating, some calcium "dependent" and some not. Alternatively, it may mean that the availability of Ca^{2+} ions to certain intracellular structures have a variable relationship to the presence or absence of extracellular calcium, at times even an inverse relationship.

CONCLUDING REMARKS

Three lines of evidence, namely, cultures of rat brain glial cells, ontogeny of guinea pig cerebral cortex, and regional variation or lack of variation in responsiveness, all seem to point to the conclusion that the major share of cyclic AMP accumulation incited by adenosine alone resides in glial elements of the central nervous system whereas the synergistic actions of adenosine are predominantly properties of neurons. Each of these lines of evidence is flawed, but somewhat different assumptions are involved in each. Verification of this conclusion will probably require the ability to detect and quantify localized accumulations of cyclic AMP at least at the cellular level. This task appears to be beyond the histochemical techniques currently available and attempts to develop semiquantitative immunohistochemical procedures are presently under way in this laboratory.

Progress in understanding the mechanisms involved in the synergistic actions of adenosine has been painfully slow. It is unlikely that further study of brain slice systems will yield important new information. This is most unfortunate since brain slices have been the only experimental system in which this phenomenon has been reliably observed, at least with regard to the regulation of cyclic AMP accumulation. It will be important to explore various dispersed cell, cultured cell, and broken cell systems for evidence of this phenomenon. However, if some of the possible mechanisms considered above (e.g., mediation by endogenous humoral agents) are actually involved, such exploration will yield negative results.

ACKNOWLEDGMENTS

The author wishes to acknowledge the collaboration of Dr. Richard Lehne and the skillful technical assistance of Mr. James Gibson, as well as support by a grant from the USPHS (NS 12764).

REFERENCES

1. Blume, A. J., Dalton, C., and Sheppard, H. (1973): *Proc. Natl. Acad. Sci. U.S.A.*, 70:3099–3102.
2. Blume, A. J., and Foster, C. J. (1975): *J. Biol. Chem.*, 250:5003–5008.
3. Bradham, L. S., Holt, D. A., and Sims, M. (1970): *Biochim. Biophys. Acta*, 201:250–260.
4. Brostrom, C. O., Huang, Y-C., Breckenridge, B. M., and Wolff, D. J. (1975): *Proc. Natl. Acad. Sci. U.S.A.*, 72:64–68.
5. Cheung, W. Y., Bradham, L. S., Lynch, T. J., Lin, Y. M., and Tallant, E. A. (1975): *Biochem. Biophys. Res. Commun.*, 68:616–625.
6. Clark, R. B., Gross, R., Su, Y. F., and Perkins, J. R. (1974): *J. Biol. Chem.*, 249:5296–5303.
7. Dobbing, J., and Sands, J. (1970): *Brain Res.*, 17:115–123.
8. Gilman, A. G., and Schrier, B. K. (1972): *Mol. Pharmacol.*, 8:410–416.
9. Harper, J., and Brooker, G. (1975): *J. Cyclic Nucleotide Res.*, 1:207–214.
10. Haslam, R. J., and Lynham, J. A. (1972): *Life Sci.*, 11(II):1143–1151.
11. Huang, M., and Daly, J. W. (1974): *Life Sci.*, 14:489–502.
12. Huang, M., Shimizu, H., and Daly, J. W. (1972): *J. Med. Chem.*, 15:462–466.
13. Kakiuchi, S., Rall, T. W., and McIlwain, H. (1969): *J. Neurochem.*, 16:485–491.
14. Londos, C., and Wolff, J. (1977): *Proc. Natl. Acad. Sci. U.S.A.*, 74:5482–5486.
15. Mah, H. D., and Daly, J. W. (1976): *Pharmacol. Res. Commun.*, 8:65–79.
16. McCarthy, K. D., and de Vellis, J. (1978): *J. Cyclic Nucleotide Res.*, 4:15–26.
17. McKenzie, S. G., and Bär, H. P. (1973): *Can. J. Physiol. Pharmacol.*, 51:190–196.
18. Perkins, J. P., and Moore, M. (1973): *Mol. Pharmacol.*, 9:774–782.
19. Perkins, J. P., Moore, M., Kalisker, A., and Su, Y-F. (1975): *Adv. Cyclic Nucleotide Res.*, 5:641–660.
20. Peters, V. B., and Flexner, L. B. (1950): *Am. J. Anat.*, 86:133–162.
21. Premont, J., Perez, M., and Bochaert, J. (1977): *Mol. Pharmacol.*, 13:662–670.
22. Rall, T. W., and Sattin, A. (1970): *Adv. Biochem. Psychopharmacol.*, 3:113–133.
23. Sattin, A., and Rall, T. W. (1970): *Mol. Pharmacol.*, 6:13–23.
24. Sattin, A., Rall, T. W., and Zanella, J. (1975): *J. Pharmacol. Exp. Ther.*, 192:22–32.

25. Schrier, B. K. (1973): *J. Neurobiol.,* 4:117–124.
26. Schultz, J., and Hamprecht, B. (1973): *Naunyn Schmiedebergs Arch. Pharmacol.,* 278:215–225.
27. Schwabe, U., and Daly, J. W. (1977): *J. Pharmacol. Exp. Ther.,* 202:134–143.
28. Shimizu, H., and Daly, J. W. (1970): *Biochim. Biophys. Acta,* 222:465–473.
29. Shonk, R. F., and Rall, T. W. (1978): *J. Neurochem. (Submitted for publication).*
30. Sturgill, T. W., Schrier, M. B. K., and Gilman, A. G. (1975): *J. Cyclic Nucleotide Res.,* 1:21–30.

Physiological and Regulatory Functions of Adenosine and Adenine Nucleotides, edited by H. P. Baer and G. I. Drummond. Raven Press, New York © 1979.

Adenosine and Cyclic Adenosine Monophosphate-Generating Systems in Brain Tissue

John W. Daly

National Institute of Arthritis, Metabolism, and Digestive Diseases, National Institutes of Health, Bethesda, Maryland 20014

Nearly a decade ago, it was observed that adenosine markedly stimulates cyclic adenosine monophosphate (cyclic AMP) formation in brain tissue (32). Subsequently, adenosine was found to inhibit the spontaneous firing of virtually all central neurons (16,25,26) and to inhibit evoked synaptic potentials in certain central pathways (22). The site and mechanism of inhibitory action of adenosine on neuronal activity and synaptic transmission are key questions with respect to the possible physiological function of adenosine in the central nervous system. At present it appears likely that adenosine-elicited accumulation of cyclic AMP at postsynaptic sites is involved in the inhibition of neuronal activity. Effects at presynaptic sites, particularly with respect to inhibition of synaptic transmission and effects via mechanisms not involving cyclic nucleotides may also pertain. Further studies are needed before the biochemistry and electrophysiology of adenosine can be definitively correlated. An overview of the current knowledge of the interactions of adenosine with cyclic AMP systems in brain tissue may provide insights into new directions and areas for such correlative research.

ADENOSINE AND CYCLIC AMP FORMATION IN BRAIN TISSUE

Adenosine activates cyclic AMP formation in brain slices (3) and in homogenates (27) presumably through interaction with an adenosine receptor coupled to adenylate cyclase. The true K_a of adenosine for such a receptor appears to be less than 10 μM. In some tissues, fat cells for example, adenosine inhibits adenylate cyclase (9). The K_I appears to be less than 10 μM. Thus adenosine receptors may have either stimulatory or inhibitory effects on adenylate cyclase, a fact that is often overlooked in considering adenosine as a stimulant only of cyclic AMP formation in brain. Adenosine also interacts with phosphodiesterases (32) and with cyclic AMP–dependent protein kinases (20), but 50% inhibition of these enzymes by adenosine requires concentrations greater than 500 and 100 μM, respectively. Thus it appears likely that the major physiological site at which adenosine influences cyclic AMP levels is an adenosine receptor coupled to adenylate cyclase. Adenosine has also been reported to enhance levels of cyclic guanosine monophosphate (cyclic GMP) in brain slices (21,30), but this effect has not yet been studied in detail.

RELEASE, METABOLISM, AND ACTION OF ADENOSINE IN BRAIN TISSUE

Activation of adenylate cyclases in brain tissue is a biochemical property exhibited by many putative neurotransmitters and adenosine (3). However, no evidence for storage of adenosine within synaptic vesicles or calcium-dependent quantal release of adenosine has been observed. Instead, it appears that enhanced levels of adenosine observed in brain tissue during electrical stimulation (28), anoxia (29), or convulsions (31) result from a steady accumulation of 5′-AMP and then adenosine as levels of adenosine triphosphate (ATP) decline under such conditions. Levels of adenosine probably increase in brain tissue whenever ATP levels decrease owing to excessive expenditure of energy. This increase in adenosine might provide a mechanism for direct inhibitory feedback or transynaptic inhibition of systems in rapidly firing neurons. Further studies are required to establish the validity of such a model. Another possible source of adenosine is, however, from ATP that is quantally released together with many neurotransmitters.

Although no evidence for storage and release of adenosine has been obtained, mechanisms for the inactivation of adenosine exist similar to those known for biogenic amines. Two relatively high-affinity enzymes metabolize adenosine. One is adenosine deaminase (K_m, 50 μM) (49) which converts adenosine to inactive inosine (18). The other enzyme is adenosine kinase (K_m, 20 μM) (45) which converts adenosine to 5′-AMP. In addition to the metabolic enzymes, a high-affinity uptake system (K_a, 1 to 5 μM) (11) for adenosine exists.

Adenosine (EC_{50}~ 30 μM) elicits accumulations of cyclic AMP in brain tissue from virtually every species and brain region that has been investigated (3). In ganglia, adenosine does *not* appear to cause accumulations of cyclic AMP. Adenosine, ATP, adenosine diphosphate (ADP), and 5′-AMP elicit similar accumulations of cyclic AMP in guinea pig cerebral cortical slices. The nucleotides appear somewhat less potent than adenosine itself. Indeed, a variety of data (18,32) suggest that the nucleotides are inactive and must be hydrolyzed to adenosine in order to activate cyclic AMP formation. This point still requires further investigation since in certain glioma cell lines, 5′-AMP appears to directly activate cyclic AMP formation (2).

Adenosine elicits an accumulation of cyclic AMP in vesicles ("microsacs") present in particulate fractions from Krebs-Ringer homogenates of brain tissue (1). In such "microsacs" from guinea pig cerebral cortex, adenosine (EC_{50}~ 10 μM) elicits a twofold accumulation of cyclic AMP in contrast to the 15- to 30-fold accumulation elicited in slices from the same tissue.

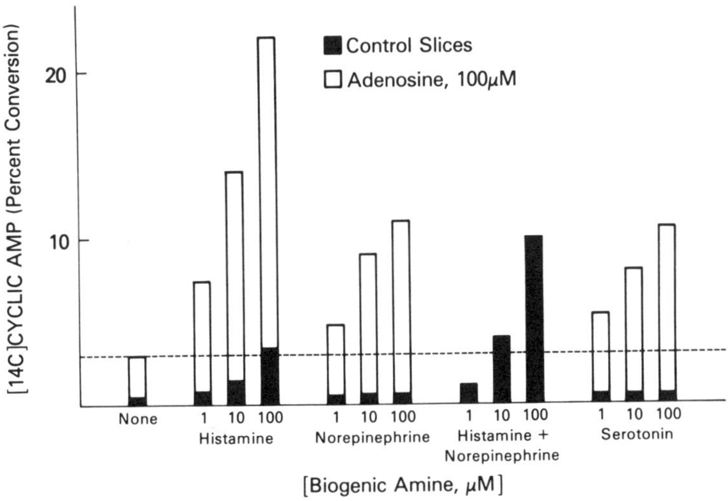

FIG. 1. Amine-elicited accumulation of cyclic AMP in adenine-labeled guinea pig cerebral cortical slices. Incubations were for 10 min in the presence of agonists. Solid bars represent response to amines. Total height of bars represents response in the presence of adenosine. (Data from ref. 14.)

Adenosine ($EC_{50}\sim$ 10 μM) has now been shown to activate adenylate cyclases in cell-free homogenates from rat striatum (27), an effect blocked by theophylline.

ADENOSINE AND BIOGENIC AMINES

One remarkable aspect of the effects of adenosine in brain slices is the marked synergistic response of cyclic AMP–generating systems to biogenic amine-adenosine combinations. Greater than additive effects on accumulation of cyclic AMP have been observed for norepinephrine-adenosine, histamine-adenosine, and serotonin-adenosine (14) (Fig. 1) but not for dopamine-adenosine. Histamine-norepinephrine combinations have greater than additive effects on cyclic AMP levels in guinea pig cerebral cortical slices. Endogenous adenosine may well be involved in the synergistic response to histamine-norepinephrine combinations (13).

Norepinephrine does not elicit a detectable accumulation of cyclic AMP in rat cerebral cortical slices until 11 days after birth (23). A marginal response to adenosine in cortical slices was observed on the fifth day after birth and this response was potentiated nearly fourfold in the presence of norepinephrine. Thus synergisms between adenosine and amines can be observed during development when responses to the biogenic amine cannot be detected.

The nature of the amine receptors involved in the synergistic interaction with adenosine has been investigated for both norepinephrine and histamine. In both guinea pig and rat cerebral cortical slices the synergistic interaction of norepineph-

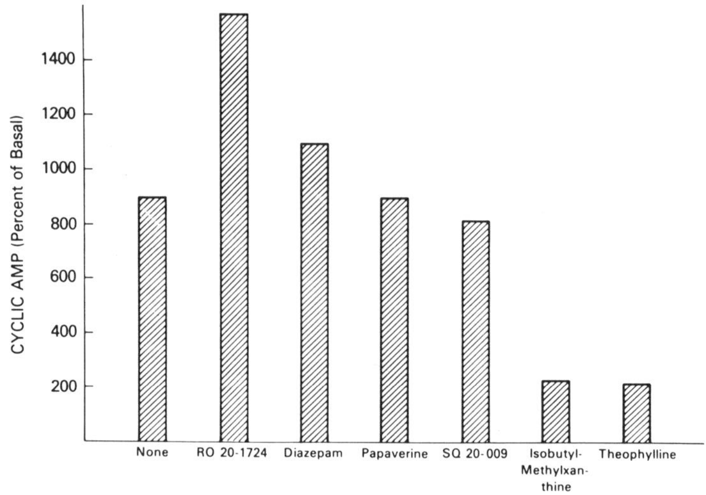

FIG. 2. Phosphodiesterase inhibitors: Effect on adenosine-elicited accumulations of cyclic AMP in adenine-labeled guinea pig cerebral cortical slices. Incubations were for 10 min in the presence of 100 μM adenosine. Phosphodiesterase inhibitors (200 μM) were added 2 min prior to adenosine. (Data from ref. 18.)

rine and adenosine appears to involve α-adrenergic receptors (24,33,38). Similarly, the synergistic interaction of histamine and adenosine appears to involve primarily H_1-histaminergic receptors (4,7).

ADENOSINE AND PHOSPHODIESTERASE INHIBITORS

Certain phosphodiesterase inhibitors markedly potentiate accumulations of cyclic AMP elicited by adenosine. These include benzyl-2-imidazolidinones such as RO 20–1724, and the structurally related phenyl-2-pyrrolidone, ZK 62711. Both classes potentiate the adenosine response about twofold (18,35,40) (Fig. 2). Diazepam causes an approximately 1.5-fold potentiation of the adenosine response, whereas papaverine and a pyrazolopyridine, SQ 20009, have little effect. Alkylxanthines such as theophylline and isobutylmethylxanthine unexpectedly cause a marked inhibition of the adenosine response (18,32). This inhibition by alkylxanthines is competitive in nature and appears unrelated to their activity as phosphodiesterase inhibitors. Thus 3-isobutyl-1-methylxanthine, which is 10-fold more potent than theophylline as a phosphodiesterase inhibitor, is nearly equipotent to theophylline as an adenosine antagonist. Current efforts to obtain an alkylxanthine with low activity as a phosphodiesterase inhibitor and high activity as an adenosine antagonist or the converse have not been completely successful. 1,3-Dibutylxanthine, however, is about 10-fold less potent as a phosphodiesterase inhibitor than isobutylmethylxanthine and about fivefold more potent than isobutylmethylxanthine as an adenosine antagonist (F. Smellie, C. Davis, and J. Daly, *unpublished data*). Conversely, 8-methylisobutylmethylxanthine is a potent phosphodiesterase inhibitor, and it is about twofold less potent than either theophylline or isobutylmethylxanthine as an adenosine antagonist.

ANALOGS OF ADENOSINE

Another approach to the discovery of a potent and specific adenosine antagonist was to investigate various adenosine analogs. In most cases modification of the ribose moiety of adenosine led to loss of agonist activity (15,18,32). Many of these analogs were, however, antagonists (Fig. 3). It remains unclear to what extent certain of these "adenosine antagonists," for example, adenine xylofuranoside, are merely general inhibitors of adenylate cyclase. In addition, certain of these "adenosine antagonists" may have weak partial agonist activity. With respect to inhibition of electrical activity of cortical neurons, an "antagonist," 2′-deoxyadenosine, did mimic the depressant effects of adenosine (25,26).

Analogs of adenosine modified in the purine moiety often retained or had enhanced agonist activity. Two examples might be cited: 2-Chloroadenosine ($EC_{50} \sim 10\ \mu M$) represents a potent adenosine analog (12) that is not a substrate for adenosine deaminase nor an effective substrate for the high-affinity uptake system. Thus for this adenosine analog incorporation into intracellular ATP pools plays no role in the stimulation of cyclic AMP formation. N^6-benzyladenosine ($EC_{50} \sim 30\ \mu M$) or N^6-phenylisopropyladenosine ($EC_{50} \sim 20\ \mu M$) represent two more potent

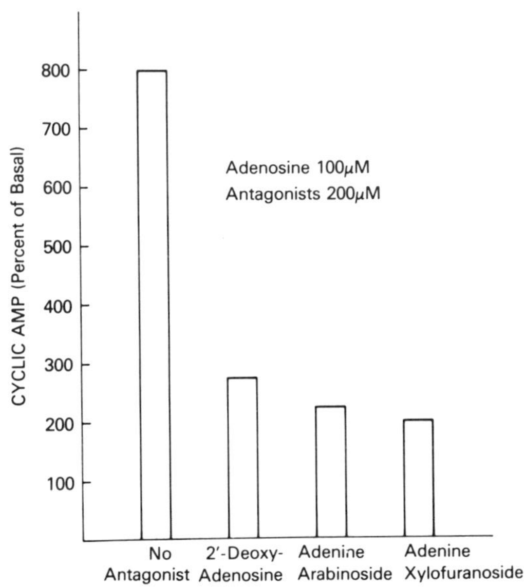

FIG. 3. Adenosine analogs: Inhibition of adenosine-elicited accumulations of cyclic AMP in adenine-labeled guinea pig cerebral cortical slices. Incubations were for 10 min in the presence of 100 μM adenosine. Analogs (200 μM) were added 2 min prior to adenosine. (Data from refs. 15 and 18.)

adenosine analogs (12,41) that are not substrates for adenosine deaminase. The latter compound exists in D- and L-enantiomers. The L-isomer is manyfold more potent than the D-isomer both with respect to eliciting accumulations of cyclic AMP in brain slices and with respect to inhibiting evoked synaptic potentials in a hippocampal slice (48).

UPTAKE AND DEAMINATION OF ADENOSINE

Inhibition of uptake of adenosine by low concentrations of dipyridamole, hexobendine, 6-nitrobenzylthioguanosine, or papaverine markedly potentiates the accumulation of cyclic AMP elicited by submaximal concentrations of exogenous adenosine (12) (Table 1). Under these conditions the EC_{50} for adenosine in brain slices is reduced about threefold from 30 μM to less than 10 μM. The adenosine-uptake inhibitors also potentiate responses to biogenic amines such as histamine or norepinephrine, apparently by inhibition of reuptake of endogenous adenosine within the brain slice. Endogenous adenosine apparently then reaches levels sufficient to synergistically potentiate aminc responses. Basal levels of cyclic AMP in brain slices are somewhat increased by dipyridamole, hexobendine, and papaverine. Although certain of the uptake inhibitors—most notably papaverine and dipyridamole—are also phosphodiesterase inhibitors it is unlikely that this property contributes significantly to the potentiation of adenosine and amine responses at the low concentrations used. Indeed, dipyridamole and papaverine at low concentrations have no effect on amine responses if theophylline is present to block adenosine receptors (12). The results point to the importance of high-affinity uptake mechanisms in the control of low concentrations of adenosine in brain tissue.

A very potent and irreversible inhibitor of adenosine deaminase, 2′-deoxy-

TABLE 1. *Adenosine-elicited accumulation of cyclic AMP in guinea pig cerebral cortical slices: effect of compounds that inhibit uptake of adenosine*

Uptake inhibitor	Inhibition of adenosine uptake (%)	Potentiation of accumulation of cyclic AMP elicited by adenosine (%)
None	0	0
Dipyridamole (10 μM)	71	300
Hexobendine (10 μM)	70	300
6-(4-Nitrobenzyl)-thioguanosine (4 μM)	46	220
Papaverine (20 μM)	48	190

Guinea pig cerebral cortical slices were incubated with 10 μM adenosine for 10 min. In separate experiments the percent inhibition of uptake of adenosine and the percent potentiation of adenosine-elicited accumulation of radioactive cyclic AMP in adenine-labeled slices were ascertained.

Data from Huang and Daly, ref. 12.

coformycin (K_I, 0.2 nM), has little effect on basal levels of cyclic AMP in rat cerebral cortical slices, but markedly potentiates responses to norepinephrine (46) (Table 2). The potentiation of the norepinephrine response by deoxycoformycin is markedly reduced by theophylline. Thus, as with adenosine-uptake inhibitors, an inhibitor of adenosine deaminase appears to cause an increase in endogenous adenosine in brain slices sufficient to potentiate amine responses. In further studies, 10 μM dipyridamole potentiated responses to amines to a somewhat greater extent than did 10 nM deoxycoformycin (P. Skolnick, Y. Nimitkitpaisan, and J. Daly, *unpublished data)*. Combinations of dipyridamole and deoxycoformycin potentiated amine responses to a somewhat greater extent than either agent alone. Thus both reuptake and deamination appear involved in maintaining low extracellular levels of adenosine in brain tissue.

ADENOSINE DEAMINASE

Based on a variety of data with inhibitors and antagonists, endogenous adenosine appears to be involved in the regulation of cyclic AMP levels in brain tissue. Adenosine deaminase added to incubation media should greatly reduce levels of endogenous adenosine in the extracellular space. Thus this enzyme provides a potentially useful tool for the investigation of roles for endogenous adenosine under a variety of conditions (see ref. 13). A summary of data on the effect of the adenosine antagonists, theophylline and 2′-deoxyadenosine, and of adenosine deaminase on stimulation of cyclic AMP formation by a variety of agents is presented in Table 3. The response to adenosine and to ATP was markedly inhibited by all agents. The inhibition of the ATP response by adenosine deaminase suggests that exogenous ATP must be hydrolyzed to adenosine prior to stimulation of cyclic AMP–generating systems. The marked inhibitions of the accumulation of cyclic AMP elicited under depolarizing conditions (veratridine, electrical stimulation) suggest an involvement of enhanced levels of extracellular adenosine in the re-

TABLE 2. *Cyclic AMP in rat cerebral cortical slices: effect of inhibition of adenosine deaminase*

Stimulant	2′-Deoxycoformycin (0.01 μM)	Cyclic AMP (pmoles/mg protein)
None	−	10.7 ± 1.4
	+	12.9 ± 1.7
Norepinephrine (30 μM)	−	77.7 ± 9.5
	+	123.0 ± 13.7
Norepinephrine + theophylline (0.5 mM)	−	44.3 ± 3.3
	+	67.0 ± 6.0

Rat cerebral cortical slices were incubated for 10 min with norepinephrine. Other agents were added 5 min prior to the norepinephrine.

Data from Skolnick et al., ref. 46.

TABLE 3. *Effect of "adenosine antagonists" on accumulation of cyclic AMP in guinea pig cerebral cortical slices*

	Percent inhibition by		
Agonist	Theophylline (200 μM)	2′-Deoxyadenosine (200 μM)	Adenosine deaminase (10 μg/ml)
Adenosine (20 μM)	98	82	90
ATP (100 μM)	62	56	84
Electrical stimulation	81	—	67
Veratridine (8 μM)	78	40	85
Glutamate (2,000 μM)	80	0	90
Histamine (100 μM)	26	15	40
Histamine-norepinephrine (100 μM)	33	40	40

Guinea pig cerebral cortical slices prelabeled with radioactive adenine were incubated for 10 min with agonists. Theophylline or 2′-deoxyadenosine were added 2 min prior to agonist, whereas adenosine deaminase was added 10 min prior to agonist.

Data from Huang et al., ref. 13, and Mah and Daly, ref. 18.

sponse of brain cyclic AMP systems to neuronal depolarization. The results with glutamate are ambiguous since although theophylline and adenosine deaminase block the glutamate response, 2′-deoxyadenosine has no effect. Further studies are needed to resolve this anomaly. Either adenosine is *not* involved in the glutamate response (see refs. 43 and 44) or glutamate releases adenosine at sites not readily accessible to either exogenous adenosine or 2′-deoxyadenosine. If adenosine is *not* involved in the glutamate response then the specificity of theophylline and adenosine deaminase as adenosine "antagonists" must be reevaluated.

Responses to amines are reduced by adenosine antagonists or adenosine deaminase (13,15) (Tables 3 and 4). The partial antagonism of the amine responses by these agents suggests that endogenous adenosine synergistically potentiates amine responses in brain slices. However, quite surprisingly, adenosine deaminase inhibits the accumulation of cyclic AMP elicited by *N*-phenylisopropyladenosine in rat cerebral cortical slices (41). Since this adenosine analog is not a substrate for the deaminase, it appears that adenosine deaminase may well have significant nonspecific effects on cyclic AMP–generating systems in brain slices (see ref. 34) or that endogenous adenosine contributes in some manner to the response to *N*-phenylisopropyladenosine in rat cortical slices (see ref. 41). A variety of evidence is not consonant with nonspecific inhibitory effects of adenosine deaminase on cyclic AMP systems. Thus adenosine deaminase does not inhibit the response to *N*-phenylisopropyladenosine in guinea pig cerebral cortical slices (41). Furthermore, the presence of the enzyme in the incubation medium *potentiates* accumulation of cyclic AMP in fat cells (10). Finally, preliminary data using adenosine deaminase irreversibly inhibited by 2′-deoxycoformycin (E. A. McNeal, C. R. Creveling, and J. Daly, *unpublished data)* provides no indication of nonspecific effects. Thus with "microsac" preparations from guinea pig cerebral cortex (see ref. 1), the presence of "inhibited adenosine deaminase" (10 μg/ml) has no effect on basal levels of cyclic AMP or on the response to adenosine or an adenosine-norepinephrine combination.

"RELEASE" OF ADENOSINE IN BRAIN SLICES

Definitive studies on the mechanism of spontaneous and evoked "release" of adenosine in brain slices have not been completed. Accumulation of cyclic AMP elicited by electrical stimulation of guinea pig cerebral and cerebellar slices appear linked to "release" of adenosine. The resultant adenosine-dependent accumulation of cyclic AMP is reduced only slightly under conditions that should inhibit calcium-dependent release of neurotransmitters, i.e., low levels of extracellular calcium ions or high concentrations of magnesium ions (50). The results strongly suggest that "release" of adenosine is not a quantal phenomenon involving calcium-dependent excitation-secretion coupling, but rather a phenomenon independent of calcium. Indeed, the absence of extracellular calcium ions causes both enhanced electrical activity and enhanced "release" of adenosine in brain slices (see refs. 39,41, and 50, and refs. therein).

CALCIUM IONS AND AMINE- AND ADENOSINE-ELICITED ACCUMULATIONS OF CYCLIC AMP IN BRAIN SLICES

Norepinephrine-elicited accumulations of cyclic AMP are *reduced* under calcium-free (EGTA) conditions in rat cerebral cortical slices, but *enhanced* in guinea pig cerebral cortical slices (36,41) (Table 5). It is primarily the α-adrenergic

TABLE 4. *Adenosine deaminase: effects on cyclic AMP in cerebral cortical slices*

Stimulant (100 μM)	Adenosine deaminase (10 μg/ml)	Cyclic AMP (pmoles/mg protein)	
		Rat	Guinea pig
None	−	12 ± 1	9 ± 2
	+	10 ± 1	7 ± 3
Histamine	−	—	35 ± 3
	+	—	22 ± 3
Norepinephrine	−	118 ± 9	33 ± 4
	+	35 ± 5	13 ± 4
N-Phenylisopropyladenosine	−	135 ± 5	131 ± 13
	+	94 ± 7	160 ± 18

Cerebral cortical slices were incubated for 10 min with stimulants. Adenosine deaminase was added 2 min prior to the stimulant.

Data from Schwabe et al., ref. 41.

TABLE 5. *Calcium ions and accumulations of cyclic AMP in cerebral cortical slices*

Stimulant (100 μM)	EGTA (2 mM)	Cyclic AMP (pmoles/mg protein) Rat	Guinea pig
None	−	17 ± 2	9 ± 2
	+	41 ± 4	18 ± 2
Norepinephrine	−	162 ± 15	33 ± 4
	+	88 ± 7	69 ± 5
Adenosine	−	119 ± 19	179 ± 19
	+	216 ± 28	309 ± 21

Cerebral cortical slices were incubated with stimulants for 10 min. The calcium ion chelator, EGTA, was added 2 min prior to the stimulant.

Data from Schwabe et al., ref. 41.

component of the norepinephrine response that is reduced under calcium-free conditions (39). Interestingly, it was also the α-component that was synergistically potentiated by adenosine (24). In calcium-free medium, adenosine deaminase completely blocks norepinephrine-elicited accumulation of cyclic AMP in both rat and guinea pig preparations. Amine responses are thus strongly dependent on both extracellular calcium ions and on adenosine. It would appear that in brain slices "release" of adenosine is enhanced when extracellular levels of calcium ions are reduced. In guinea pig cerebral slices synergistic amine-adenosine interactions then predominate, whereas in rat cerebral cortical slices the dependence of the amine response on extracellular calcium ions predominates (see ref. 41).

Adenosine-elicited accumulation of cyclic AMP is greatly potentiated under calcium-free conditions in *both* rat and guinea pig cerebral cortical slices (36,41) (Table 5). Furthermore, adenosine-elicited accumulation of cyclic AMP in guinea pig olfactory cortical slices is markedly inhibited by high concentrations of calcium (17).

The interactions of adenosine and calcium ions appear to be consistent with the following model: Extracellular calcium ions are normally inhibitory to adenosine responses. Reductions in extracellular calcium ions increase membrane excitability which leads to an increase in efflux of adenosine from the hyperactive neurons. Adenosine alone or in consort with amines then activates cyclic AMP formation. Amine-elicited accumulations of cyclic AMP probably occur only under conditions in which either extracellular calcium or adenosine are present.

This model conforms to a feedback control system in which reduction of calcium leads to a release of adenosine, thereby assuring that amine receptors remain in a responsive, perhaps hyperresponsive, state during neuronal hyperactivity. Cyclic AMP generated in response to adenosine and/or amines may then play a role in reducing neuronal activity, thereby preventing continued hyperactivity. Normal levels of calcium ions would be restored during this period of reduced neuronal activity.

DESENSITIZATION OF AMINE RECEPTORS: ROLE OF ADENOSINE

The "desensitization" of amine receptors coupled to adenylate cyclase has been studied extensively in many types of cells (see refs. 8,19, and 42). Mechanisms appear to differ in different systems. In guinea pig cerebral cortical slices histamine and/or norepinephrine elicit only one accumulation of cyclic AMP; after washing the preparation it appears refractory to restimulation by the amine (36,37). The presence of exogenous adenosine or endogenous adenosine "released" under calcium-free conditions restores the responsiveness of the amine-receptor adenylate cyclase complex. An intervening incubation with adenosine does not restore responsiveness of the amine-sensitive systems. The mechanisms involved in this phenomenon have not been elucidated. In rat cerebral cortical slices, cyclic AMP–generating systems did not appear, in at least one study, to become refractory after a prior stimulation with norepinephrine (47).

DESENSITIZATION OF ADENOSINE RECEPTORS

In guinea pig cerebral cortical slices, adenosine elicits sequential stimulations of cyclic AMP–generating systems with no loss in responsiveness (37). However, preliminary data with "microsacs" from guinea pig cerebral cortex (see ref. 1) suggests that endogenous adenosine present during prolonged incubations causes a marked "desensitization" of the adenosine receptors (E. T. McNeal, C. R. Creveling, and J. Daly, *unpublished data*). For example, in a representative experiment the response to adenosine obtained during 10-min assay periods decreased about 60% as a result of a prolonged 90-min preincubation. When adenosine deaminase was present during the 90-min preincubation period the response to adenosine decreased only 6%. 2′-Deoxycoformycin was added prior to the 10-min assay period in order to inhibit the adenosine deaminase. Responses to norepinephrine and, to a lesser extent, histamine were reduced when adenosine deaminase had been present during short 10-min preincubation periods (E. T. McNeal, C. R. Creveling, and J. Daly, *unpublished data*). Reduction in endogenous adenosine in the preincubation period appeared to have resulted in reduced synergistic interactions of amine and endogenous adenosine during the 10-min assay period.

Further study on desensitization and other aspects of the biochemistry of adenosine would be greatly facilitated if a radioactive ligand for the adenosine receptor were available. Radioactive 2-chloroadenosine has been investigated as such a ligand but definitive evidence for specific binding was not obtained (Y. Nimitkitpaisan, P. Skolnick, and J. Daly, *unpublished results*). Other agents including certain antagonists may prove to be more satisfactory as ligands for the adenosine receptors.

ALTERATIONS IN ADENOSINE-SENSITIVE CYCLIC AMP-GENERATING SYSTEMS *IN VIVO* IN BRAIN

The responsiveness of norepinephrine-sensitive cyclic AMP–generating systems undergoes compensatory changes as a result of alterations in noradrenergic input (5). Possible adaptive responses after alterations in adenosine input have not been systematically investigated. However, although treatment of guinea pigs with reserpine does not increase the responsiveness of norepinephrine-sensitive cyclic AMP–generating systems in cerebral cortical slices, this treatment does enhance the magnitude of adenosine responses about twofold (6). Synergistic responses to adenosine-norepinephrine combinations are also increased. To determine if this unexpected alteration in adenosine responses represents an alternative means of enhancing amine responsiveness after drug treatment or lesions will require further investigation.

SUMMARY

The accumulation of cyclic AMP elicited in brain tissue by adenosine can be potentiated by certain phosphodiesterase inhibitors, blocked by alkylxanthines and certain adenosine analogs, and potentiated by agents that inhibit uptake or deamination of adenosine. The response to adenosine is greatly enhanced in the presence of low extracellular levels of calcium ions and is synergistically augmented by biogenic amines such as norepinephrine, histamine, and serotonin. Adenosine deaminase reduces extracellular levels of adenosine in brain tissue and thereby reduces the magnitude of accumulation of cyclic AMP elicited by biogenic amines. In certain brain preparations adenosine-sensitive cyclic AMP–generating systems appear to undergo considerable "desensitization" owing to endogenous adenosine. Enhanced responsiveness of cyclic AMP systems to adenosine in brain slices have been noted after at least one drug treatment, namely with reserpine. The biochemical literature on adenosine and cyclic AMP under various conditions thus provides extensive data for correlative studies on the electrophysiological effects of adenosine in the central nervous system.

REFERENCES

1. Chasin, M., Mamrak, F., and Samaniego, S. G. (1974): *J. Neurochem.,* 22:1031–1038.
2. Clark, R. B., Su, Y.-F., Gross, R., and Perkins, J. P. (1974): *J. Biol. Chem.,* 249:5296–5303.
3. Daly, J. (1977): *Cyclic Nucleotides in the Nervous System.* Plenum Press, New York.
4. Daly, J. W., McNeal, E. T., and Creveling, C. R. (1978): *A. N. Richards Symposium on Histamine Receptors.* Spectrum Publications, New York *(in press).*
5. Dismukes, R. K., and Daly, J. W. (1976): *J. Cyclic Nucleotide Res.,* 2:321–336.
6. Dismukes, R. K., Ghosh, P., Creveling, C. R., and Daly, J. W. (1976): *Exp. Neurol.,* 52:206–215.
7. Dismukes, K., Rogers, M., and Daly, J. W. (1976): *J. Neurochem.,* 26:785–790.
8. Drummond, A. H., Baguley, B. C., and Staehelin, M. (1977): *Mol. Pharmacol.,* 13:1159–1169.
9. Ebert, R., and Schwabe, U. (1973): *Naunyn Schmiedebergs Arch. Pharmacol.* 278:247–259.
10. Fain, J. N. (1973): *Mol. Pharmacol.,* 9:595–604.
11. Hopkins, S. V., and Goldie, R. G. (1971): *Biochem. Pharmacol.,* 20:3359–3365.

12. Huang, M., and Daly, J. W. (1974): *Life Sci.,* 14:489–503.
13. Huang, M., Gruenstein, E., and Daly, J. W. (1973): *Biochim. Biophys. Acta.,* 329:147–151.
14. Huang, M., Shimizu, H., and Daly, J. (1971): *Mol. Pharmacol.,* 7:155–162.
15. Huang, M., Shimizu, H., and Daly, J. W. (1972): *J. Med. Chem.,* 15:462–466.
16. Kostopoulos, G. K., and Phillis, J. W. (1977): *Exp. Neurology,* 55:719–724.
17. Kuroda, Y., Saito, M., and Kobayashi, K. (1976): *Proc. Jpn. Acad.,* 52:86–89.
18. Mah, H. D., and Daly, J. W. (1976): *Pharmacol. Res. Commun.,* 8:65–79.
19. Mickey, J. V., Tate, R., Mullikin, D., and Lefkowitz, R. J. (1976): *Mol. Pharmacol.,* 12:409–419.
20. Miyamoto, E., Kuo, J. F., and Greengard, P. (1969): *J. Biol. Chem.,* 244:6395–6402.
21. Ohga, Y., and Daly, J. W. (1977): *Biochim. Biophys. Acta.,* 498:46–60.
22. Okada, Y., and Kuroda, Y. (1975): *Proc. Jpn. Acad.,* 51:491–494.
23. Perkins, J. P., and Moore, M. (1973): *Mol. Pharmacol.,* 9:774–782.
24. Perkins, J. P., Moore, M. M., Kalisker, A., and Su, Y-F. (1975): *Adv. Cyclic Nucleotide Res.,* 5:641–660.
25. Phillis, J. W., Kostopoulos, G. K., and Limacher, J. J. (1974): *Can. J. Physiol. Pharmacol.,* 52:1226–1229.
26. Phillis, J. W., Kostopoulus, G. K., and Limacher, J. J. (1975): *Eur. J. Pharmacol.,* 30:125–129.
27. Premont, J., Perez, M., and Bockaert, J. (1977): *FEBS Lett.,* 75:209–212.
28. Pull, I., and McIlwain, H. (1972): *Biochem. J.,* 130:975–981.
29. Rubio, R., Berne, R. M., Bockman, E. L., and Curnish, R. R. (1975): *Am. J. Physiol.,* 228:1896–1902.
30. Saito, M. (1977): *Biochim. Biophys. Acta,* 498:316–325.
31. Sattin, A. (1971): *J. Neurochem.,* 18:1087–1096.
32. Sattin, A., and Rall, T. W. (1970): *Mol. Pharmacol.,* 6:13–23.
33. Sattin, A., Rall, T. W., and Zanella, J. (1975): *J. Pharmacol. Exp. Ther.,* 192:22–32.
34. Schmidt, M. J., Thornberry, J. F., and Molloy, B. B. (1977): *Brain Res.,* 121:182–189.
35. Schultz, J. (1974): *J. Neurochem.,* 22:685–690.
36. Schultz, J. (1975): *J. Neurochem.,* 24:495–501.
37. Schultz, J., and Daly, J. W. (1973): *J. Biol. Chem.,* 248:860–866.
38. Schultz, J., and Daly, J. W. (1973): *J. Neurochem.,* 21:573–579.
39. Schwabe, U., and Daly, J. W. (1977): *J. Pharmacol. Exp. Ther.,* 202:134–143.
40. Schwabe, U., Miyake, M., Ohga, Y., and Daly, J. W. (1976): *Mol. Pharmacol.,* 12:900–910.
41. Schwabe, U., Ohga, Y., and Daly, J. W. (1978): *Naunyn Schmiedebergs Arch. Pharmacol.,* 302:141–151.
42. Shear, M., Insel, P. A., Melmon, K. L., and Coffino, P. (1976): *J. Biol. Chem.,* 251:7572–7575.
43. Shimizu, H., Ichishita, H., and Mizokami, Y. (1975): *J. Cyclic Nucleotide Res.,* 1:61–67.
44. Shimizu, H., Ichishita, H., and Umeda, I. (1975): *Mol. Pharmacol.,* 11:866–873.
45. Shimizu, H., Tanaka, S., and Kodama, T. (1972): *J. Neurochem.,* 19:687–698.
46. Skolnick, P., Nimitkitpaisan, Y., Stalvey, L., and Daly, J. W. (1978): *J. Neurochem.,* 30:1579–1582.
47. Skolnick, P., Schultz, J., and Daly, J. W. (1975): *J. Neurochem.,* 24:1263–1265.
48. Smellie, F., Daly, J. W., Dunwiddie, T., and Hoffer, B. (1978): *Pharmacologist,* 20:231.
49. Trams, E. G., and Lauter, C. J. (1975): *Biochem. J.,* 152:681–687.
50. Zanella, J., Jr., and Rall, T. W. (1973): *J. Pharmacol. Exp. Ther.,* 186:241–251.

Physiological and Regulatory Functions of Adenosine and Adenine Nucleotides, edited by H. P. Baer and G. I. Drummond. Raven Press, New York © 1979.

Biochemical Characterization of Adenosine Receptors in the Brain

Hirotoshi Shimizu

Nippon Roche Research Center, Kajiwara, Kamakura, Japan 247

In the past decades various biogenic amines including catecholamines, serotonin, histamine, and acetyl choline have been studied extensively as neurotransmitters in the peripheral and central nervous systems. Adenosine, which also possesses an amino group, was not regarded as a member of these neuromodulators or neurotransmitters until 1967 when Sattin and Rall (5) reported that addition of adenosine or adenosine monophosphate (AMP) to brain slices caused 100- to 500-fold increases in their cyclic AMP content. This effect of adenosine was soon confirmed by Shimizu et al. (11), and it was demonstrated at the same time that this action of adenosine was not merely the result of an increased supply of the precursor ATP but the result of the increased catalytic activity of the adenylate cyclase.

Sattin and Rall (6) and Daly and co-workers (2,7,10,12) have further studied the adenosine-elicited accumulation of cyclic AMP in brain slices and the findings may be summarized as follows: (a) The maximal elevation of cyclic AMP was achieved in the concentration range of 0.1 to 0.5 mM, with a half-maximal effect of 0.05 mM. (b) Analogs of adenosine with a small substitution at position 2 or N^6 (as the secondary amine) of the purine moiety are agonists whereas modification at position 1 of the purine moiety or at position 2′ or 3′ of the ribose moiety usually abolishes activity. 5′-Phosphorylated analogs are active, but are probably dephosphorylated prior to interacting with the enzyme. (c) Deoxyribose (2′ or 3′) and methylxanthines are fairly selective adenosine antagonists. (d) The effect of biogenic amines or glutamate on cyclic AMP elevation in incubated brain slices is potentiated in the presence of adenosine. (e) Binding of adenosine to the cyclase-coupled receptor is likely to be reversible since removal of adenosine from the incubation medium lowers cyclic AMP levels, and readdition of adenosine causes an elevation of the same magnitude as before. (f) In contrast to cyclic AMP elevation elicited with depolarizing agents, the adenosine-elicited cyclic AMP elevation is Ca^{2+} independent and not inhibited by cocaine. (g) Adenosine-stimulated cyclic AMP formation utilizes as the substrate the ATP compartment also used for cyclic AMP formation elicited by other biogenic amines.

In addition to these biochemical findings, our electronmicroscopical examination of human brain slices preincubated with [^{3}H]adenosine revealed that the ^{3}H incorporated is localized in nerve terminals with much higher density than in cell bodies and glia (8). Shimizu et al. (4) also noticed that the brain slices preincubated with

[^{3}H]adenosine spontaneously release the nucleoside into the medium during further incubation and that the rate of the release was nearly doubled by application of depolarizing stimuli. The last finding was later demonstrated with a more elegant technique by McIlwain's group (4).

All these findings are suggestive of a possible presence of a purinergic nerve mechanism in the brain (8), the physiological role of which remains to be established.

Following reports on the occurrence of adenosine-sensitive adenylate cyclase in brain slices, various groups have searched for such systems in other tissues as well as in other neuronal preparations. Indeed, platelets, thymocytes, cardiac muscle slices, and adrenal tumor cells are among the tissues in which adenosine-sensitive adenylate cyclase has been found. In these tissues or cells the structure-activity profiles of adenosine analogs are essentially similar. With broken cell particulate preparations of glial cells, a half-maximal stimulation of the cyclase has been achieved with 3 to 5 μM adenosine (1). This concentration contrasts with the higher half-maximal concentrations (50 μM) observed with the brain slice preparations, in which access of adenosine to its receptor is perhaps impaired anatomically or influenced by enzyme action.

ENZYMES INVOLVED IN ADENOSINE METABOLISM

Adenosine is deaminated mainly in homogenates of brain because of high adenosine deaminase levels. However, [^{14}C]adenosine injected into the carotid artery or incubated with brain slices was mostly phosphorylated to [^{14}C]adenine nucleotides (13). Thus, in the process of uptake of adenosine through the cell membrane, phosphorylation occurs in preference to deamination, and nucleotide synthesis might provide increased substrate levels for adenylate cyclase (10) (Fig. 1). Hydrolysis of the glycosidic bond between adenine and ribose by adenine phosphoribosyl transferase was not observed by us. When brain slices were incubated with [2,8-^{3}H, U-^{14}C]adenosine for 3 to 6 min, there was no significant difference in the ^{3}H/^{14}C ratios between the precuorsor adenosine and the phosphorylated metabolites isolated from the incubated slices (13).

The overall results point out the importance of adenosine kinase in the process of metabolism, and hence for the regulatory function of adenosine in the intact tissue. Little was known about adenosine kinase of brain tissue until 1972, when partial purification of the kinase from guinea pig brain was achieved (13). In the course of ammonium sulfate fractionation the highest activity was found in the fraction between 50 and 75% saturation. The K_m for adenosine was of the order of 20 μM, which is close to the half-maximal concentration for the phosphorylation-coupled uptake of adenosine into incubated brain slices. It is still uncertain whether or not a specific permeation factor is involved in the adenosine uptake in brain cells, but available data are in agreement with a facilitated diffusion mechanism, whereby the phosphorylation by adenosine kinase is rate limiting.

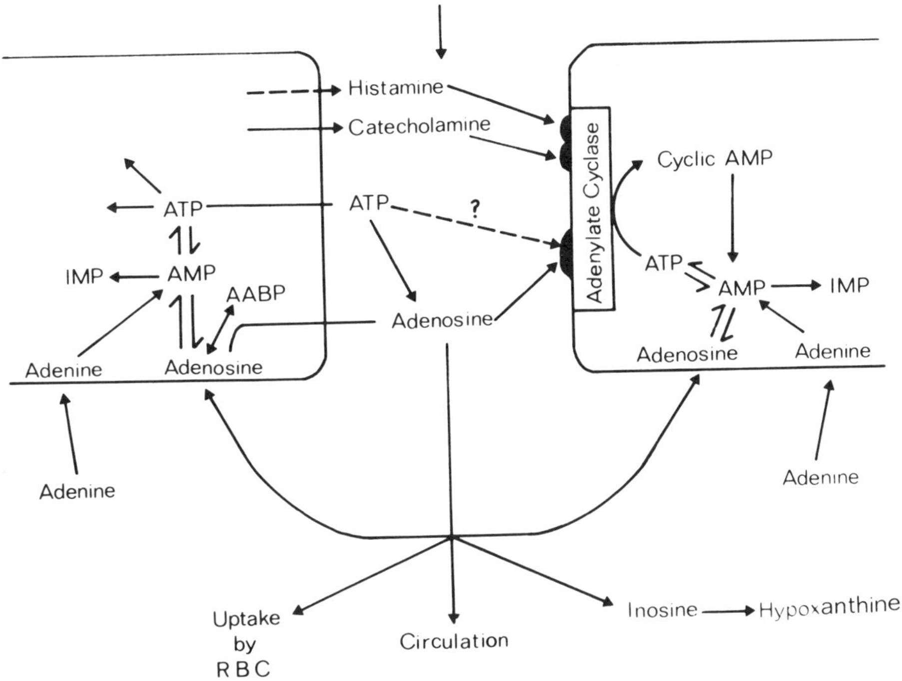

FIG. 1. Fate of adenosine.

ADENOSINE-BINDING PROTEINS

In addition to the three enzymes for adenosine metabolism and the postulated receptor of the adenylate cyclase system, at least one class of high-affinity adenosine-binding proteins has been demonstrated in liver (14,15) and red blood cells (16). These proteins also bind cyclic AMP, but do not inhibit the catalytic subunit of the cyclic AMP–dependent protein kinase. Although there are still conflicting reports on the characteristics of the cyclic AMP/adenosine–binding proteins, various adenine compounds compete with the binding of [^{3}H]-cyclic AMP and/or [^{3}H]adenosine and hence this type of protein is often designated as "adenine analog–binding protein" or AABP. A physiological role of AABP is not yet known, and significant amounts of AABP have been found only in liver and erythrocytes (14).

During the past few years we have looked for adenosine-binding proteins in the brain. At first we chose [^{3}H]-N^6-phenyladenosine as the ligand, because this adenosine analog is resistant to the adenosine-metabolizing enzymes but still stimulates cerebral cyclic AMP formation. Despite an extensive search for a particulate fraction from brain that specifically binds the adenosine analog, we have not been successful studying membranes from cerebral neocortex of either rat or guinea pig.

Further attempts were made using the 9,000 × g supernatant of whole rat brain homogenates and [^{3}H]adenosine as a ligand. The supernatant proteins were fractionated by ammonium sulfate and DEAE-cellulose column chromatography as summarized in Fig. 2. Each column fraction was checked for adenosine-binding activity by means of Millipore membrane filtration. The binding assays contained 0.5 ml protein solution (20 mM Tris-HCl, pH7.5) plus 0.05 ml [^{3}H]adenosine (5 × 10^5 dpm, 7 × 10^{-12} moles) in H_2O. After incubation for 45 min at 25° C. The mixture was filtered through 0.45 μm Millipore HA filters.

Usually 4 or 5 [^{3}H]adenosine peaks (15 nM) were eluted from the column during the linear gradient (Fig. 3). Peaks 3 and 4 were eluted at salt concentrations of about 0.15 M and 0.3 M, respectively. The two peaks were further purified separately on a column of DEAE Sephadex. The resultant Peak 4 (P-4) was virtually free from adenosine kinase, deaminase, and nucleosidase, and had an apparent K_a for adenosine of approximately 80 nM. The P-4 protein also binds [^{3}H]-cyclic AMP to almost the same extent as [^{3}H]adenosine.

The effect of adenine and its analogs on the binding of [^{3}H]adenosine (15 nM) to P-4 protein was investigated (Table 1). Adenylic acid, cyclic AMP, and 2-chloroadenosine at 1.5 μM showed more than 60% inhibition. Weaker inhibitory activities were found for N^6-phenyladenosine and 1-methylisoguanosine whereas

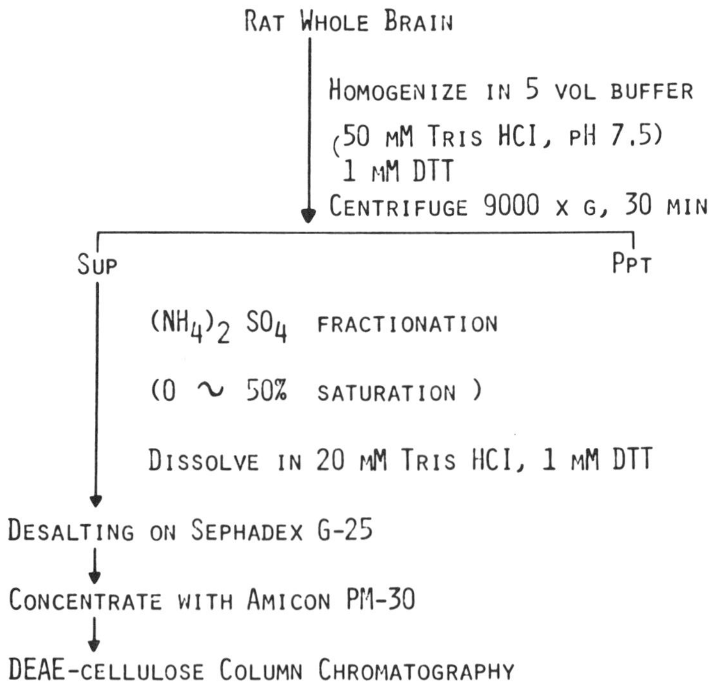

FIG. 2. Diagram of preparation for adenosine binding proteins.

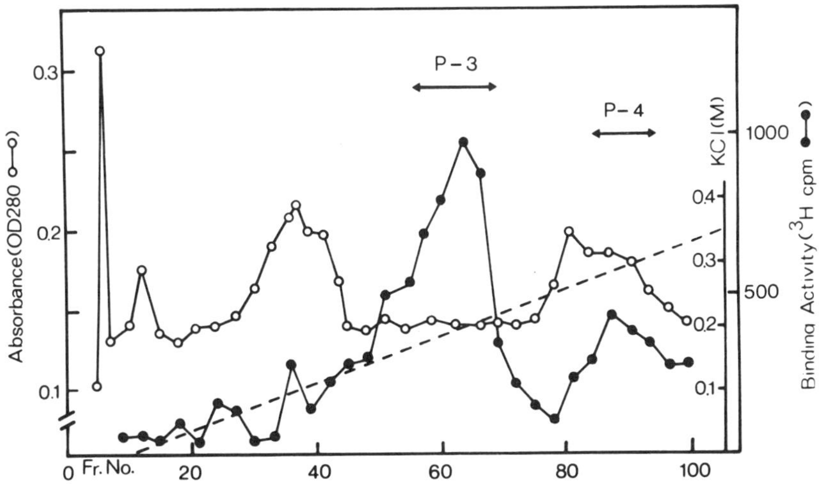

FIG. 3. Elution profile of adenosine binding proteins from a DEAE-cellulose column. Eight rat brains were pooled, homogenized, and fractionated as shown in Fig. 2. Approximately 25 ml of the concentrate filtered through Amicon PM-30 Diaflo membrane was applied onto a 2.5 × 27 cm column of DEAE-cellulose previously equilibrated with 20 mM Tris-HCl, pH 7.5, containing 1 mM dithiothreitol. Proteins were then eluted with a gradient of KCl increasing linearly up to 400 mM. Fractions, 15 ml each, were collected and protein (○) and the binding activity (●) were assayed in each fraction.

adenine, inosine, guanosine, and theophylline were virtually not inhibitory. Binding of [^{3}H]-cyclic AMP to P-4 protein was also inhibited by adenosine to a significant extent. These properties indicate that P-4 is distinct from the regulatory subunit of cyclic AMP–dependent protein kinase.

TABLE 1. *Effect of adenine and its analogs on the binding of adenosine to P-4 protein*

Compound tested	Inhibition
Adenine	—
5′-AMP	++
3′,5′-Cyclic AMP	++
Inosine	—
Guanosine	—
2-Chloroadenosine	++
N^6-Phenyladenosine	+
Theophylline	—

Competitive binding activity was measured at both 1.5 and 3.0 μm and is expressed on a scale of "—" (no activity) to "++" relative to the activity of nonlabeled adenosine, which was "+++".

The results of competitive inhibition studies indicate rather that P-4 protein is similar to AABP. However, Sugden and Corbin (14) were unable to find a significant amount of AABP in rat brain. Since there are considerable differences in the spectrum of AABP inhibitors between liver (14) and erythrocyte (3), brain may contain AABP that differs from these as well.

CONCLUSION

The possible physiological importance of adenosine in various mammalian tissues has been recognized. In view of the high activity of adenosine in elevating brain cyclic AMP, the nucleoside probably plays an important role in central neuronal transmission. Synthesis, release, degradation, and reuptake of adenosine occur in the brain, and adenosine kinase is likely to play a pivotal role in the regulation of its extra- and intracellular levels. A high-affinity adenosine-binding protein was found in rat brain. This protein, revealing high affinity also to cyclic AMP, may resemble the so-called "adenine analog–binding protein" found in liver and erythrocytes. Understanding the biological significance of this and related adenosine binding proteins may contribute to an understanding of the role of adenosine as a modulator of neurotransmission in the brain.

REFERENCES

1. Clark, R. B., and Seney, M. N. (1976): *J. Biol. Chem.,* 251:4236–4249.
2. Daly, J. (1977): In: *Cyclic Nucleotides in the Nervous System,* pp. 108–116. Plenum Press, New York.
3. Olsson, R. A. (1978): *Biochemistry,* 17:367–375.
4. Pull, I., and McIlwain, H. (1972): *Biochem. J.,* 126:965–973.
5. Sattin, A., and Rall, T. W. (1967): *Fed. Proc.,* 26:707.
6. Sattin, A., and Rall, T. W. (1970): *Mol. Pharmacol.,* 6:13–23.
7. Schultz, J., and Daly, J. W. (1973): *J. Biol. Chem.,* 248:843–852.
8. Shimizu, H. (1973): In: *Fourth Int. Mtg. Int. Soc. Neurochem.,* pp. 88–89.
9. Shimizu, H., Creveling, C. R., and Daly, J. (1970): *Proc. Natl. Acad. Sci. U.S.A.,* 65:1033–1040.
10. Shimizu, H., and Daly, J. (1970): *Biochem. Biophys. Acta,* 222:465–473.
11. Shimizu, H., Daly, J. W., and Creveling, C. R. (1969): *J. Neurochem.,* 16:1609–1619.
12. Shimizu, H., Ichishita, H., and Odagiri, H. (1974): *J. Biol. Chem.,* 249:5955–5962.
13. Shimizu, H., Tanaka, S., and Kodama, T. (1972): *J. Neurochem.,* 19:687–698.
14. Sugden, P. H., and Corbin, J. D. (1976): *Biochem, J.,* 159:423–437.
15. Ueland, P. M., and Døskeland, S. O. (1977): *J. Biol. Chem.,* 252:677–686.
16. Yuh, K-C., M., and Tao, M. (1974): *Biochemistry,* 13:5220–5226.

Physiological and Regulatory Functions of Adenosine and Adenine Nucleotides, edited by H. P. Baer and G. I. Drummond. Raven Press, New York © 1979.

Regulation of Mouse Neuroblastoma Adenylate Cyclase by Adenosine

Arthur J. Blume

Department of Physiological Chemistry and Pharmacology, Roche Institute of Molecular Biology, Nutley, New Jersey 07110

There is now ample evidence to indicate that adenosine and/or adenosine nucleotides serve as inhibitory neurotransmitters or neuromodulators (7,8). In addition, as was shown with the adrenergic inhibitory neurotransmitter dopamine (13,14, 21), exogenous adenosine mediates elevations in intracellular cyclic adenosine monophosphate (cAMP) (10,11,16,18–20). Thus it seems possible that adenosine produces its inhibitory effects through an alteration of neuronal cAMP metabolism, and a simple neuronal system responding to exogenous adenosine with changes in both cAMP metabolism and in plasma membrane properties would be of interest. The work described below is a summary of the efforts made in this laboratory to elucidate the manner in which mouse neuroblastoma tissue culture cells alter their cAMP metabolism in response to exogenous adenosine.

METHODS

Studies with intact, viable mouse neuroblastoma cells were carried out with clonal line NS20 or NIE (1). Cells were grown and analyzed for intracellular cAMP content as described previously (2,3). Assays for adenylate cyclase in homogenates of NS20 cells were performed according to the method of Salomon et al. (17) using [^{32}P]–adenosine triphosphate ([^{32}P]ATP) or [^{32}P]AMP-P(NH)P with 50 mM Tris/maleate buffer pH 7.4, 5 to 10 mM $MgCl_2$, 0.5 mM cAMP, 0.3 mM Ro 20-1724 [a phosphodiesterase (PDE) inhibitor], and a nucleotide regenerating system composed of 20 mM sodium creatine phosphate and 24 units/ml creatine phosphokinase. Cell homogenates were prepared and adenylate cyclase assayed as described previously (4,6).

STUDIES WITH INTACT CELLS

Mouse neuroblastoma cells incubated in the presence of certain inhibitors of cyclic nucleotide PDE respond to exogenous adenosine (or 2-chloroadenosine) with a significant increase in intracellular cAMP content (Table 1). Addition of adenosine (or 2-chloroadenosine) to these cells in the absence of a PDE inhibitor

TABLE 1. *Specificity of the adenosine elevation of intracellular cAMP concentration of NS20 cells*

Additions	PDE inhibitor present (pmoles cAMP/mg protein)		
	None	Ro 20-1724 (0.7 mM)	IBMX (0.5 mM)
None	40	40	45
Adenosine	64	865	729
2-Chloroadenosine	60	822	693
Guanosine	44	44	47
Inosine	43	42	46
AMP	40	140	n.d.
ADP	41	60	n.d.
ATP	43	62	n.d.

Cells were placed into buffer at 22° C with the above compounds and cAMP content determined 50 min later. Final concentration of all additions was 200 μM.n.d. = Not determined. IBMX = isobutylmethylxanthine.

leads to a less than 1.5-fold increase in cAMP content. With the PDE inhibitor 4-(3-butoxy-4-methoxybenzyl)-2-imidazolidinone (Ro 20-1724) present, the clonal cell line NS20 responded to adenosine with a rapid, large, and reversible increase in cAMP content. Within 1 min increases were measurable (half-maximal response after 9 min) and reached a steady state with a 16- to 22-fold increase over basal levels to approximately 1,700 pmoles/mg protein. The intracellular cAMP content

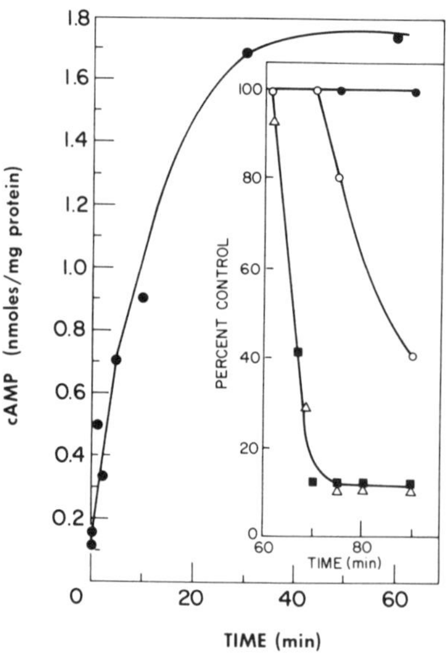

FIG. 1. Time course of elevation of NS20 intracellular cAMP content by adenosine and reversibility of the adenosine effect. NS20 cells, pretreated for 3 hr with 0.7 mM Ro 20-1724, were washed and placed at 22° C in buffer with 200 μM adenosine and 0.7 mM Ro 20-1724. **Insert:** After 60 min, adenosine (○), Ro 20-1724 (■), adenosine and Ro 20-1724 (Δ), or neither (●) were removed by rapid washing (~1 min) with buffers minus the appropriate compounds and then resuspended at 22° C. One hundred percent control = cAMP content before any washes (i.e., 1,730 pmoles/mg protein). (Part of the data from ref. 3, with permission.)

returned to nearly original levels on removal of either adenosine or PDE inhibitor (Fig. 1). However, we have observed that not all inhibitors of the soluble PDE activity of NS20 cells potentiated the elevation in cAMP content owing to exogenous adenosine (2). A plausible explanation for this finding is that membrane permeability differs with respect to PDE inhibitors.

Studies with NS20 and NIE clonal neuroblastoma cells showed that both had a similar adenosine-specific receptor linked to cAMP metabolism. This receptor exhibited an apparent affinity for adenosine of 2 μM. 2-Chloroadenosine could replace adenosine whereas guanosine, inosine, adenine, adenosine diphosphate (ADP), and ATP could not (Table 1). Although there was a small response to 5′-AMP, it could have resulted from as little as 1% contamination of the nucleotide with adenosine.

Theophylline blocked the response of NS20 (or NIE) cells to adenosine (or 2-chloroadenosine) in an apparently competitive fashion (Fig. 2). Theophylline did not block the elevation of intracellular cAMP produced in these cells by exogenous prostaglandin. Therefore, adenosine receptors must exist interacting specifically with agonists and antagonists.

The elevation of the cAMP content in NS20 cells by exogenous adenosine was also blocked by certain cholinergic agonists and antagonists as well as by valinomy-

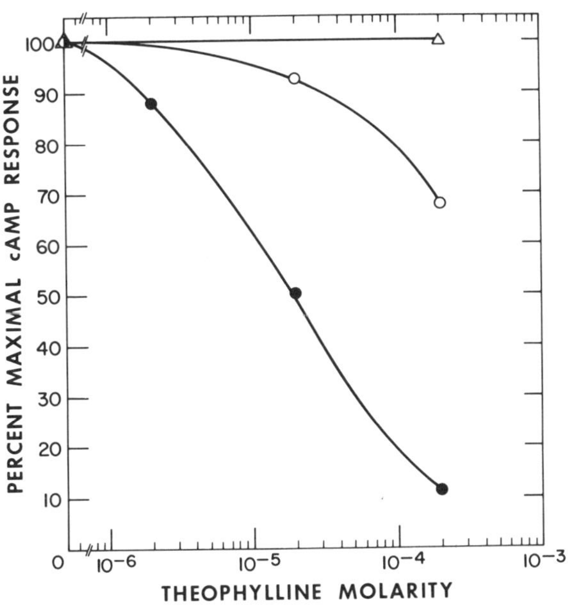

FIG. 2. Competitive nature of the theophylline inhibition of the *in vivo* elevation of cAMP produced by adenosine. Cells incubated at 22° C with 0.7 mM Ro 20-1724 plus adenosine at 2 μM (•), 20 μM (○), or 200 μM (Δ) and increasing theophylline concentrations for 50 min. Maximal cAMP content (i.e., in the absence of theophylline) was 571 (•), 645 (○), and 1,098 (Δ) pmoles cAMP/mg protein, respectively. (From ref. 3, with permission.)

TABLE 2. *Blockers of the elevation of intracellular NS20 cAMP content by adenosine*

Agent	pmoles cAMP/mg protein ± SD	Percent inhibition of elevation ± SD
None	60 ± 5	—
Adenosine	758 ± 84	0 ± 11
Adenosine + acetylcholine	187 ± 50	84 ± 4
Adenosine + decamethonium	194 ± 26	83 ± 2
Adenosine + acetate	701 ± 40	8 ± 6
Adenosine + choline	650 ± 75	20 + 10
Adenosine + valinomycin	108 ± 7	93 ± 7
Adenosine + A23187	60 ± 6	100 ± 6
Adenosine + X537A	151 ± 2	87 ± 2
Adenosine + veratridine	656 ± 7	6 ± 1
Adenosine + theophylline	69 ± 9	90 ± 10

Cells pretreated with Ro 20-1724 (700 μM) for 3 hr were placed in buffer with Ro 20-1724 at 22° C and the compounds listed. Cyclic AMP content was determined 50 min later. Final concentrations: Adenosine, theophylline, and veratridine: 200 μM; acetylcholine, acetate, and choline: 1 μM; decamethonium: 5 μM; valinomycin; 100 nM; A23187 and X537A: 100 μM.
From ref. 2, with permission.

cin (a specific ionophore for K^+), A23187 (an ionophore for the divalent cations Ca^{2+} and Mg^{2+}), and X537A (a nonspecific ionophore for both mono- and divalent cations like Na^+, K^+, Ca^{2+}, and Mg^{2+}) (Table 2) (2). Although these agents did not significantly alter the basal cellular cAMP content, they blocked the effects of prostaglandin on cAMP levels. The action of the cholinergic compounds was shown to be mediated through a muscarinic cholinergic receptor (2). The blocking action of the ionophores and the fact that depolarizing cholinergic antagonists, but not nondepolarizing cholinergic antagonists, blocked cAMP elevation (2) suggest that some event related to membrane depolarization (i.e., some ion translocation) might be responsible for preventing adenosine elevations in intracellular cAMP content. However, since both prostaglandin and adenosine responses were similarly blocked by this group of compounds, this type of control over cAMP metabolism is not mediated directly through the adenosine receptor.

Our studies with intact cells have thus revealed that mouse neuroblastoma cells respond to adenosine via specific receptors coupled to a control mechanism of cAMP metabolism. As increases in cAMP content were seen after inhibition of PDE activity, this effect would appear to be mediated by stimulation of cAMP synthesis. Indeed, activation of NS20 adenylate cyclase by adenosine *in vitro* could be demonstrated (see next section). Last, it was established that neuronal cells in culture, capable of generating action potentials (9,15), have receptors for adenosine and acetylcholine and that the latter can regulate the cAMP response to adenosine.

STUDIES ON CELL-FREE NS20 HOMOGENATES

When adenosine or 2-chloroadenosine were incubated with adenylate cyclase in homogenates of NS20 cells, stimulation of enzyme activity was observed (16,17).

The increase in activity ranged between 2.7- and 4.5-fold over basal values and no evidence of any inhibitory effects was observed with up to 200 μM nucleoside. Adenosine and 2-chloroadenosine stimulated activity in a noncooperative manner with apparent K_as of 67.6 and 6.7 μM, respectively, and apparently increased the maximum velocity of the enzyme without changing the K_m for ATP (Fig. 3).

Analysis of the specificity of the direct stimulation of NS20 adenylate cyclase by nucleosides matched the specificity observed for the nucleoside response of intact NS20 cells (Table 3). Activation of adenylate cyclase *in vitro* did not occur with guanosine or inosine and, as indicated above, did not occur with ATP. A number of nucleosides with minor modifications in the 2 and 8 positions on the purine moiety were found to be agonists. Compounds modified with sugar moieties or with drastic changes in the base moiety were devoid of agonist action. Theophylline appeared to be a specific adenosine antagonist, i.e., it did not interfere with prostaglandin (PG) activation of the cyclase and did not modify basal activity. When tested directly against 2-chloroadenosine, theophylline acted as a competitive inhibitor (Fig. 4). It was also apparent that nucleosides such as 2′- and 5′-deoxyadenosine acted as inhibitors of basal, PGE- and 2-chloroadenosine–stimulated adenylate cyclase. This group of compounds is therefore similar to that recently described by Londos and Wolff (12) which is not blocked by theophylline and appeared to involve an inhibitory site common to all adenylate cyclases studied to date.

Studies directed at elucidating the mechanism of stimulation of neuroblastoma adenylate cyclase by the theophylline-sensitive adenosine receptors have shown that although the maximal velocity of adenylate cyclase is increased by NaF, the guanine nucleotide 5′-guanylylimidodiphate [GMP-P(NH)P], as well as by 2-chloro-

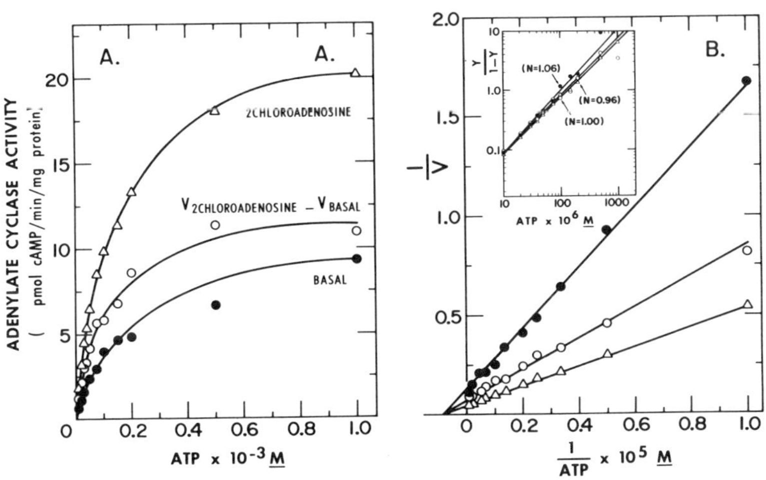

FIG. 3. Comparison of the requirement for ATP for basal and 2-chloroadenosine–stimulated NS20 adenylate cyclase. Hill plots are given in the insert to panel **B** (N = Hill coefficient). (From ref. 6, with permission.)

TABLE 3. *Effect of various nucleotides on the adenylate cyclase activity of NS20 homogenates*

Agent	Molarity ($\times 10^{-6}$)	V_{agent}/V_{basal}	Percent inhibition of stimulation owing to: 2-Chloroadenosine (15 μM)	PGE$_1$ (1.4 μM)
None		1.0		
Adenosine	50	2.5		
2-Chloroadenosine	50	5.0		
2-Hydroxyadenosine	50	3.5		
8-Methylaminoadenosine	50	2.1		
7-Deazaadenosine	200	1.2	<5	<5
2-Deoxy-AMP	200	1.1	<5	<5
2-Deoxy-D-ribose	200	1.1	<5	<5
Adenine-9-β-D-xylopyranoside	200	0.9	<5	<5
Inosine	200	1.0	≤5	≤5
Guanosine	200	1.1	≤5	≤5
Cytosine	200	1.2	≤5	≤5
Guanine	200	1.0	≤5	≤5
Tolazoline	200	1.2	≤5	≤5
Yohimbine	200	1.4	≤5	≤5
AMP	200	0.8	19	≤5
Adenine	200	1.0	22	9
Theophylline	200	1.0	60	≤5
2′-Deoxyadenosine	200	0.6	48	55
5′-Deoxyadenosine	200	0.6	60	70

Substrate in all cases was 1 mM [^{32}P]ATP, and the reaction velocities (at 30° C) (V) are given as pmoles cAMP produced/min/mg protein. For tests of inhibitory effects, stimulators (i.e., 2-chloroadenosine or PGE$_1$) and the compound to be tested were added together. V_{agent}/V_{basal} = ratio of stimulated to nonstimulated basal activity.

From ref. 4, with permission.

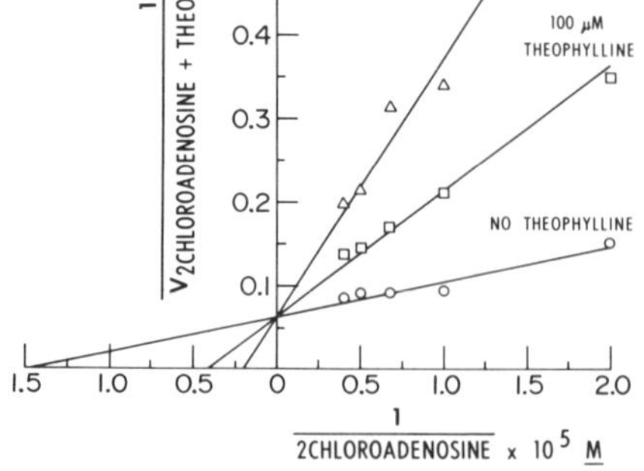

FIG. 4. Competitive inhibition by theophylline of 2-chloroadenosine stimulation of NS20 adenylate cyclase. *In vitro* assays were performed at 30° C with 1 mM [^{32}P]ATP as substrate. V = enzyme velocity as pmoles/min/mg protein. $V_{theophylline}$ was equal to the basal velocity of the enzyme. (From ref. 4, with permission.)

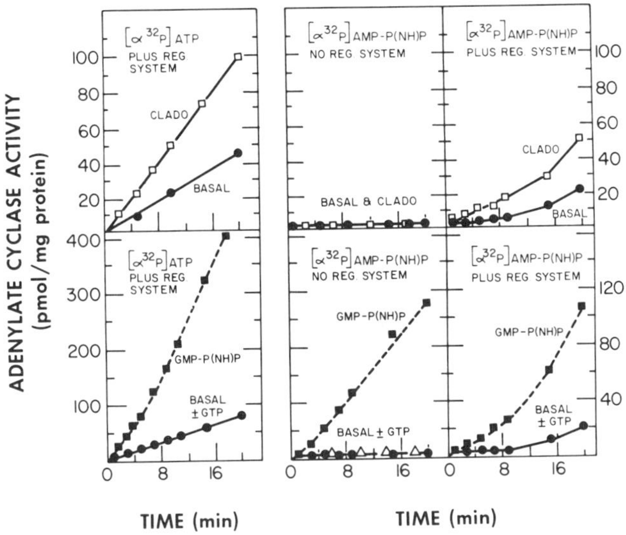

FIG. 5. Difference in responsiveness of NS20 adenylate cyclase to 2-chloroadenosine versus GMP-P(NH)P. Assays were performed at 30° C and contained 6 mM $MgCl_2$, 0.5 mM cAMP, 0.3 mM Ro 20-1724, 50 mM Tris/maleate buffer pH 7.4, 0.1, mM substrate [$^{\alpha 32}$P] or [^{32}P]AMP-P(NH)P) and, where indicated, a nucleotide triphosphate–regenerating system (20 mM creatine phosphate and 24 units/ml creatine phosphokinase), 50 μM 2-chloroadenosine, 50 μM GMP-P(NH)P, or 100 μM GTP.

adenosine or adenosine, these three stimulatory effects can be distinguished (Fig. 5) (5,6). In the standard enzyme assay that contains ATP as substrate and a nucleotide triphosphate regenerating system, NaF and 2-chloroadenosine activation is rapid, being complete at the earliest times assayed (i.e., 1 min). In contrast, activation by GMP-P(NH)P is slow, taking up to 12 min for completion. With AMP-P(NH)P as a substrate and in the absence of nucleotide regenerating system, activation by NaF and GMP-P(NH)P persists and the rate of activation by GMP-P(NH)P was increased. In contrast, under these conditions no activation by 2-chloroadenosine is demonstrable. Addition of a nucleotide regenerating system to the AMP-P(NH)P assay system reestablishes stimulation of the enzyme by 2-chloroadenosine and the extended lag time before maximal activation by GMP-P(NH)P.

Although clear differences exist between the stimulatory processes with the guanine nucleotide and 2-chloroadenosine, we have observed an interdependence between these two processes. Further, a similar interrelationship apparently exists with PGE_1 and GMP-P(NH)P activation. 2-Chloroadenosine is observed to shorten the time required for full activation of the enzyme by GMP-P(NH)P in an assay system containing ATP or AMP-P(NH)P as substrate provided there is a nucleotide regenerating system present (5). In addition, the ability of guanosine diphosphate (GDP) (as measured by its IC_{50} dose) to block activation of the enzyme by GMP-P(NH)P is decreased by 2-chloroadenosine (Fig. 6).

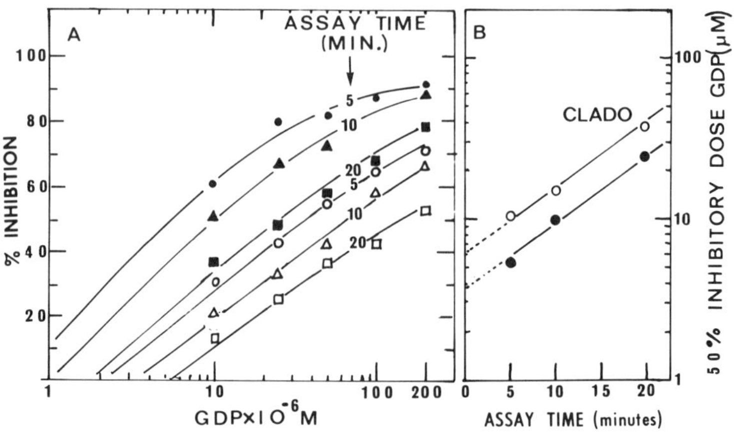

FIG. 6. 2-Chloroadenosine–mediated decrease in sensitivity of NS20 adenylate cyclase to inhibition by GDP. **A:** Enzyme activity was measured at 30° C and pH 7.4 with 5 mM $MgCl_2$, and 0.1 mM [^{32}P]AMP-P(NH)P as substrate in the absence of a nucleotide-regenerating system, with and without 50 μM 2-chloroadenosine for 5, 10, and 20 min. **B:** IC_{50} Dose of GDP calculated for the 5-, 10-, and 20-min reaction times. (●) No 2-chloroadenosine; (○), 2-chloroadenosine present. (From ref. 5, with permission.)

CONCLUSIONS

The data reviewed in this chapter clearly indicate that mouse neuroblastoma cells in culture have specific adenosine receptors linked to adenylate cyclase. In order to observe a significant increase in the accumulation of intracellular cAMP after addition of adenosine, the presence of an inhibitor of PDE is essential. Theophylline, which is itself known as a PDE inhibitor, blocks elevation of cAMP content by adenosine, both *in vivo* and *in vitro,* as an apparently competitive adenosine receptor antagonist.

The profile of action of adenosine and other nucleosides on adenylate cyclase *in vitro* parallels that observed with cAMP stimulation on the level of the intact cell. Adenosine action requires the presence of guanosinetriphosphate (GTP) and adenosine shortens the time required for maximal stimulation of the enzyme by the hydrolysis-resistant GTP analog GMP-P(NH)P. In addition, adenosine decreases the ability of GDP to block GMP-P(NH)P access to the important guanine nucleotide regulatory site of cyclase.

REFERENCES

1. Amano, T., Richelson, E., and Nirenberg, M. (1972): *Proc. Natl. Acad. Sci. U.S.A.,* 69:258–263.
2. Blume, A. J., Chen, C., and Foster, C. J. (1977): *J. Neurochem.,* 29:625–632.
3. Blume, A. J., Dalton, C., and Sheppard, H. (1973): *Proc. Natl. Acad. Sci. U.S.A.,* 70:3099–3102.
4. Blume, A. J., and Foster, C. J. (1974): *J. Biol. Chem.,* 250:5003–5008.
5. Blume, A. J., and Foster, C. J. (1976): *J. Biol. Chem.,* 251:3399–3404.
6. Blume, A. J., and Foster, C. J. (1976): *J. Neurochem.,* 26:305–311.
7. Burnstock, G. (1972): *Pharmacol. Rev.,* 24:509–581.

8. Burnstock, G. (1978): In: *Cell Membrane Receptors for Drugs and Hormones: A Multidisciplinary Approach,* edited by L. Bolis and R. W. Straub, pp. 107–118. Raven Press, New York.
9. Catterall, W. A., and Nirenberg, M. (1973): *Proc. Natl. Acad. Sci. U.S.A.,* 70:3759–3763.
10. Huang, M., Shimizu, H., and Daly, J. W. (1972): *J. Med. Chem.,* 15:462–466.
11. Kakiuchi, S., Rall, J. W., and McIlwain, H. (1969): *J. Neurochem.,* 16:485–491.
12. Londos, C., and Wolff, J. (1977): *Proc. Natl. Acad. Sci. U.S.A.,* 74:5482–5486.
13. McAfee, D. A., and Greengard, P. (1972): *Science,* 178:310–312.
14. McAfee, D. A., Schorderel, M., and Greengard, P. (1971): *Science,* 171:1156–1158.
15. Nelson, P. G. (1973): In: *Tissue Culture of the Nervous System,* edited by G. Sato, pp. 135–138. Plenum Press, New York.
16. Pull, I., and McIlwain, H. (1972): *Biochem. J.,* 127:91.
17. Salomon, Y., Londos, C., and Rodbell, M. (1974): *Anal. Biochem.,* 58:541–548.
18. Sattin, A., and Rall, T. W. (1970): *Mol. Pharmacol.,* 6:13–23.
19. Schultz, J., and Daly, J. W. (1973): *J. Biol. Chem.,* 248:843–852.
20. Shimizu, H., and Daly, J. W. (1972): *Eur. J. Pharmacol.,* 17:240–252.
21. Siggins, G. R., Oliver, A. P., Hoffer, B. J., and Bloom, F. E. (1971): *Science,* 171:192–194.

Physiological and Regulatory Functions of Adenosine and Adenine Nucleotides, edited by H. P. Baer and G. I. Drummond. Raven Press, New York © 1979.

Effects of Adenosine on Adenylate Cyclase in Broken Cell Preparations of Central Nervous System

*Joël Premont, **Jean-Pol Tassin, **Gérard Blanc, and *Joël Bockaert

**Laboratoire de Physiologie Cellulaire, and Groupe NB-INSERM U 114, College de France, 75231 Paris Cedex 05, France*

Adenosine is very potent in raising cyclic adenosine monophosphate (cyclic AMP) levels in brain slices (6,22). In intact tissues this could involve one or several of the following mechanisms: (a) increase of the substrate pool, adenosine triphosphate (ATP), for adenylate cyclase (23); (b) inhibition of cyclic AMP breakdown (14); (c) release of compounds able to increase cyclic AMP accumulation such as Ca^{2+} (4); and (d) activation of an adenylate cyclase directly through binding to an adenosine receptor. The following evidence suggests that this last mechanism exists in brain slices: Adenosine increases cyclic AMP accumulation even in the presence of specific uptake inhibitors (8), and 2-chloroadenosine is as active as adenosine (25). However, two observations suggest that the other mechanisms could also participate in cyclic AMP accumulation. When brain slices (24) or some cell lines (14) are exposed to adenosine in the presence of a second stimulatory hormone, such as histamine or norepinephrine, the resulting cyclic AMP accumulation is more than additive. In addition, adenosine increases cyclic AMP accumulation in brain slices of all areas of brain of all species tested.

Good evidence for an adenosine receptor would be the demonstration of an adenosine-sensitive adenylate cyclase in cell-free preparation of brain. Although such an enzyme has been demonstrated in some cerebral cell lines (2,17) and other tissues (13), it has not been found previously in brain. In this report, we describe and characterize extensively an adenosine-sensitive adenylate cyclase present in homogenates of several rat brain regions. This enzyme is shown to have characteristics similar to those of other neurotransmitter- or hormone-sensitive adenylate cyclases.

EXPERIMENTAL PROCEDURE

Homogenate Preparation

Male Sprague-Dawley rats (350 to 500 g) were killed by decapitation. Their brains were either dissected on a cooled plate (4° C) or frozen on a Leitz Wetzler microtome stage refrigerated at −7° C. Frozen brains were used to study the topographical distribution of adenylate cyclases. In this case microdiscs were punched out from a series of frozen slices in different brain areas following the method of Palkovits (16). Tissues were homogenized in 2 mM Tris-maleate, pH 7.2, and 2 mM EGTA (ethylene-glycol-bis(aminoethyl ether)-N,N′-tetraacetic acid).

Adenylate Cyclase Assay

Adenylate cyclase activity was measured by the conversion of [α^{32}P]ATP to cyclic [^{32}P]AMP. The reaction mixture (final volume, 100 μl) contained 25 mM Tris-maleate (pH 7.2); 0.25 mM ATP, 1 mM $MgSo_4$ (except where otherwise stated), 0.4 mM EGTA (added with the homogenate), 0.2 mg/ml of creatine kinase, 20 mM sodium creatine phosphate, 0.1 mM papaverine, and, where indicated, 0.4 IU/ml adenosine deaminase.

The reaction was initiated by adding the homogenate (20 μl). After 2 min at 30° C, 1 μCi of [α^{32}P]ATP and 0.001 μCi of [^{3}H]-cyclic AMP were added and the reaction was allowed to proceed for 5 min at 30° C except where otherwise stated. Cyclic AMP production was measured as described by Salomon et al. (21).

RESULTS AND DISCUSSION

Inhibition of Adenylate Cyclase of Rat Striatal Homogenates by Methylxanthines and Adenosine Deaminase

Our experimental design eliminated, or allowed for correction of, phosphodiesterase interference in adenylate cyclase assays. Thus the observed reduction (60 to

TABLE 1. *Effects of cyclic AMP, methylxanthines, and adenosine deaminase on adenylate cyclase of rat striatal homogenates*

Addition	Concentration (mM)	Adenylate cyclase activity (pmoles/min/mg protein)
None		310.4 ± 22.6
Cyclic AMP	1	339.2 ± 21.0
Theophylline	0.6	130.0 ± 10.8
Caffeine	1.0	163.4 ± 3.6
	5.0	126.8 ± 6.2
Isobutylmethylxanthine	0.1	197.8 ± 9.8
	1.0	118.2 ± 3.6
Adenosine deaminase (0.4 IU/ml)		175.4 ± 2.8
Adenosine deaminase + theophylline	0.6	122.0 ± 9.0

Results are the means ± SEM of three determinations.

70%) of enzyme activity in the presence of methylxanthines was unrelated to phosphodiesterase inhibition and, instead, could result from inhibition of endogenous adenosine action (Table 1). Indeed, methylxanthines are known to block the adenosine-induced cyclic AMP accumulation in brain slices (22). Our hypothesis was further strengthened by the observation that adenosine deaminase (ADA) was also able to reduce adenylate cyclase activity in striatal homogenates and that the effects of ADA and theophylline were not additive (Table 1).

Specificity of the Adenosine "Receptor" Implicated in Adenylate Cyclase Stimulation in Striatal Homogenates

Figure 1 shows the effect of three deamination-resistant analogs of adenosine on adenylate cyclase activity measured in the presence of ADA. 2-Chloroadenosine, N[6]-phenylisopropyladenosine, N[6]-dimethyl- and N[6]-methyladenosine were found to be full agonists of the system, i.e., in their presence adenylate cyclase activity was equal to that obtained with 10 μM adenosine in the absence of ADA (Fig. 1 and Table 2). Since 6-mercaptoadenosine and inosine were inactive one can conclude that an amino group at position 6 of the purine ring is essential for activity, as has been found in brain slices (8). 5′-Deoxyadenosine is a partial agonist of the enzyme whereas 2′- and 3′-deoxyadenosine were noncompetitive inhibitors. In this cell-free preparation in the presence of ADA, phosphorylated derivatives of adenosine, 5′-AMP, ADP, and ATP were also inactive. In brain slices they are active (8) but their effect could be indirect since they may be rapidly hydrolyzed to adenosine. Theophylline was a competitive inhibitor of the stimulation by 2-chloroadenosine (18) with a K_I of 20 μM (Table 2). Using this value and the dose inhibition curves of the adenylate cyclase by theophylline determined in the presence of several concentrations of adenosine (18), it was possible to estimate the affinity of adenosine for its receptor (0.5 μM) (Table 2).

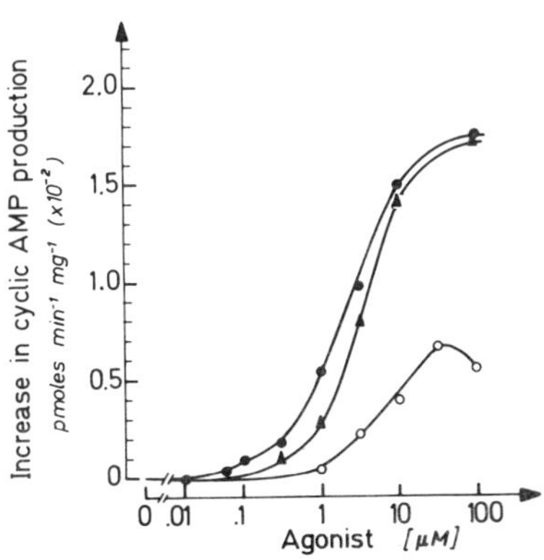

FIG. 1. Dose activation curves for three adenosine derivatives resistant to ADA. The reaction was conducted in the presence of ADA (0.4 IU/ml). In the absence of adenosine derivatives, adenylate cyclase activity was 175.4 ± 2.8 pmoles/min/mg protein (N=3). (-•-), 2-chloroadenosine; (-▲-) N[6]-phenylisopropyladenosine; (-○-), 5′-deoxyadenosine.

TABLE 2. *Chemical specificity of adenosine receptors implicated in the stimulation of the striatal adenylate cyclase*

Receptor	Maximal stimulation (%)	K_a app (μM)	K_I app (μM)
Adenosine [a]	100	0.5	
5′-AMP	0	—	
ADP [b]	0	—	
ATP	0	—	
Inosine	0	—	
Guanosine	0	—	
2-Chloroadenosine	100	1.2	
N^6-Phenylisopropyladenosine	100	3.2	
N^6-Dimethyladenosine	100	28.0	
N^6-Methyladenosine	120	85.0	
6-Mercaptopurine riboside	0	—	
5′-Deoxyadenosine	36	9	
3′-Deoxyadenosine [a]	Noncompetitive inhibition	—	
2′-Deoxyadenosine [a]	Noncompetitive inhibition	—	
Theophylline	Competitive inhibition		20

The maximal stimulation of each compound is expressed as percent of the stimulation obtained with adenosine.

K_a app=concentration giving half-maximal stimulation.

K_I app=apparent inhibition constant of adenosine-sensitive adenylate cyclase.

[a] These compounds were tested in the absence of ADA.

[b] This compound was tested in the absence of an ATP-regenerating system.

Origin of Adenosine in the Adenylate Cyclase Assay

The adenosine concentration in striatal homogenates was 10 ± 2.5 μM (N = 4) corresponding to a final concentration of 2 μM in the assay (18). This concentration

TABLE 3. *Evidence for hydrolysis of unlabeled ATP to adenosine*

First period (2 min)	Second period (3 min)	Final incubation (5 min)	Adenylate cyclase activity (pmoles/min/mg)	Percent
ATP	None	[α^{32}P]ATP	310	100
ATP + ADA	None	[α^{32}P]ATP	175.2	56
ATP + ADA + coformycin	None	[α^{32}P]ATP	306.8	99
ATP + ADA	Coformycin	[α^{32}P]ATP	288.4	93

The experiment was divided into three periods: Two preliminary incubations of 2 and 3 min and a final incubation period of 5 min with a tracer amount of [α^{32}P]ATP. The concentrations of ADA, 2-chloroadenosine, theophylline, ATP, and coformycin were 0.4 IU/ml, 50 μM, 0.5 mM, 0.25 mM, and 1 μM, respectively.

was high enough to fully activate the adenosine-sensitive adenylate cyclase. However, after washing the particulate fractions by centrifugation, the adenylate cyclase activity was still inhibited by ADA (18), suggesting that adenosine could also be formed from ATP hydrolysis during the assay. This was confirmed by the reactivation of adenosine-sensitive adenylate cyclase after blocking the action of ADA present in the assay using coformycin, a highly potent inhibitor of ADA (Table 3).

Effect of ATP and Mg^{2+}: Evidence for a Biphasic Stimulatory and Inhibitory Effect of Adenosine on Striatal Adenylate Cyclase

Effect of ATP and Mg^{2+} on the Stimulatory Effect

In most adenylate cyclase systems, hormones or neurotransmitters increased the reaction velocity without causing a change in the affinity of the enzyme for the substrate ATP-Mg. This was also the case for adenosine-sensitive adenylate cyclase (J. Premont et al., *in preparation*). In the presence of adenosine the enzyme was less susceptible to inhibition by free ATP. In the presence of low concentrations of ATP (0.01 mM) the adenylate cyclase activities measured in absence or presence of dopamine or 2 chloroadenosine increased as the function of Mg^{2+} concentration. Indeed, as in other systems, the enzyme activity continued to increase for Mg^{2+} concentrations in excess of that required to give a maximal concentration of ATP-Mg (Fig. 2). This could be taken as evidence for an allosteric activator role of free Mg^{2+} as proposed by several investigators working on hormone- or neurotransmitter-sensitive adenylate cyclase (1,10). Indeed, the Lineweaver-Burk

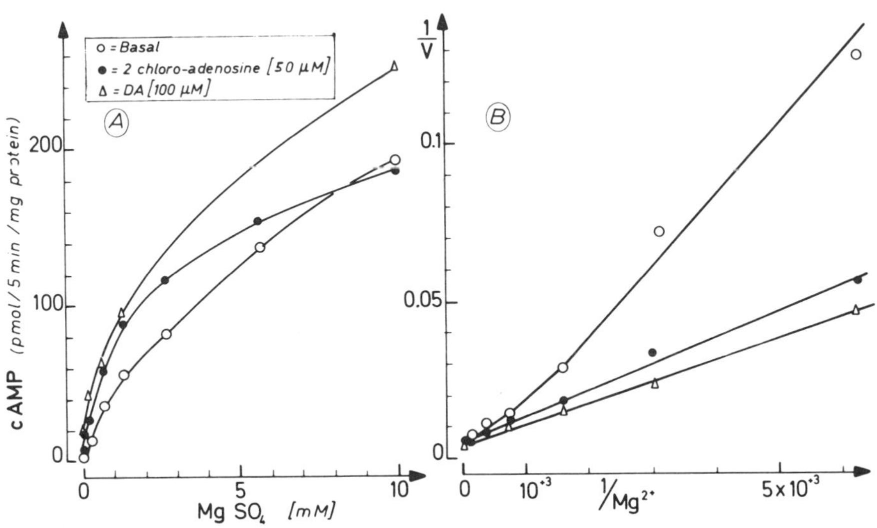

FIG. 2. Effect of Mg^{2+} concentration on adenosine-sensitive adenylate cyclase. The ATP concentration was kept constant (0.01 mM). The right part of the figure shows the Lineweaver-Burk plot of the data. Adenosine deaminase was present (0.4 IU/ml).

plot of the basal adenylate cyclase activity measured as a function of Mg^{2+} concentration was nonlinear (Fig. 2). On the contrary, in the presence of 2 chloroadenosine and dopamine, the same plot gives a straight line (Fig. 2). Increasing the molar ratio of added magnesium to added ATP from 2 to 1,000 decreased the fold stimulation by dopamine or 2 chloroadenosine. However, in the presence of 10 mM Mg^{2+} there was still a stimulation by dopamine whereas 2-chloroadenosine was slightly inhibitory (Fig. 2).

Effect of ATP and Mg^{2+} on the Inhibitory Effect

Starting from this last observation and from reports showing that in some systems adenosine has a biphasic stimulatory and inhibitory effect (13) we have further investigated the effect of adenosine and adenosine analogs on striatal adenylate cyclase at different Mg^{2+} concentrations. The adenosine analog may be divided into three groups differing in their effect on this enzyme (J. Premont et al., *in preparation*). The first group (adenosine, 2-chloroadenosine, 5'-deoxyadenosine) when tested at 100 μM was stimulatory at 1 mM Mg^{2+} and inhibitory at 10 mM Mg^{2+} concentrations. The second group (2'-deoxyadenosine 3'-deoxyadenosine) gave inhibition at all Mg^{2+} concentrations tested. The third group (N^6-phenylisopropyladenosine, N^6-methyl- and N^6-dimethyladenosine) was stimulatory at all Mg^{2+} concentrations. The inhibition was dependent on the concentration of Mg^{2+} and was maximal at 10 mM. At 10 mM Mg^{2+} the inhibition by purine nucleotides was dose dependent and saturable (Fig. 3). When producing an inhibition, the main effect of adenosine analogs was a reduction of the maximal velocity; the affinity of the enzyme for its substrate being only slightly changed (J. Premont et al., *in preparation*).

The adenosine site triggering the inhibitory effect is different from the adenosine site implicated in adenylate cyclase stimulation since (a) it does not have the same

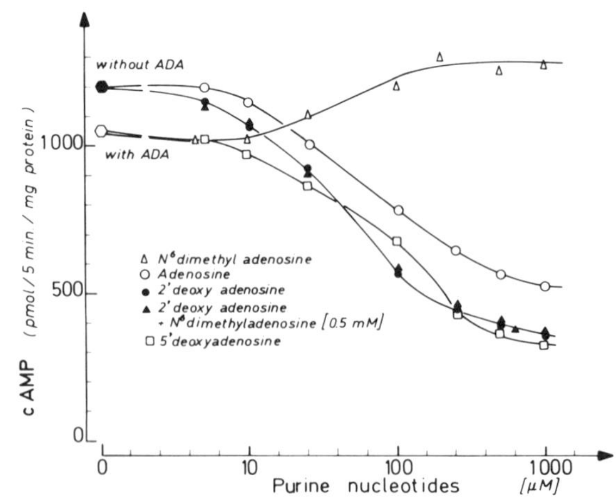

FIG. 3. Effect of some adenosine derivatives on the striatal adenylate cyclase measured at high Mg^{2+} concentration (10 mM). ADA, 0.4 IU/ml. The ATP concentration was 0.5 mM. The incubation time was 10 min.

chemical specificity and (b) N^6-dimethyladenosine (0.5 mM), an adenosine analog that is stimulatory only (even when tested at 1 mM), did not overcome the inhibitory effect of 2′-deoxyadenosine (Table 3). These results are in agreement with those of Londos and Wolff (13) on peripheral tissues.

It is interesting to note that purine nucleotides only partially inhibited the "basal" adenylate cyclase (Fig. 3). Although inhibition by adenosine or adenosine derivatives is obtained under nonphysiological conditions (high Mg^{2+} and nucleotide concentrations), it is a specific and saturable phenomenon. Thus one can speculate that in striatal homogenates part of the adenylate cyclase is stimulated by an unknown endogenous compound or more likely by an unknown mechanism that can be antagonized only under nonphysiological conditions. Furthermore, it is possible that other divalent cations such as Mn^{2+} could have the same effect as Mg^{2+} but at lower concentrations (12).

Subcellular Distribution of Striatal Adenosine-Stimulated Adenylate Cyclase: Evidence for a Guanosine Triphosphate Requirement for Its Stimulation

Subfractionation of striatal homogenate showed that most of the adenosine-sensitive adenylate cyclase was present in the nuclear and mitochondrial subfractions (Table 4). The high recovery in the nuclear pellet (1,000 × *g)* resulted mainly from the gentle homogenization procedure employed. Homogenization was performed by a motor driven (100 rpm) Teflon-glass homogenizer (4 strokes) that gave a low-speed pellet composed of nuclei but also of much cell debris. More drastic homogenization resulted in a loss of adenosine stimulation of adenylate cyclase. The mitochondrial subfraction, containing nerve endings, was further subfractionated according to Jones and Matus (9). The highest specific activity of basal and 2-chloroadenosine–sensitive adenylate cyclase was found in synaptic membranes. In this fraction the enzyme stimulation by 2-chloroadenosine was very small in the absence of guanosine triphosphate (GTP) (15%) and was increased by this nucleotide (52%) (Table 4) (20). A GTP requirement was also evident when comparing the percent recovery of total cyclic AMP production owing to 2-chloroadenosine obtained in the presence (79.8%) and absence (49.8%) of GTP (Table 4). Myelin and F3 fractions contained, respectively, 3% and 50% of the 2-chloroadenosine–sensitive enzyme found in the synaptic fraction. The 2 chloroadenosine–sensitive adenylate cyclase of the F3 fraction was not sensitive to GTP, probably because of the high content of GTP in mitochondria. Thus adenosine-sensitive adenylate cyclase, like basal- (7), dopamine- (5), or histamine-sensitive (10) adenylate cyclases was localized mainly in synaptic membranes.

Topographical Distribution of Adenosine-Stimulated Adenylate Cyclase in Rat Brain and Striatum: Relationship with Ca^{2+}- and Dopamine-Sensitive Adenylate Cyclases

A highly significant stimulation of adenylate cyclase activity by 2-chloroadenosine was found in posterior cerebellar cortex, olfactory lobe, tuber-

TABLE 4. *Distribution of adenosine-sensitive adenylate cyclase activity in subcellular fractions of striatum*

Fraction	Protein (mg)	Adenylate cyclase activity: Specific activity (pmoles/5 min/mg protein)				Total cyclic AMP production due to 2-chloroadenosine (% of homogenate)	
		Basal		+ 2-chloroadenosine			
		−GTP	+ GTP	− GTP	+ GTP	−GTP	+ GTP
Primary Subfractions							
Starting homogenate	37.4	640	466	1,306	829	100	100
P_1 Nuclear (1,000 × *g* pellet)	14.6	692	570	1,367	1,042	39	50
S_1 (1,000 × *g* supernatant)	24.7	323	295	815	535	48	43
P_2 Mitochondrial (10,000 × *g* pellet)	13.2	433	553	913	961	25	39
S_2 Microsomal (10,000 × *g* supernatant)	10.5	76	81	135	134	2.4	4.3
Subfractions of P_2							
F_1 (myelin)	1.14	52	176	90	235	0.18	0.5
F_2 (mainly synaptic membranes)	4.02	832	1,082	959	1,646	2.0	16.4
F_3 (mainly mitochondria)	4.56	286	341	638	603	6.2	8.6
Total recovery (%)	93					49.8	79.8

Fractionations were performed according to Jones and Matus (9) modified as follows. Initial homogenization of striatum was in 9 volumes of 0.30 M sucrose, 2 mM Tris-maleate (pH 7.2) to 2 mM EGTA (pH 7.2), using a glass homogenizer with a Teflon pestle at 100 rpm (4 strokes). The crude mitochondrial pellet was lysed in 2 mM Tris-maleate (pH 7.2) for 30 min. The lysate was made up to 34% (wt/wt sucrose by addition of the appropriate volume of 48% (wt/wt) sucrose and laid on the bottom of the tube. It was overlaid by two phases of 28.5% (wt/wt) and 10% (wt/wt) sucrose. Sucrose solutions were made up in 2 mM Tris-maleate (pH 7.2), 2 mM EGTA (pH 7.2). The sedimentation flotation density gradient was centrifuged at 60,000 × *g* for 110 min. F_1 was collected at 10% to 28.5% interphase; F_2 at 34% to 28.5%, and F_3 was the pellet. The different fractions were collected, diluted in large volume of 2 mM Tris-maleate (pH 7.2) to 2 mM EGTA (pH 7.2), and centrifuged at 100,000 × *g* for 30 min.

TABLE 5. *Regional distribution of adenosine-sensitive adenylate cyclase in rat brain*

Brain area	Adenylate cyclase activity (pmoles/min/mg protein) ADA	ADA + 2-chloroadenosine (10 μM)
Frontal cerebral cortex	55.4 ± 0.7	52.3 ± 2.1
Parietal cerebral cortex	60.0 ± 2.6	63.4 ± 2.4
Anterior cerebellar cortex	62.0 ± 4.0	67.6 ± 0.2
Central cerebellar cortex	79.7 ± 3.0	86.6 ± 1.70
Posterior cerebellar cortex	59.4 ± 0.9	70.3 ± 1.80[a]
Olfactory lobe	56.7 ± 0.8	66.7 ± 1.30[a]
Tuberculum olfactorium	27.4 ± 1.6	42.5 ± 1.20[b]
Substantia nigra (A_9)	52.1 ± 0.6	52.8 ± 1.65
Ventral tegmental area (A_{10})	37.0 ± 0.1	25.8 ± 2.0
Hippocampus	17.0 ± 2.0	18.1 ± 0.6
Hypothalamus	57.9 ± 1.0	48.1 ± 1
Rostral striatum	206.0 ± 5.0	373.0 ± 11 [b]
Caudal striatum	169.0 ± 6	277.0 ± 10[b]
Nucleus accumbens	116.0 ± 1	267.0 ± 5[b]
Globus pallidus	41.5 ± 0.5	80.0 ± 0.6[b]

The different areas were dissected from a series of frozen slices (−7°C) and localized according to the Atlas of Köning and Klippel (11). Results are the means ± SEM of three or four experiments.
[a] Significance, $p<0.001$.
[b] Stimulation, $p<0.001$ (Student's *t*-test).

culum olfactorium, striatum, nucleus accumbens, and globus pallidus. No stimulation was found in cerebral cortex, the anterior and central cerebellar cortex, substantia nigra, ventral tegmental area, hippocampus, and hypothalamus. Furthermore, within the striatum, the distribution of adenosine-sensitive adenylate cyclase was heterogeneous and very similar to that of dopamine-sensitive adenylate cyclase (ref. 3; and J. Premont et al., *in preparation).* On the contrary, the distribution of β-adrenergic–sensitive adenylate cyclase in this structure was found to be homogeneous (3). These results and the observation that adenosine-sensitive adenylate cyclase disappeared after injection of kainic acid, a substance known to induce neuronal degeneration (15) in striatum (J. Premont et al., *in preparation),* suggest that adenosine-sensitive adenylate cyclase is localized on specific neurons rather than on glial cells.

The most striking difference between results obtained in brain slices on cyclic accumulation induced by adenosine and our results on adenosine-sensitive adenylate cyclase in cell-free preparations is that in this system some brain areas, in particular cerebral cortex, did not contain this enzyme. This was also true for guinea pig cerebral cortex (J. Premont et al., *in preparation),* in slices of which an adenosine-induced accumulation of cyclic AMP is known to be very marked (22). This absence of adenylate cyclase stimulation in cerebral cortex homogenates was obtained no matter what Mg^{2+} concentration was tested in either fresh or frozen tissues. Adenosine was also unable to potentiate the norepinephrine stimulation of the adenylate cyclase in homogenates of the frontal cerebral cortex (J. Premont et al., *in preparation)* or dopamine stimulation in striatal homogenates. Thus it is possible that in cortical brain slices the adenosine effect does not involve the direct stimulation of an adenylate cyclase. Indeed, in the cortex adenosine-induced potentiation of neurotransmitter-sensitive adenylate cyclases is most important, being absent in striatal slices (26). However, we cannot exclude the possibility that the adenosine-sensitive adenylate cyclase from cerebral cortex is more labile during tissue homogenization than that of other structures. Further work is needed to clarify this point.

The activities of dopamine-, adenosine-, and Ca^{2+}-sensitive adenylate cyclases were additive (18). This might suggest that their receptors are localized on different cell types since no clear report exists on the additivity of hormone effects on a single cell type.

CONCLUSION

Adenosine stimulated adenylate cyclase in homogenates of certain brain areas with characteristics similar to other neurotransmitters:

1. The stimulation was mediated through a ''receptor'' highly specific for adenosine.
2. Adenosine increased the maximal reaction velocity of adenylate cyclase without altering the K_m for its substrate ATP-Mg.
3. Adenosine increased the apparent affinity of the Mg^{2+} allosteric site of the enzyme as reported for other neurotransmitters (10,19).
4. Guanosine triphosphate is required in order to obtain optimal stimulation.
5. The highest total and specific activities were observed in fractions enriched in nerve endings whereas adenylate cyclase activity was very low in cytosol or myelin fractions.
6. The topographical distribution is heterogeneous indicating that the adenosine stimulation is not a general property of brain adenylate cyclase.

These results support the hypothesis that adenosine may be a neurotransmitter in brain.

ACKNOWLEDGMENTS

This work was supported by LRA 219 from the Centre National de la Recherche Scientifique and by CLR 76.4.04.16 and ATP 58-78-90 from the Institut National de la Recherche Medicale.

REFERENCES

1. Birnbaumer, L., Pohl, S. L., and Rodbell, M. (1969): *J. Biol. Chem.,* 244:3468–3476.
2. Blume, A. J., and Foster, C. J. (1976): *J. Neurochem.,* 26:305–311.

3. Bockaert, J., Premont, J., Glowinski, J., Thierry, A. M., and Tassin, J. P. (1976): *Brain Res.,* 107:303–315.
4. Bradham, L. S., Holt, D. A., and Sims, M. (1970): *Biochem. Biophys. Acta,* 201:250–260.
5. Clement-Cormier, Y. C., Parrish, R. G., Petzold, G. L., Kebabian, J. W., and Greengard, P. (1975): *J. Neurochem.,* 25:143–149.
6. Daly, J. W. (1976): *Life Sci.,* 18:1349–1358.
7. De Robertis, E., De Lores Arnaiz, G. R., Alberici, M., Butcher, R. W., and Sutherland, E. (1967): *J. Biol. Chem.,* 242:3487–3493.
8. Huang, M., and Daly, J. W. (1974): *Life Sci.,* 14:489–503.
9. Jones, D. H., and Matus, A. I. (1974): *Biophys. Biochem. Acta,* 356:276–287.
10. Kanof, P. D., Hegstrand, L. R., and Greengard, P. (1977): *Arch. Biochem. Biophys.,* 182:321–334.
11. König, J. F. R., and Klippel, R. A. (1963): *The Rat Brain—A Stereotaxic Atlas.* Raven Press, New York.
12. Londos, C., and Preston, M. S. (1977): *J. Biol. Chem.,* 252:595–5956.
13. Londos, C., and Wolff, J. (1977): *Proc. Natl. Acad. Sci. U.S.A.,* 74:5482–5486.
14. Maguire, M. E., Surgill, T. W., Anderson, H. J., Minna, J. D., and Gilman, A. G. (1975): In: *Advances in Cyclic Nucleotide Research,* Vol. 5, edited by G. I. Drummond, P. Greengard, and G. A. Robison, pp. 699–718. Raven Press.
15. McGeer, E. G., Innanen, V. T., and McGeer, P. L. (1976): *Brain Res.,* 118:356–358.
16. Palkovits, M. (1973): *Brain Res.,* 59:449–450.
17. Penit, J., Huot, J., and Jard, S. (1976): *J. Neurochem.,* 26:265–273.
18. Premont, J., Perez, M., and Bockaert, J. (1977): *Mol. Pharmacol.,* 13:662–670.
19. Premont, J. Tassin, J. P., Thierry, A. M., and Bockaert, J. (1976): In: *Advances in Biochemical Psychopharmacology,* Vol. 15, edited by E. Costa and E. Giacobini, pp. 347–356. Raven Press, New York.
20. Rodbell, M., Lin, M. C., Salomon, Y., Londos, C., Harwood, J. P., Martin, B. R., Rendell, M., and Berman, M. (1975): In: *Advances in Cyclic Nucleotide Research,* Vol. 5, edited by G. I. Drummond, P. Greengard, and G. A. Robison, pp. 3–29. Raven Press, New York.
21. Salomon, Y., Londos, C., and Rodbell, M. (1974): *Anal. Biochem.,* 58:541–548.
22. Sattin, A., and Rall, T. W. (1970): *Mol. Pharmacol.,* 6:13–23.
23. Schultz, J., and Daly, J. W. (1973): *J. Biol. Chem.,* 248:843–852.
24. Skolnick, P., and Daly, J. W. (1977): In: *Cyclic 3′,5′ Nucleotides: Mechanisms of Action,* edited by H. Cramer and J. Schultz, pp. 289–315. Wiley and Sons, London.
25. Sturgill, J. W., Schrier, B. K., and Gilman, A. G. (1975): *J. Cyclic Nucleotide Res.,* 1:21–30.
26. Wilkening, D., and Makman, M. H. (1975): *Brain Res.,* 92:522–528.

Physiological and Regulatory Functions of Adenosine and Adenine Nucleotides, edited by H. P. Baer and G. I. Drummond. Raven Press, New York © 1979.

Action of Adenosine on Adenylate Cyclase

Constantine Londos, J. Wolff, and Dermot M. F. Cooper

National Institute of Arthritis, Metabolism, and Digestive Diseases, National Institutes of Health, Bethesda, Maryland 20014

It is now well established that cyclic adenosine monophosphate (cyclic AMP) levels in many cell types can be modified by low concentrations of adenosine, and sufficient data has been published to permit certain generalizations regarding this nucleoside effect: (a) Adenosine increases cyclic AMP levels in many tissues including brain slices (33,37), platelets (14), lymphocytes (41), bone cells (26), ventricular myocardium (16), epidermis (18), and a wide variety of cultured and transformed cell lines (3,6,24,42). (b) With the use of blockers of adenosine uptake, it has been shown that these adenosine effects are mediated by an external cell surface receptor in several tissues (6,12,14,16,42). (c) In nearly all cases, methylxanthines block this action of adenosine at the cell surface receptor; theophylline appears to be effective with all tissues examined (3,5,12,14,16–18,24,33,40,42). (d) A number of purine-modified adenosine derivatives, such as N^6-methyladenosine (12,16), N^6-phenyladenosine (16), 2-aminoadenosine (17), and 2-chloroadenosine (12,14,16,17,24,40,42), mimic this effect of the natural nucleoside.

On the other hand, compounds modified in the ribose moiety, such as 2′-deoxyadenosine (12,16,17,24,33) and 9-β-D-arabinofuranosyladenine (17), do not mimic the stimulatory effects of adenosine and, in fact, depress cyclic AMP levels in some cells in which adenosine has the opposite effect (12,17). There is one important exception to the above summary, the rat adipocyte. In this cell the surface receptor for adenosine and purine-modified analogs mediates *depression* of cellular cyclic AMP levels; this phenomenon is dealt with later in this chapter.

METHODS AND MATERIALS

The adenylate cyclase assay method and sources of materials used therein have been described (32), as have the sources of adenosine analogs (23) and preparations of membranes from rat liver (29), rat adipocyte (44), and Leydig tumor cell (42). Assay conditions are described in the legends to the figures and tables.

For the modified adenylate cyclase assay, [α^{32}P]2′-deoxy-adenosine triphosphate (ATP) was purchased from the International and Chemical Nuclear Corp. Initially, recovery of [^{32}P]2′-deoxy-cyclic AMP was monitored with [^{3}H]2′-deoxy-cyclic AMP prepared enzymatically by reacting [^{3}H]2′-deoxy-ATP with partially purified adenylate cyclase from *Bordetella pertussis* (15) kindly provided by Dr. Erik

Hewlett, of the National Institutes of Health. Subsequently, [^{3}H]2′-deoxy-cyclic AMP was supplied by ICN.

TWO SITES FOR ADENOSINE ACTION ON ADENYLATE CYCLASE

At the time we became interested in the actions of adenosine there were several reports showing that the nucleoside produces a biphasic effect on adenylate cyclase activity in membrane preparations from several different cell types (2,7,13,27). The typical pattern was activation of cyclase by low concentrations and inhibition by higher adenosine concentrations. Effects at the lower but not at the higher concentrations were inhibited by methylxanthines, which led to the suggestion by Haslam and Lynham (13) that these opposing actions of adenosine are mediated by different sites. Moreover, Blume and Foster (4) had shown that several purine-modified adenosine analogs stimulated adenylate cyclase activity in neuroblastoma homogenates, but that 2′-deoxyadenosine inhibited the stimulatory effects of both 2-chloroadenosine and prostaglandin E_1, providing further evidence for two sites for adenosine action on the enzyme.

There were two adenylate cyclases that did not exhibit the biphasic response to adenosine, that from rat liver in which the only apparent effect of adenosine was inhibition (22), and that from the Leydig tumor cell in which the only apparent effect was activation (42). A comparison of the actions of a number of adenosine analogs revealed that these two adenylate cyclases contain rather different sites for nucleoside action (23).

Table 1 lists the apparent K_a and K_I values for adenosine and adenosine analogs for activation of the Leydig cell cyclase and inhibition of the liver cyclase, respectively. The analogs are grouped into four categories: those that activate the Leydig cell enzyme but have no effect on the liver enzyme (Class A), those that inhibit the liver but have either no effect or are weakly effective on the Leydig cell enzyme (Class B), those that act on both cyclases (Class C), and those that act on neither (Class D). Note that Class A is comprised of purine-modified analogs. Based on the requirement for integrity of the ribose moiety, the site through which adenosine and its analogs increase activity of the Leydig cell enzyme is termed the R site, and based on the requirement for integrity of the purine moiety the site through which adenosine and its analogs inhibit the liver cyclase is termed the P site.

The activation by adenosine and R-type analogs of the Leydig cell enzyme is antagonized by theophylline. Also, the R-type analogs are those that mimic the cyclic AMP–elevating action of adenosine in cells in which the nucleoside has been shown to act at an external cell surface receptor (see first two paragraphs of this chapter). Thus, the R site appears to be the cell surface adenosine receptor.

On the other hand, the available information suggests that the P site is on the internal surface, associated with a regulatory unit of the catalytic component of the cyclase. Haslam and Rosson (14) have shown that a blocker of adenosine uptake prevents the depression of platelet cyclic AMP levels caused by high concentrations of adenosine. Second, the methylxanthines have no effect on the inhibition of the

TABLE 1. *A comparison of the effects of adenosine and adenosine analogs on adenylate cyclases from liver and I-10 Leydig cells*

Test compound	Leydig cell $K_{a(app)}$	Liver $K_{I(app)}$
Adenosine	7 μM	24 μM
Class A		
2-Methyladenosine	30	NI
2-Aminoadenosine	35	NI
N^6-Methyladenosine	40	NI
N^6-Phenylisopropyladenosine	40	NI
9-β-D-Ribofuranosyl purine	35	NI
Formycin	200	NI
Class B		
2′,5′-Dideoxyadenosine	NA	0.7
2′-Deoxyadenosine	700	70
2′,3′-Isopropylideneadenosine	NA	200
9-β-D-Arabinofuranosyl adenine	NA	25
9-β-D-Xylofuranosyl adenine	800	25
Class C		
2-Chloroadenosine	8	240
3′-Deoxyadenosine	30	250
5′-Deoxyadenosine	30	6
Class D		
7-Deazaadenosine	NA	NI
8-Azaadenosine	NA	NI
8-Bromoadenosine	NA	NI
9-β-L-Ribofuranosyl adenine	NA	NI

Values shown for liver, $K_{I(app)}$, are the concentrations of compounds required to inhibit activity by 50%; NI indicates that no inhibition was observed at the highest concentration tested which was at least 100 μM. Values shown for the Leydig cell enzyme, $K_{a(app)}$, are the concentrations required for half-maximal activation; NA indicates that no activation was observed.

liver cyclase by P-type analogs. Finally, with both the hepatic (21) and thyroid (43) cyclases the P site remains intact on solubilization of membranes with neutral detergents, a procedure that usually destroys interactions between cell surface receptors and the catalytic component of adenylate cyclase. However, further studies are required before the location of the P site can be determined precisely.

With the use of the analogs shown in Table 1 it was determined that the adenylate cyclases from adrenal tumor cell and human platelets possess both P and R sites (23). We have found subsequently that cyclases from the following sources contain both P and R sites: turkey erythrocytes, mouse splenic-lymphocytes, fetal liver cells, and rat brain cortex. Of particular interest is the fact that so many cells contain an R-type receptor for activation of adenylate cyclase, indicating that this receptor may be as ubiquitous as the beta adrenergic receptor for catecholamines which also activates a wide variety of adenylate cyclase systems.

IMPORTANCE OF ASSAY CONDITIONS IN DETECTION OF P AND R ADENOSINE SITES

In some cells the demonstration of both P and R sites proved difficult. Thus we reported initially that the rat liver cyclase has no R site for activation, and that the Leydig tumor cell enzyme has no P site for inhibition (23). We have now been able to show that extensive manipulation of assay conditions may be required to reveal either P or R sites. These are discussed below.

In view of the data showing that 2-chloroadenosine increases cyclic AMP levels in transformed rat liver cells (24), we thought it unusual that rat liver adenylate cyclase was not activated by either adenosine or R site agonists (22,23). No activation was observed even when the ATP concentration in the assay medium was lowered to reduce the amount of adenosine formed as a result of nucleotide hydrolysis by membrane-bound enzymes (22). In the course of examining the adipocyte adenylate cyclase (see below), which is extraordinarily sensitive to adenosine, an assay system was developed using (α^{32}P)2′-deoxy-ATP which had been shown to serve as a substrate of hormone-sensitive adenylate cyclase from fat cells (1). Since metabolism of this substrate leads to the formation of 2′-deoxyadenosine, which is a weak R site effector (Table 1), the effects of low concentrations of adenosine on adenylate cyclase activity can be determined. With

TABLE 2. *Effects of adenosine, methylxanthines, and adenosine deaminase on rat liver adenylate cyclase with 2′-deoxy-ATP as substrate*

Adenosine (μM)	Adenylate cyclase[a]	1 U/ml Adenosine deaminase + N^6-phenylisopropyladenosine (μM)	Adenylate cyclase[a]
0	93	0	52
0.08	109		
0.16	117	0.022	56
0.64	137	0.08	71
2.5	155	0.75	102
10.0	160	2.2	114
Theophylline (μM)		6.7	118
20.0	73	20	141
100	56	60	135
500	43		
MIX (μM)			
20	68		
100	52		
500	38		

Adenylate cyclase was assayed in a medium containing 0.1 mM 2′-deoxy-ATP, 0.1 mM 2′-deoxy-cyclic AMP, 1 μCi (α^{32}P)2′-deoxy-ATP, 5 mM $MgCl_2$, 10 μM GTP, 25 U/ml creatine phosphokinase, 5 mM creatine phosphate, 1 mM dithiothreitol, 30 mM Tris-HCl, pH 7.4, and approximately 0.1 mg/ml liver membrane protein. The reaction was stopped after 8 min at 30° C.

[a] Activity is expressed as pmoles cyclic AMP/mg protein/8 min.

this substrate (Table 2) adenosine activates the "basal" activity of hepatic cyclase at concentrations well below 1 μM, and the effect is maximal at 2.5 μM. If ATP had been the substrate the "basal" activity would have been in the vicinity of the maximum attained here and stimulation by added adenosine would not have been observed. The fact that adenosine deaminase and low concentrations of theophylline or 1-methyl-3-isobutylxanthine (MIX) depress control activity (Table 2) indicates that the membranes also contribute some adenosine to the assay medium. If account is taken of this contaminating adenosine, the K_a for activation of the liver cyclase would be approximately 0.1 μM. Note also that N^6-phenylisopropyladenosine, an R site effector resistant to adenosine deaminase, stimulates activity in the presence of deaminase with an apparent K_a value of 0.1 μM. The latter approach has been employed by Premont et al. (30) to show that the adenylate cyclase in brain striatal membranes is activated by adenosine analogs.

The meaning of adenylate cyclase activity in membrane preparations in the absence of hormone, known as basal activity, has been the subject of considerable speculation (28). As noted by Clark and Seney (7), basal activity in membranes is higher, with respect to hormone-stimulated activity, than is the ratio of unstimulated to stimulated cyclic AMP levels in intact cells. These investigators surmised that much of the basal activity could result from contaminating adenosine in the assay medium. The data in Table 2 show that as much as 75% of the basal activity of hepatic membranes can be attributed to the action of adenosine. A similar conclusion can be drawn from the data of Premont et al. (30) on the adenylate cyclase in striatal homogenates. Given the widespread occurrence of receptors sensitive to low concentrations of adenosine and the considerable contamination of cyclase reaction mixtures by the nucleoside, it is likely that a substantial portion of basal activity in many membranes results from adenosine stimulation.

One further point to be made is that activation of the liver cyclase by adenosine and R site effector analogs is absolutely dependent on the addition of guanosine triphosphate (GTP) to the assay medium. This requirement for GTP is not surprising in view of the fact that adenosine, like hormones, seems to act via a cell surface receptor, and the GTP requirement for hormonal stimulation of adenylate cyclases is well documented (31). Synergism between guanine nucleotides and adenosine has been observed with several other adenylate cyclases (7,36,42).

A characteristic feature of P site–mediated inhibition is that the sensitivity to adenosine is increased by agents that increase adenylate cyclase activity, such as divalent cations (5,22,23,27,39), hormones (19,22,23,25), and cholera toxin (45). As shown in Table 3, 2′,5′-dideoxyadenosine, a potent P site effector, does not inhibit the Leydig cell cyclase in the presence of 1.5 mM Mg^{2+}. If, however, Mn^{2+} is combined with the Mg^{2+}, activity increases and the adenosine analogs become inhibitory, again showing the critical nature of the assay conditions. Thus the Leydig cell cyclase contains a P site that exhibits the characteristic cation dependence, but that has a lower apparent affinity for 2′,5′-dideoxyadenosine than is seen with other cyclase systems (10,22,45).

TABLE 3. *Cation dependence for inhibition of Leydig tumor cell adenylate cyclase by 2′,5′-dideoxyadenosine*

2′,5′-Dideoxyadenosine (μM)	1.5 mM Mg^{2+}	1.5 mM Mg^{2+} + 0.2 mM Mn^{2+}
None	147	652
40	125	469
100	140	340
200	138	320
400	135	269

The assay medium contained 0.1 mM ATP, 2 μCi [α^{32}P]ATP, 0.33 mM cyclic AMP, 30 μM Gpp (NH)p, 0.1% crystalline bovine serum albumin, 30 U/ml creatine phosphokinase, 6.7 mM creatine phosphate, 25 mM Tris-HCl, pH 7.7, 33 μg/ml membrane protein, and divalent cations as indicated in the table. The reaction was stopped after 10 min at 37° C. Activity is expressed as pmol cyclic AMP/mg protein/10 min.

ADENOSINE ACTION ON ADIPOCYTE ADENYLATE CYCLASE

An interesting exception to the above generalizations regarding the actions of R and P site effector analogs is the rat adipocyte. Briefly, adipocytes extrude adenosine which then inhibits the actions of hormones that increase cyclic AMP levels (11,34,35). Adenosine, in the low-nanomolar range, inhibits both the increased cyclic AMP and resultant lipolysis caused by these hormones (8,10, 11,34,35,38). However, three aspects of adenosine action suggest that its effect is mediated by an R-type site, that is, by the site that usually mediates activation of adenylate cyclase: (a) although both P site analogs, such as 2′,5′-dideoxyadenosine and R site analogs, such as N^6-phenylisopropyladenosine, are effective in decreasing cellular cyclic AMP levels, the latter are far more potent (9,11,28); (b) adenosine appears to act at a cell surface receptor, as determined by studies with blockers of adenosine uptake (8,9); and (c) methylxanthines antagonize the actions of adenosine (8,11). However, despite repeated attempts by others (10,38), adenosine was found to be only a weak inhibitor of the fat cell adenylate cyclase in membrane preparations, and this inhibition was neither blocked by methylxanthines nor mimicked by those analogs, such as N^6-phenylisopropyladenosine, which are highly potent in decreasing fat cell cyclic AMP levels (9,11,38). On the other hand, 2′,5′-dideoxyadenosine inhibited the cyclase (10,38). It was concluded that the more sensitive adenosine site had been lost during membrane preparation (10,38).

The findings using analogs on intact cells suggested the presence of both P and R sites in adipocyte membranes; thus we set out to define assay conditions that would permit observation of both sites *in vitro*. Recently, we have found that adenosine and R site analogs are highly potent inhibitors of adenylate cyclase in purified rat adipocyte membranes (20). This was accomplished through careful control of the assay medium GTP concentration. It is known that GTP exerts a biphasic stimulatory and inhibitory effect on hormone-stimulated (catecholamine, adrenocorticotropic hormone, etc.) adenylate cyclase activity in fat cell membranes (44); a typical curve is shown in Fig. 1 with isoproterenol as the activating hormone. It is seen that R site effectors, such as N^6-phenylisopropyladenosine, have no effect on hormone-stimulated activity on the ascending, but clearly inhibit on the descending phase of the GTP curve. That is, R site effectors inhibit activity in concert with the inhibitory GTP action. Moreover, the actions of these analogs are competitively antagonized by theophylline, as shown in Fig. 2. The data in Fig. 2 compare the effects of theophylline on the inhibitory actions of N^6-methyladenosine and 2′,5′-dideoxyandenosine, R and P site effectors, respectively. Theophylline antagonizes the inhibition by N^6-methyladenosine (upper panel); the K_I for the methylxanthine was calculated to be 1 μM from a Schild plot of these data. In other experiments, MIX was found to be approximately fivefold more potent, and caffeine approximately fivefold less potent than theophylline. In contrast, the inhibition produced by 2′,5′-dideoxyadenosine is not affected by theophylline.

Another feature that distinguishes between R and P site effects is the level of inhibition achieved. R site effectors inhibit to an intermediate plateau of activity, as seen in the upper panel of Fig. 2, whereas P site effectors abolish activity, as shown in the lower panel.

In addition to the two N^6-substituted adenosine analogs shown in Fig. 1, other R

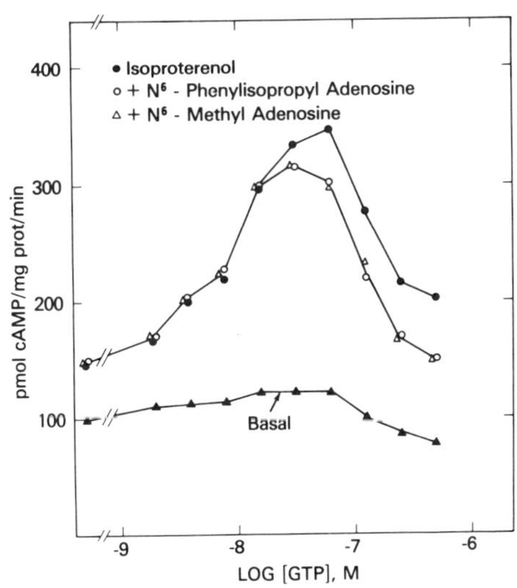

FIG. 1. Effects of N^6-phenylisopropyladenosine and N^6-methyladenosine on isoproterenol stimulated fat cell adenylate cyclase activity as a function of GTP concentration. The reaction mixture was as described in the legend to Table 4. Basal activity without hormone, (▲); 1 μM isoproterenol alone, (•); isoproterenol + 50 μM N^6-phenylisopropyladenosine, (○); and isoproterenol + 50 μM N^6-methyladenosine, (Δ). Basal activity was slightly inhibited by the two N^6-substituted analogs in this experiment (not shown). Note that a fairly constant increment of hormone-stimulated activity was inhibitied by the analogs. In other experiments it was found that the percent inhibition of the hormone response increased as the hormone concentration was lowered, and reached a maximum of approximately 70% when the hormone concentration was sufficient to double the basal activity. However, there was always a finite increment of activity elicited by isoproterenol that could not be inhibited by R site–reactive adenosine analogs.

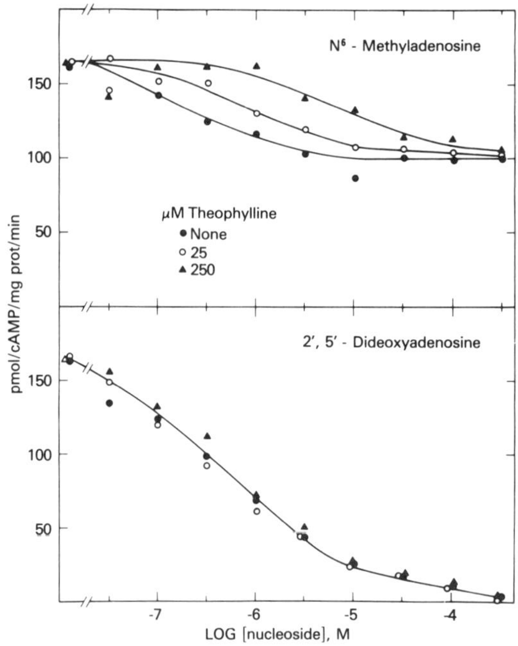

FIG. 2. Effects of theophylline on the inhibition of fat cell adenylate cyclase activity by N[6]-methyladenosine *(upper panel)* and 2′,5′-dideoxyadenosine *(lower panel)*. The reaction mixture was as described in the legend to Table 4 with the inclusion of 10 μM GTP.

site effectors (Class A, Table 1) also inhibit isoproterenol-stimulated activity only in the presence of "inhibitory" concentrations of GTP (Table 4). Note that although activity in the presence of 0.5 μM GTP is higher than that seen in the absence of nucleotide, this concentration is inhibitory with respect to peak activity (see Fig. 1). On the other hand, P site effectors (class B, Table 1) are inhibitory both in the absence and presence of GTP (Table 4). Moreover, the inhibition produced by R site analogs listed in Table 4 is antagonized by theophylline, whereas the inhibition by P site effectors is unaltered by the methylxanthine (data not shown).

It is concluded, therefore, that the fat cell cyclase contains both P and R sites for adenosine actions, and that both sites mediate inhibition of the enzyme. These sites are distinguished by adenosine analogs, the extent of inhibition, GTP dependence, and methylxanthine antagonism. The unique feature of the fat cell system is that R site effectors *inhibit,* whereas in all other cyclases examined, these analogs activate.

Under the assay conditions noted in the legend to Fig. 1, the apparent K_I values (nanomolar) for several adenosine analogs are: N[6]-propyladenosine, 5; 2-chloroadenosine, 10; N[6]-cyclohexyladenosine, 10; N[6]-phenylisopropyladenosine, 10; and N[6]-methyladenosine, 200. These values correlate closely with those found for the actions of these same analogs on the intact cell (38).

In the fat cell cyclase studies discussed above, adenosine deaminase was included

TABLE 4. *Guanosine triphosphate dependence for inhibition of fat cell adenylate cyclase by R but not P site adenosine analogs*

Analogs	Adenylate cyclase activity (pmoles cyclic AMP/mg protein/min) No GTP	+ 0.5 μM GTP
Experiment #1: P site analogs		
Control	160	188
1 μM 9-β-D-Xylofuranosyladenine	66	92
30 μM 2′-Deoxyadenosine	102	127
6 μM 9-β-D-Arabinofuranosyladenine	88	113
Experiment #2: R site analogs		
Control	101	158
30 μM 9-β-D-Ribofuranosylpurine	91	94
5 μM 2-Methyladenosine	107	109

Adenylate cyclase was assayed in a medium containing 0.1 mM ATP, 1 μCi (α^{32}P)ATP, 0.1 mM cyclic AMP, 0.1 U/ml purified calf duodenal mucosa adenosine deaminase (Sigma), 4 mM $MgCl_2$, 2 mM creatine phosphate, 25 U/ml creatine phosphokinase, 0.1% crystalline bovine serum albumin, 30 mM Tris-HCl, pH 7.4, 0.1 μM isoproterenol, and 10 μg/ml purified fat cell membrane protein. The reaction was stopped after 30 min at 24° C.

in the assay medium to remove contaminating adenosine. With the use of 2′-deoxy-ATP as substrate, and in the absence of deaminase, the K_I value for adenosine was found to be 10 nM. Thus the extraordinary sensitivity to adenosine by the intact adipocyte is observed *in vitro* with adenylate cyclase in purified plasma membranes.

It is concluded that this demonstration of an R site that (uniquely) mediates inhibition of the fat cell cyclase explains some of the potent actions of adenosine and methylxanthines on adipocyte metabolism.

CONCLUSIONS

The membranes of most cells contain two different sites, designated P and R, through which adenosine modulates adenylate cyclase activity. Each site has properties that distinguish it from the other, and the information accumulated to the present time is consistent with the classification shown in Table 5. When only one site can be demonstrated, we assume, as a working hypothesis, that conditions for revealing the other have not been found. Moreover, a substantial fraction of "basal" adenylate cyclase activity is likely to result from adenosine present under usual assay conditions.

From information obtained with adenosine analogs and methylxanthines, it would appear that it is the R site that mediates some of the biological actions elicited by low extracellular adenosine concentrations. Inasmuch as the R site appears to be on the external cell surface, it is not surprising that activation of adenylate cyclase, the phenomenon observed with most membranes examined, proceeds via a GTP-

TABLE 5. *Adenylate cyclase adenosine sites*

Characteristics	P Site	R Site
Analogs accepted	Ribose modified	Purine modified
Methylxanthines	No effect	Antagonize
Divalent cations (Mg^{2+}, Mn^{2+})	Enhance	No effect
Location	Internal surface	External surface
Effect	Inhibition	Activation (fat cell inhibition)

dependent process. The one exception, the fat cell, provides an example of the complexity of adenylate cyclase. In this cell adenosine acts at a cell surface R receptor to *inhibit,* again by a GTP-mediated process, the activation of adenylate cyclase by hormones that also act at cell surface receptors to activate the enzyme by a second GTP-mediated process.

There are no biological actions yet of adenosine that can be attributed to P site effects. However, at this site, which is associated with all adenylate cyclases we have examined, the enzyme can be inhibited by micromolar concentrations of adenosine. Given the intracellular location of the P site, the increased sensitivity to inhibition by adenosine when the enzyme is activated, and the likelihood of the occurrence of micromolar adenosine concentrations within the cell, it is not unreasonable to suggest a role for the P site in mediating feedback inhibition of the enzyme.

REFERENCES

1. Bär, H. P., and Hechter, O. (1969): *Biochim. Biophys. Acta,* 192:141–144.
2. Birnbaumer, L., Nakahara, T., and Yang, P. C. (1974): *J. Biol. Chem.,* 249:7857–7866.
3. Blume, A. J., Dalton, C., and Sheppard, H. (1973): *Proc. Natl. Acad. Sci. U.S.A.,* 70:3099–3102.
4. Blume, A. J., and Foster, C. J. (1975): *J. Biol. Chem.,* 250:5003–5008.
5. Blume, A. J., and Foster, C. J. (1976): *J. Neurochem.,* 26:305–311.
6. Clark, R. B., Gross, R., Su, Y. F., and Perkins, J. P. (1974): *J. Biol. Chem.,* 249:5296–5303.
7. Clark, R. B., and Seney, M. N. (1976): *J. Biol. Chem.,* 251:4239–4296.
8. Ebert, R., and Schwabe, U. (1973): *Arch. Pharmacol.,* 278:247–259.
9. Fain, J. N. (1973): *Mol. Pharmacol.,* 9:595–604.
10. Fain, J. N., Pointer, R. H., and Ward, W. F. (1972): *J. Biol. Chem.,* 247:6866–6872.
11. Fain, J. N., and Weiser, P. B. (1975): *J. Biol. Chem.,* 250:1027–1034.
12. Green, R. D., and Stanberry, L. R. (1977): *Biochem. Pharmacol.,* 26:37–43.
13. Haslam, R. J., and Lynham, J. A. (1972): *Life Sci.,* 11:1143–1154.
14. Haslam, R. J., and Rosson, G. M. (1975): *Mol. Pharmacol.,* 11:528–544.
15. Hewlett, E., and Wolff, J. (1976): *J. Bacteriol.,* 127:890–898.
16. Huang, M., and Drummond, G. (1976): *Biochem. Pharmacol.,* 25:2713–2719.
17. Huang, M., Shimizu, H., and Daly, J. W. (1972): *J. Med. Chem.,* 5:462–466.
18. Iizuka, H., Adachi, K., Halprin, K., and Levine, V. (1976): *Biochim. Biophys. Acta,* 444:685–693.
19. Johnson, R. A., and Welden, J. (1977): *Arch. Biochem. Biophys.,* 183:216–227.
20. Londos, C., Cooper, D. M. F., Schlegel, W., and Rodbell, M. (1978): *Proc. Natl. Acad. Sci. U.S.A.,* 75:5362–5366.
21. Londos, C., Lad, P. M., Nielsen, T. B., and Rodbell, M. (1978): *J. Supramol. Struct. (in press).*
22. Londos, C., and Preston, M. S. (1977): *J. Biol. Chem.,* 252:5951–5956.
23. Londos, C., and Wolff, J. (1977): *Proc. Natl. Acad. Sci. U.S.A.,* 74:5482–5486.

24. Maguire, M. E., Sturgill, T. W., Anderson, H. J., Minna, J. D., and Gilman, A. G. (1975): *Adv. Cyclic Nucleotide Res.,* 5:699–718.
25. McKenzie, S. G., and Bär, H. P. (1973): *Can. J. Physiol. Pharmacol.,* 51:190–196.
26. Peck, W. A., Carpenter, J., and Messinger, K. (1974): *Endocrinology,* 94:148–154.
27. Peck, W. A., Carpenter, J. G., and Schuster, R. J. (1976): *Endocrinology,* 99:901–909.
28. Perkins, J. P. (1973): *Adv. Cyclic Nucleotide Res.,* 3:1–64.
29. Pohl, S. L., Birnbaumer, L., and Rodbell, M. (1971): *J. Biol. Chem.,* 246:1849–1856.
30. Premont, J., Perez, M., and Bockaert, J. (1977): *Mol. Pharmacol.,* 13:662–670.
31. Rodbell, M., Lin, M. C., Salomon, Y., Londos, C., Harwood, J. P., Martin, B. R., Rendell, M., and Berman, M. (1975): *Adv. Cyclic Nucleotide Res.,* 5:3–29.
32. Salomon, Y., Londos, C., and Rodbell, M. (1974): *Anal. Biochem.,* 58:541–548.
33. Sattin, A., and Rall, T. W. (1970): *Mol. Pharmacol.,* 6:13–23.
34. Schwabe, U., and Ebert, R. (1974): *Arch. Pharmacol.,* 282:33–44.
35. Schwabe, U., Ebert, R., and Erbler, H. C. (1973): *Arch. Pharmacol.,* 276:133–148.
36. Sevilla, N., Tolkovsky, A. M., and Levitzki, A. (1977): *FEBS Lett.,* 81:339–341.
37. Shimizu, H., and Daly, J. (1970): *Biochim. Biophys. Acta,* 222:465–473.
38. Trost, T., and Stock, K. (1977): *Arch. Pharmacol.,* 299:33–40.
39. Weinryb, I., and Michel, I. M. (1974): *Biochim. Biophys. Acta,* 334:218–225.
40. Wilkening, D., and Makman, M. H. (1975): *Brain Res.,* 92:522–528.
41. Wolberg, G., Zimmerman, T. P., Hiemstra, K., Winston, M., and Chu, L. C. (1975): *Science,* 187:957–959.
42. Wolff, J., and Cook, G. H. (1977): *J. Biol. Chem.,* 252:687–693.
43. Wolff, J., Londos, C., and Cook, G. H. (1978): *Arch. Biochem. Biophys.,* 191:161–168.
44. Yamamura, H., Lad, P. M., and Rodbell, M. (1977): *J. Biol. Chem.,* 252:7964–7966.
45. Zenser, T. V. (1976): *Proc. Soc. Exp. Biol. Med.,* 152:126–129.

Physiological and Regulatory Functions of Adenosine and Adenine Nucleotides, edited by H. P. Baer and G. I. Drummond. Raven Press, New York © 1979.

Conformational Component of Structure-Activity Relationships in Nucleosides

Douglas L. Miles, Daniel W. Miles, and Henry Eyring

Department of Chemistry, University of Utah, Salt Lake City, Utah 84112

The ability of an enzyme to catalyze a molecular transformation depends on the intrinsic properties of potential substrates. Modifications of a nucleoside substrate produces analogs with differing intrinsic properties. The consequences of molecular modification can be conveniently divided into two intramolecular effects, namely, the electronic character and the conformational properties of the nucleoside analog. The intramolecular changes that accompany molecular modification of a nucleoside may, in turn, produce alterations in intermolecular effects. For example, the behavior of a nucleoside analog at an enzyme active or regulatory site may be changed. Other interactions important to the expression of biological effects such as solubility and membrane transport may be affected.

The information accumulated for a wide range of nucleoside analogs (6) suggests that altering intramolecular properties results in meaningful alterations of intermolecular interactions, particularly those involving enzyme catalysis. The complexity of the task of deriving correct interpretations from available data is indicated by the observations that some drastic alterations of a nucleoside may produce little change in its biochemical properties whereas minor and conservative changes can completely reverse the biochemical character of a nucleoside (6). The availability of a wide range of nucleoside analogs and of many studies of their biological properties provides the initial information essential to molecular level descriptions of the relationships between molecular properties and the biological effects expressed.

The apparent conformational specificity of some purine nucleoside interconversions and cyclic adenosine monophosphate (cyclic AMP) related enzymes are discussed. These investigations indicate the importance of conformational properties of nucleosides and nucleotides in directing the proper three-dimensional positioning of substrate binding and reactive entities at the active site of an enzyme during the course of enzyme-substrate interaction. The conformational distribution of purine analogs is strongly influenced by interactions between the nitrogen at the N3 position and the -CH group at the C8 position with the ribose moiety. The nature of these interactions is described as well as their potential modification by other factors.

A description of the conformational and electronic properties required by an enzyme of a potential substrate is of considerable value in describing and predicting the biological effects of nucleoside analogs. The ability of nucleoside analogs to act as disease-controlling therapeutic agents with a high degree of specificity or as modulators of physiological processes can potentially be predicted from a consideration of their conformational and electronic properties.

CONFORMATIONAL ENERGY CALCULATIONS

The Iterative Extended Hückel Theory (IEHT) was used to calculate the conformational and electronic properties of the nucleoside analogs discussed (34). The IEHT method is held to be a reliable semiempirical approach for doing conformational energy calculations (16). For the conformational and electronic properties presented here the coordinates used are based on the X-ray coordinates of adenosine 5′-phosphate (28). Where necessary standard bond lengths and angles were used (33).

The conformational definitions used are described in Fig. 1. In Figs. 2 and 3 are calculations of 8-azaadenosine, 8-azainosine, inosine, and 2-azainosine. For each of these analogs a comparison can be made with results previously published for adenosine, 2-fluoroadenosine, and 2-azaadenosine (16). The same structural framework, i.e., that of adenosine (28), was used for each analog. Modifications of the adenine moiety generated each analog. This approach minimizes the effects of coordinate variations from one set of X-ray coordinates to another. In Fig. 2 are presented our IEHT calculations of the conformational properties of 8-azaadenosine and 8-azainosine, which are compared with that of 2-fluoroadenosine. For each of

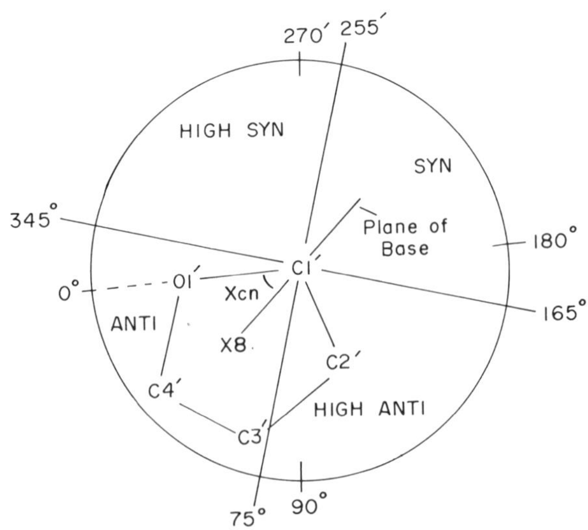

FIG. 1. Definition of glycosidic conformational ranges. A dihedral angle of 0° corresponds to a cis planar arrangement of OI′-CI′-NI-C8 for purines.

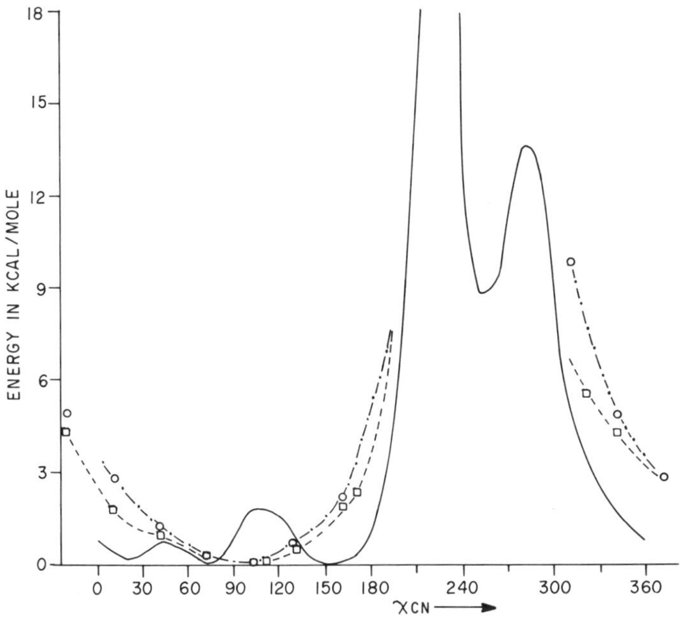

FIG. 2. Conformational energy calculations for 2-fluoroadenosine (—), 8-azadenosine (---), and 8 azainosine (-·-·-·).

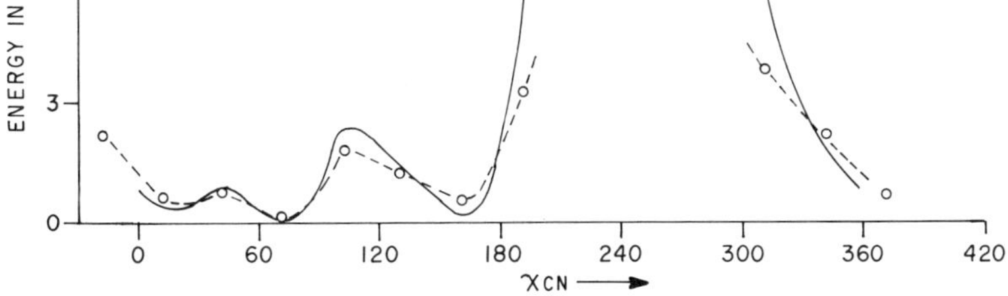

FIG. 3. Conformational energy calculations for inosine (—) and 2-azainosine (-----).

these analogs there is a shift toward greater stability in the high anti conformation compared with adenosine (16). The calculations in Fig. 2 support the concept that purine analogs in which the -CH group at the 8 position is substituted by a nitrogen are shifted toward a greater preference for high anti orientation. There is substantial evidence of such a shift in the relative populations of the anti versus the high anti conformation (39). In this respect, purine analogs with a nitrogen at the 8 position appear to resemble pyrimidine analogs such as 6-azauridine in which a similar shift toward greater relative stability of the high anti conformational range is indicated by both theoretical and experimental results (37). The conformational energy results presented here differ from those of others (35) as well as previous calculations of ours (16) on 8-azaadenosine that show a retention of the preference for an anti orientation similar to that of adenosine, which is considered to result from the choice of coordinates or crystal structures used in the computations and not to reflect the conformational effect of either the 8-azapurine (16,35) or 6-azapyrimidine substitutions (25). This indicates the importance of using comparable molecular coordinates to demonstrate the conformational effects of various modifications. This approach eliminates factors such as a modified sugar puckering that can mask the conformational effect of a molecular modification. This approach, however, fails to incorporate the changes in bond angles and bond lengths produced by such a molecular modification. We have previously discussed the possible errors introduced by this approach (18). For the calculations presented here these types of errors are not considered qualitatively important.

The results presented in Figs. 2 and 3, compared with the calculated electronic properties shown in Table 1, indicate that a decrease in net atomic charge at the N3 position stabilizes the high anti conformational range. However, other base ribose

TABLE 1. *Net atomic charges at some key positions for various purine nucleoside derivatives*

Molecule	Net atomic charges			
	C6	N1	N3	N7
Adenosine	0.086	−0.186	−0.170	−0.184
2-Fluoroadenosine	0.111	−0.156	−0.143	−0.180
2-Azaadenosine	0.108	−0.150	−0.134	−0.180
8-Azaadenosine	0.082	−0.193	−0.165	−0.154
8-Azainosine	0.121	0.004	−0.144	−0.140
2-Azainosine	0.115	0.019	−0.095	−0.173
7-Deazaadenosine	0.056	−0.209	−0.180	
1-Deazaadenosine	0.058		−0.192	−0.192
2,6-Diaminopurine ribonucleoside	0.064	−0.213	−0.187	−0.194
Inosine	0.102	−0.008	−0.152	−0.181
Guanosine	0.080	−0.032	−0.189	−0.200

Net atomic changes at the C6, N1, N3, and N7 positions calculated using the IEHT method.

interactions tend to minimize this stabilization. Thus inosine and 2-azainosine are conformationally similar in the anti and high anti regions, in spite of their differing net atomic charge densities at the N3 position. The same effect is seen in the comparison of the conformational properties and N3 net atomic charges for 8-aza-adenosine and 8-azainosine. For each of these analogs, as well as the 2-fluoro-adenosine and 2-azaadenosine analogs reported previously (16) there is a stabilization of the high anti conformational energy region compared with adenosine.

CONFORMATIONAL EFFECTS OF MOLECULAR MODIFICATIONS

Conformation and N3 Electronic Properties

It has been demonstrated recently that a reduction in net atomic charge at the N3 position of purines alters the conformational properties of nucleosides (16,17). The high anti conformational range is stabilized for adenosine analogs such as 2-fluoroadenosine and 2-azaadenosine (16). For both analogs a significant reduction in electron density at the N3 position appears to account for both the theoretically and experimentally demonstrated shift of conformer population to high anti and syn conformational regions. Similarly, it might be expected that nucleosides such as guanosine or nucleoside analogs such as 1-deazaadenosine and 2,6-diaminopurine ribonucleoside, which have an increased electron density at the 3 position (see Table 1) will have an increased anti conformer population if hydrogen bonding between N3 and the 5′-hydroxyl group is not an important factor in stabilizing the syn conformation. Since these differences are retained in nucleotides or cyclic nucleotides, perhaps one factor contributing to the dissimilar biological properties of cyclic AMP compared with cyclic guanosine monophosphate (cyclic GMP) may be a greater "anti preference" by cyclic GMP. Several enzymes that operate on nucleosides appear to be sensitive to the electronic properties at the N3 position of purines. These enzymes include inosine monophosphate (IMP) dehydrogenase, purine nucleoside phosphorylase, adenosine kinase, deoxycytidine kinase, adenosine deaminase, adenylate cyclase, and protein kinase II. In contrast to suggestions that the N3 nitrogen is involved in enzyme binding, we propose that the N3 nitrogen may often alter the glycosidic conformational preferences.

Inosine Monophosphate Dehydrogenase

The enzyme IMP dehydrogenase is inhibited by analogs such as ribavirin that can apparently exist in the high anti conformation (19). Ribavirin derivatives that are not stable in this orientation are not inhibitors of this enzyme (18,19). Their inactivity may represent both their failure to be readily phosphorylated by adenosine kinase, which appears to require a high anti orientation (20), or to effectively displace IMP at the active site of the enzyme IMP dehydrogenase which also appears to prefer substrates that can readily exist in a high anti conformation. Of further interest is the finding that 8-azainosine is predominantly converted to guanine nucleotides rather than to adenine nucleotides (2). By comparing the conformational energy profile of inosine in Fig. 3 with that of 8-azainosine in Fig. 2 it can be seen that between 90 and 140° kcal 8-azainosine is predicted to be more stable than inosine. This region is expected to be unstable for inactive ribavirin analogs such as 5-methylribavirin (27). Perhaps an orientation in this range is specified by IMP dehydrogenase.

Another factor that may influence the metabolic fate of 8-aza-IMP is the modification in net atomic charge at the N7 position. By referring to Table 1 it can be seen that the reduction in net atomic charge at this position is significant. Inosine monophosphate is apparently insensitive to this property, but the competing enzyme, adenylosuccinate synthetase may be more sensitive to atomic charge variations at the N7 position.

A finding of some interest is that 6-azauridine, which is known to have its glycosidic orientation shifted toward a preference for the high anti conformation because of substantial repulsive interactions between the N6 nitrogen and the 01′ oxygen (37) produces an accumulation of adenine nucleotides at the expense of guanine nucleotides (10). Although this point has not been investigated to our knowledge, it may be that 6-azauridine, like ribavirin, is an inhibitor of IMP dehydrogenase. 8-Azainosine has been shown to resemble 6-azauridine in its ability to inhibit the conversion of orotic acid to uridine nucleotides (3,31). It appears that 8-azapurines as well as purine nucleosides like adenosine can mimic the effects of pyrimidine analogs such as 6-azauridine as inhibitors of pyrimidine enzymes. This similarity is proposed to be related to their potentially similar glycosidic properties. It also appears that pyrimidine analogs such as 6-azauridine can inhibit purine-utilizing enzymes such as IMP dehydrogenase because of their conformational resemblance to purine substrates such as inosine or purine analogs such as 8-azainosine. The possible inhibition of IMP dehydrogenase by 6-azauridine would indicate that, like ribavirin, a major modification of the substrate to produce an inhibitory analog is effective as long as functional groups necessary for the binding of the substrate are retained and the active conformation for enzyme action is not energetically unfavorable. Thus 6-azauridine, ribavirin, 8-azainosine, inosine, and adenosine have in common the essential conformational characteristics to give them similar biological effects with respect to certain enzymes of purine and pyrimidine metabolism. This interchangeability of purine and pyrimidine nucleosides represents a regulatory option in which pyrimidines regulate purine metabolism and purine nucleosides regulate pyrimidine metabolism. This can occur because of their similar conformational and functional group specificity.

Adenosine Kinase

We have previously proposed that a high anti orientation is required of adenosine or adenosine analogs by this enzyme (20). Adenine arabinoside (ara-A) is a poor substrate for adenosine kinase if it is a substrate at all (36). The slow conversion of ara-A to ara nucleotides, which can be accomplished by deoxycytidine kinase (12) and deoxyadenosine kinase (5), has occasionally been interpreted to suggest that ara-A is a substrate for adenosine kinase. However, to our knowledge there is no evidence in the literature that the isolated adenosine kinase enzyme has the capacity to act on this analog. Based on the inactivity of ara-A and 2′-deoxyadenosine as substrates for adenosine kinase it has been proposed that a trans relationship between the base moiety and a hydroxyl group at the 2′ position was required by the enzyme (4). A possible exception to this rule is that 2-fluoro-2′-deoxyadenosine has been reported to be a substrate for this enzyme (30–38). However, the possibility remains that enzymes such as deoxycytidine kinase may also accomplish this phosphorylation in addition to or in place of adenosine kinase.

Another interesting investigation was the molecular orbital study by Kaneti (11) who found that there was a good correlation between the electronic indices at the N3 position and the rates of phosphorylation for most adenosine analogs. Thus the lower the electron density at N3 the more rapid the enzyme reaction. Electronic properties at other locations in the purine moiety were reported to be not important (11).

As an alternative to the trans rule proposed by Bennett and Hill (4) and the formation of an enzyme substrate complex proposed by Kaneti (11) we have considered this information from a conformational point of view. Thus the high anti orientation proposed to be required by the enzyme adenosine kinase was found to be unstable for the inactive ara-A analog and for 2′-deoxyadenosine. However, the conformational effect of removing the 2′-OH group is considered to be effectively counterbalanced by certain substituents at the 2 position of the purine moiety. Thus the reported activity of 2-fluoro-2′-deoxyadenosine (30,38), compared with the lack of activity of 2′-deoxyadenosine, may reflect the ability of the fluoro substituent to stabilize the high anti orientation near 160° kcal that may be specified by the enzyme (20). The 2-fluoro modification may also produce this effect by failing to act as an effective substrate for the competing enzyme, adenosine deaminase, or by accelerating the activity of the analog as a substrate for deoxycytidine kinase (12).

The increased activity of adenosine analogs such as 8-azaadenosine and tubercidin (20) as substrates for adenosine kinase that show increased stability in the high anti conformation also indicates a possible requirement for a high anti substrate orientation by the adenosine kinase enzyme.

The correlation between increases in adenosine kinase activity with reductions in N3 electronic properties as well as the activity of analogs that don't easily fit this correlation (11) supports a conformational interpretation in which the postulated active high anti conformation is specified by the enzyme (16,20).

Purine Nucleoside Phosphorylase

Hypoxanthine arabinoside (ara-H) is apparently not an effective substrate for purine nucleoside phosphorylase in man since it is substantially excreted in the urine as a nucleoside (14). Similarly, in mice labeled ara-A is excreted in the urine as ara-H (7). This and other information (26) is in conflict with the assumption that ara-H is effectively cleaved to give the free base (8,13). The slow conversion reported of labeled ara-H to labeled nucleotides (8,13) may be attributable to the ability of enzymes such as deoxycytidine kinase to slowly phosphorylate ara-H (12). However, the findings that labeled ara-A is converted primarily to nucleotides such as IMP and AMP (8,13) have been interpreted to suggest that cleavage of the glycosidic bond does occur. However, either inversion of configuration followed by cleavage or direct phosphorylation of either ara-adenine or ara-H followed by inversion could account for these results. If arabinosyl-l-phosphate is found not to be able to replace ribose-1-phosphate generally, as is the case with fish muscle purine nucleoside phosphorylase, (40) then cleavage of the glycosidic bond would favor conversion of labeled ara-H to normal nucleotides. It would also result in an irreversible reaction so that if there is even a slow cleavage of the glycosidic bond of ara-H it would be unlikely that 88 or 97% recoveries of labeled ara-H in man (14) would occur after a 24-hr period.

For purine nucleoside phosphorylase, which operates on both nucleosides and 2′-deoxy nucleosides, there can be no doubt that a 2′-OH group is not essential for the enzyme reaction (3,29). Thus the conformational effect of a 2′-OH group cis to the purine moiety could rule out the active conformation. There is little chance that a 2′-OH group, which is essentially internal to the substrate, as in ara-H will interfere with the enzyme reaction through intermolecular interactions. Thus, with limited experimental data available, it appears that either the high anti orientation or the opposite high syn conformation is required by the enzyme. The reduced activity of 2-aza analogs of both inosine and 2′-deoxyinosine (3,29) may possibly be related to their expected increased stability in an inactive syn conformation compared with inosine and 2′-deoxyinosine. The anti and high anti conformations are about equally probable (see Fig. 3) for inosine or 2-azainosine. From Fig. 3 it may be suggested that the syn and high syn conformations might be more stable for the 2-aza nucleosides or deoxynucleosides. Thus inosine or 2′-deoxyinosine with the high anti orientation probably being more populated would be expected to be more active than the 2-aza analogs which are expected to shift their conformational preference toward a greater inactive conformational population.

Deoxycytidine Kinase

The proposed conformational specificity of deoxycytidine kinase has been described previously (16) and is currently under further investigation. It appears that the increased activity of 2-fluoro-2′-deoxyadenosine compared with 2′-deoxyadenosine and 2-amino-2′-deoxyadenosine (12) is another illustration of the ability of a reduction in net atomic charge at the N3 position to stabilize a postulated active conformation (16) and increase the rate of the enzyme reaction.

Adenylate Cyclase

Adenosine appears to be able to activate adenylate cyclase under conditions in which ara-A has no effect (41). This suggests another hypothesis that adenosine receptors that activate adenylate cyclase require that the nucleoside is bound in a high anti orientation (21). Two distinct adenosine reactive binding sites have been proposed by Londos and Wolff (15). They suggested that one site, which may require a high anti oriented adenosine analog, will activate adenylate cyclase whereas the other type of site that may be specific for an anti conformation in an adenosine analog will suppress adenylate cyclase activity (15). It may be found that the ability of substituents at the C2 position in purines to affect adenylate cyclase activity is related, in part, to their abilities to modify electronic properties at the N3 position in purine nucleosides. However, it appears necessary to go beyond a mere calculation of electronic properties at the N3 position since complex interactions created by modifications in the purine or ribose moieties can, in some cases, dominate or mask the effect of changing the electron density at the N3 position. Effects of this type are shown in Figs. 2 and 3.

Protein Kinase II

A recent investigation has shown that there is a good relationship between the electronic properties at the N3 position of cyclic AMP analogs and their abilities to activate protein kinase II (24). Instead of the N3 position participating in binding to the regulatory subunit of the protein kinase as proposed, it may be that the variations in activity can be related to the conformational effects produced by those variations. Under this interpretation it appears that an anti orientation is required of cyclic AMP analogs if they are to stimulate cyclic AMP–dependent protein kinase II. A comparison of the CNDO/2 calculations of N3 electronic properties (24) and our IEHT calculations as predictors of the activity of various cyclic AMP analogs is shown in Fig. 4. The considerable activity of various 8-substituted derivatives of cyclic AMP may suggest a syn orientation as the active conformation (23). However, the syn conformation, as defined in Fig. 1, appears to place the N3 nitrogen in a position that is virtually inaccessible to an enzyme entity.

Conformation and Modifications at the 8 Position of Purines

It has generally been accepted that substitutions of bulky groups at the 8 position of purines results in a preference for a syn conformation. A recent crystallographic investigation has shown that 8-aminoadenosine and 8-methylaminoadenosine monophosphates are stable in the anti conformation (9). A subsequent theoretical investigation supported this finding but held that an 8-methyl derivative would result in a preference for the syn conformation (32). Theoretical and experimental results from this laboratory indicate that 8-methyl purine nucleoside derivatives should also be regarded as being distributed between syn and anti orientations as well as certain regions of the high anti conformational range (22). In particular, circular dichroism studies indicate that whereas 3-deazaadenosine, benzimidazole, and 8-chlorobenzimidazole prefer a syn conformation in solution the 8-methyl derivative of benzimidazole prefers the anti conformation. Thus interpretations that assume that a methyl or amino group at the 8 position rule over an anti orientation or

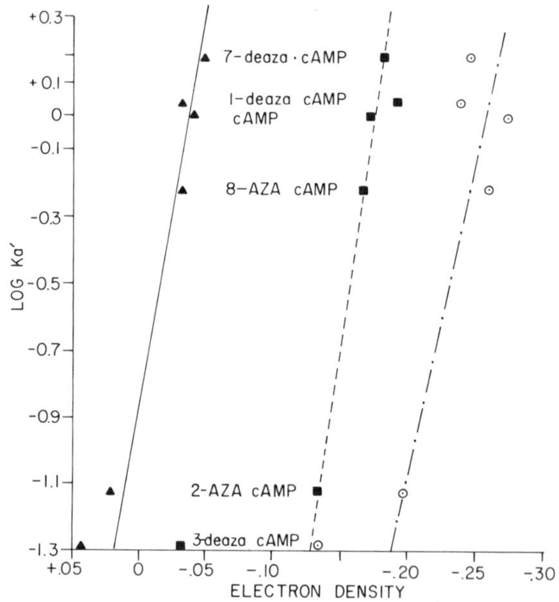

FIG. 4. Plots of log K_a' for cyclic AMP–dependent protein kinase from bovine brain versus CNDO/2 calculation of net atomic charge at N3 (⊙-⊙-⊙), CNDO/2 calculation of net atomic charge at N3 minus the electron density in the P_z orbital at N3 (▲-▲-▲) as recently reported. For comparison, an IEHT calculation of N3 net atomic charge (■-■-■) is included.

certain regions in the high anti conformational range are in error. Just as hydrophobic interactions create a preferred anti conformation for 8-methylbenzimidazole in aqueous solution, the environmental factors existing at an enzyme's active site or a cell surface receptor may modify the conformational state of a substrate-enzyme or substrate-receptor complex. The analog 8-bromoadenosine can potentially exist in the anti conformation. In a crystalline complex of 8-bromo-ADP-ribose with alcohol dehydrogenase, the 8-bromoadenosine moiety is in the anti conformation (1). Some enzymes can apparently stabilize an anti orientation for nucleoside derivatives with various bulky 8-substituents (1,9). This stabilization may also occur with 8-substituted cyclic AMP analogs as activators of protein kinase II (23) and would suggest an anti conformation being required by this enzyme of cyclic AMP derivatives.

CONCLUSIONS

The intrinsic conformational and electronic properties of nucleoside analogs can be correlated with their observed biological effects. It appears that the electronic properties at the N3 position in purines exert important effects on glycosidic orientation. Modulation of this effect occurs with 2-azainosine and tubercidin and indicates that other factors besides N3 electronic properties can counteract or augment the stabilization of the high anti conformation generally observed with a reduction in net atomic charge at the N3 position. Modifications in the five-membered ring of purines also exert important conformational effects. Both of these factors deserve further investigation since minor modifications can radically alter the conformational character of nucleosides. To characterize the relationships between molecular modification and the biochemical effect of the modification it appears necessary to calculate the conformational energy profile for each analog in spite of the good correlations, in some cases, between N3 electronic properties and activity. Results so far suggest that such calculations, if done with a comparable set of atomic coordinates for the analogs under investigation, will account for the observed correlations as well as the exceptions found with strictly electronic correlations.

Similarly, substitutions at the 8-position must be viewed with caution because of the apparent ability of some enzymes to impose an anti orientation on certain purine derivatives.

Molecular orbital calculations of nucleoside conformation and nucleoside electronic structure can facilitate the characterization of enzyme or receptor specificity provided the activity of an appropriate set of analogs has been experimentally determined. The ability of nucleoside analogs to affect physiological processes can hopefully be described and predicted using theoretical descriptions of their inherent conformational and electronic properties.

ACKNOWLEDGMENTS

The authors wish to thank the National Institutes of Health for their support through grant #GM 12862-13 and the University of Utah Research Fund for their assistance.

REFERENCES

1. Abdallah, M. A., Biellmann, J. F., Nordstrom, B., and Branden, C. I. (1975): *Eur. J. Biochem.*, 50:475–481.
2. Bennett, L. L., Jr., and Allan, P. W. (1976): *Cancer Res.*, 36:3917–3923.
3. Bennett, L. L., Jr., Allan, P. W., Carpenter, J. W., and Hill, D. L. (1976): *Biochem. Pharmacol.*, 25:517–521.
4. Bennett, L. L., Jr., and Hill, D. J. (1975): *Mol. Pharmacol.*, 11:803–808.
5. Bennett, L. L., Jr., Montgomery, J. A., Brockman, R. W., and Shealy, Y. F. (1978): *Adv. Enzyme Regul. (in press)*.
6. Bloch, A. (1975): *Ann. N. Y. Acad. Sci.*, 255:576–596.
7. Brink, J. J., and LePage, G. A. (1964): *Cancer Res.*, 24:1042–1049.
8. Drach, J. C., Bus, J. S., Schultz, S. K., and Sandberg, J. N. (1974): *Biochem. Pharmacol.*, 23:2761–2767.
9. Evans, F. E., and Kaplan, N. O. (1976): *J. Biol. Chem.*, 251:6791–6797.
10. Janeway, C. M., and Cha, S. (1977): *Cancer Res.*, 37:4382–4388.
11. Kaneti, J. (1973): *J. Theor. Biol.*, 38:169–179.
12. Krenitsky, T. A., Tuttle, J. V., Kosalka, G. W., Chen, I. S., Beacham, L. M. III, Rideout, J. L., and Elion, G. B. (1976): *J. Biol. Chem.*, 251:4055–4061.
13. LePage, G. A. (1970): *Can. J. Biochem.*, 48:75–78.
14. LePage, G. A., Khaliq, A., and Gottlieb, J. A. (1973): *Drug Metab. Dis.*, 1:756–759.

15. Londos, C., and Wolff, J. (1977): *Proc. Natl. Acad. Sci. U.S.A.,* 74:5482–5486.
16. Miles, D. L., and Eyring, H. (1978): Int. J. of Quantum Chemistry: Quantum Biology Symposium 5, 173–190. John Wiley, N.Y.
17. Miles, D. W., Townsend, L. B., Redington, P., and Eyring, H. (1977): *Proc. Natl. Acad. Sci. U.S.A.,* 73:2384–2387.
18. Miles, D. L., Miles, D. W., and Eyring, H. (1977): *Biochim. Biophys. Acta,* 518:17–30.
19. Miles, D. L., Miles, D. W., Redington, P. K., and Eyring, H. (1976): *Proc. Natl. Acad. Sci. U.S.A.,* 73:4257–4260.
20. Miles, D. L., Miles, D. W., Redington, P. K., and Eyring, H. (1977): *J. Theor. Biol.,* 67:499–514.
21. Miles, D. L., Miles, D. W., and Eyring, H. (1977): *Proc. Natl. Acad. Sci. U.S.A.,* 74:2194–2198.
22. Miles, D. W., Townsend, L. B., Miles, D. L., and Eyring, H. (1978): (in press.)
23. Miller, J. P. (1977): In: *Cyclic 3′,5′-Nucleotides: Mechanisms of Action,* edited by H. Cramer and J. Schultz, pp. 77–105. Wiley and Sons, New York.
24. Miller, J. P., Christensen, L. F., Andrea, T. A., Meyer, R. B., Jr., Kitano, S., and Mizuna, Y. (1978): *J. Cycl. Neucl. Res.,* 4:133–144.
25. Mitra, C., and Saran, A. (1978): *Biochim. Biophys. Acta,* 518:193–204.
26. Montgomery, J. A., Elliott, R. D., and Thomas, H. J. (1975): *Ann. N. Y. Acad. Sci.,* 255:292–305.
27. Naik, S. R., Witkowski, J. T., and Robins, R. K. (1974): *J. Heterocycl. Chem.,* 11:57–61.
28. Neidle, S., Kuhlbrandt, W., and Achari, A. (1976): *Acta Crystallogr. Sect. B.,* 32:1850–1855.
29. Parks, R. E., Jr., and Agarwal, R. P. (1972): In: *The Enzymes,* Vol. 7, edited by P. D. Boyer, pp. 483–514. Academic Press, London.
30. Parks, R. E., Jr., and Brown, P. R. (1973): *Biochemistry,* 12:3294–3302.
31. Pasternak, C. A., and Handschumacher, R. E. (1959): *J. Biol. Chem.,* 234:2992–2997.
32. Pohorille, A., Perahia, D., and Pullman, B. (1978): *Biochim. Biophys. Acta,* 517:511–516.
33. Pople, J. A., and Gordon, M. (1967): *J. Am. Chem. Soc.,* 96:5276–5288.
34. Rein, R., Clarke, G. A., and Harris, F. E. (1970): In: *Quantim Aspects of Heterocyclic Compounds in Chemistry and Biochemistry,* Vol. 2, pp. 86–117. Israel Academy of Science and Humanities, Jerusalem.
35. Saran, A., Mitra, C., and Pullman, B. (1978): *Biochim. Biophys. Acta,* 517:255–264.
36. Schnebli, H. P., Hill, D. L., and Bennett, L. L., Jr. (1967): *J. Biol. Chem.,* 242:1997–2004.
37. Schwalbe, C. H., and Saenger, W. (1973): *J. Mol. Biol.,* 75:129–143.
38. Senft, A. W., and Crabtree, G. W. (1977): *Biochem. Pharmacol.,* 26:1847–1855.
39. Sprang, S., Scheller, R., Rohrer, D., and Sundaralingam, M. (1978): *J. Am. Chem. Soc.,* 100:2867–2872.
40. Tarr, H. L. A. (1958): *Can. J. Biochem. Physiol.,* 36:517–530.
41. Zimmerman, T. P., Rideout, J. L., Wolberg, G., Duncan, G. S., and Elion, G. B. (1976): *J. Biol. Chem.,* 251:6757–6766.

Transport and Metabolism of Adenosine

Physiological and Regulatory Functions of Adenosine and Adenine Nucleotides, edited by H. P. Baer and G. I. Drummond. Raven Press, New York © 1979.

Adenosine-Binding Proteins in Dog Heart

R. A. Olsson, R. B. Vomacka, and D. G. Nixon

Suncoast Heart Association Cardiovascular Research Laboratory, Department of Internal Medicine, University of South Florida, College of Medicine, Tampa, Florida 33620

The cardiac adenosine pool is thought to be almost entirely extracellular as a consequence of the high adenosine deaminase and adenosine kinase activities of cardiac muscle (3,18). This conventional view of cardiac adenosine metabolism is based on direct estimates of the activities of these enzymes in cardiac muscle extracts (2,4,16) and on experiments that show that exogenous adenosine is very efficiently incorporated into the cardiac adenine nucleotide pool (11,12,15,19). This view is challenged, however, by experiments that show that dipyridamole, an inhibitor of adenosine transport, fails to prevent the accumulation of adenosine in hypoxic hearts but does prevent the release of this nucleoside and its degradation products into the venous effluent. This experiment has been interpreted as evidence for an intracellular pool of adenosine (6,13). More direct evidence for this possibility was recently obtained by Frick and Lowenstein (7), who perfused rat hearts with mixtures of [^{3}H]adenosine and [^{14}C]adenosine 5′-monophosphate ([^{14}C]5′-AMP), then switched to purine-free perfusate to wash the exogenous purines out of the interstitial space, and subsequently analyzed the hearts for adenine derivatives. Although [^{3}H]adenosine was promptly washed out on switching to purine-free buffer, the cardiac concentration of [^{14}C]adenosine resulting from the hydrolysis of [^{14}C]5′-AMP remained elevated, its inaccessibility to washout from the interstitium giving evidence for an intracellular location.

This evidence supporting an intracellular adenosine pool was reported as we were studying the properties of several proteins in heart muscle that bind adenosine. This chapter concerns preliminary observations on two of these proteins.

METHODS

Ventricular muscle obtained from pentobarbital-anesthetized dogs was homogenized in 0.01 M $KHPO_4$, pH 7, containing 1 mM Na_2-ethylenediaminetetra-acetic acid, and solid $(NH_4)_2SO_4$ (313 g/liter) was added to the supernatant obtained by centrifugation. The precipitated protein was collected by centrifugation, dissolved in 0.1 M $KHPO_4$, pH 7.0, and the adenosine-binding activity present was estimated by incubating serially diluted aliquots of this solution with [2,8-^{3}H]adenosine for 4 hr at 37° C. Unbound adenosine was removed by treating the mixture with a charcoal-dextran suspension; the mixture was diluted with buffer, centrifuged, and protein-bound ^3H activity in an aliquot of the supernatant was estimated by liquid scintillation spectrometry.

Initial attempts at fractionating the adenosine-binding proteins employed an LKB gel slab isoelectric focusing apparatus and ampholytes giving a pH gradient between 4 and 8. At the end of each run the gel was sliced into 15 sections, transferred to plastic chromatography columns and the protein eluted with 0.1 M $KHPO_4$, pH 7.0. In later experiments the proteins were purified by ion exchange and gel filtration chromatography. These techniques have not attained purification to homogeneity, but the degree of purification is sufficient to permit the studies described below.

RESULTS

Adenosine Binding Activity of Heart Muscle

The amount of [^3H]adenosine bound in the 50% $(NH_4)_2SO_4$ precipitate was proportional to the amount of protein added below a protein concentration of 200 μg/ml. The binding capacity was calculated as the product of the total protein in this precipitate (g) and the binding activity (moles [^3H]adenosine bound/g) divided by the original weight of heart muscle (g). In four preparations the estimated activity ranged between 0.45 and 0.54 nmoles/g ventricular muscle.

Isoelectric Focusing

[^3H]adenosine-binding activity was found in a single peak at pH 5.7, but the activity in this peak accounted for no more than 15% of the activity originally applied. However, mixing material having a pI of ~ 4.5 with an aliquot of each of the other fractions revealed a second peak of activity between pH 5.5 and 6.1. The total binding activity in the two peaks averaged 70% of that applied. The three factors identified in this experiment were designated operationally as:

1. Intrinsic (I): The activity having a pI of 5.7 that binds adenosine in the absence of any additional factors.
2. Latent (L): The activity having a pI between 5.5 and 6.1 that is manifest only after the addition of the material of pI ~ 4.5.
3. Stimulator (S): The material of pI ~ 4.5 that stimulates [^3H]adenosine binding by material focusing between pH 5.5 and 6.1.

We have subsequently prepared all three factors by ion exchange and gel filtration chromatography, employing "stimulator" prepared by isoelectric focusing as a detection reagent for "latent" factor and vice versa.

All three factors are heat- and trypsin-labile, suggesting that they are proteins. [^3H]Adenine, [^3H]-cyclic AMP and [^3H]5′-AMP do not bind to I or to mixtures of L and S, nor is the binding of [^3H]adenosine to I or to mixtures of L and S inhibited by these unlabeled purines, suggesting that the binding reaction is specific for adenosine.

Interaction of L, S, and [^{3}H]Adenosine

The rate and extent of [^{3}H]adenosine binding in mixtures of L and S depends individually on the concentration of each reactant. When the three reactants are mixed simultaneously, the rate at which adenosine binds first accelerates, then slows as the reaction approaches equilibrium. Because this sort of time course could reflect the dependence of the overall reaction on the concentration of some type of intermediate complex, we evaluated the effect of preincubating mixtures of two of the reactants, i.e., L + Ado (where Ado = adenosine), S + Ado, or L + S, on the course of the reaction initiated when the omitted reactant was subsequently added. The rate of adenosine binding was compared to that in a control mixture prepared by adding all the reactants simultaneously. As shown in Fig. 1, the preincubation of S + Ado substantially accelerated adenosine binding when L was added later. This degree of acceleration was unique to the S + Ado mixture, which is evidence that the binding reaction proceeds via an intermediate S·Ado complex. It was not possible to demonstrate this complex when charcoal-dextran was used to remove unbound [^{3}H]adenosine from the reaction mixture or when the complex was separated from free adenosine by gel filtration chromatography, apparently because the complex dissociates rapidly. However, confirmatory evidence for this complex was obtained by chromatography on gel columns equilibrated with buffers containing [^{3}H]adenosine (Fig. 2). Similar experiments failed to demonstrate an L·Ado complex. Thus it appears that adenosine binding to these proteins proceeds by a two-step reaction:

$$\text{S} + \text{Ado} \rightleftharpoons \text{S·Ado}$$
$$\text{L} + \text{S·Ado} \rightleftharpoons \text{L·S·Ado}$$

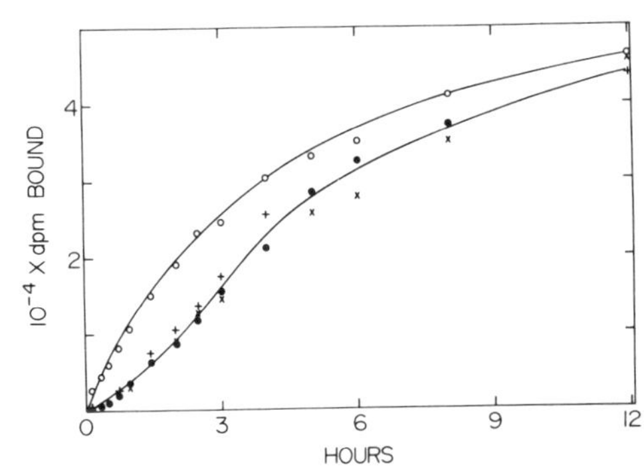

FIG. 1. A mixture containing (per ml) 27 μg S, 156 μg L, and 7 pmoles [^{3}H]Ado (500,000 dpm) in 50 mM Tris-HCl, pH 7.5, was incubated at 37° C (+). Other mixtures containing S + [^{3}H]Ado (○), L + [^{3}H]Ado (•), or L + S (×) were preincubated for 1 hr and the omitted reactant added. Periodically duplicate 0.1 ml aliquots were withdrawn and treated with 0.025 ml of a charcoal-dextran suspension, diluted with 1.00-ml ice-cold 50 mM Tris-HCl, pH 7.5, centrifuged at 4,800 × *g* for 5 min and 0.7 ml of supernatant counted. **Abscissa:** Time (hr). **Ordinate:** [^{3}H]Adenosine bound (dpm/ml). Some of the symbols (○, ×, or +), when superimposed, have been omitted for clarity.

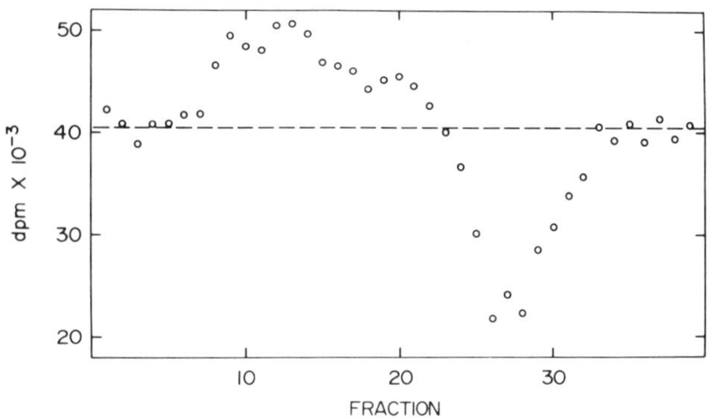

FIG. 2. A mixture of 75 μg S and 140,000 dpm [^{3}H]Ado in 0.25 ml 50 mM Tris-HCl, pH 7.5, was incubated for 2 hr at 37° C. A 0.2 ml aliquot was applied to a 0.5 × 10 cm column of Sephadex G-25 equilibrated with 50 mM Tris-HCl, pH 7.5, containing 560,000 dpm [^{3}H]Ado/ml. The column was eluted with this buffer as five drop fractions were collected and counted for ^{3}H activity. Each data point represents the average activity in three consecutive fractions.

These experiments also show that because the S·Ado complex dissociates so readily, the use of charcoal-dextran to separate free and bound adenosine reflects formation of the L·S·Ado complex (Reaction 2) rather than binding of adenosine per se (Reaction 1).

L·S·[^{3}H]Ado Complex

The rate of complex formation was identical in 50 mM Tris-HCl, pH 7.5, to that when 0.1 M $KHPO_4$, pH 7.0, in 0.2 to 0.5 M NaCl, was used as a buffer. As shown in Table 1, however, complex formation was inhibited by several substances that could interfere with hydrophobic and/or hydrogen bond formation (9), suggesting that the forces governing the interaction of these proteins may be primarily nonionic in character. Once formed, the L·S·[^{3}H]Ado complex appears to be stable. [^{3}H]Adenosine was not displaced when the complex was incubated for 3 hr at 37° C in buffers of pH as low as 4.0, or at pH 7.5 in the presence of charcoal-dextran, adenosine deaminase (6 IU/ml), adenosine (0.5 mM), or any of the reagents listed in Table 1 that inhibit formation of the complex.

The size of the L·S·[^{3}H]Ado complex was estimated on a calibrated column of Bio Gel A5m (Fig. 3). A peak of ^{3}H activity emerged in the void volume and this was followed by a "trail" of activity that extended to nearly half the included volume. This result suggests that the L·S·[^{3}H]Ado complex may aggregate to form polymers of indefinite size:

$$n\ \mathrm{L{\cdot}S{\cdot}Ado} \rightleftharpoons (\mathrm{L{\cdot}S{\cdot}Ado})_n$$

Ultimately these polymers may attain Stokes radii equivalent to that of a spherical protein having a molecular weight ≥ 5 million. This estimate of the molecular size of the polymer awaits independent confirmation by ultracentrifugation.

TABLE 1. *Inhibition of L·S·[³H]Ado complex formation*

Addition	Final concentration	cpm/0.7 ml	Percent inhibition
None	—	2,537	0
Sucrose	10%	474	81
Glycerol	20%	1,225	52
Dextran T-70	5%	880	65
Polyvinylpyrollidone	5%	2,231	12
Urea	0.5 M	663	74
Urea	1 M	250	90

Mixture of L, S, and [^{3}H]adenosine in 50 mM Tris-HCl, pH 7.5, containing the substances listed below were incubated 1 hr at 37° C. Duplicate 100 μl aliquots of each mixture were withdrawn at 15-min intervals, treated with 25 μl charcoal-dextran suspension and diluted with 1 ml cold 50 mM Tris buffer. After centrifugation for 5 min at 4,800 × *g*, 0.7 ml aliquots of the supernatant were counted. Data refer to the values at 1 hr.

Light Scattering

We monitored Rayleigh light scattering during the incubation of S and adenosine and also following the addition of L, as a means of examining the possibility that polymerization begins with the interaction of S and adenosine rather than as a consequence of the formation of the L·S·Ado complex. Since light scattering is an index of the weight average molecular weight (24), any increase in this variable signifies protein aggregation. In this experiment, light scattering was monitored for 1 hr after the addition of [^{3}H]adenosine to a solution of S and monitoring was continued following the addition of L. These measurements were compared to those in a second cuvette in which L, S, and [^{3}H]adenosine were mixed simultaneously.

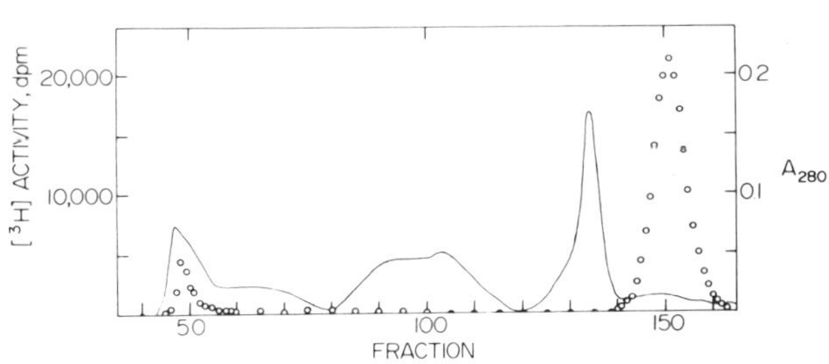

FIG. 3. A mixture of 19 μg L, 18 μg S, 0.2 mg bovine serum albumin, and 650,000 dpm [^{3}H]Ado in 0.5 ml 0.1 M $KHPO_4$, pH 7.0, in 0.2 M NaCl was incubated at 37° C for 16 hr and then applied to a 0.9 × 60 cm column of Bio Gel A5m equilibrated with the $KHPO_4$/NaCl buffer. The column was eluted with this buffer as five drop fractions were collected and counted for ^{3}H activity. The L and S used in this experiment had been purified previously on this column. L activity emerging in fractions 90 to 110 and S activity emerging in fractions 135 to 150 were used in this experiment. V_0 (Blue Dextran) was at fraction 46; V_t ([^{3}H]Ado) was at fraction 151. The A_{280} peak at fraction 136 is the result of the BSA added as an internal standard. For clarity the ^{3}H activity of every fifth fraction was plotted between fractions 60 and 140. Identical results were obtained when 50 mM Tris-HCl, pH 7.5, was substituted for $KHPO_4$/NaCl buffer.

The rate of L·S·[^{3}H]Ado complex formation was estimated by the charcoal-dextran method on samples drawn periodically from each of the two reaction mixtures. Figure 4 shows that light scattering did not change with time in the mixture of S and adenosine. The more rapid formation of L·S·[^{3}H]Ado complex and increase in light scattering produced by preincubating S and adenosine is evidence that the S·Ado complex indeed formed under these conditions. Because we could not monitor the angular dependence of the scattered light in this experiment, it is not possible to estimate the molecular weight of the product. However, the increase in light scattering is strong evidence that the interaction of L and S produced a product whose molecular weight is greater than that of either protein alone.

Biological Activity of L and S

[^{3}H]-Labeled adenine, adenosine, inosine, AMP, and cyclic AMP were recovered unchanged when these purines were incubated with S, L, or a mixture of S and L. Certain erythrocyte proteins that bind adenosine exhibit S-adenosylhomocysteinase activity (10). To see if the proteins under study exhibited this activity, L, S, and a combination of L and S were incubated with [^{14}C]adenosine and S-adenosylhomocysteine for 1 hr at 37° C and the products were separated on DEAE cellulose thin-layer plates developed with 50 mM Tris-HCl, pH 7.5. All of the ^{14}C activity applied was recovered in the adenosine spot, indicating that no conversion to S-adenosylhomocysteine had occurred. Further, L, S, and the L·S·[^{3}H]Ado complex did not bind to columns of adenosine deaminase covalently coupled to CNBr-activated agarose. These experiments show that these cardiac

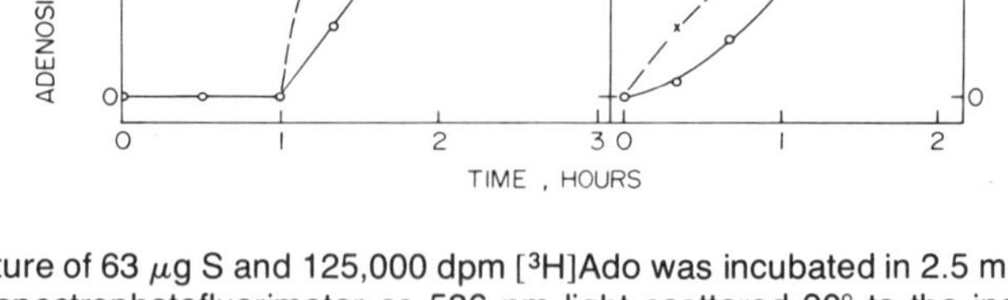

FIG. 4. A mixture of 63 μg S and 125,000 dpm [^{3}H]Ado was incubated in 2.5 ml 50 mM Tris-HCl, pH 7.5, in a spectrophotofluorimeter as 536 nm light scattered 90° to the incident beam was monitored. After 1 hr L was added to a final concentration of 280 μg/ml and monitoring of scattered light continued. At intervals duplicate 0.05 ml aliquots of the mixture were withdrawn and treated as described in the legend to Fig. 1. A second reaction mixture containing identical amounts of S, L, and [^{3}H]Ado was prepared by mixing the three reactants simultaneously. Light scattering in the two reaction mixtures and in a standard solution of 0.3 mg BSA/ml was monitored sequentially by rotating the cell holder of the spectrophotofluorimeter. **Panel A:** Preincubation of S + [^{3}H]Ado. **Panel B:** Simultaneous mixture of S, L, and [^{3}H] Ado. Δ light scattered is expressed relative to the BSA control.

adenosine-binding proteins are different than the adenosine deaminase–complexing proteins found in a variety of cells (1,22,26). Finally, since these proteins do not bind cyclic AMP, they differ from the adenine analog–binding proteins isolated from erythrocytes (25) and hepatocytes (23).

DISCUSSION

The experiments of Frick and Lowenstein (7) described above indicate that the cardiac adenosine pool has an intracellular compartment in addition to the adenosine in the interstitial space. At least some of the adenosine in the intracellular compartment could be bound to the adenosine-binding proteins identified in the present study. From the rough estimates of the number of binding sites per gram of heart muscle, it appears that adenosine in the intracellular compartment could account for a significant fraction of the total cardiac adenosine pool.

What is the biological function of adenosine in an intracellular compartment? Our preliminary results have not identified an activity associated with any of these proteins other than the capacity to bind adenosine, so that the biological function of these proteins and the adenosine complexes they form remains unclear. However, the observation that the binding of adenosine to the ''stimulator'' protein seems to promote its interaction with ''latent'' protein to form an indefinite polymer suggests that adenosine may play a role in protein-protein interactions. This appears to be an example of ligand-induced polymerization (14), a phenomenon of great importance in metabolic control (8).

One practical consequence of the compartmentation of the cardiac adenosine pool is that it may complicate the interpretation of experiments that test the adenosine hypothesis by comparing total cardiac adenosine levels, rather than the adenosine concentration in the cardiac interstitium, with coronary vascular resistance. Employing a radioligand binding assay (17) we have found that the adenosine content of normally perfused left ventricular muscle in open-chest dogs averages 1.1 nmole/g wet weight (20). If this amount of adenosine were confined to the interstitial space, whose volume is ~ 0.25 ml/g (5), the concentration would be about 4 μM in the vicinity of the coronary resistance vessels. This concentration would be expected to cause maximum coronary vasodilation, which clearly was not the case in these experiments. The discrepancy suggests that alternative methods, such as the pericardial superfusion technique employed by Berne and co-workers (21), may reflect the adenosine concentration at the coronary resistance vessels more accurately than the concentration estimated by direct analysis of muscle samples.

ACKNOWLEDGMENTS

This research was supported by a Grant-In-Aid from the Suncoast Chapter, Florida Affiliate, American Heart Association.

REFERENCES

1. Akedo, H., Nishihera, H., Shinikai, K., and Komatsu, K. (1970): *Biochim. Biophys. Acta,* 302:429–442.
2. Berne, R. M., and Rubio, R. (1974): *Circ. Res.,* 35 (Suppl III): III-109–III-118.
3. Berne, R. M., and Rubio, R. (1974): *Adv. Cardiol.,* 12:303–317.
4. Berne, R. M., Rubio, R., Dobson, J. B., Jr., and Curnish, R. R. (1971): *Circ. Res.,* 28 (Suppl I):I115–119.
5. Frank, J. S., and Langer, G. A. (1974): *J. Cell. Biol.,* 60:586–601.
6. Frick, G. P., and Lowenstein, J. M. (1976): *J. Biol. Chem.,* 251:6372–6378.
7. Frick, G. P., and Lowenstein, J. M. (1978): *J. Biol. Chem.,* 253:1240–1244.
8. Frieden, C. (1971): *Annu. Rev. Biochem.,* 40:653–696.
9. Gordon, J. A., and Jencks, W. P. (1963): *Biochemistry,* 2:47–57.
10. Hershfield, M. S. (1978): *Fed. Proc.,* 37:1466.
11. Hopkins, S. V., and Goldie, R. G. (1971): *Biochem. Pharmacol.,* 20:3359–3365.
12. Jacob, M. I., and Berne, R. M. (1960): *Am. J. Physiol.,* 198:322–326.
13. Kubler, W., and Spieckermann, P. G. (1970): *J. Mol. Cell. Cardiol.,* 1:351–377.
14. Levitzki, A., and Koshland, D. E. (1972): *Biochemistry,* 11:247–252.
15. Liu, M. S., and Feinberg, H. (1971): *Am. J. Physiol.,* 220:1242–1248.
16. Olsson, R. A. (1970): *Circ. Res.,* 26:301–306.
17. Olsson, R. A., Davis, C. J., Gentry, M. K., and Vomacka, R. B. (1978): *Anal. Biochem.,* 85:132–138.
18. Olsson, R. A., and Patterson, R. E. (1976): *Prog. Mol. Subcell. Biol.,* 4:228–248.
19. Olsson, R. A., Snow, J. A., Gentry, M. K., and Frick, G. P. (1972): *Circ. Res.,* 29:767–778.
20. Olsson, R. A., Vomacka, R. B., and Nixon, D. G. (1978): *Fed. Proc.,* 37:418.
21. Rubio, R., and Berne, R. M. (1969): *Circ. Res.,* 25:407–415.
22. Schrader, W. P., and Stacy, A. R. (1977): *J. Biol. Chem.,* 252:6409–6415.
23. Sugden, P. H., and Corbin, J. D. (1976): *Biochem. J.,* 159:423–437.
24. Tanford, C. (1961): *Physical Chemistry of Macromolecules.* New York, Wiley & Sons.
25. Yyh, K.-C. M., and Tao, M. (1974): *Biochemistry,* 13:5220–5226.
26. Van Der Weyden, M. B., and Kelley, W. N. (1976): *J. Biol. Chem.,* 251:5448–5456.

Physiological and Regulatory Functions of Adenosine and Adenine Nucleotides, edited by H. P. Baer and G. I. Drummond. Raven Press, New York © 1979.

Adenosine Transport

Alan R. P. Paterson

Cancer Research Unit (McEachern Laboratory), University of Alberta, Edmonton, Alberta T6G 2H7, Canada

The processes involved in the cellular uptake[1] of adenosine and other nucleosides are becoming understood through studies of the kinetics of uptake and metabolism of these compounds. Nitrobenzylthioinosine (NBMPR) [2] and related compounds are potent and specific inhibitors of nucleoside transport[1]; this chapter is concerned with the use of these inhibitors as probes of the adenosine uptake process. Adenosine uptake has been investigated mainly with cell preparations from animals (erythrocytes, lymphocytes, macrophages) or with cultured cells, although definitive contributions have also been made with intact tissues.

NUCLEOSIDE TRANSPORT BY ERYTHROCYTES

The entry of adenosine and other nucleosides into many types of animal cells is mediated by nucleoside-specific elements of the plasma membrane; adenosine enters human erythrocytes to a significant extent by diffusion (29) and only by a diffusion process in erythrocytes of some sheep (36). The existence of a nucleoside transport mechanism was clearly discernible in studies of uridine and thymidine fluxes across the plasma membrane of human erythrocytes; neither permeant is metabolized in these cells and the kinetic characteristics of their fluxes are attributable to a facilitated diffusion transport process with some contribution from simple diffusion (7,16,26). On the basis of a careful examination of uridine fluxes in human erythrocytes, Cabantchik and Ginsburg (5,11) have concluded that the kinetics of these fluxes are consistent with a simple carrier mechanism which is asymmetric in that kinetic parameters of efflux differ from those of influx. The nucleoside transport mechanism of erythrocytes accepts as "substrates" both purine and pyrimidine ribonucleosides and deoxyribonucleosides; this wide specificity was apparent in the ability of various nucleosides, including adenosine, to promote the efflux of uridine or thymidine from erythrocytes preloaded with either nucleoside (8,16).

[1] The terms "uptake" and "uptake mechanism (or process)" refer to the intracellular accumulation of permeant and its metabolites. These terms include both transport and metabolic events involved in nucleoside uptake. "Transport" refers only to the mediated passage of permeant across the plasma membrane.

[2] Abbreviations: Nitrobenzylthioinosine (NBMPR), 6-[(4-nitrobenzyl)thio]-9-β-D-ribofuranosylpurine; dipyridamole (Persantine), 2,6-bis-(diethanolamino)-4,8-diperidinopyrimido-(5,4-*d*) pyrimidine; tricyclic nucleoside, 3-amino-1,5-dihydro-5-methyl-1-β-D-ribofuranosyl-1,4,5,6,7,8-pentaazaacenaphthylene.

NITROBENZYLTHIOINOSINE, A POTENT AND SPECIFIC INHIBITOR OF NUCLEOSIDE TRANSPORT

Studies of NBMPR binding to human erythrocytes indicate that the inhibitor was bound tightly but reversibly to erythrocyte membrane sites, the occupancy of which resulted in loss of transporter function (6,25). Human erythrocytes possess about 10^4 NBMPR binding sites and the dissociation constant of the bound inhibitor is about 1 nM (6); mouse erythrocytes have similar NBMPR binding properties (9). It has been demonstrated that adenosine uptake by erythrocytes has a mediated component for which saturation kinetics can be shown (29,31,36) and that the mediated component is eliminated in the presence of NBMPR (36). Nitrobenzylthioinosine does not inhibit the uptake of nucleobases, sugars, and amino acids in erythrocytes and HeLa cells (18,20,21). Thus NBMPR is a specific, tightly bound inhibitor of nucleoside transport, and, as such, is useful as a "transport reagent" enabling, for example, demonstration of the transport component of nucleoside uptake processes in cells that metabolize nucleoside permeants. Nitrobenzylthioinosine is the best-studied member of a group of S^6-substituted 6-thiopurine nucleosides which are inhibitors of nucleoside transport (18,20,21,23).

Dipyridamole, long recognized as a potent inhibitor of nucleoside transport (see, for example, refs. 4,27, and 29), also eliminates the mediated component of adenosine uptake by erythrocytes (29,36); however, dipyridamole may also modify the activity of other membrane transport systems (4). Evidence that dipyridamole and NBMPR may bind to the same sites on the nucleoside transport mechanism is discussed below.

TRANSPORT AND METABOLISM COMPONENTS OF THE NUCLEOSIDE UPTAKE PROCESS

On entry into cells, nucleosides are metabolized and thereby removed from transport equilibria; the principal metabolites formed from low (physiological) concentrations of nucleoside permeants are nucleoside 5′-phosphates, which are retained within cells. In this way, influent nucleoside molecules are "trapped" intracellularly as nucleotide derivatives. If the trapping component of a nucleoside uptake process is slower than the mediated entry component, as the uptake process proceeds intracellular free permeant will accumulate, and permeant outflow will occur. Ultimately, a steady state condition will be achieved in which the net rate of uptake will be that of the metabolizing process. These circumstances may have been realized in recent studies of transport activities in cultured cells for thymidine and deoxycytidine in which K_m and $V_{\max}$ values for transport greatly exceeded (more than 10-fold) those of the kinases by which the metabolism of these permeants is initiated (15,27,28). The foregoing arguments indicate that uptake rates will provide a measure of transport kinetics when nucleoside permeants are trapped as metabolites only if they are initial rates; this has been generally recognized, but the rapidity with which initial rates decline to steady state rates has not been appreciated. Thus, when the kinetics of nucleoside uptake are measured under assay conditions with relatively long (minutes) intervals of uptake, the kinetic parameters obtained may represent those of the rate-limiting metabolic step and not those of the transport step. This comment, which reiterates an earlier warning (4), applies to many observations reported on nucleoside uptake rates. Thus the existence of nucleoside transport activities with high K_m and high capacity (V_{max}) that operate in conjunction with nucleoside kinase activities having K_m and V_{max} values one to two orders of magnitude lower, makes apparent the need for a careful approach to the measurement of initial rates, if uptake kinetics are to be used to measure rates of nucleoside transport.

Rapid sampling techniques (5,18,32,33,35) which allow the assay of nucleoside uptake during very brief intervals (seconds) have enabled measurement of initial uptake rates. Using cell monolayers on glass coverslips to permit rapid sampling, Taube and Berlin (33) showed that rabbit polymorphonuclear leukocytes possess a saturable nucleoside transport mechanism which, on the basis of kinetic criteria, appears to accept a variety of nucleoside permeants. The K_m value for adenosine in this system was about 10 μM. Strauss et al. (32) have demonstrated with rapid sampling the presence in mouse spleen lymphocytes of two saturable systems for adenosine transport with K_m values of 12 and 400 μM; the permeant specificities of these systems were not defined. Pearson et al. (24) have also demonstrated the presence of two discrete processes (K_m values of 3 μM and 250 μM) that mediate adenosine uptake by pig aortic endothelial cells in monolayer culture; dipyridamole was a potent inhibitor of the low K_m process.

With oil layer centrifugation techniques for the isolation of cells after very brief intervals of exposure to permeant, Plagemann and co-workers (15,35) have shown that adenosine triphosphate (ATP) -depleted cultured cells (Novikoff hepatoma) possess a transport mechanism that mediates the entry of thymidine with K_m values about 100-fold higher than values observed previously when slower methods were applied to cells that phosphorylated the permeant. Other nucleosides competed with thymidine for entry and caused countertransport in the ATP-depleted cells (27); these observations established the existence of a facilitated diffusion system of broad specificity, resembling those demonstrated in erythrocytes (16) and leukocytes (33). Similar findings have been reported for deoxycytidine transport in the same cell system (28).

In summary, the existence in cultured cells of a facilitated diffusion mechanism of broad specificity for the transport of nucleosides has been shown. In other cell types, adenosine uptake appears to be mediated by two mechanisms distinguishable by K_m values that differ by 10-fold or more; adenosine uptake processes with the higher K_m values (100 to 400 μM) may be mediated by the facilitated diffusion mechanism of broad specificity mentioned above. In the second type of adenosine transport mechanism, K_m values for uptake are about 10 μM; saturability and inhibition by dipyridamole and NBMPR have been shown, but the process has not been identified as a facilitated diffusion mechanism.

ADENOSINE TRANSPORT BY HeLa CELLS IN MONOLAYER CULTURE

A technique with replicate monolayer cultures of HeLa cells, which permits assays of permeant uptake during intervals as brief as 10 sec, has provided an approach to the measurement of initial rates of uptake from low concentrations (less than 10 μM of adenosine (18). The rate of adenosine uptake in this system was a saturable function of concentration (K_m about 2.5 μM); in the presence of 5 μM NBMPR, the saturable uptake was eliminated, revealing a diffusional component of adenosine entry. At low NBMPR concentrations, inhibition of adenosine uptake was competitive (18). Under the conditions of the uptake assay, the influent adenosine was converted, immediately and almost entirely, to nucleotide metabolites, primarily ATP, adenosine diphosphate, and adenosine monophosphate. The time course of adenosine uptake was essentially that of formation of the adenosine phosphates, and the time course of the latter was linear and extrapolated through time zero.

NITROBENZYLTHIOINOSINE INHIBITION OF NUCLEOSIDE TRANSPORT IN CELLS OTHER THAN ERYTHROCYTES

When NBMPR and related inhibitors of nucleoside transport have been used to probe the nucleoside transport process in cells that metabolize nucleosides, the findings that emerged have supported the general features of the uptake process discerned from kinetic studies. Nitrobenzylthioinosine acts as a potent general inhibitor of nucleoside uptake in a wide variety of cells through interference with the transport step of the uptake process. The uptake of nucleobases, sugars, and amino acids by erythrocytes and HeLa cells is not affected by these inhibitors under conditions in which nucleoside transport is blocked (18,20,21), nor is the phosphorylation of uridine, thymidine, or adenosine in cell extracts inhibited (17,20; A. R. P. Paterson and C. E. Cass, *unpublished data*), nor is the deamination of adenosine inhibited (17) by these compounds.

Various types of cultured cells, including HeLa cells (13), L5178Y mouse lymphoblasts, RPMI 6410 human lymphoblastoid cells (22), and MCT hamster cells (10), possess sites, evidently on the plasma membrane, to which NBMPR binds reversibly but with high affinity. Because occupancy of such sites is directly related to inhibition of nucleoside uptake (6), they are on, or associated with, the nucleoside transport mechanism. Our assay for the site-specific binding of NBMPR measures cell-associated NBMPR (labeled with ^{35}S or ^{3}H) that is displaceable by an excess of an NBMPR homolog, nitrobenzylthioguanosine (13). Mass law analysis of data for the site-specific binding of NBMPR to HeLa cells showed that a single type of receptor was involved and that at 37° C, HeLa cells possess about 1.3×10^5 sites at which the dissociation constant for NBMPR was 0.6 nM. Similar high-affinity binding sites of a single type are also present on L5178Y cells (A. R. P. Paterson and C. E. Cass, *unpublished data),* NBMPR occupancy of these sites results in inhibition of nucleoside uptake; this is illustrated for adenosine uptake by HeLa cells in Fig. 1. Similarly shaped concentration-response curves have been

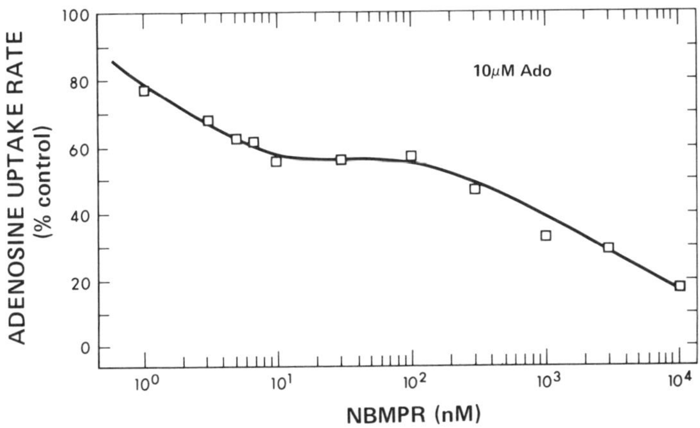

FIG. 1. Nitrobenzylthioinosine inhibition of adenosine (Ado) uptake by HeLa cells. Replicate monolayer cultures were equilibrated at 20° C for 20 min with serum-free culture medium (18) with and without (controls) NBMPR at the indicated concentrations. Portions of cell-equilibrated medium at each NBMPR concentration were then added to lyophilized residues of [2-^{3}H]adenosine, achieving final concentrations of 10 μM; these solutions were then used to assay adenosine uptake rates by the same monolayer cultures with which they had been equilibrated. Assays of adenosine uptake, performed at 20° C in triplicate, employed uptake intervals of 10 sec begun and ended as described in detail previously (18); monolayers were solubilized and assayed for ^{3}H content by conventional liquid scintillation methods.

found for NBMPR inhibition of the uptake by HeLa cells of guanosine, thymidine, and uridine (A. R. P. Paterson and C. E. Cass, *unpublished data*).

The NBMPR concentration-effect relationship is obviously complex; thus, partial inhibition of adenosine uptake occurred in the range of NBMPR concentrations at which the saturable, high-affinity binding of NBMPR took place. Substantial inhibition of the remaining transport activity required concentrations of NBMPR 1,000-fold higher. Eilam and Cabantchik (10) have also observed that after saturation of high-affinity binding sites for NBMPR on MCT cells, a substantial portion of uridine uptake activity remained active. The kinetic properties of the latter system were interpreted in terms of an allosteric carrier model with NBMPR binding sites independent of the permeation sites; the allosteric model was favored over an alternative model involving two transport systems with different sensitivities to NBMPR.

ADENOSINE ANALOGS

"Protection" experiments with NBMPR have afforded a means of exploring the specificity of the nucleoside transport mechanism; the low specificity of the transporter with respect to the base portion of the permeant is made apparent in this approach. We showed in earlier experiments that L5178Y cells, when cultured in medium containing NBMPR, proliferated in the presence of otherwise toxic concentrations of several nucleoside analogs, evidently because cellular uptake of the latter was reduced by the transport inhibitor (34). Table 1 summarizes similar experiments in which cultured lymphoblastoid cells, RPMI 6410, were protected by NBMPR from inhibitory concentrations of various cytotoxic adenosine analogs; it is implicit in such protection that entry of the analog into cells is mediated by the nucleoside transport mechanism. In similar experiments, dipyridamole, a potent inhibitor of adenosine transport (see below) protected RPMI 6410 cells against various cytotoxic nucleosides.

The compounds listed in Table 1 are adenosine analogs in the sense that they are substrates for adenosine kinase (see references). It is apparent in these data that both the kinase and the nucleoside transport mechanism accept a variety of heterocyclic structures as the aglycone portion of their substrates. In other studies, we have shown that cultured cells are also protected by NBMPR from various cytotoxic pyrimidine nucleosides (34; A. R. P. Paterson and C. E. Cass, *unpublished data)*.

The foregoing observations demonstrate that "transportability" of various nucleoside analogs is a determinant of cytotoxicity. The experiment of Table 2 suggests that this principle may have *in vivo* applicability. These data show that otherwise lethal doses of tubercidin were tolerated by mice when NBMPR was administered simultaneously. In similar experiments, NBMPR protected mice against the toxicity of nebularine (purine riboside). It would appear that the pharmacokinetics or the *in vivo* distribution of the toxic agents may be changed by the coadministration of NBMPR; these possibilities are under test.

TABLE 1. *Nitrobenzylthioinosine protection of RPMI 6410 cells against cytotoxic analogs of adenosine*

Analog (μM)	Cell proliferation rate [a] (% of control) +NBMPR [b]	+NBMPR [b]	Source [c]	Ref.
2-Azaadenosine (1.0) [d]	−13	98	A	1
8-Azaadenosine (3.0) [d]	21	91	A	2
Carbocyclic adenosine (5.0) [d]	−10	84	B	2
2-Fluoroadenosine (0.5) [d]	30	85	A	30
Formycin (6.0) [d]	−18	98	C	14
6-Methylthioinosine (0.2)	−1	72	D	30
Nebularine (1.0)	9	88	D	14
Pyrazofurin (0.3)	41	71	E	12
Tricyclicnucleoside (0.01) [d]	−1.0	98	F	3
Tubercidin (0.1) [d]	−27	88	D	14
Xylosyladenine (300)	13	59	G	14

[a] Population doublings in 72 hr as a percentage of rates in cultures without additives (3.3 to 3.8 doublings in 72 hr).

[b] This concentration did not significantly affect cell proliferation rates. 5 μM NBMPR.

[c] Sources: (A): J. A. Montgomery; (B): Y. F. Shealy; (C): H. Umezawa; (D): commercial (E): M. J. Sweeney; (F): L. B. Townsend; and (G): G. A. Le Page.

[d] Date from ref. 22.

TABLE 2. *Nitrobenzylthioinosine protection against toxicity of tubercidin toward BDF_1 female mice*

Day	Number of mice alive after treatment[a] at these dosages:				
	Tubercidin (mg/kg)	4	6	4	6
	NBMPR (mg/kg)	0	0	40[b]	40[b]
1–4		6	6	6	6
5		5	5	6	6
6		5	3	6	6
7		4	1	6	6
8		1	0	6	6
9–20		0	0	6	6

[a] Five i.p. injections at 24-hr intervals.
[b] Given simultaneously with tubercidin.

DIPYRIDAMOLE

Vasodilation by dipyridamole has been linked repeatedly to the ability of this compound to inhibit transport of adenosine (17). The potent inhibition by dipyridamole of adenosine uptake (during intervals of 10 sec) by HeLa monolayer cells is illustrated in Fig. 2; an IC_{50} value of 0.2 μM was found. The NBMPR concentration-effect relationships for the inhibition of adenosine transport (Fig. 1) are more complex than those for dipyridamole; the reasons for these differences are not known.

Experiments with HeLa cell monolayers have shown that [G-^{3}H]NBMPR at

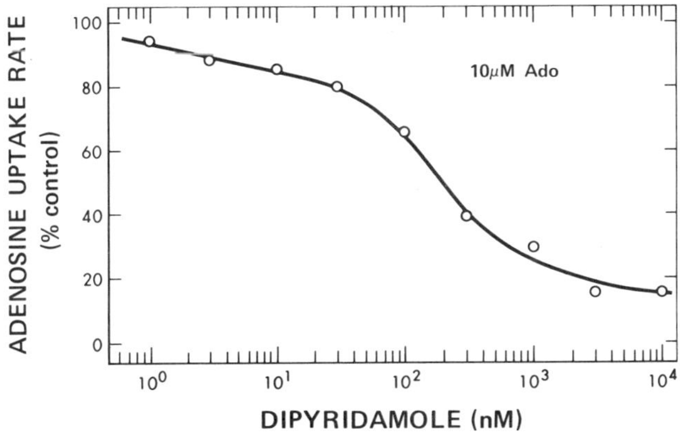

FIG. 2. Dipyridamole inhibition of NBMPR binding by HeLa cells. Replicate monolayer cultures were incubated for 20 min at 20° C in medium containing 1.0 nM [G-^{3}H]NBMPR (Moravek Biochemicals, Los Angeles) and the indicated concentrations of dipyridamole; after a single wash, the cellular ^{3}H content was determined, as specified previously (18). Ado, adenosine.

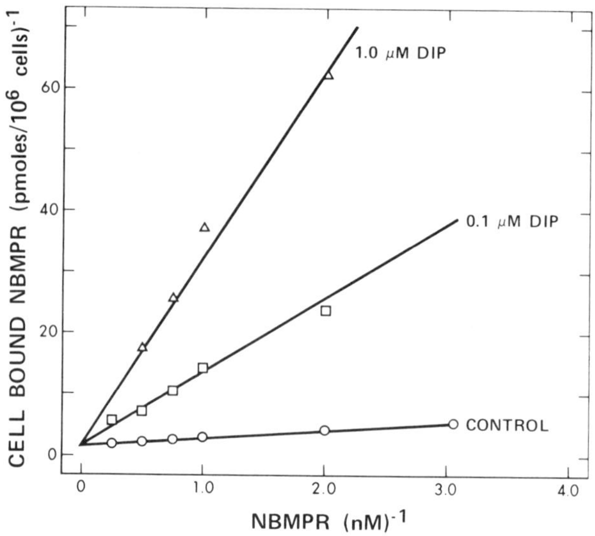

FIG. 3. Dipyridamole (DIP) inhibition of NBMPR binding by HeLa cells. Replicate monolayer cultures were incubated at 20° C serum-free medium (18) containing graded concentrations of [G-^{3}H]NBMPR without (○) and with dipyridamole (0.1 μM, □; 1.0 μM, Δ); after incubation for 20 min, monolayers were washed and both cells and medium were assayed for ^{3}H content.

equilibrium with the high-affinity cellular binding sites was displaced from these sites by 25 μM dipyridamole (19); the displacement proceeded rapidly at 37° C and more slowly at 20° C, as did the previously reported displacement of bound [^{35}S]NBMPR by nitrobenzylthioguanosine (13). This result suggested that dipyridamole may bind to the NBMPR sites. The data of Fig. 3 indicate that dipyridamole is a potent, apparently competitive inhibitor of the high-affinity, site-specific binding of NBMPR to HeLa cells. The dissociation constant for dipyridamole in this experiment, calculated assuming competition with NBMPR at the latter's binding sites, was about 30 μM, a value 40-fold higher than that of NBMPR. However, the obvious differences between the concentration-effect relationships for the two inhibitors (Figs. 1 and 2) would seem to argue against a simple competition between them for the same binding sites; explanations for these anomalies are not presently available.

ACKNOWLEDGMENTS

This work was supported by the National Cancer Institute of Canada and the Medical Research Council of Canada. Also acknowledged are helpful discussions with Dr. Carol E. Cass during preparation of this paper.

REFERENCES

1. Bennett, L. L., Jr., Allan, P. W., Carpenter, J. W., and Hill, D. L. (1976): *Biochem. Pharmacol.*, 25:517–521. *Pharmacol.*, 27:233–241.

2. Bennett, L. L., Jr., and Hill, D. L. (1975): *Mol. Pharmacol.,* 11:803–808.
3. Bennett, L. L., Jr., Smithers, D., Hill, D. L., Rose, L. M., and Alexander, J. A. (1978): *Biochem. Pharmacol.,* 27:233–241.
4. Berlin, R. D., and Oliver, J. M. (1975): *Int. Rev. Cytol.,* 42:287–336.
5. Cabantchik, Z. I., and Ginsburg, H. (1977): *J. Gen. Physiol.,* 69:75–96.
6. Cass, C. E., Gaudette, L. A., and Paterson, A. R. P. (1974): *Biochim. Biophys. Acta,* 345:1–10.
7. Cass, C. E., and Paterson, A. R. P. (1972): *J. Biol. Chem.,* 247:3314–3320.
8. Cass, C. E., and Paterson, A. R. P. (1973): *Biochim. Biophys. Acta,* 291:734–746.
9. Cass, C. E., and Paterson, A. R. P. (1975): *Biochem. Pharmacol.,* 24:1989–1993.
10. Eilam, Y., and Cabantchik, Z. I. (1977): *J. Cell. Physiol.,* 92:185–202.
11. Ginsburg, H., and Cabantchik, Z. I. (1977): *J. Gen. Physiol.,* 70:679–680.
12. Jakubowski, A., Lehman, C., Moyer, J., and Handschumacher, R. A. (1977): *Proc. Am. Assoc. Cancer Res.,* 18:217.
13. Lauzon, G. J., and Paterson, A. R. P. (1977): *Mol. Pharmacol.,* 13:883–891.
14. Lindberg, B., Klenow, H., and Hansen, K. (1967): *J. Biol. Chem.,* 242:350–356.
15. Marz, R., Wohlheuter, R. M., and Plagemann, P. G. W. (1977): *J. Supramol. Struct.,* 6:433–440.
16. Oliver, J. M., and Paterson, A. R. P. (1971): *Can. J. Biochem.,* 49:262–270.
17. Olsson, R. A., Snow, J. A., Gentry, M. K., and Frick, G. P. (1972): *Circ. Res.,* 31:767–778.
18. Paterson, A. R. P., Babb, L. R., Paran, J. H., and Cass, C. E. (1977): *Mol. Pharmacol.,* 13:1147–1158.
19. Paterson, A. R. P., Lau, E. Y., and Cass, C. E. (1979): *Mol. Pharmacol. (in press).*
20. Paterson, A. R. P., Naik, S. R., and Cass, C. E. (1977): *Mol. Pharmacol.,* 13:1014–1023.
21. Paterson, A. R. P., and Oliver, J. M. (1971): *Can. J. Biochem.,* 49:271–274.
22. Paterson, A. R. P., Yang, S., Lau, E. Y., and Cass, C. E. (1979): *Mol. Pharmacol. (in press).*
23. Paul, B., Chen, M. F., and Paterson, A. R. P. (1975): *J. Med. Chem.,* 18:968–973.
24. Pearson, J. D., Carleton, J. S., Hutchings, A., and Gordon, J. L. (1978): *Biochem. J.,* 170:265–271.
25. Pickard, M. A., Brown, R. R., Paul, B., and Paterson, A. R. P. (1973): *Can. J. Biochem.,* 51:666–672.
26. Pickard, M. A., and Paterson, A. R. P. (1972): *Can. J. Biochem.,* 50:704–705.
27. Plagemann, P. G. W., Marz, R., and Erbe, J. (1976): *J. Cell. Physiol.,* 89:1–18.
28. Plagemann, P. G. W., Marz, R., and Wohlheuter, R. M. (1978): *Cancer Res.,* 38:978–989.
29. Roos, H., and Pfleger, K. (1972): *Mol. Pharmacol.,* 8:417–425.
30. Schnebli, H. P., Hill, D. H., and Bennett, L. L., Jr. (1967): *J. Biol. Chem.,* 242:1997–2004.
31. Schrader, J., Berne, R. M., and Rubio, R. (1972): *Am. J. Physiol.,* 223:159–166.
32. Strauss, P. R., Sheehan, J. M., and Kashket, E. R. (1976): *J. Exp. Med.,* 144:1009–1021.
33. Taube, R. A., and Berlin, R. D. (1972): *Biochim. Biophys. Acta,* 255:6–18.
34. Warnick, C. T., Muzik, H., and Paterson, A. R. P. (1972): *Cancer Res.,* 32:2017–2022.
35. Wohlheuter, R. M., Marz, R., Graff, J. C., and Plagemann, P. G. W. (1976): *J. Cell. Physiol.,* 89:605–612.
36. Young, J. D. (1978): *J. Physiol.,* 277:325–339.

Physiological and Regulatory Functions of Adenosine and Adenine Nucleotides, edited by H. P. Baer and G. I. Drummond. Raven Press, New York © 1979.

Regulation of Adenosine Metabolism

J. Frank Henderson

Cancer Research Unit, McEachern Laboratory, and Department of Biochemistry, University of Alberta, Edmonton, Alberta T6G 2H7, Canada

The ways in which adenosine synthesis and metabolism are regulated in animal cells and tissues are very far from being understood, and indeed, the considerable complexity of adenosine metabolism has become appreciated only relatively recently. This chapter considers not only what is presently known about the regulation of adenosine metabolism, but also emphasizes the questions that remain to be answered, and the considerable variation that exists among biological systems with respect to adenosine metabolism and its regulation will also be dealt with. This subject is discussed from two points of view: first, the regulation of the metabolism of exogenous adenosine, and second, the regulation of the metabolism of endogenous adenosine.

EXOGENOUS ADENOSINE

Adenosine metabolism and its regulation are often thought of simply in terms of exogenous adenosine, the adenosine supplied to an experimental system by the investigator. In this case the only major pathways of metabolism to be considered are deamination of adenosine to inosine, and its phosphorylation to adenylate. The simplest approach to the question of the regulation of adenosine metabolism, therefore, is to consider the relative rates of deamination and phosphorylation as they are influenced (a) by the intracellular concentration of adenosine, (b) by the kinetic properties—especially the Michaelis constants—of adenosine deaminase and adenosine kinase, and (c) by the relative total activities of these two enzymes.

The intracellular concentration of adenosine is determined (a) by the amount of adenosine supplied to the cell or tissue, the effective volume of the system, and the rate of supply, (b) by the rate of metabolism (i.e., deamination) in the extracellular compartment, (c) by the rate of entry of adenosine into the cells, and (d) by the rate of metabolism within the cells.

The extracellular metabolism of adenosine is especially influenced by the adenosine deaminase activity of serum, and there is considerable species variation in this property. Thus 10% mouse serum deaminated 2% of 100 μM adenosine in 30 min. at 37° C, dog serum also 2%, moose serum 7%, goat serum 9%, guinea pig serum 17%, and fetal calf serum 24% (M. M. Johnson and J. F. Henderson, *unpublished data)*. Because the adenosine deaminase activity of horse serum is so much less than that of fetal calf serum, the former is much preferred in tissue culture studies involving adenosine (23).

Kinetic Properties

The Michaelis constants reported for adenosine kinase are generally lower than those for adenosine deaminase. Thus those for adenosine kinase are in the range of 0.4 to 1.6 μM adenosine (5,7,10,12,14), whereas most of the values reported for adenosine deaminase are in the range of 10 to 60 μM; however, some Michaelis constants for adenosine deaminase go as low as 1 μM and as high as 400 μM (27). The Michaelis constants for the small–, intermediate–, and large–molecular weight forms of adenosine deaminase have been shown to be virtually the same (17,25).

Michaelis constants reported for adenosine kinase for adenosine triphosphate (ATP) range from 10 to 200 μM (5,7,10). Adenosine kinase has also been found to be inhibited by high concentrations of adenosine (7,12), with 50% inhibition obtained at 100 μM adenosine in one study (7).

Adenosine kinase is also subject to product inhibition, with a K_I for adenosine diphosphate (ADP) of 60 μM, and a K_I for 6-methylthiopurine ribonucleoside 5′-phosphate, an analog of adenylate, of 160 μM (10). Inosine acts as a product inhibitor of adenosine deaminase, with K_I values reported between 0.6 and 1.6 mM (25).

Enzyme Amounts

The total activities of adenosine deaminase and adenosine kinase vary greatly both among tissues and among species. Thus adenosine deaminase activities in the spleens of different animals have been reported to be as follows: cat 88.5, rat 17.6, mouse 12.4, rabbit 11.6, dog 11.0, guinea pig 3.1 (2). Adenosine kinase activities in the hearts of different species are: rat 48, guinea pig 46, cow 29, cat 23, rabbit 19, man 18, dog 15, swine 4 (5).

When adenosine deaminase activities were measured in different tissues in one species, man, the following results were obtained: Duodenum 4.26, stomach 1.83, small intestine 1.70, spleen 1.19, pancreas 0.51, lymph nodes 0.46, liver 0.33, heart 0.18, and brain 0.12 (17). In the rhesus monkey adenosine kinase activity varied among tissues as follows: Liver 1,200, spleen 810, lymph node 530, kidney 500, testis 360, lung 270, small intestine 220, erythrocytes 200, brain 79, heart 66, skeletal muscle 63, skin 38 (11).

Adenosine Cleavage

Finally, it should be noted that adenosine can not only be deaminated and phosphorylated in animal cells, but also phosphorylyzed to adenine and ribose-1-P. This reaction can be catalyzed by purine nucleoside phosphorylase (K_m for hypoxanthine, 20 to 30 μM; K_m for adenine, 300 to 400 μM) (28), but evidence has been presented that a specific adenosine phosphorylase may also exist in animal cells (K_m for adenine, 20 to 30 μM) (6); this picture is complicated, however, by a recent report that adenosine phosphorylase activity may be associated with mycoplasma infection (24). Quantitatively, the cleavage of adenosine to adenine (or adenosine synthesis from adenine plus ribose-1-P, for this reaction is freely reversible) is a minor reaction, and relative rates of the synthesis of adenosine, inosine, and guanosine from adenine, hypoxanthine, and guanine (plus ribose-1-P) in extracts of rat liver were found to be 0.53, 239, and 190, respectively (6).

Regulation in Intact Cells

Although the principal enzymes of adenosine metabolism have been studied in considerable detail, it is not at all certain to what extent their kinetic properties and total amounts are the sole or even major determining factors in the regulation of the relative rates of deamination and phosphorylation of adenosine. In no single case have the properties of isolated enzymes and the relative rates of alternative pathways of adenosine metabolism in intact cells been thoroughly compared, and in most studies intracellular concentrations have not been held constant during the entire experimental period.

A careful study of adenosine metabolism in human lymphocytes stimulated with phytohemagglutinin has shown a number of differences between enzyme properties in extracts and the actual metabolism of adenosine in intact cells (23). Thus the specific activity of adenosine deaminase in lymphocytes was decreased approximately 50% following stimulation, whereas its total activity increased 1.7-fold. The specific activity of adenosine kinase was unchanged, whereas its total activity increased 2.9-fold; this corresponds to the increase in total protein per culture.

In contrast to these small changes in enzyme amounts, the deamination of adenosine in intact stimulated lymphocytes increased sixfold, and phosphorylation increased 12-fold; it was suggested that these increases may be caused mainly by increased entry of adenosine into the cells. It is of interest that the total rate of adenosine metabolism by intact lymphocytes was never greater than 25% of the total enzyme activities measured in extracts.

In stimulated lymphocytes the relative rates of deamination and phosphorylation of adenosine varied with adenosine concentration in a manner that would be expected from the Michaelis constants of the two enzymes; thus phosphorylation predominated at adenosine concentrations less than 5 μM, whereas deamination was the major pathway at higher adenosine concentrations. What is of particular interest, however, is that in unstimulated lymphocytes this ratio did not vary with adenosine concentration, and deamination always markedly predominated (9:1). The factors that regulate the selective rates of deamination and phosphorylation under these conditions are not known.

The ratio of phosphorylation to deamination also varies in intact erythrocytes of different species, even under conditions of adenosine saturation. Some of these ratios are as follows: Mouse 0.024, Syrian hamster 0.190, rat 0.398, moose 0.726, goat 0.958, mule deer 1.22 (M. M. Johnson and J. F. Henderson, *unpublished data*). The relationship of these data to enzyme properties has not yet been studied.

Four patterns of adenosine metabolism can be distinguished on the basis of the relative rates of deamination and phosphorylation as a function of adenosine concentration. (a) Deamination predominates at all adenosine concentrations; this has been observed in unstimulated human peripheral lymphocytes (23) and in sheep erythrocytes (21). (b) Phosphorylation predominates at all adenosine concentrations; so far, this would appear to be the case only in adenosine deaminase–deficient human cells. (c) Phosphorylation predominates at low adenosine concentrations and deamination predominates at high adenosine concentrations; this occurs in Ehrlich ascites tumor cells (21) and in phytohemogglutinin-stimulated human lymphocytes (23). (d) Deamination predominates at low adenosine concentrations and phosphorylation predominates at high adenosine concentration; this pattern has not been observed, and would be expected to be uncommon.

Conclusion

In view of the cell and species differences in adenosine metabolism that have been noted to date, it would seem necessary to characterize each individual experimental system in terms of rates of extracellular metabolism and the relative rates of phosphorylation and deamination at a series of adenosine concentrations, as well as to determine what the adenosine concentration actually is in each instance, and to evaluate, if possible, the role of phosphorolysis in the overall pattern of adenosine metabolism.

ENDOGENOUS ADENOSINE

The regulation of the metabolism of endogenous adenosine is far more complicated than that of exogenous adenosine. Thus, in addition to the questions of the relative kinetic properties and total activities of adenosine deaminase and adenosine kinase already discussed, there are a number of other questions that must be asked regarding endogenous adenosine. For example, what is the intracellular concentration of adenosine in different cells and under various conditions, at what rate is adenosine formed, by what reactions is it formed, and how are these reactions regulated? Furthermore, the different functions of the several adenosine-forming reactions should be considered, as well as variation among cell types and among species with respect to each of these matters.

It seems correct to say that normal intracellular concentrations of adenosine are not known for any mammalian cell, but that these are probably quite low. Similarly, it seems correct to say that in no cell type or experimental system is the rate of adenosine formation accurately known; in this case it is not even possible to state that this is a relatively high or a relatively low rate.

Pathways of Adenosine Synthesis

Adenosine can potentially be synthesized via a number of different pathways: (a) The dephosphorylation of 5′-adenylate; (b) from S-adenosylhomocysteine formed in the process of methylation; (c) from 3′-adenylate; (d) from adenosine end groups in ribonucleic acid (RNA); and (e) by the reaction of adenine and ribose-1-P. Little is known about several of these pathways and some have not been thought of seriously from the point of view of adenosine synthesis.

The synthesis of adenosine from adenine and ribose-1-P probably makes a negligible contribution to total adenosine synthesis, since the availability of both adenine and ribose-1-P would seem to be quite limited, and total enzyme activities appear to be low.

Presumably, also, the cleavage of RNA terminal groups to form free adenosine is not a major pathway; however, the potential contribution of this pathway has not been determined. Breakdown of RNA to 3′-adenylate, and the subsequent dephosphorylation of the latter, may be a major source of adenosine, at least in some systems. The RNA could be derived from the diet, produced by cell death and turnover, or intracellular RNA cleaved in the course of normal function and metabolism. However, the question of the relative rates of hydrolysis of RNA to 3′-nucleotides and to 5′-nucleotides appears to be still unresolved, and indeed, may vary among cell types and experimental and physiological situations.

It is more certain that adenosine is formed from S-adenosylhomocysteine when the latter is produced from S-adenosylmethionine in the course of methylation reactions. This may be a very important pathway of adenosine synthesis, and, presumably, the rate at which methylation occurs in different experimental systems could be determined.

Dephosphorylation of Adenylate

Most attention has been traditionally placed on the formation of adenosine via the dephosphorylation of 5′-adenylate. This precursor, however, may have several different sources, including the breakdown of RNA (derived from the diet, from cell turnover, and in the course of turnover of intracellular RNA), the adenine ribonucleotide pool, and coenzymes (e.g., 3′,5′-cyclic adenylate) that form 5′-adenylate in the course of their own metabolism. In addition, there is evidence for compartmentation of adenine nucleotides in some systems; thus, in guinea pig heart, both adenosine and cyclic adenylate appear to be formed from one small pool of adenine nucleotides (20).

Although adenosine formation from 5′-adenylate is undoubtedly important in cells whose energy metabolism is impaired, it is not at all certain that this is a quantitatively major pathway under other conditions. It is important to realize that, despite many assumptions, there is really very little evidence on this point.

The regulation of 5′-adenylate dephosphorylation involves a number of considerations. First, the enzyme or enzymes responsible are not known with certainty.

Although it is often assumed that 5′-nucleotidase is involved, it must be remembered that both alkaline phosphatase and acid phosphatase can dephosphorylate adenylate as well; in addition, there are both specific and nonspecific 5′-nucleotidases in some systems, as well as other nonspecific phosphatases. In some animal tumor cells alkaline phosphatase appears to be particularly important (26).

Cellular and intracellular compartmentation of enzymes that can dephosphorylate 5′-adenylate also occur (8). Thus, in addition to cell-to-cell variation in the total activities of the several dephosphorylating enzymes, 5′-nucleotidase in some cells occurs on the outer surface of the cell membrane, and the extent and mechanism of access of this enzyme to its substrate is not entirely clear. Other cells have a soluble, intracellular 5′-nucleotidase in addition to, or even instead of, the membrane form. In addition, localization of acid phosphatase in lysosomes has been well established and some cells have a lysosomal nucleotidase as well. Finally, considerable species variation in 5′-nucleotidase activity has been demonstrated, and in one study using heart the following relative activity results were obtained: Rat 100, guinea pig 33, mouse 17, sheep 16, beef 3, pig 1.5, rabbit 1 (1).

Concentrations of adenylate are normally quite low in cells with unimpaired energy metabolism; 40 to 80 nmoles/g in Ehrlich ascites tumor cells (15) and 100 to 130 nmoles/g in guinea pig heart (20) are representative values. The Michaelis constant for adenylate for the 5′-nucleotidase of Ehrlich ascites tumor cells is 67 μM, whereas a value of 18 μM has been reported for the heart enzyme (8).

5′-Nucleotidase is strongly inhibited by several purine and pyrimidine nucleoside triphosphates. The K_I for inhibition of adenylate dephosphorylation by ATP by the Ehrlich ascites tumor cell enzyme is 12 μM (18), whereas for the rat heart 5′-nucleotidase this value is 1.8 μM (1). Since ATP concentrations in cells are typically 2,000 to 5,000 μM, this enzyme should be strongly inhibited in viable cells under all conditions if these concentrations of ATP have access to the enzyme.

Studies of adenine nucleotide catabolism in Ehrlich ascites tumor cells incubated with 2-deoxyglucose indicate that the dephosphorylation of adenylate to adenosine is influenced more by the rate of adenylate formation from ADP and ATP than by the actual concentration of adenylate. Thus, although the concentration of adenylate varies considerably during the incubation, adenosine accumulation increases steadily with time as well as with adenylate production. At adenylate levels greater than 200 μM, however, concentration may also affect the rate of dephosphorylation (15).

Alternative Pathways of Adenylate Metabolism

Finally, the dephosphorylation of adenylate to adenosine must be considered in relationship to the alternative pathways of metabolism of adenylate, principally (a) phosphorylation to ADP and ATP, and (b) deamination to inosinate. Certainly in cells supplied with ample oxygen and/or glucose, phosphorylation to ATP is the major net pathway of adenylate metabolism. Studies of Ehrlich ascites tumor cells incubated with radioactive adenine under optimum conditions have shown that the relative rates of phosphorylation, deamination, and dephosphorylation were 2,800:40:2, respectively (16,21,22).

Even when adenine nucleotide breakdown is accelerated under conditions of impaired energy metabolism, alternative pathways of adenylate metabolism may still be important. Thus, whereas dephosphorylation is the major route (92%) of adenylate metabolism in human erythrocytes incubated with 2-deoxyglucose (P. Burridge and J. F. Henderson, *unpublished data),* less than 20% of adenylate is metabolized via dephosphorylation in Ehrlich ascites tumor cells incubated with 2-deoxyglucose, the remaining being deaminated to inosinate (16).

Functions of Pathways of Adenosine Formation

Another way to approach the problem of the regulation of adenosine formation and metabolism is to consider the actual or potential functions of the various alternative pathways. These include (a) digestion of dietary RNA, (b) degradation of RNA from dead cells, (c) turnover of intracellular RNA (of which there are, of course, several species), (d) the turnover and renewal of the adenine nucleotide pool, (e) methylation reactions using S-adenosylmethionine, (f) the conversion of adenylate to inosinate via the so-called "adenosine cycle," (g) the transport of purines from one tissue to another via erythrocytes, (h) energy metabolism, and, finally, (i) other functions that may be called "regulatory" in nature. Some of these functions and pathways have already been discussed.

An additional function of adenosine formation that has been proposed (9) is to provide a pathway of synthesis of inosinate from adenylate that bypasses the adenylate deaminase reaction. The so-called adenosine cycle (adenylate → adenosine → inosine → hypoxanthine → inosinate) may indeed be important in some microorganisms, but studies of several types of mammalian cells have failed to provide evidence for its physiological functioning (3). (It is true, of course, that cells such as human erythrocytes do convert appreciable amounts of exogenous adenosine to inosinate.)

There is evidence, however, that adenosine may be produced by tissues such as liver, move out of the hepatocytes, and be rapidly picked up by erythrocytes, in which it is phosphorylated to adenine nucleotides, converted to inosinate, or both. Then, in other tissues, adenosine together with inosine and hypoxanthine may be formed, leave the erythrocyte, and be used for nucleotide formation by the recipient tissues (13). This quasinutritional role of adenosine formation has been demonstrated in animal systems, but its quantitative importance is not known.

The formation of adenosine from adenylate, together with the deamination of adenylate to inosinate, are sensitive indicators of energy metabolism, i.e., the sum of reactions involved in the phosphorylation of ADP to ATP. That adenosine (and inosinate) are produced from adenine nucleotides when oxygen tension decreases, when glucose supply is impaired, or under the influence of inhibitors of glycolysis or uncouplers of oxidative phosphorylation, etc., has been demonstrated in many systems (16,20,21).

Finally, many other real or potential "regulatory" functions of adenosine have been demonstrated, many of which are associated with cyclic adenylate functions. However, the links between these regulatory functions and the metabolic pathways that produce adenosine have not yet been completely elucidated.

Conclusions

In summary, it need only be said that there are many very important questions about the regulation of adenosine metabolism—particularly in regard to endogenous adenosine—that remain to be answered.

ACKNOWLEDGMENTS

The preparation of this chapter was supported by the National Cancer Institute of Canada, and the original work presented was supported by the National Cancer Institute and the Medical Research Council.

REFERENCES

1. Baer, H.-P., Drummond, G. I., and Duncan, E. L. (1966): *Mol. Pharmacol.*, 2:67–76.
2. Brady, T. G., and O'Donovan, C. I. (1965): *Comp. Biochem. Physiol.*, 14:101–120.
3. Brox, L. W., and Henderson, J. F. (1976): *Can. J. Biochem.*, 54:200–202.
4. Daddona, P. E., and Kelley, W. N. (1977): *J. Biol. Chem.*, 252:110–115.
5. DeJong, J. W. (1977): *Arch. Int. Physiol. Biochim.*, 85:557–569.
6. Divekar, A. Y. (1976): *Biochim. Biophys. Acta*, 422:15–28.
7. Divekar, A. Y., and Hakala, M. T. (1971): *Mol. Pharmacol.*, 7:663–673.
8. Drummond, G. I., and Yamamoto, M. (1971): In: *The Enzymes*, 3rd ed., edited by P. D. Boyer, 4:337–354.
9. Green, H., and Ishii, K. (1972): *J. Cell. Sci.*, 11:173–177.
10. Henderson, J. F., Mikoshiba, A., Chu, S. Y., and Caldwell, I. C. (1972): *J. Biol. Chem.*, 247:1972–1975.
11. Krenitsky, T. A., Miller, R. L., and Fyfe, J. A. (1974): *Biochem. Pharmacol.*, 23:170–172.
12. Leibach, T. K., Spiess, G. I., Neudecker, T. J., Peschke, G. J., Puchwein, G., and Hartmann, G. R. (1971): *Hoppe Seylers Z. Physiol. Chem.*, 352:328–344.
13. Lerner, M. H., and Lowy, B. A. (1974): *J. Biol. Chem.*, 249:959–967.
14. Lindberg, B., Klenow, H., and Hansen, K. (1967): *J. Biol. Chem.*, 242:350–356.
15. Lomax, C. A., Bagnara, A. S., and Henderson, J. F. (1975): *Can. J. Biochem.*, 53:231–241.
16. Lomax, C. A., and Henderson, J. F. (1973): *Cancer Res.*, 33:2825–2829.
17. Ma, P. F., and Magers, T. A. (1975): *Int. J. Biochem.*, 6:281–286.
18. Murray, A. W., and Friedricks, B. (1969): *Biochem. J.*, 111:83–89.
19. Schnebli, H. P., Hill, D. L., and Bennett, L. L., Jr. (1967): *J. Biol. Chem.*, 242:1997–2004.
20. Schrader, J., and Gerlach, E. (1976): *Pfluegers Arch.*, 367:129–135.
21. Snyder, F. F., and Henderson, J. F. (1973): *J. Biol. Chem.*, 248:5899–5904.
22. Snyder, F. F., and Henderson, J. F. (1973): *J. Cell Physiol.*, 82:349–361.
23. Snyder, F. F., Mendelsohn, J., and Seegmiller, J. E. (1976): *J. Clin. Invest.*, 58:654–666.
24. Thomas, M. A., Shipman, C., Jr., Sandberg, J. N., and Drach, J. C. (1977): *In Vitro*, 13:502–509.
25. Van der Weyden, M. B., and Kelley, W. N. (1976): *J. Biol. Chem.*, 251:5448–5456.
26. Wolpert, M. K., Dombe, S. P., Brown, J. E., Sznycer, E., Agarwal, K. C., and Sartorelli, A. C. (1971): *Cancer Res.*, 31:1620–1626.
27. Zielke, C. L., and Suelter, C. H. (1971): In: *The Enzymes*, 3rd ed., edited by P. D. Boyer, 4:47–78.
28. Zimmerman, T. P., Gersten, N. B., Ross, A. F., and Miech, R. P. (1971): *Can. J. Biochem.*, 49:1050–1054.

Physiological and Regulatory Functions of Adenosine and Adenine Nucleotides, edited by H. P. Baer and G. I. Drummond. Raven Press, New York © 1979.

Ecto-Enzymes Concerned with Nucleotide Metabolism

J. F. Manery and E. E. Dryden

Department of Biochemistry, University of Toronto, Toronto, Ontario M5S 1A8 Canada

Plasma membrane (PM) enzymes whose active sites face the external medium rather than the cell cytoplasm have been designated ecto-enzymes. An ecto-enzyme is a true PM enzyme, but not all true PM enzymes are ecto-enzymes. Some, like $(Na^{+}+K^{+})Mg^{2+}$ATPase, span the membrane having a site on both the external and internal surfaces; others, like adenylate cyclase, probably have only an internal site. There is a paucity of reports dealing with the "ecto" position of enzymes in whole cells compared with the current spate of publications describing enzymes associated with isolated PMs. This is deplorable because only when the enzyme is undisturbed in its natural position on or in the PM of the intact cell can its enzymic characteristics truly reflect their contributions to the function of the whole cell.

A warning is needed against the temptation to ascribe to all cell surfaces the ecto position of a certain enzyme because this position has been firmly established for one cell type. The use of acetylcholinesterase as a marker for the external surface of the erythrocyte (21,24) is so commonplace and so useful in interpreting erythrocyte membrane data that it is referred to as a typical ecto-enzyme. It has not been found, however, on guinea pig leukocytes (8), on muscle cell surfaces (14), or on isolated muscle PM (J. F. Manery, E. E. Dryden, G. Madapallimattam, and C. Brotman, *unpublished data*). In addition, this enzyme on the erythrocyte ghost is so sensitive to membrane perturbations (by detergents, lipases, etc.) that it, too, fails to satisfy some of the desired criteria for ecto-enzymes (21).

Rigorous ideal criteria to be satisfied before defining an enzyme on an intact cell as an ecto-enzyme have been documented by DePierre and Karnovsky (8,10), Trams and Lauter (36), and Newby et al. (23). These authors have used specific examples to help investigators interpret the "strictures" and "caveats" (10) imposed by these criteria, because it is almost impossible to obtain from most cells completely satisfactory experimental data. The criteria include: (a) the enzyme acts on substrates supplied extracellularly; (b) the substrate does not penetrate the cells; (c) the products are located in the extracellular medium; (d) there is no release of the ecto-enzyme or of known intracellular enzymes into the surrounding medium; (e) cells are shown to remain intact by demonstrating retention of their normal cell constituents; (d) if the enzyme is confined to the PM in an ecto position homogenization should not increase its activity; (f) inhibition (or stimulation as in the case of some lectins) by large nonpenetrating molecules provides good evidence of an ecto-enzyme.

Some of the difficulties encountered in attempting to satisfy these criteria are listed, along with suggestions for coping with them. (a) The presence of nonspecific phosphatase activity can be assessed by using substrates not attacked by the ecto-enzyme under study. (b) Methodology makes it difficult to obtain straight-line initial rates. (c) The products of the reaction may not remain in the extracellular medium because they are further metabolized by other enzymes, or because they enter the cells. The time period should be as short as possible so that the amounts of substrate degraded and products formed still show stoichiometry. (d) Choice of a short time period ensures minimal degradation of cells, a particular hazard when using cultured cells. (e) The total enzyme activity of the cell may not be confined to the PM. In only two known cell types, guinea pig polymorphonuclear leukocytes (8) and intact rat adipocytes (23), does the ecto-5′-nucleotidase comprise all of the 5′-adenosine monophosphate phosphohydrolase (5′-AMPase) of the cell. In most cases, however, the ecto-enzyme contributes the largest proportion (50 to 100%) of the total activity. (f) Appreciating the desirability of finding the same quantitative parameters for the enzyme on the intact cell and in the isolated PM, one must accept the fact that preparation of the PM may affect the enzyme activity. Acceptance of qualitative similarities, such as enzyme specificities and the action of accelerators and inhibitors, is legitimate for identifying a presumptive ecto-enzyme. (g) If further breakdown of products releasing P_i, or exit of P_i from the cell interior, cannot be minimized sufficiently by a short period of incubation, the use of a labeled substrate [γ^{32}P]ATP, [^{32}P]AMP, etc., permits identification of the source of the measured P_i.

In this chapter the following ecto-enzymes[1] are described: (Ca^{2+}, Mg^{2+})-ATPase, 5′-AMPase (5′-nucleotidase), AMP deaminase, ρNPPase, and PDE. The classification as "ecto" is most firmly established for the first four of these. The evidence is strong and convincing that the external membrane of unicellular organisms, e.g., ameba (32) and yeast (29), and the PM of free-swimming cells (leukocytes, lymphocytes, etc.), the PM of cells in culture and of cells of organized tissues like muscle and adipose tissue all contain ecto-enzymes. Because of the limitations of space a complete review of the literature is impossible. I have described the ideal criteria for ecto-enzymes and the difficulties, imposed by using viable living cells, that are encountered satisfying the criteria. Accepting good presumptive evidence for the ecto-enzyme I have selected representative papers describing the ecto-enzyme and a few in which a new dimension has been added to the discovery of the enzyme; these include a new probe to test their ecto-position as well as their characteristics on embryonic and transformed cells. The reader is

[1] (Ca^{2+}, Mg^{2+})adenosine triphosphate γ-phosphohydrolase (ATPase) (EC 3.6.1.3); 5′-adenosine monophosphate phosphohydrolase (5′-AMPase) same as 5′-nucleotidase (EC 3.1.3.5); adenosine monophosphate deaminase or aminohydrolase (AMP deaminase) (EC 3.5.4.6); ρ-nitrophenylphosphatase (ρNPPase) (EC 3.1.3.1); phosphodiesterase (PDE) (EC 3.1.4.1); (sodium + potassium)-stimulated magnesium adenosine triphosphatase [(Na^++K^+)Mg^{2+}ATPase] (EC 3.6.1.4).

referred to these papers for background references. Perhaps the most glaring omission is the vast literature on ecto-enzymes of platelets.

MUSCLE

Small intact muscles of the hindleg of the frog provide excellent material for studying ecto-enzymes. Over many years we have demonstrated that frog muscles are biologically stable if they are carefully dissected to minimize injury and allowed to recover from handling by soaking overnight at 0 to 4° C in a frog Ringer's solution buffered with Tris-HCl (RT) and containing a high K^+ concentration (6 mM instead of 2.5, the concentration in frog plasma). During incubation at 22° C in a balanced salt solution with pure oxygen in the gas phase they retain their excitability properties and electrolyte concentrations, and respire at a constant rate over many hours. The special advantages for the investigation of ecto-enzymes are as follows: (a) the similarity of contralateral muscles such that muscles from one leg provide good controls for muscles from the other leg; (b) the low level (if any) of a nonspecific ecto-phosphatase; (c) the virtual absence of P_i leakage out of the cells or entry into the cells under the incubation conditions; (d) the remarkable stability of the ecto-enzymes even after hours of multiple incubations.

In the early 1960s we found, almost by accident, a system of nucleotide-converting enzymes on the external surface of frog muscles (11). The enzymes faced outward. After 4 hr the concentration of ATP in the medium was reduced from 3 mM to about 0.4 mM and the products were recovered almost stoichiometrically in accord with a reaction sequence involving the enzyme ATPase, adenylate kinase, and 5′-AMP deaminase [ATP→inosine monophosphate (IMP) + NH_3 + 2 P_i].

Figure 1 shows the time course of the conversion of ATP (3 mM) to IMP through adenosine diphosphate (ADP) and AMP (11). From 0.3 to 0.4 g of muscle in 2 ml of RT reduced the concentration of added nucleotide (3 mM) to about one-half in less than 1 hr. The decrease in conversion rate after the first hour was demonstrated to be caused by the low substrate concentration and not by the accumulation of products (16). The ATPase activity was not mediated by the $(Na^+ + K^+)Mg^{2+}$ATPase since no inhibition by ouabain was observed, nor were adenylate kinase (EC 2.7.4.3) or AMP deaminase affected by ouabain, although in every case direct analyses of muscle revealed inhibition by the glycoside of Na^+ and K^+ transport (11). Ethacrynic acid, a powerful diuretic, also inhibits isolated muscle $(Na^+ + K^+)Mg^{2+}$ATPase (J. F. Manery, E. E. Dryden, G. Madapallimattam, and C. Brotman, *unpublished data*) and Na^+ and K^+ transport in frog muscle regardless of the substrate supplied (27). Using [^{14}C]-labeled substrates and separation by paper chromatography it became clear that this drug exerted no inhibition on any of the three enzymes (27).

In view of these data it was somewhat disconcerting to observe (Fig. 2) a marked decrease in the production of IMP from ATP, ADP, and AMP if Na^+ and K^+ were replaced in the medium by choline (16). Figure 2D provides the explanation. Obviously, the absence of Na^+ and K^+ had no inhibitory effect whatsoever on the

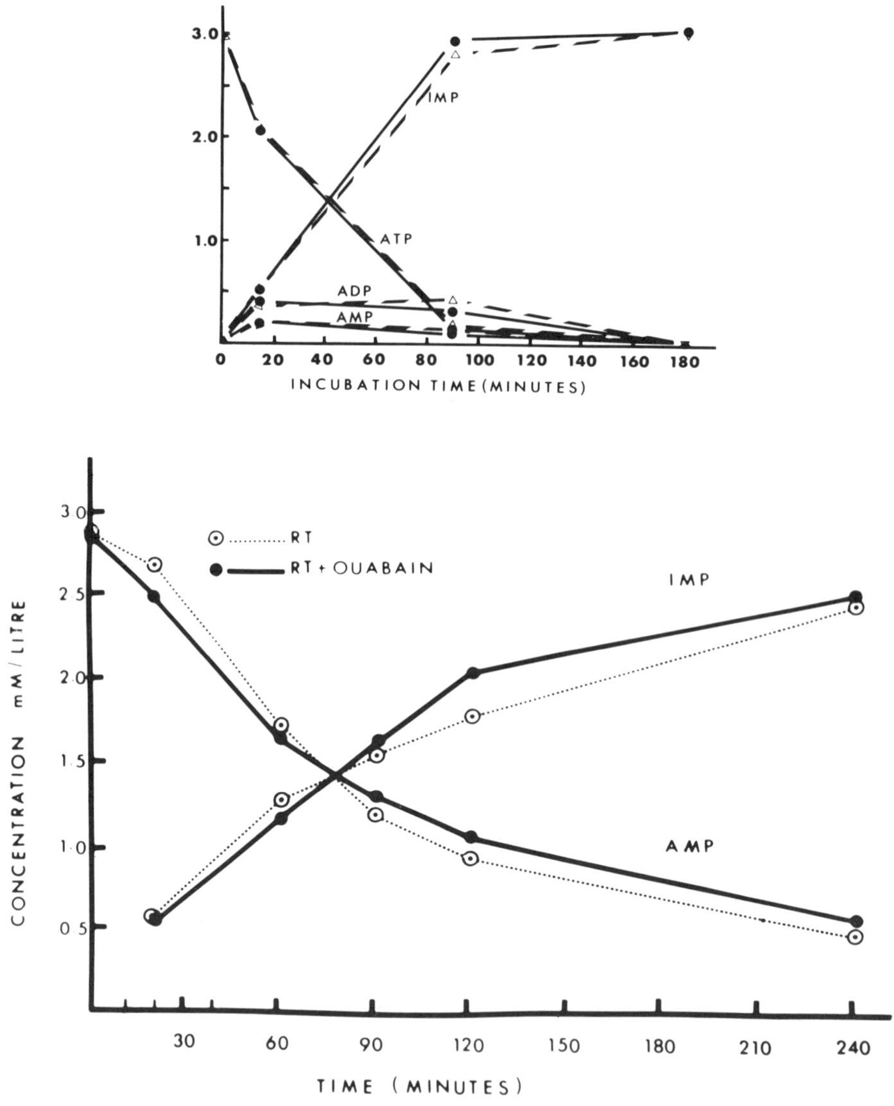

FIG. 1. Time course of conversion of ATP → IMP in media surrounding intact frog muscles. Matched muscles were used, one of a pair being the control, the other exposed to 1 μM ouabain (⊙-----⊙, *bottom*) or to 1 mM ethacrynic acid (Δ———Δ, *top*). After overnight soaking (see text) muscles were incubated with or without the drug in RT containing (mM): ATP 3.0; NaCl 106.7; KCl 6.0; $CaCl_2 \cdot 2H_2O$ 1.26; $MgSO_4$ 1.18; Tris(hydroxymethyl)aminomethane 13.2, pH 7.4, 245 mOsm. Nucleotide concentrations in the media were determined by ultraviolet spectral analyses in the ouabain experiment; in the ethacrynic acid experiment [^{14}C]ATP was supplied and the concentration of the products estimated by liquid scintillation spectrometry after chromatographic separation. (Data from ref. 11 and ref. 27, with permission.)

production of P_i, which is a product of the reactions catalyzed by ATPase and adenylate kinase. Using 16 paired incubations it was proved conclusively that the lack of Na^+ and K^+ caused a decrease in the conversion of AMP to IMP that was highly significant at $p = 0.01$ using the t-test for paired variates. Muscle analyses for Na^+ and K^+ showed that the enzyme responded to the extracellular, not the intracellular, concentrations of these cations. Attention is drawn to the low, almost negligible, amounts of P_i found in the first hour of incubation with AMP, which demonstrate the absence of the nonspecific phosphatase frequently found on the surface of other cells (see below).

It was demonstrated many times that there was no leakage of ATPase or AMP deaminase and only a slight leakage of adenylate kinase from muscles into the media during the 1- to 4-hr incubation periods. These data were obtained by incubating ATP, ADP, or AMP in the medium from which a muscle had been removed after a long incubation. Most impressive was the tenacity with which these ecto-enzymes were held on the muscle surface. Data in ref. 16 show how reproducible the conversion of AMP to IMP can be during successive incubations. In these experiments muscles were incubated for four successive periods of 4 hr each, three during the first day and the fourth on the following day. Muscles were left in a large volume of RT for 20 min between the first three incubations and overnight before the last incubation. There was no leakage of the enzyme into these media.

Characteristics of the ecto-ATPase were studied in some detail (J. F. Manery, E. E. Dryden, and G. Madapallimattam, *unpublished data)*. The average activity over the first 20 min provided a good approximation to initial rates, since the release of P_i from ATP was almost linear during this period. Table 1 shows that there was no measurable loss of P_i from the muscle during the first 20 min, that divalent cations are essential, and that Ca^{2+} is not preferred over Mg^{2+}. Divalent cadmium and trivalent lanthanum at 1 mM inhibited the ATPase in the presence of Ca^{2+} (2 mM) suggesting that Ca^{2+}ATP and Mg^{2+}ATP are the true enzyme substrates. Mersalyl also inhibited this ecto-enzyme. Mersalyl (Salyrgan) is a sulfhydryl inhibitor known to inhibit actomyosin ATPase. Neither oligomycin, dinitrophenol, nor ouabain had any effect, showing that the ATPase activity was not mitochondrial in origin or a part of the $(Na^+ + K^+)Mg^{2+}$ATPase. The ecto-ATPase hydrolyzes other trinucleotides showing a preference for ATP.

Initially, AMPase (5′-nucleotidase) was alleged to be a PM marker because it was found in liver PM. The enzyme activity had been either very low (2) or difficult to find (33) in the isolated PM of skeletal muscle until Woo and Manery (43) showed that the activity was readily detectable in the Boegman et al. preparation (2) if μM rather than mM concentrations of either AMP or IMP were used; it is also present in the PM of rabbit skeletal muscle (J. F. Manery, E. E. Dryden, and G. Madapal-

FIG. 2. (See pages 328–329.) The effects of $Na^+ + K^+$ on the time course of conversion of adenine nucleotides to IMP and on changes in the P_i concentrations in the media. The label indicates the nucleotide (3mM) initially provided as substrate. The medium was RT (without $CaCl_2 \cdot 2H_2O$) as shown in Fig. 1. Choline chloride was substituted for $Na^+ + K^+$, mOsm of this medium being 228. (Data from ref. 16, with permission.)

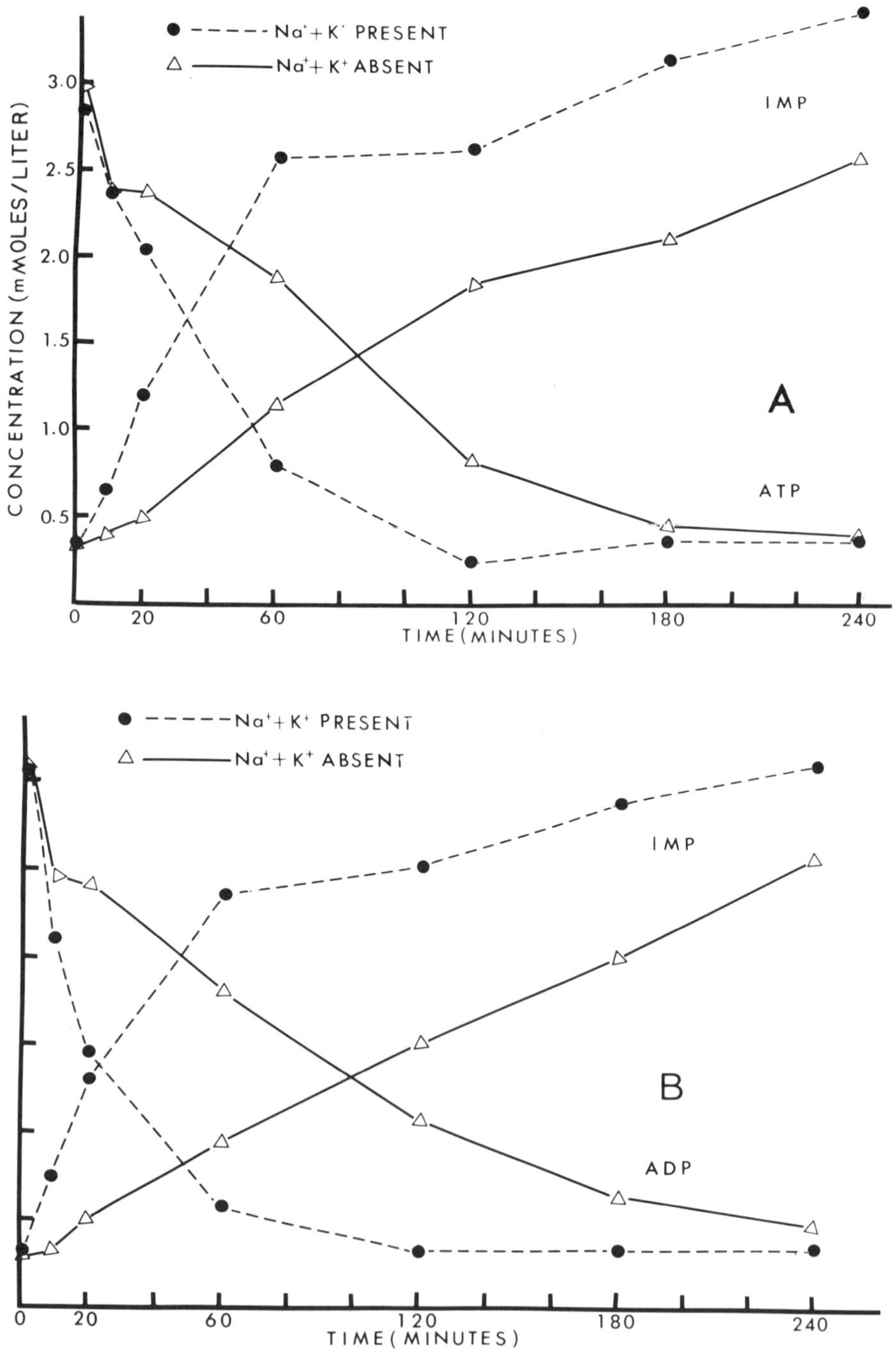

Na⁺+K⁺ PRESENT
Na⁺+K⁺ ABSENT
CONCENTRATION (mMOLES/LITER)
3.0
2.5
2.0
1.5
1.0
0.5
IMP
A
ATP
0
20
60
120
180
240
TIME (MINUTES)
Na⁺+K⁺ PRESENT
Na⁺+K⁺ ABSENT
IMP
B
ADP
0
20
60
120
180
240
TIME (MINUTES)

FIG. 2A and B

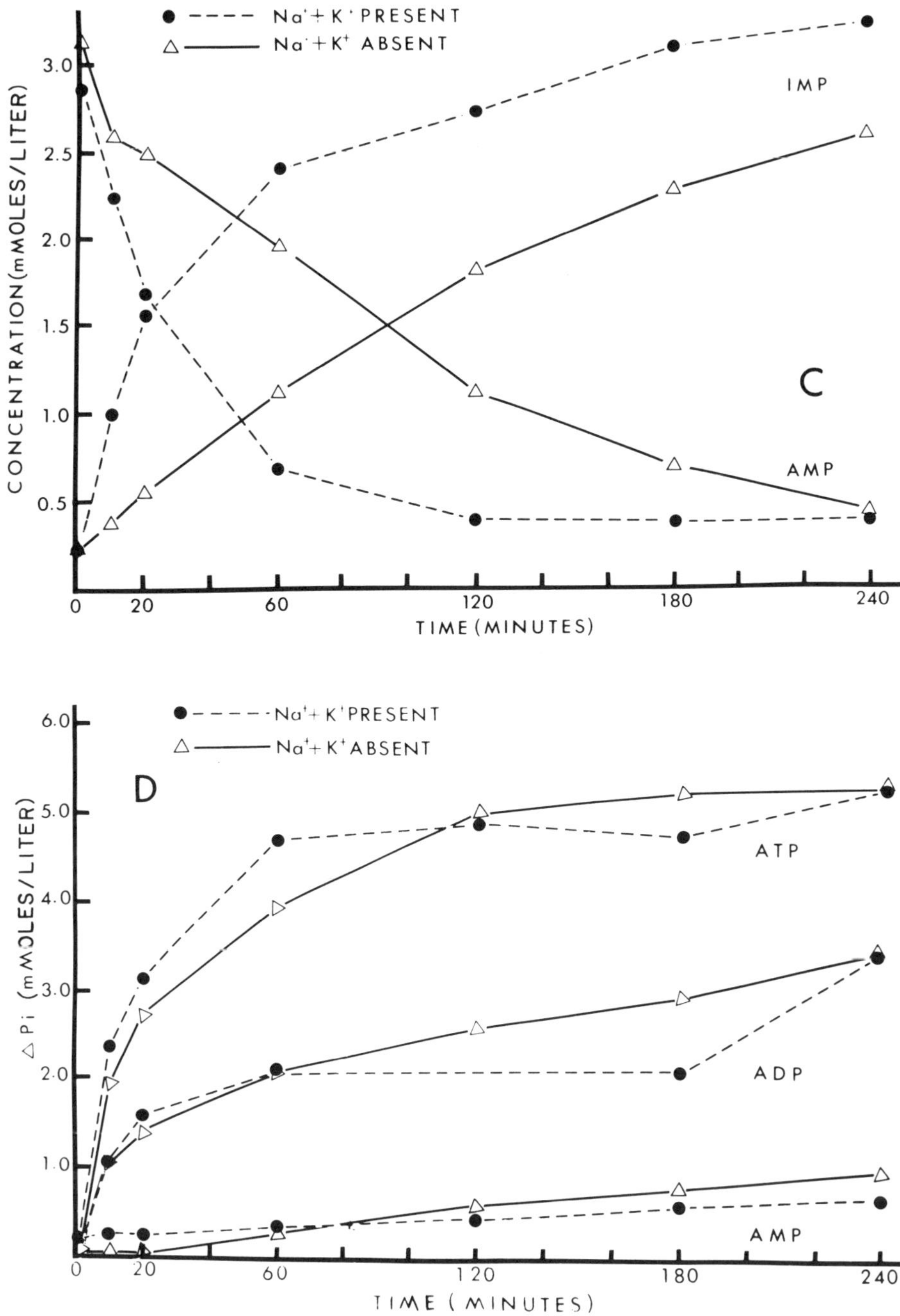
Na⁺+K⁺ PRESENT
Na⁺+K⁺ ABSENT
IMP
AMP
C
CONCENTRATION (mMOLES/LITER)
TIME (MINUTES)
0
20
60
120
180
240
0.5
1.0
1.5
2.0
2.5
3.0
Na⁺+K⁺PRESENT
Na⁺+K⁺ABSENT
D
ATP
ADP
AMP
△ Pi (mMOLES/LITER)
TIME (MINUTES)
1.0
2.0
3.0
4.0
5.0
6.0

FIG. 2C and D

TABLE 1. *Effect of various conditions on initial rates of muscle ecto-ATPase*

Conditions[a]	Effect (% of control)[b]
0 Ca^{2+}, 2 mM Mg^{2+}	+8
0 Ca^{2+}	−18
0 Ca^{2+}, 2 mM EDTA	−65
1 mM Cd^{2+}	−28
1 mM La^{3+}	−56
1 mM Mersalyl	−27
Oligomyosin or DPN or ouabain	0
Control, 30 muscles, μmoles P_i/g/20 min = 7.5 ± 0.04	

Media were analyzed for P_i after incubation of muscles for 20 min at 20° C. The control muscles were in RT containing (mM); ATP 2.0; NaCl 106.7; KCl 6.0; $CaCl_2$•2.0; Tris buffer 13.2, pH 7.4. Matched muscles were always used, one of a pair being incubated in the control RT and the other in the same medium with the deletions and additions indicated.

[a] No ATP.

[b] No P_i.

limattam, *unpublished data)*. Using substrate concentrations from 2 to 10 μM, an active ecto-5′-nucleotidase was easily demonstrated (Fig. 3) on intact frog muscles (43). It was firmly bound because no activity was evident in media from which muscles had been removed after previous incubation. The figure shows that AMP is deaminated to IMP and after an initial buildup, both nucleotides are converted by the nucleotidase to adenosine and inosine; part of the inosine formed is attributed to the activity of an adenosine deaminase. For the data of part B, Ca^{2+} and Mg^{2+} were omitted from the medium and 1 mM ethylenediametetraacetate (EDTA) was added. No nucleotidase activity is evident, only that of AMP deaminase showing the complete dependence of the ecto-5′-nucleotidase on divalent cations.

Again the absence of a nonspecific phosphatase was confirmed by showing that the presence of ρNPP and of 3′-AMP had little or no effect on either the time course of the degradation of IMP by the ecto-enzyme of the intact cell, or on the 5′-nucleotidase activity of the isolated PM (43). In addition, the enzyme in both preparations showed activity toward other 5′-nucleotides. This was demonstrated by the abolition of measurable breakdown of a [^{14}C]-5′-nucleotide when the enzyme was exposed to the labeled nucleotide in the presence of nonradioactive IMP, cytidine monophosphate (CMP), guanosine monophosphate (GMP), or AMP. These experiments, in which we showed that the activity of the muscle surface 5′-nucleotidase could be demonstrated only at μM substrate concentrations, should militate against the general tendency to ignore differences in PM enzymes among species, tissue, or cell type.

Finally, an active PDE activity that catalyzes the breakdown of cyclic adenosine 3′,5′ monophosphate (cAMP) to 5′-AMP was found on the muscle surface (42).

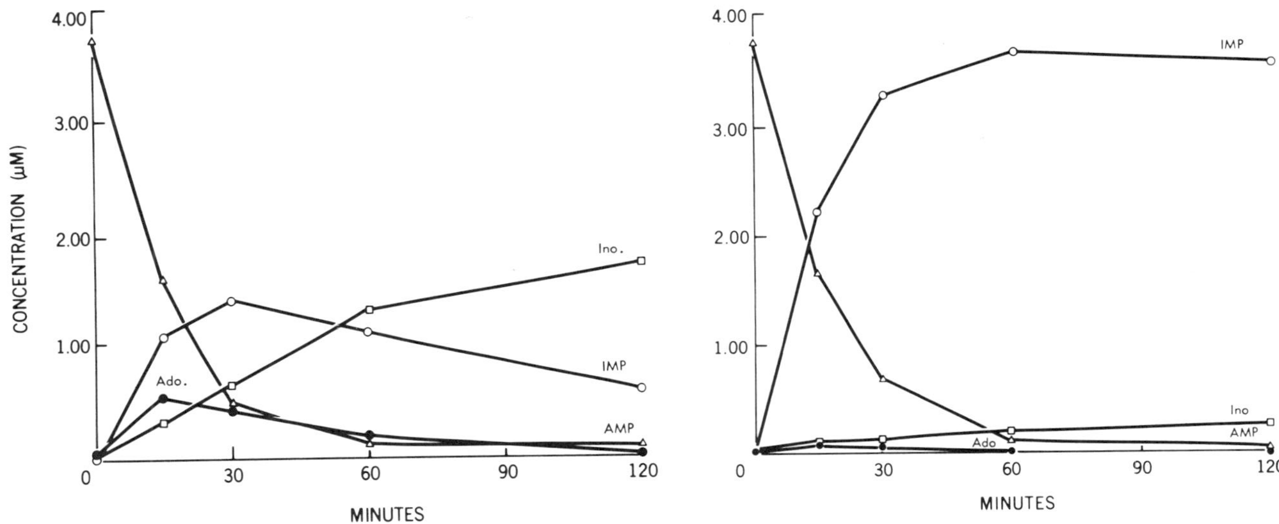

FIG. 3. Degradation of AMP (3.9 μM) in media surrounding intact muscles. The initial substrate was [^{14}C]AMP and the products were separated and identified by chromatography. **Left:** Muscles in complete RT. **Right:** Muscles in Ca^{2+}-, Mg^{2+}-free RT containing EDTA (1 mM). (Data from ref. 43, with permission.)

"Leakage" experiments and successive incubations revealed that over 85% of the enzyme remained bound to the muscle. Kinetic studies of the ecto-enzyme *in situ* demonstrated two apparent K_m values that were not too dissimilar from those reported for purified or partially purified PDEs from other cells (brain, heart, skeletal muscle, frog bladder, and kidney) (see ref. 42 for references). Furthermore, insulin stimulated this enzyme *in situ* at low extracellular concentrations of cAMP (2 to 17 μM). The authors speculated about the significance of this stimulation (42).

After this work was completed an early (1957), short report of an ecto-ATPase on the surface of isolated rat diaphragm came to our attention (17) in which it was concluded that the surface of rat diaphragm exhibited ecto-ATPase activity.

It has been difficult to reconcile the characteristics of muscle surface enzymes that we have studied so intensively and observed so consistently with those described for bullfrog and toad sartorii by Sakai et al. (30). They reported the absence of AMP deaminase and the activation of the ecto-ATPase by Na^+ and K^+ as well as its inhibition by ouabain, concluding that this ATPase is the Na^+ and K^+ transport enzyme. One striking difference in the preparation of the intact muscle is the length of time given for its recovery from handling or from the injury imposed during dissection. In our laboratory, not only are dissections done only by experienced dissectors but muscles with the smallest visible injury are discarded, and during the long recovery period of 16 to 24 hr fibers with undetectable injury heal or become extracellular space. The Japanese group allowed only 30 min to 1 hr for recovery. In a later paper (15) Kayama accepted the possibility of some slight ecto-AMP deaminase activity but alleged that this was evidence of damage and enzyme leakage from the cell interior. There is no reason to preclude the possibility of AMP deaminase activity located inside the cell as well as outside (see ref. 31 for enzymes that are both ecto- and endo-enzymes). In our opinion, our data showing the consistency of AMP deaminase activity after long and successive incubations and the complete absence of the enzyme in these media attest to the ecto position of this enzyme on uninjured fibers. The absence of 5′-nucleotidase on both the intact muscle and the isolated sarcolemma (15) can probably be explained by failure to use μM substrate concentrations (43).

Cultured hamster cardiac cells have recently been shown to possess an ecto-ATPase, dependent on Ca^{2+} or Mg^{2+} and not inhibited by ouabain (5). Its active center is exposed to the environment since the products ADP and P_i can be measured in the medium. The enzyme showed kinetic properties somewhat different from those of myofibrillar myosin, which does not rule out the possibility that myosin-like proteins located in the PM might be associated with the ecto-ATPase. In various cultured cell types myosins appear to be integral components of the PM (for further discussion see section below titled Biological Significance of Ecto-Enzymes).

Attention should be paid to the distinctive morphology of the muscle sarcolemma when searching for the location of muscle ecto-enzymes. Also, caution should be exercised in interpreting the extent of the effect on the ecto-ATPase of certain agents during measurement of initial rates. Nayler (22) pointed out that the PM (about 9 nm thick) was a small integral part of the sarcolemmal complex. This complex is made up of (a) a diffuse reticulum of collagen fibrils, (b) an extracellular protein-polysaccharide coating (the basement membrane) that may be 50 to 100 nm thick or more, and (c) the PM, which is contiguous with the cytoplasm and appears to extend as tubular invaginations, the "T tubules," which branch, ramify, and anastomose throughout the cell. The protein-polysaccharide layer protects the PM from the extracellular phase and, because of their large molecular weight, bulky shape, charge, or other parameters, some substances may be retarded from reaching the PM as they penetrate this thick layer.

ADIPOCYTES AND MAST CELLS

An active ecto-5′-nucleotidase with a K_m of 43 μM was demonstrated in 1975 by Newby et al. (23) on the external surface of isolated fat cells of the rat. Extracellular AMP was hydrolyzed producing stoichiometric amounts of P_i and adenosine. The intactness of the cells was judged by their normal ATP concentrations and their retention of lactate dehydrogenase. When a small amount of nucleotidase activity was found in the medium a parallel increase in lactate dehydrogenase was also observed. Antibodies raised to whole liver cells and fat cells of the rat inhibited the 5′-nucleotidase of the intact adipocyte. The nucleotidase of a particulate fraction (the PM) showed the same characteristics as the enzyme on the intact cell, namely the same K_m, hydrolysis of all 5′-mononucleotides but not other phosphate compounds, inhibition by low concentrations of ATP, etc.

Mast cells, which store histamine and release it in the presence of ATP and Ca^{2+}, contain a Ca^{2+}- or Mg^{2+}-stimulated ecto-ATPase (6). The enzyme resembles the actomyosin ATPase and not the Ca^{2+}-transport ATPase. Chelation of the cations by EDTA and EGTA inhibited the enzyme, whereas ouabain had no effect. For these studies on the whole cell the complication owing to histamine release was overcome by the use of deoxyglucose, a known inhibitor of this release. Under these conditions the Ca^{2+}- and Mg^{2+}-activated ATPase activities were the same in both whole cell and homogenate. Little or no 5′-nucleotidase activity was detected using 5 mM AMP as substrate. After sucrose gradient separation of the homogenate fractions, the ATPase and low nucleotidase activities were both found in the same light fraction; this did not contain the intracellular lactate dehydrogenase.

LEUKOCYTES AND LYMPHOCYTES

In many ways free-swimming cells are easier to work with than cells of organized tissues. Well-documented studies have shown that guinea pig polymorphonuclear leukocytes will hydrolyze exogenous ATP, AMP, and ρNPP (8,10). The three enzymes involved could not be washed off in 10 washes with 50 to 100 volumes of a Krebs-Ringer solution (8). Disruption of the cells by homogenization (8) did not increase the AMPase activity, showing that the total cell nucleotidase was in the ecto position. Activities of ATPase and ρNPPase increased twofold on cell disruption, suggesting that the PM acted as a permeability barrier to ATP and ρNPP.

Substrates (ATP and AMP) were labeled with [^{32}P]phosphate so that the source of the product could be identified. To prove that the substrates were cleaved outside the cells and not inside and products transported out, the cells were first loaded with [^{33}P]phosphate. After hydrolyzing the [^{32}P]-labeled substrate, virtually all of the ^{33}P label (87 to 95%) remained inside the cells and less than 1% of the ^{32}P label was found inside. Such unequal mixing of the two isotopes inside the cells provides ample proof that the ^{32}P label did not enter the leukocytes (8). A diazonium salt of sulfanilic acid believed not to penetrate the cells markedly inhibited the ecto-enzymes ATPase, AMPase, and ρNPPase, but not the cytoplasmic enzyme lactate dehydrogenase; the reagent, however, effectively inhibited the dehydrogenase in sonicated cell preparations.

Finally, the AMPase was proved to be an ecto-5′-nucleotidase because it hydrolyzed the four mononucleotides (9), the affinity order being as follows: AMP≥uridine monophosphate (UMP)>CMP>GMP>IMP. In addition, substrate competition studies demonstrated that there were three separate ecto-enzymes, the ρNPPase being the least specific. To eliminate its effect on the ATPase 5 mM ρNPP was routinely added to an ATPase assay. Important for our purpose was the complete lack of response of the ecto-ATPase to $Na^+ + K^+$ or ouabain showing that the enzyme was unrelated to $(Na^+ + K^+)Mg^{2+}$ATPase. The ATPase was not stimulated by 2,4-dinitrophenol (DNP) or inhibited by oligomycin. Species differences must be appreciated (10).

Medzihradsky et al. (20), using isolated rat leukocytes, confirmed many of the characteristics described above for guinea pig leukocytes and extended the research to include the effects of drugs. Rat leukocytes also possess an active ecto-ATPase (K_m for ATP = 0.3 mM). There was little or no effect on this MgATPase of oligomycin, Ca^{2+}, or DNP. Nonspecific phosphatase activity was excluded. The ouabain-inhibitable, $(Na^+ + K^+)$-activated portion was negligible (about 7%). That the cells were not leaky was demonstrated by the normal concentrations of Na^+ and K^+ found inside the cells. Of real interest was the finding (19) of almost complete inhibition of the ecto-ATPase by phenothiazine tranquilizers and tricyclic antidepressants. The potency of the drugs as enzyme inhibitors correlated well with the percentage inhibition of the phagocytic activity of the cells, which suggests a possible relationship of the ecto-ATPase to contractile elements.

Mouse splenic lymphocytes were studied by Pommier et al. (26) with respect to the sidedness of the PM marker enzymes $(Na^+ + K^+)Mg^{2+}$ATPase, Mg^{2+}ATPase, and 5′-nucleotidase. Microsomes, whole cells, and homogenates were investigated. The same level of MgATPase and 5′-nucleotidase was observed in intact cells and in homogenates, which is evidence of ecto-enzymes. In contrast, there was no ouabain-inhibitable $(Na^+ + K^+)$-sensitive MgATPase activity observable until the cells were disrupted, showing that this ATPase faces inward. The lectin concanavalin A (Con A), a large molecule that does not cross PMs, markedly stimulated a MgATPase activity catalyzed by intact cells. Although the 5′-nucleotidase activity is very low in these cells, the lectin appeared to inhibit it slightly. Inhibition of this enzyme by Con A has been observed many times (see refs. 1,3 and 26).

The existence of ecto-ATPase and ecto-AMPase in various cell types is now so well established that it is not surprising that investigators are searching for differences in the characteristics of these surface enzymes between normal and malignant cells. Recently it has been shown by Weiss and Sachs (40) that rat myeloid leukemic cells had a lower ecto-ATPase than normal early myeloid cells, granulocytes, and macrophages. Malignant transformation of the normal early myeloid cells appeared to be associated with a block in the amount of ecto-ATPase formed. Moreover, of the three groups of leukemic cells tested for their response to a differentiation-inducing protein, the group that showed no differentiation had the highest ecto-ATPase activity.

In contrast, ecto-AMPase activity was lower in normal cells than in myeloid leukemic clones. The clone with the highest ecto-AMPase activity was not the one with the highest ecto-ATPase. Also, induction of differentiation did not change the level of AMPase in either normal or leukemic cells. Clearly, the two enzymes are regulated independently.

MAMMARY GLAND AND TUMORS

Having previously investigated the 5′-nucleotidase and Mg^{2+}ATPase of partially purified mammary PMs of the rat, in 1976 Carraway and co-workers (3) compared the enzymes of the PM with those of the intact cell and of solubilized membranes using a mammary ascites adenocarcinoma (13762). A low K_m (25 μM) 5′-nucleotidase, the inhibition of which by Con A was reversed by α-methylmannoside, was found in the isolated PM of normal and ascites cells and in the intact 13762 ascites cells. Careful studies showed that the Con A inactivation was occurring at the cell surface. The authors noted that the inhibition exhibited no cooperativity in the intact cells or the isolated membrane envelopes but showed substantial cooperativity when the membranes were in a vesicular form or solubilized in deoxycholate. They suggested that ecto-enzymes may act as "reporter" groups for membrane alterations if a change in cooperativity of the enzyme-lectin interaction was observed.

CELLS OF CENTRAL NERVOUS SYSTEM AND OTHER CULTURED CELLS

Applying the elegant micro-techniques of Holger Hydén to neurons carefully dissected by hand from Deiter's nucleus in the rabbit brain, in 1962 Hydén and Cummins (7) demonstrated beyond question the Mg^{2+}-ecto-ATPase on glial cells; the ATPase was unaffected by Na^+, K^+, or ouabain. Using intact neurons, on the other hand, no surface Mg^{2+}ATPase could be detected. When the neurons were homogenized, an active $(Na^+ + K^+)Mg^{2+}$ATPase activity was revealed; this enzyme was firmly bound to a membrane fraction isolated by a mini-homogenizer.

In a careful study by Trams and Lauter (36) of ecto-enzymes on intact cells of several cultured cell lines (HeLa; KB; human, mouse, and guinea pig hepatocyte; mouse neuroblastoma) nonspecific ecto-Mg^{2+}ATPase and ecto-5′-nucleotidase activities were identified using many of the ideal criteria described above. The neuroblastoma cells exhibited an ecto-Mg^{2+}ATPase that was unaffected by ouabain. In contrast, adenylate cyclase, although clearly a PM enzyme, faced the cytoplasm and not the external medium.

Although the authors (36) endorse the possibility that an ecto-ATPase and ecto-5′-nucleotidase characterize eukaryotic cells, they stress the disparity in specific enzyme activity that can exist between different cell types, and readers should be cautioned against generalizations. Finally, it may be of some interest that when HeLa cells were exposed to [^{14}C]AMP there was evidence of ecto-deaminases for adenosine and AMP (see ref. 16 for muscle ecto-deaminase). A later paper from this laboratory (37) provides evidence for ecto-ATPase and 5′-nucleotidase on several cultured cell lines derived from the central nervous system; these were mouse neuroblastoma, neonatal Syrian hamster astrocytes, human astrocytoma, and primary mouse astroblasts.

It will be recalled that Cummins and Hydén (7) found an active ecto-ATPase on glial cells. More recently, Stefanović et al. (35) published the first report of nearly stoichiometric production of extracellular adenosine and P_i from exogenous AMP by intact cultured nerve cells (C6 rat glioma cells). More than 60% of the total 5′-nucleotidase activity of the cell was in the ecto position. In only two cases has the total nucleotidase activity of the cell been reported to reside solely in the PM [adipocytes (23) and leukocytes (8)]. Differential labeling of the P_i in the cellular and intracellular compartments using [^{33}P]phosphate and [^{32}P]-substrates, respectively, showed convincingly that AMP did not enter the cell and that it was hydrolyzed by an AMPase facing outward (35). The enzyme hydrolyzed other 5′-nucleotides, was inhibited by Con A, and displayed precise Michaelis-Menten kinetics *in situ;* there was no nonspecific phosphohydrolase activity.

A significant extension of this work was described in 1977 (34). It is known that contact with nonneuronal cells enhances differentiation of dissociated neurons and favors their maintenance in long-term cultures. The authors studied the ecto-enzymes of cells during coculture and after reisolation. The cells studied showed a Ca^{2+}ATPase activity that was consistently higher than the Mg^{2+}ATPase activity. Normal hamster astroblasts (clonal line NN) and mouse neuroblastoma cells (clonal line Ml) originally have low ecto-ATPase activity. When cocultured for 7 days the specific activity of the enzyme increased two to four times. The greatest increase was in the Ca^{2+}ATPase activity. The mouse neuroblasts reisolated from coculture with the hamster glial cells after 7 days and after 2 months showed a continued increase with time in ATPase activity above their original activity; they showed a higher V_{max} and a higher K_m. The higher ecto-ATPase was maintained for more than 20 replications. Mouse neuroblastoma cells isolated after coculture with chick fibroblasts had Ca^{2+}- and Mg^{2+}-dependent ATPase activities that were four times higher than in the original cell line. In contrast, 5′-nucleotidase activity of cocultured or reisolated neuroblasts and astroblasts did not differ from that of the original cells. For the purposes of this chapter it is significant that the neuronal-glial interaction manifests itself by a marked increase in surface enzyme activity. Thus attention is focused on the response of ecto-enzymes to external influences on the cell surface. These and similar observations should generate greater efforts to understand the biological importance of ecto-enzymes.

BIOLOGICAL SIGNIFICANCE OF ECTO-ENZYMES

The PM of cells is important for at least three cell functions, one related to transport, one to cell contact, and one to cell deformity. It has been suggested (3) that the cell surface ecto-enzymes might be used as reporter groups for assessing membrane changes. The advantages are: (a) Assays are done in intact cells without cell disruption or membrane isolation; and (b) properties of enzymes *in situ* can be measured consistently and quantitatively, including specific activities, kinetic parameters, allosteric or cooperative effects, and inhibition or activation by external agents. Evidence has been presented above showing that cell surface enzymes are sensitive to changes in environment (34,40), and thus they may become more useful probes of membrane function than has hitherto been suspected. This chapter describes how to use ecto-enzymes but does not go beyond mere conjecture with respect to their true physiological function.

Beneficial consequences of the activities of these ecto-enzymes can be readily cited if a teleological approach to the problem of their function is acceptable. Cells are generally impermeable to molecules like ATP; their degradation to nucleotides and nucleosides that enter cells readily provides substrates for intracellular syntheses. Adenosine is vasodilatory and the possible role of ecto-5′-AMPase in the regulation of blood flow will be apparent from other chapters in this volume. Adenosine triphosphate is also a powerful vasodilator but its effect must be transient because of the active ecto-ATPase on the surface of the large mass of skeletal muscle, in addition to the plasma ATPase and the ecto-ATPase on other cells. Whatever the reason, the ATP concentration is very low in human plasma. Using a bioassay, Forrester (12) found 10 μM; with other methods workers failed to detect any ATP.

Adenosine has been implicated as a neurohumoral agent in brain and implicated in hormonal stimulation of lipolysis in fat cells (see ref. 23 for cited literature). The transient increase of membrane permeability induced by cytoplasmic ATP that may be translocated to the membrane has been related to brain seizures in a discussion by Rosenblatt et al. (28). A possible role of AMP and adenosine at the cell surface as effectors of the adenylate cyclase that is exclusively on the inside of the PM has been suggested (35). Unfortunately, the exact assignment of physiological roles to the ecto-enzymes described here is not possible at present.

The apparent similarity of the ecto-ATPases to muscle actomyosin ATPase suggests the intriguing possibility that ecto-ATPases contribute to the function of the PM that is related to cell contact and cell deformability. The free-swimming cells and organisms, vertebrate cells in organized tissues, and those grown in culture described above have a common characteristic: they display motility, deformability, and ameboid movement. Moreover, fibrous proteins resembling muscle actomyosin, especially that of smooth muscle, now appear to be integral components of the cell surface membranes of various cell types including cultured cell lines (5,13,41).

Platelets also contain an active ecto-(Ca^{2+},Mg^{2+})ATPase (4) believed to be associated with thrombosthenin, the contractile protein component of the platelet PM. Platelet ecto-ATPase is firmly bound, uninhibited by ouabain (4,18,39), but inhibited by the sulfhydryl reagents which decrease induced platelet aggregation (18). Proteases degrade myosin molecules to two products, the heavy meromyosin globular head with ATPase activity capable of reacting with actin, and a rod-like fibrous protein devoid of ATPase activity. There is clear EM evidence that platelet actin is located in the cytoplasm beneath the PM but the localization of myosin, either totally in the cytoplasm or partially on the outside of the PM, is unclear (25,38). The reader is referred to ref. 38 since a detailed discussion of the voluminous literature on platelets is beyond the scope of this chapter. It will suffice to say that earlier immunological techniques used to locate the ecto-ATPase and contractile proteins have been criticized because the antisera may have contained contaminant antibodies (13,38).

The same problems in the preparation of pure, specific antibodies are encountered using cultured cells, as discussed by Willingham et al. (41). These authors (41) prepared antisera to myosin from mouse L929 cells, mouse uterus, and rabbit skeletal muscle. The L cells reacted by agglutination only to the antiserum against L cell myosin. Immunofluorescence was also highly specific against this antisera and seemed to locate the myosin in cultured nonmuscle cells on the PM surfaces. A specific myosin antibody is needed that will inhibit both the ecto-ATPase of the intact cell and cell motility, or any other function mediated by the contractile proteins. Pertinent to this suggestion is the observation (13) that antibodies to smooth muscle myosin from chicken gizzard, known to inhibit aggregation of embryonic cells, reduced the (Ca^{2+},Mg^{2+})-ATPase of the isolated PM of mouse liver cells which contain contractile proteins; antibodies to skeletal muscle myosin had no effect. Mersalyl (0.5 mM), which inhibits actomyosin ATPase, likewise arrested the ATPase of the isolated mouse liver PM (13).

Thus, although there is widespread, if indirect, evidence that the ecto-(Ca^{2+},Mg^{2+})ATPase is associated with the myosin-like protein of the PM, there is no proof that the ATPase of the heavy meromyosin head is situated so that the active site faces outward. The speculation is intriguing but definitive experiments have not yet been done.

ACKNOWLEDGMENTS

Grateful acknowledgment is expressed to the Medical Research Council of Canada and to the Muscular Dystrophy Association of Canada for the support of our research program.

REFERENCES

1. Abney, E. R., Evans, W. H., and Parkhouse, R. M. E. (1976): *Biochem. J.*, 159:293–299.
2. Boegman, R. J., Manery, J. F., and Pinteric, L. (1970): *Biochim. Biophys. Acta*, 203:506–530.

3. Carraway, K. L., Fogle, D. D., Chesnut, R. W., Huggins, J. W., and Carraway, C. A. C. (1976): *J. Biol. Chem.,* 251:6173–6178.
4. Chambers, D. A., Salzman, E. W., and Neri, L. L. (1967): *Arch. Biochem.,* 119:173–178.
5. Coetzee, G. A., and Gevers, W. (1977): *Biochem. J.,* 164:645–652.
6. Cooper, P. H., and Stanworth, D. R. (1976): *Biochem. J.,* 156:691–700.
7. Cummins, J., and Hydén, H. (1962): *Biochim. Biophys. Acta,* 60:271–283.
8. DePierre, J. W., and Karnovsky, M. L. (1974): *J. Biol. Chem.,* 249:7111–7120.
9. DePierre, J. W., and Karnovsky, M. L. (1974): *J. Biol. Chem.,* 249:7121–7129.
10. DePierre, J. W., and Karnovsky, M. L. (1976): In: *In Vitro Methods Cell-Mediated Tumor Immunology,* edited by B. R. Bloom and J. R. David, pp. 353–358. Academic Press, New York.
11. Dunkley, C. R., Manery, J. F., and Dryden, E. E. (1966): *J. Cell. Physiol.,* 68:241–248.
12. Forrester, T. (1969): *J. Physiol. (Lond.),* 200:53P–54P.
13. Garnett, H. M., Kemp, R. B., and Gröschel-Stewart, U. (1976): *Arch. Biochem. Biophys.,* 172:419–424.
14. Headon, D. R., Keating, H., and Barrett, E. J. (1974): In: *Methodological Developments in Biochemistry,* Vol. 4, edited by E. Reid, pp. 279–291. Longmans, London.
15. Kayama, Y. (1977): *Jikeikai Med. J.,* 24:101–111.
16. Manery, J. F., Riordan, J. R., and Dryden, E. E. (1968): *Can. J. Physiol. Pharmacol.,* 46:537–547.
17. Marsh, J. B., and Haugaard, N. (1957): *Biochim. Biophys. Acta,* 23:204–205.
18. Mason, R. G., and Saba, S. R. (1969): *Am. J. Pathol.,* 55:215–223.
19. Medzihradsky, F., Bole, G. G., Jr., Nork, L. M., and Lin, H. (1976): *Res. Commun. Chem. Pathol. Pharmacol.,* 14:277–285.
20. Medzihradsky, F., Lin, H., and Marks, M. J. (1975): *Life Sci.,* 16:1417–1428.
21. Moore, R. B., Manery, J. F., and Dryden, E. E. (1978): *Prog. Clin. Biol. Res.,* 20:51–67. Alan R. Liss, Inc., New York.
22. Nayler, W. G. (1977): In: *Mammalian Cell Membranes,* Vol. 3, edited by G. A. Jamieson and D. M. Robinson, pp. 147–171. Butterworths, London.
23. Newby, A. C., Luzio, J. P., and Hales, C. N. (1975): *Biochem. J.,* 146:625–633.
24. Niday, E., Wang, C. S., and Alanpovic, P. (1977): *Biochim. Biophys. Acta,* 469:180–193.
25. Pollard, T. D., Fujiwara, K., Handin, R., and Weiss, G. (1977): *Ann. N. Y. Acad. Sci.,* 283:218–239.
26. Pommier, G., Ripert, G., Azoulay, E., and Depieds, R. (1975): *Biochim. Biophys. Acta,* 389:483–494.
27. Riordan, J. R., Manery, J. F., Dryden, E. E., and Still, J. S. (1972): *Can. J. Physiol. Pharmacol.,* 50:432–444.
28. Rosenblatt, D. E., Lauter, C. J., and Trams, E. G. (1976): *J. Neurochem.,* 27:1299–1304.
29. Rothstein, A., and Meier, R. (1948): *J. Cell. Comp. Physiol.,* 32:77–95.
30. Sakai, R., Sato, T., Kayama, Y., and Sakai, T. (1973): *Jikeikai Med. J.,* 20:143–158.
31. Schrier, S. L. (1977): *Blood,* 50:227–237.
32. Sells, B. H., Six, N., and Brachet, J. (1961): *Exp. Cell Res.,* 22:246–256.
33. Severson, D. L., Drummond, G. I., and Sulakhe, P. V. (1972): *J. Biol. Chem.,* 247:2949–2958.
34. Stefanović, V., Ciesielski-Treska, J., and Mandel, P. (1977): *Brain Res.,* 122:313–323.
35. Stefanović, V. Mandel, P., and Rosenberg, A. (1976): *J. Biol. Chem.,* 251:3900–3905.
36. Trams, E. G., and Lauter, C. J. (1974): *Biochim. Biophys. Acta,* 345:180–197.
37. Trams, E. G., Lauter, C. J., and Banfield, W. G. (1976): *J. Neurochem.,* 27:1035–1042.
38. Vroman, L., and Leonard, E. F. (editors) (1977): *Ann. N. Y. Acad. Sci.,* 283:1–560.
39. Wang, T. Y., Hussey, C. V., Sasse, E. A., and Hause, L. L. (1977): *Am. J. Clin. Pathol.,* 67:528–532.
40. Weiss, B., and Sachs, L. (1977): *J. Cell. Physiol.,* 93:183–188.
41. Willingham, M. C., Ostlund, R. E., and Pastan, I. (1974): *Proc. Natl. Acad. Sci. U.S.A.,* 71:4144–4148.
42. Woo, Y-T., and Manery, J. F. (1973): *Arch. Biochem. Biophys.,* 154:510–519.
43. Woo, Y-T., and Manery, J. F. (1975): *Biochim. Biophys. Acta,* 397:144–152.

Electrophysiological Studies

Physiological and Regulatory Functions of Adenosine and Adenine Nucleotides, edited by H. P. Baer and G. I. Drummond. Raven Press, New York © 1979.

Role of Adenosine and Adenine Nucleotides in Central Nervous Function

J. W. Phillis, G. K. Kostopoulos, J. P. Edstrom, and S. W. Ellis

Department of Physiology, College of Medicine, University of Saskatchewan, Saskatoon, Saskatchewan, S7N OWO, Canada

Adenosine derivatives have been proposed as transmitters at synaptic junctions in both peripheral and central nervous (CNS) systems. The possibility that adenosine triphosphate (ATP) or a related purine may be the transmitter at various peripheral autonomic junctions has been reviewed by Burnstock (3), and more than 20 years have elapsed since Holton and Holton (11) suggested that 5′-ATP might be the transmitter for dorsal root afferent fibers.

There have now been a number of studies showing that adenosine and its nucleotides are released from nervous tissues during activity (2,12,16,24, 25,28,31,34,36,37). Furthermore, evidence has accumulated indicating that such released adenosine could have a presynaptic inhibitory action on the release of other transmitters such as acetylcholine (ACh) and noradrenaline (8,9,35) from peripheral and central nerve terminals. Although Feldberg and Sherwood (6) described sedative actions of intraventricularly administered adenosine and 5′-ATP in cats, potential interest in possible inhibitory actions of iontophoretically applied noncyclic adenine nucleotides on central synaptic transmission was discouraged by preliminary reports that 5′-ATP was without action on neurons in the spinal cord (4) and had excitatory actions on cells in the brainstem and on cerebellar Purkinje cells (7,29).

The discovery in 1974 that adenosine and various adenine nucleotides were potent depressants of the firing rate of rat cerebral cortical neurons (23) initiated a series of iontophoretic studies into the actions of purine and pyrimidine nucleotides on central neurons in various regions of the brain of rats anesthetized with nitrous oxide and methoxyflurane. These studies were subsequently extended to include agents that purportedly inhibit the enzymatic degradation of adenosine, block the uptake of adenosine, or block the adenosine receptor. This chapter presents a summary of the major findings of these studies.

ACTIONS OF ADENOSINE AND ADENINE NUCLEOTIDES ON CEREBRAL CORTICAL NEURONS

Depression

The most pronounced action of the adenine derivatives on deep (800 to 1,400 μM) spontaneously firing cortical neurons was to decrease the rate of spontaneous discharge. This depression had a rapid onset, occurring with a latency of only a few seconds, and a duration dependent on both the magnitude of the depression and the duration of application of adenosine or the adenine nucleotide. Adenosine and/or adenosine 5′-monophosphate (5′-AMP) were tested on several hundred deep spontaneously firing neurons and in almost every instance depressed the characteristic bursting discharge of these neurons. Many of the neurons were identifiable by antidromic activation as corticospinal neurons and the action of the adenine derivatives on these was similar to that on unidentified deep cortical neurons. The depressant activity of 5′-AMP was compared with that of the potent inhibitory monocarboxylic amino acid, γ-amino butyric acid (GABA). When tested on 21 unidentified spontaneously active neurons, for similar ejection times, applications of 5′-AMP by currents of 15 nA were found to be as effective as currents of 10 nA passed through the GABA barrel. On some neurons, removal of the "holding" current on the 5′-AMP barrel allowed sufficient nucleotide to diffuse out of the electrode to depress spontaneous firing.

The effects of adenosine and its nucleotides on various types of chemically evoked firing were less clear-cut. Many of the deep spontaneously active cortical neurons are excited by ACh. A consistent finding was that applications of the adenine derivatives that were able to depress spontaneous firing had little effect on the ACh-evoked increase in firing rate. Larger applications of the adenine derivative usually depressed the ACh-evoked activity (Fig. 1). Applications of L-glutamate were frequently used to excite more superficial, quiescent, cortical neurons. This excitation was found to be somewhat resistant to depression by adenosine or adenine nucleotides.

The adenine nucleoside and nucleotides were also found to be rather ineffective at

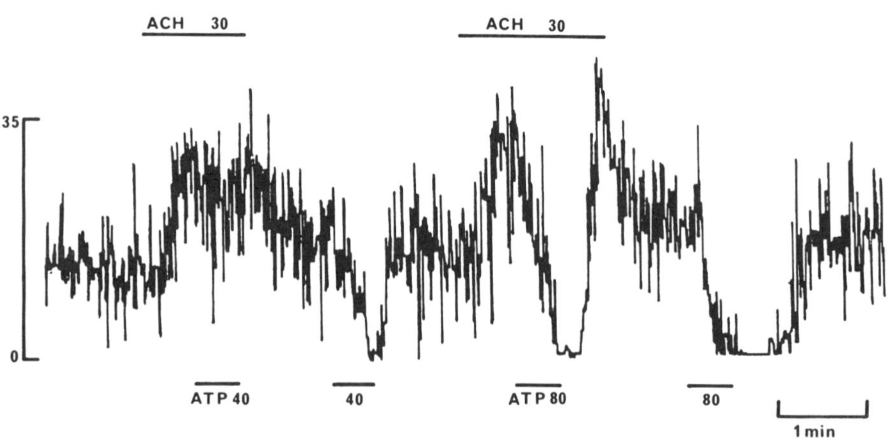

FIG. 1. Rate meter recording of neuronal firing. Acetylcholine (ACh) applied twice by currents of 30 nA excited this deep spontaneously active cerebral cortical neuron. Adenosine 5′-triphosphate (40 nA) depressed the spontaneous discharge but had little effect on the ACh-evoked firing. Adenosine 5′-triphosphate (80 nA) depressed both ACh-evoked and spontaneous firing.

depressing the invasion of the soma of corticospinal neurons by antidromically propagated action potentials. In a few instances, the passage of large currents (100 to 200 nA) through the adenosine or 5′-AMP barrel resulted in a failure of spike invasion. This effect was difficult to reproduce consistently, even on the same neuron.

The depressant actions of 5′-AMP were studied by intracellular recording from nine cortical neurons. The most striking effects were a hyperpolarization, accompanied by a decrease in spontaneously occurring and evoked (by stimulation of the adjacent cortical surface) synaptic potentials (Fig. 2) (5). These effects occurred in the absence of any significant change in the cell's input resistance, measured with de- and hyperpolarizing pulses. In eight of the nine cells, direct intracellular stimulation could produce action potentials before, during, and after the application of 5′-AMP. In three of these it was possible to accurately measure the threshold for action potential generation during this sequence. There was no significant change in threshold during 5′-AMP application, showing that 5′-AMP probably does not inhibit cell firing by altering the excitability of its membrane.

The mechanism of adenosine and adenine nucleotide depression of neuronal firing is still by no means clear but the findings obtained with both intra- and extracellular recording suggest that their principal action may be some form of interference with synaptic transmission. An inhibition of the presynaptic excitation-transmitter release coupling appears to be the most logical explanation and would be consistent with observations made at peripheral cholinergic and adrenergic nerve terminals. Other possibilities include a block of postsynaptic receptors and/or inhibition of the coupling between the receptors and the conductance mechanisms linked to them. In smooth muscle tissues there is evidence that adenosine may have both pre- and postsynaptic actions in the same tissue, and such a possibility cannot presently be excluded in the CNS.

Lekić (18) has recently presented evidence favoring a presynaptic action of 5′-AMP at motor axon collateral–Renshaw cell synapses in the spinal cord. Adenosine 5′-monophosphate depressed synaptic activation of Renshaw cells by afferent impulses in the cholinergic motor axon collaterals but did not affect their responses to iontophoretically applied ACh.

Excitation

Adenosine 5′-triphosphate-evoked excitations of central neurons have previously been described by Galindo et al. (7) and Siggins et al. (29) and were observed in our studies on cerebral cortical neurons. Such effects were most frequently demonstrable on quiescent or slowly firing neurons, and instances were observed in which 5′-ATP had both excitant and depressant effects on the same neuron. Adenosine 5′-triphosphate-evoked excitation usually occurred after a very brief latency, and firing ceased immediately once the application ceased. On many spontaneously firing neurons it was possible to demonstrate that 5′-ATP had an initial brief excitant action followed by a phase of depression. This apparent duality

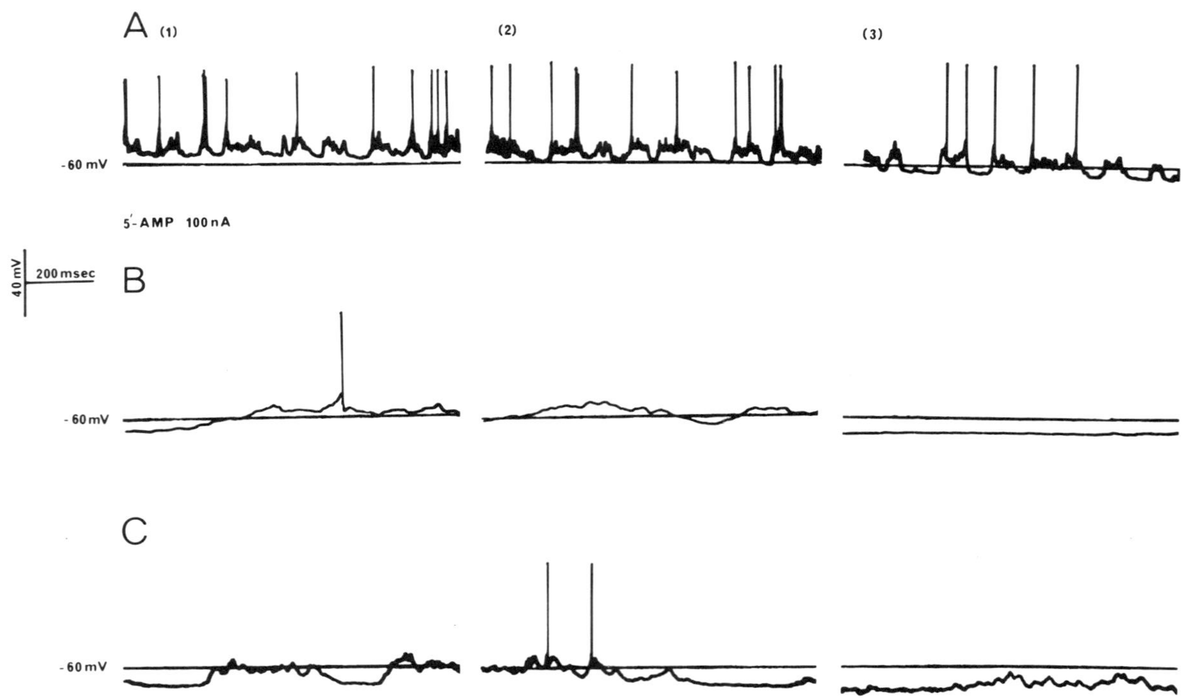

FIG. 2. Intracellular records from a corticospinal neuron. The sets of three records in **A, B,** and **C** were recorded before, during, and after, respectively, the application of 5′-AMP (100 nA) from an extracellular electrode barrel attached to the intracellular electrode (tip separation 50 μ). The calibration line on each recording indicates a membrane potential level of −60 mV. During the application of 5′-AMP the neuron was hyperpolarized and the spontaneously occurring excitatory postsynaptic potentials disappeared. Recovery began to occur after about 1 min when the 5′-AMP application was terminated and was complete 5 min later.

of action of 5′-ATP frequently resulted in its appearing to have a slower onset and weaker depressant action than 5′-AMP and adenosine. The neuron illustrated in Fig. 3 was unusual in that over a period of time its rate of spontaneous firing altered quite markedly. During the period of time when it was nearly quiescent, 5′-ATP had a clear-cut excitatory action on this neuron, but when it was spontaneously active, similar applications of the nucleotide had a depressant action.

Comparable excitant actions were observed with adenosine 5′-tetraphosphate. A reduction in the length of the 5′-phosphate chain reduced excitant activity. Thus adenosine 5′-diphosphate (5′-ADP) had a weaker excitant action, and such effects were only infrequently observed with 5′-AMP and cyclic 3′,5′-AMP. It should be stressed that since most of our studies used deep spontaneously active cortical cells, the likelihood of observing any excitant effects of the adenine nucleotides was minimized.

Because adenine nucleotides chelate divalent metalic cations, it has been suggested that removal of calcium from the extracellular space could account for the excitant actions of 5′-ATP (7,29). In an attempt to test this possibility, Mg^{2+} or Ca^{2+} ions were ejected simultaneously with 5′-ATP from another barrel of the multiple barrel micropipette. Application of either cation antagonized the excitant actions of 5′-ATP. One experiment of this type is illustrated in Fig. 4 in which it is apparent that applications of Mg^{2+} prevented the excitations evoked by brief pulses of 5′-ATP but not those evoked by L-glutamate. Mg^{2+} was preferred in these experiments because it has less of a direct depressant effect on neuronal excitability than does Ca^{2+}. Adenosine 5′-triphosphate has a higher affinity for Mg^{2+} than Ca^{2+} and would, therefore, saturate at a lower Mg^{2+} concentration (1). Results of this type suggested that the excitation evoked by iontophoretically applied 5′-ATP could be a result of Ca^{2+} chelation. However, on a different preparation, the isolated

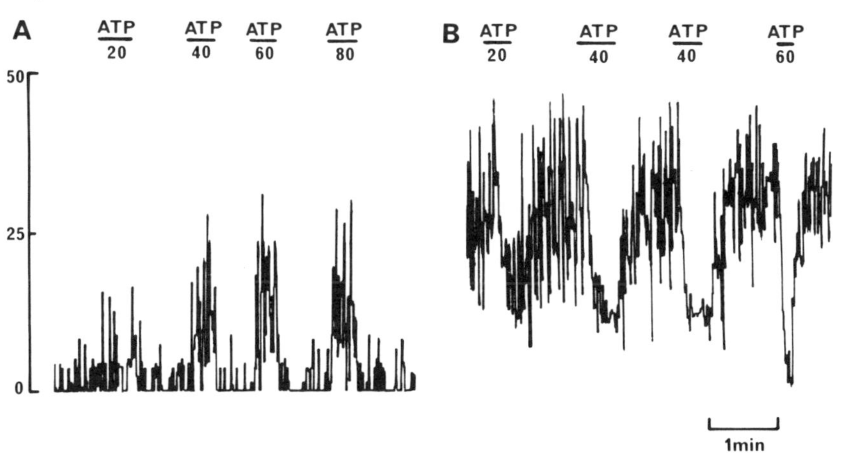

FIG. 3. This cerebral cortical neuron was excited by 5′-ATP at a time when spontaneous activity was low **(A)** and was subsequently depressed by 5′-ATP when its firing rate had increased **(B).**

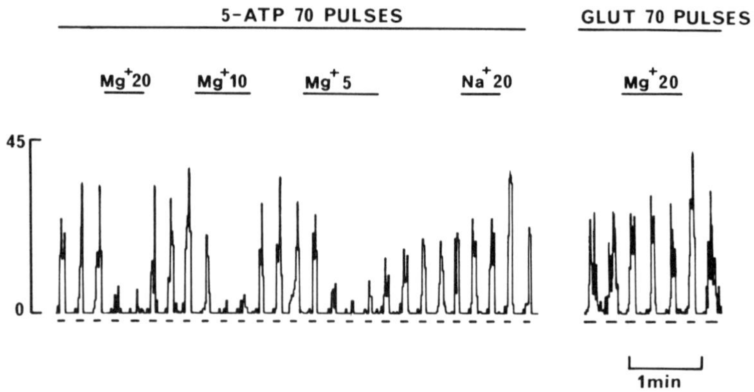

FIG. 4. This cortical neuron was excited by pulses of 5′-ATP (70 nA) or L-glutamate (70 nA). The periods of application of the excitants are indicated by the black lines under the rate meter recording. Mg^{2+} (20, 10, and 5 nA) applications greatly reduced the excitant effects of 5′-ATP but did not alter those of L-glutamate. An application of Na^{+} (20 nA) had no effect on the 5′-ATP–evoked excitations.

hemisected toad spinal cord, we have obtained evidence that 5′-ATP can have an excitant action unrelated to Ca^{2+} chelation. Perfusion of this preparation with a Ca^{2+}-5′-ATP–containing solution was observed to cause excitant effects comparable to perfusion with a Na^{+}-5′-ATP solution (see section below titled Actions of Adenine Derivatives on Other Regions of the CNS).

STRUCTURE—ACTIVITY RELATIONSHIPS

Depressant Activity

An extensive range of purine and pyrimidine nucleotides has been tested by iontophoretic application onto identified corticospinal and other deep spontaneously firing neurons in the rat sensory motor cerebral cortex. Various adenine nucleotides and adenosine were observed to have a pronounced depressant action on the spontaneous firing of these neurons. A weak depressant action was noted when guanine nucleotides were tested, and little or no effect was observed with the cytidine, inosine, thymidine, uridine, and xanthine nucleotides.

A more detailed list of the adenine nucleotides tested on rat cerebral cortical neurons is presented in Table 1. The 2′-, 3′- and 5′-AMPs, 5′-ADP, 5′-ATP, adenosine 2′, 3′-cyclic monophosphate, adenosine 5′-imidodiphosphate, adenosine 2′, 5′-diphosphate, and adenosine 3′,5′-diphosphate were all highly potent depressants comparable with adenosine in activity. Substitution of halogen atoms in position 2 of the purine ring of adenosine gave analogs that were either significantly more potent than adenosine (2-chloro- and 2-fluoroadenosine) or that, in the case of 2-bromoadenosine, had a very prolonged action. 2-Aminoadenosine, 2-hydroxyadenosine, and 2-*p*-methoxyphenyladenosine were also relatively powerful depres-

TABLE 1. *Structure-activity relationships for adenosine analogs*

Substance	Depressant activity on cerebral cortical neurons
Adenine	*[a]
Adenosine	****
Inosine	**
Purine derivatives of adenosine	
2-Chloroadenosine	*****
2-Fluoroadenosine	*****
2-Bromoadenosine	**
2-Aminoadenosine	***
2-Hydroxyadenosine	***
2-*p*-Methoxyphenyladenosine	***
N^6-Phenyladenosine	**
8-Bromoadenosine	0
Ribose derivatives of adenosine	
2′-Deoxyadenosine	**
3′-Deoxyadenosine	**
5′-Deoxyadenosine	***
Adenine xylofuranoside	*
Adenosine N^1-oxide	*
Adenosine 5′-sulfate	0
Adenosine 5′-mononicotinate	***
7-Deazaadenosine (tubercidin)	**
Mononucleotides and dinucleotides	
Adenosine 5′-monophosphate	****
Adenosine 5′-diphosphate	****
Adenosine 5′-triphosphate	****
Adenosine 5′-tetraphosphate	****
Adenosine 2′-monophosphate	****
Adenosine 2′,5′-diphosphate	****
Adenosine 3′-monophosphate	****
Adenosine 3′,5′-diphosphate	****
Adenosine 3′,5′-cyclic monophosphate	**
Adenosine 2′,3′-cyclic monophosphate	****
Adenosine 5′-imidodiphosphate	****
Homoadenosine 6′-phosphonic acid	0
α,β-Methylene 5′-ATP	*
β,γ-Methylene 5′-ATP	***
α,β-Methylene 5′-ADP	*
2′-Deoxyadenosine 5′-monophosphate	**
2′-Deoxyadenosine 3′-monophosphate	**
Nicotinamide adenine dinucleotide phosphate	***
Coenzyme A	***

[a] 0=no depressant activity; 1–5 asterisks, increasingly potent depressive effects.

sants. With bromine in the 8 position there was a complete loss of activity. Cyclic 3′,5′-AMP had considerably less depressant activity than the other adenosine phosphates. Adenosine 5′-sulfate and purine riboside were devoid of activity.

The structure-activity relationships described above and in Table 1 indicate that depressant activity is dependent on the presence of a 6-amino group on the adenine moiety and a ribose group. Substitution at the 2 position may enhance activity, but at the 8 position (8-bromoadenosine) it is detrimental to activity. Substitution of a phosphate chain at the 5′ position is acceptable but not of a sulfate group. Unsubstituted 2′- and 3′-hydroxy groups are also necessary for the manifestation of full agonist potency.

Examples of the depressant actions of several adenosine analogs are presented in Figs. 5 and 6. Nicotinamide adenine dinucleotide phosphate (NADP) (20 and 10 nA) caused a pronounced depression of the spontaneous firing of the neuron illustrated in Fig. 5. The 2-*p*-methoxyphenyladenosine derivative (CV-1674) (30 and 20 nA) also depressed firing. The records in Fig. 6 were obtained by summing the firing frequencies recorded before, during, and immediately following three successive applications of each compound using Ortec signal averager and histogram modules. The 2′,5′- and 3′,5′-adenosine diphosphates were both active agonists, and, when applied with the same current (26 nA) as 5′-AMP, caused comparable amounts of depression. Coenzyme A and adenosine 3′,5′-diphosphate, a compound likely to result from the breakdown of coenzyme A, had longer durations of action than 5′-AMP and adenosine 2′,5′-diphosphate. Coenzyme A required a larger application (40 nA) current to elicit a clear depression. 2′-Deoxyadenosine 3′-monophosphate, a naturally occurring inhibitor of adenylate cyclase in mammalian cells (26), also depressed the firing of this neuron, indicating that it has some agonist activity. This substance was then applied for 16 min during the latter 4 min of which the three summed responses to 5′-AMP shown in Fig. 6

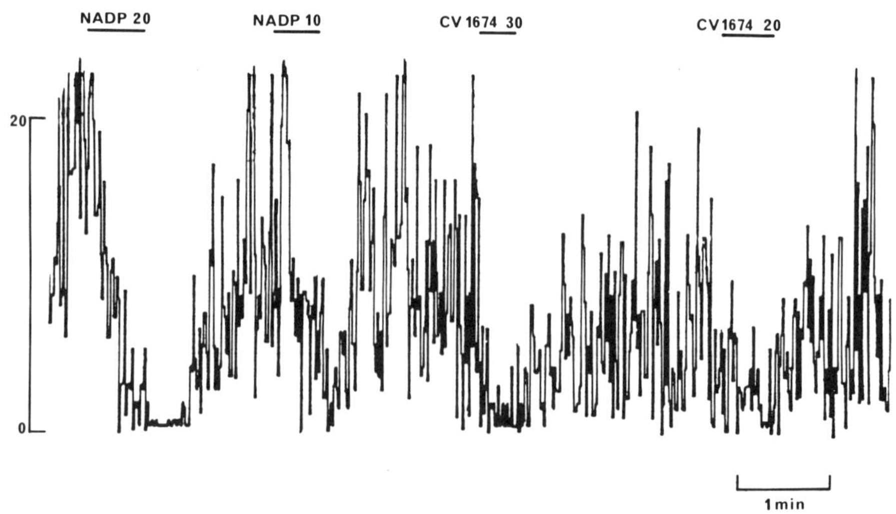

FIG. 5. Rate meter record of a cerebral cortical neuron showing depressant actions of nicotinamide adenine dinucleotide phosphate (NADP) (20 and 10 nA) and 2-*p*-methoxyphenyladenosine (CV-1674) (30 and 20 nA) on spontaneous firing.

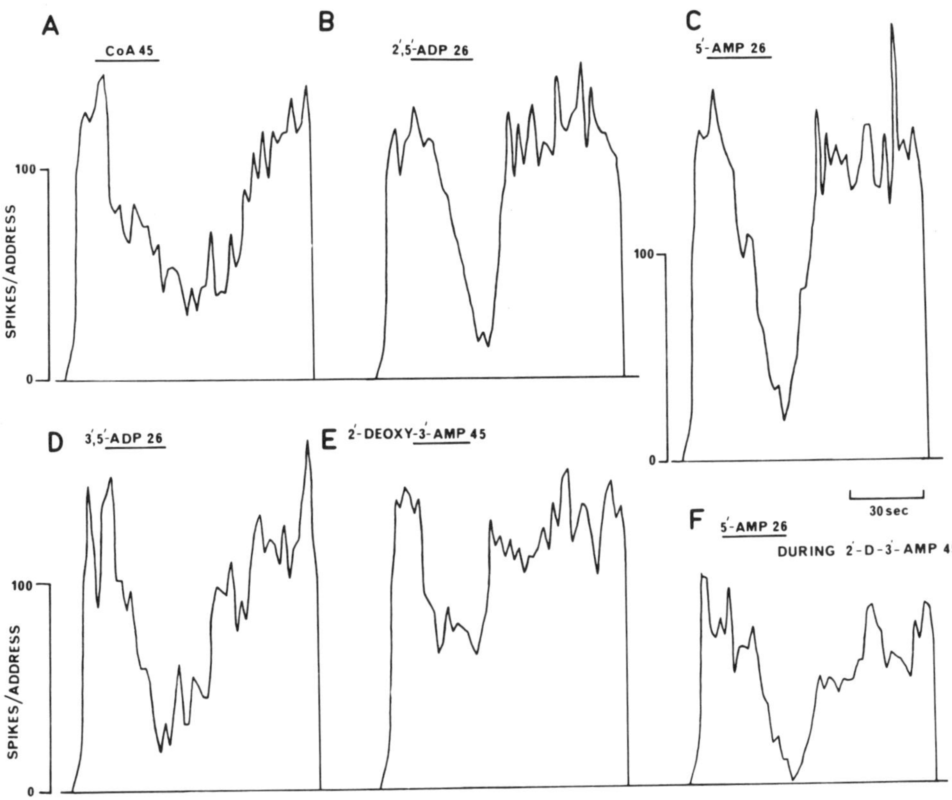

FIG. 6. Summed records of the firing of a cerebral cortical neuron during three successive applications of various adenine derivatives. Coenzyme A (CoA) (45 nA), 2′,5′-adenosine diphosphate (2′,5′-ADP) (26 nA), 3′-5′-adenosine diphosphate (3′,5′-ADP) (26 nA), 5′-AMP (26 nA), and 2′-deoxyadenosine 3′-monophosphate (45 nA) depressed spontaneous firing. The record in **F** was obtained during a continuous application of 2′-deoxyadenosine 3′-monophosphate (45 nA). The first of the three 5′-AMP applications started 11 min after the onset of the 2′-deoxyadenosine 3′-monophosphate application.

were recorded. It is apparent that the 2′-deoxyadenosine 3′-monophosphate application did not affect the action of 5′-AMP.

An issue of considerable controversy has been whether prior degradation of the adenine nucleotides to adenosine is necessary before they can activate the receptor, or if the receptor will accept a variety of differing polyphosphate side chains on the 5′ position. This problem has been investigated with methylene isosteres of 5′-AMP, -ADP, and -ATP. The phosphorous methylene bond is considered to be stable to metabolic transformations involving either hydrolysis or phosphate transfer (38). The α,β isosteres of ADP and ATP were found to be very weak depressants of cerebral cortical neurons whereas the β,γ analog of ATP possessed quite pronounced depressant activity, as did 5′-adenylimidodiphosphate (AMP-PNP). This finding suggests that even though the β,γ bond in the ATP analog is stable, cleavage could still occur at the α,β position yielding the easily metabolized compound, 5′-AMP. However, when the bond in the α,β position is stabilized the compounds were obviously more resistant to hydrolysis to adenosine. The methylene analog of 5′-AMP, homoadenosine 6′-phosphonic acid, had no depressant activity on cerebral cortical neurons, suggesting an absolute requirement for dephosphorylation to adenosine for expression of depressant activity. Similar findings have been made on brain slices, in which the ability of adenosine and ATP to elicit cyclic AMP formation is only weakly reproduced by the α,β- and β,-γ-methylene isosteres of ATP (19). Some caution must be exercised, however, in accepting these conclusions regarding the activities of the methylene isosteres. The space-filling characteristics of the modified polyphosphate chains will differ from those of the mono-, di- and triphosphates, and hence some steric distortion will be present (38). The different steric and charge distribution characteristics of the modified phosphate chains could directly influence binding of these derivatives to receptor sites and thus affect their agonist potency. Further evidence that adenine nucleotides act on adenosine receptors after their dephosphorylation to adenosine has been obtained by Mah and Daly (19) who have shown that the stimulant action of exogenous 5′-ATP and cyclic 3′,5′-AMP on endogenous cyclic AMP formation in brain slices is absent in the presence of added adenosine deaminase. Adenosine 5′-triphosphate and cyclic AMP are not substrates for this enzyme, so their ability to stimulate cyclic AMP formation must depend on hydrolysis to adenosine prior to receptor activation.

Of the other purine and pyrimidine nucleosides tested, the purines guanosine 5′-monophosphate (5′-GMP), guanosine 5′-diphosphate (5′-GDP), guanosine 5′-triphosphate (5′-GTP), and cyclic 3′,5′-GMP were observed to have depressant actions on the spontaneous firing of many neurons, and inosine, inosine 5′-monophosphate (5′-IMP), and inosine 5′-triphosphate (5′-ITP) had a weak depressant action on some neurons. Xanthosine and its nucleotides and the pyrimidine nucleosides and nucleotides were without effect on most cells.

Excitant Actions

The marked spontaneous activity of the majority of cortical neurons on which the various compounds were tested precluded a precise study of the excitant actions of the purine and pyrimidine nucleotides (see section above titled Excitation). It is clear, however, that the triphosphates of all the purines and pyrimidines tested can have an excitant action on cerebral cortical neurons. This effect usually had a short latency of onset and declined rapidly once the application had ceased. Diphosphates had less pronounced excitant effects and monophosphates (including cyclic 3′,5′-AMP and cyclic 3′,5′-GMP) had weak excitant actions on some neurons.

Cyclic 3′,5′-guanosine monophosphate is of some interest in that it has been proposed as the intracellular second messenger for the muscarinic actions of ACh. Both excitant and depressant actions of extracellularly applied cyclic GMP have been reported in previous investigations on cerebral cortical neurons and cerebellar Purkinje cells (10,20,32), and it is likely that the ratio of depressant actions to excitant actions in any particular series of experiments will be determined by such technical factors as the choice of anesthetic and the level of spontaneous activity of the neurons tested.

POTENTIATION OF ADENOSINE AND 5′-AMP

Inactivation or removal of extracellular adenosine as factors in the termination of adenosine-induced depression of cell firing has been studied using a series of compounds that either inhibit the enzyme adenosine deaminase or inhibit uptake of adenosine from the extracellular space.

The role of adenosine inactivation by deamination to the relatively inert metabolite inosine has been assessed with two potent inhibitors of adenosine deaminase, erythro-9-(2-hydroxy-3-nonyl) adenine and deoxycoformycin. Erythro-9-(2-hydroxy-3-nonyl) adenine potentiated the depressant actions of adenosine and 5′-AMP and frequently caused a long-lasting reduction in the rate of firing of cortical neurons. Recovery in such instances frequently took up to 1 hr. With smaller doses of the inhibitor it was possible to demonstrate a potentiation of the action of 5′-AMP or adenosine in the absence of any alteration in the basal firing frequency of the neuron. Similar effects occurred with deoxycoformycin which potentiated the actions of 5′-AMP and 5′-ATP and frequently caused a prolonged reduction in the spontaneous discharge frequency.

Uptake of adenosine into guinea pig cerebral cortical slices is inhibited by a series of compounds including hexobendine, papaverine, dipyridamole, diazepam, and 2-hydroxy-5-nitrobenzylthioguanosine (13,19). When these agents were applied iontophoretically onto cerebral cortical neurons they potentiated the depressant effects of adenosine and 5′-AMP, suggesting a similar block of uptake in the intact cortex and that its uptake plays an important role in terminating the action of adenosine.

All five uptake inhibitors had a depressant action on cortical neurons if applied with sufficiently large currents. When they were used with low nondepressing currents the depressant actions of adenosine and 5′-AMP were potentiated. Dipyridamole was administered by intravenous injection as well as by iontophoretic ejection and at doses of 0.1 mg/kg caused a clear-cut potentiation of 5′-AMP–evoked depressions. Intravenous injection of 1 mg/kg of dipyridamole resulted in a reduction in the rate of firing of cortical neurons and some further potentiation of the effect of 5′-AMP. Diazepam was also administered by intravenous injection and at a dose of 0.1 mg/kg potentiated the actions of 5′-AMP. Higher doses of diazepam reduced the rate of firing of neurons and further potentiated 5′-AMP.

Lidoflazine reportedly potentiates the actions of adenosine on the heart by inhibiting its uptake. When applied iontophoretically onto cortical neurons, it potentiated the depressant effects of 5′-AMP and, with larger ejection currents reduced the rate of spontaneous firing.

The importance of uptake and deamination to inosine as mechanisms for terminating the actions of extracellularly released adenosine and the adenine nucleotides are clearly demonstrated by these results. It is tempting to suggest that the depression of spontaneous firing of cortical neurons evoked by all of these compounds may have been a result of an accumulation of endogenous adenosine, whether released directly as adenosine or formed by the dephosphorylation of adenine nucleotides.

ANTAGONISM BY METHYLXANTHINES

The methylxanthines caffeine, theophylline, and 3-isobutyl-l-methylxanthine block the stimulatory effects of adenosine on cyclic 3′,5′-AMP formation in brain slices (19,27).

Attempts to test these agents by iontophoretic application were hindered by their

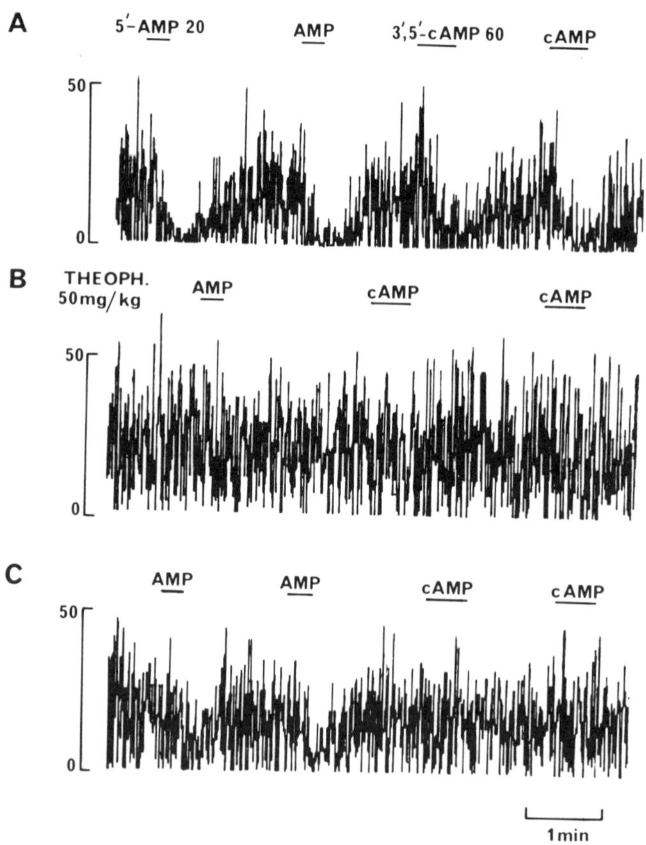

FIG. 7. Theophylline antagonism of 5′-AMP (20 nA) and cyclic 3′,5′-AMP (60 nA) depression of the spontaneous firing of a corticospinal neuron. Theophylline (50 mg/kg) was administered intravenously after the control nucleotide responses had been recorded. Trace **B** starts 7 min after the theophylline injection when both 5′-AMP and 3′,5′-cyclic AMP depressions were abolished. Trace **C** was recorded 75 min after theophylline when the 5′-AMP responses were beginning to recover; cyclic 3′,5′-AMP was still blocked.

sparing solubility in water, although some antagonism of 5′-AMP was observed with iontophoretically applied caffeine. Aminophylline is a more soluble form of theophylline and contains ethylene diamine. Unlike theophylline, aminophylline has a potent depressant action on cortical neurons that is mimicked by ethylene diamine (21) and should not, therefore, be used as a substitute for theophylline.

Theophylline, caffeine, and isobutylmethylxanthine were tested by intravenous administration in doses of 10 to 100 mg/kg. In these doses, they cause an initial transient fall in blood pressure followed by a return to control level or a long-lasting but small increase in pressure. The immediate effect on cortical neurons was often an increase in the rate of spontaneous firing that returned to control firing levels after several minutes. Initially, during this period of enhanced cell firing, the depressant effects of the adenine derivatives sometimes appeared to be potentiated. Within a few minutes, however, the actions of iontophoretically applied adenosine, 5′-AMP, 5′-ATP, or cyclic 3′,5′-AMP were noticeably reduced and maximal blocking activity was attained 8 to 12 min after the administration of the methylxanthine (Fig. 7). The antagonism could invariably be overcome if larger amounts of adenosine or adenine nucleotide were applied. Partial recovery from the antagonism could be observed from 90 min to 3 hr after the administration of the methylxanthine. Theophylline was also tested as an antagonist of 5′-GMP and 5′-GTP–evoked depressions of cortical neurons in 7 rats. In 2 animals some antagonism was evident, but in the others doses of theophylline up to 100 mg/kg failed to reduce the depressant effects of the guanine nucleotides. The possibility of a theophylline-insensitive action of guanine nucleotides thus becomes a definite consideration.

Isobutylmethylxanthine was a less effective antagonist than theophylline and caffeine. Iontophoretically applied isobutylmethylxanthine was often observed to potentiate the depressant effects of ATP. This compound inhibits the uptake of adenosine into brain slices as well as blocking the adenosine receptor (13). The variable results obtained when it is administered intravenously as an antagonist may result from these two conflicting actions.

ACTIONS OF ADENINE DERIVATIVES ON OTHER REGIONS OF THE CNS

Further insight into the phenomenon of purinergic depression of neuronal firing has been sought by examining the responses in other regions of the rat brain to iontophoretically applied adenosine, 5′-AMP, or other adenine nucleotides. Cerebellar Purkinje cells were examined in the dorsal layers of the vermis. These neurons are spontaneously active and their firing was consistently depressed by adenosine, 5′-AMP, 5′-ADP, 5′-ATP, 3′-AMP, and cyclic 2′,-3′-AMP (14). Cyclic 3′,5′-AMP was a much weaker depressant, reducing the firing rate in 37% of the cells tested and exciting a further 37%.

Depressant effects of adenosine and 5′-AMP on neuronal firing have also been observed in the hippocampus, caudate nucleus, thalamus, superior colliculus, and olfactory bulb (15). Biphasic responses to 5′-AMP (but not to adenosine) were observed in the olfactory bulb and caudate nucleus, with an initial excitation followed by an inhibition.

The effects of various purines and pyrimidines have been ascertained on the isolated hemisected toad spinal cord (22). This preparation offers the advantages that drugs can be applied to it in known concentrations and that it is possible to alter the ionic composition of the perfusion solution. Both hyperpolarizing and depolarizing actions of purine and pyrimidine nucleotides were recorded from the dorsal and ventral roots of this preparation. Adenosine and its 5′-mono-, di-, and triphosphates were able to cause a hyperpolarization of dorsal root terminals and motoneurons. This effect was not evident in unstimulated preparations or in those in which dorsal root stimulation elicited weak dorsal and ventral root responses. The threshold concentration for adenosine and 5′-AMP was of the order of 100 μM. Hyperpolarization of both roots by the adenine derivatives occurred without any diminution in the amplitude of the dorsal root stimulation–evoked dorsal and ventral root potentials. There was, however, some reduction in the duration of the evoked potentials. The most likely explanation for these actions of the adenine derivatives is that they suppress activity in those neuronal circuits in the toad spinal cord responsible for the late after-discharge following dorsal root stimulation. They do not affect the depolarization of motoneurons and dorsal root terminals by short latency pathways (either mono- or multisynaptic).

Inosine nucleotides had a weak hyperpolarizing action on dorsal and ventral root fibers suggesting a weak activation of the adenosine receptor. Guanine nucleotides (but not guanosine itself) also hyperpolarized dorsal root terminals and motoneurons in repetitively stimulated preparations. This hyperpolarization was associated with a reduction in both the amplitude and duration of the dorsal root–evoked dorsal and ventral root potentials; and at high concentrations (5 to 10 mM) there was almost total suppression of these synaptically evoked potentials.

Theophylline (100 μM) antagonized the adenine derivative–evoked hyperpolarization, but in most preparations had little or no effect on the guanine nucleotide–evoked responses. Adenine and guanine derivatives may, therefore, act at different membrane sites in the amphibian spinal cord as well as in the cerebral cortex.

The triphosphates and, to a lesser extent, the diphosphates of purines and pyrimidines tested (adenine, inosine, uridine, cytidine, thymidine, xanthosine, 2′-deoxyadenosine) depolarized dorsal and ventral roots of the toad spinal cord. Small depolarizing responses could also be observed when the monophosphates were applied at high (10 μM) concentrations. Uridine 5′-triphosphate was the most active compound in this respect and had an action that lasted for several minutes after it had been washed out of the bath. Adenosine 5′-triphosphate had a somewhat shorter duration of action as did cytidine 5′-triphosphate (5′-CTP) and 5′-ITP. Depolarizing responses to the purine and pyrimidine nucleotides were not antagonized by theophylline. Indeed, the application of theophylline to preparations on which adenine nucleotides were being tested usually resulted in either the conversion of hyperpolarizing into depolarizing responses or in an enhancement of depolarizing responses.

Purine and pyrimidine nucleotides are known to chelate calcium ions and with the high concentrations of nucleotide tested in these experiments, removal of calcium from the perfusion solution could account for the observed excitant actions. It has been possible to eliminate this as an explanation for the action of 5′-ATP, since applications of the dicalcium salt of 5′-ATP had comparable excitant actions to the sodium salt. This result does not, however, eliminate the possibility that the nucleotides may act by chelating some other metal.

When tested on isolated spinal cord preparations in which synaptic transmission had been abolished by perfusion with a Ca^{2+}-free solution containing 10 mM Mg^{2+}, adenosine had no effect on potentials in either dorsal or ventral roots. Adenosine 5′-monophosphate, 5′-ATP, and uridine 5′-triphosphate (5′-UTP) had a depolarizing action on dorsal root terminals in such preparations but did not affect motoneuronal membrane potentials. It is likely, therefore, that these compounds depolarize motoneurons in normal preparations by an action on interneurons in the spinal cord, but have a direct effect on afferent terminals.

CONCLUSIONS

The structure-activity relationships displayed in Table 1 are remarkably consistent with those resulting from biochemical investigations on the stimulation of cyclic AMP formation in cerebral cortical slices by adenosine analogs. Adenosine elicits an accumulation of cyclic AMP in brain slices by stimulation of an extracellular adenosine receptor (13,19,27). Adenosine's action is shared by the adenine nucleotides, which in brain are thought to undergo extracellular hydrolysis to adenosine in order to exert their effects, since the actions of 5′-ATP and cyclic AMP on cyclic AMP formation are almost completely blocked by added adenosine deaminase. The observation that adenosine can elicit an accumulation of cyclic AMP from prelabeled stores of intracellular ATP provides further evidence for the presence of an adenosine receptor, activation of which can enhance cyclic AMP levels. Further evidence for the concept of a receptor comes from the finding that adenosine analogs, such as 2-chloroadenosine, stimulate an accumulation of cyclic AMP but do not serve as precursors for the cyclic nucleotide then formed. An extracellular location of the receptor can be deduced from the results obtained using adenosine uptake inhibitors that potentiate the response to extracellular adenosine while concomittantly inhibiting its uptake into the tissue. This extracellular receptor is competitively blocked by the methylxanthines theophylline, caffeine, and 3-isobutyl-l-methylxanthine.

The similarities between the agonist potencies of various adenosine analogs observed in our experiment and the biochemical studies mentioned above suggest that in both instances a similar adenosine receptor is involved. Further support for a one-receptor hypothesis has been forthcoming from studies with the isolated guinea pig olfactory cortex in which depression of evoked synaptic potentials has been correlated with an enhancement of cyclic AMP levels (17).

An interesting difference between the electrophysiological and biochemical responses to stimulation of the adenosine receptor has been revealed by our finding of agonist actions of 2′- and 3′-deoxyadenosine. In cerebral cortical slices, these substances do not elicit the formation of cyclic AMP but rather act as antagonists toward adenosine (19). 5′-Deoxyadenosine is a partial agonist in cerebral cortical slices. The antagonism of adenosine-evoked formation of cyclic AMP by these compounds is thought to be the result of a direct inhibitory action of these compounds at an inhibitory site on the adenylate cyclase and it is possible that for the 2′- and 3′-deoxyadenosines this inhibitory action precludes any manifestation of a stimulant action on the adenosine receptor. In the present experiments all three compounds as well as 2′-deoxyadenosine 5′-monophosphate, and 2′-deoxyadenosine 3′-monophosphate had depressant actions and did not antagonize adenosine action, suggesting that adenylate cyclase may not be involved in the pharmacological actions of adenosine and its analogs. Alternatively it is possible that there are two adenosine receptors. One mediates cyclic AMP formation and is unresponsive to 2′- and 3′-deoxyadenosine; the other mediates the pharmacological effects of adenosine and is activated by the 2′- and 3′-deoxyadenosines. A further possibility that must be considered is that the 2′ and 3′-deoxyadenosines may be able to penetrate cell membranes more readily in isolated brain slices than in *in vivo* neurons. The inhibition of adenylate cyclase manifested in the isolated slice preparations may not occur in neurons in the intact cortex, allowing these compounds to exert a measurable effect on the adenosine receptor on the external surface of the cell plasma membrane.

The evidence presented in this chapter is consistent with the suggestion that the pharmacological actions of adenosine may result from a depression of transmitter release from presynaptic nerve terminals, possibly as a result of a reduction of Ca^{2+} influx during invasion of the nerve terminal by an action potential. The site(s) of accumulation of cyclic AMP are less well defined although both neuronal and glial elements appear to be involved.

Adenosine and its nucleotides are released from the intact cerebral cortex (34,37) but their actual site of release remains to be elucidated. Adenine nucleotides can be released by isolated brain synaptosomal preparation (16,36) suggesting that their release in the intact brain may originate from presynaptic nerve terminals. Adenosine 5′-triphosphate is packaged with ACh and norepinephrine in cholinergic and sympathetic nerve terminals and release at such sites would accompany the release of the primary transmitter. Release of 5′-ATP as a cotransmitter may also occur from terminals in the CNS, in addition to release from putative purinergic nerves.

A negative feedback role of 5′-ATP has been postulated for skeletal motor nerve terminals (30) and for the adrenergic nerves in blood vessels (33). It is conceivable that purinergic feedback regulation is a common feature at many nerve terminals. Adenosine release from postsynaptic sites may also occur during intense activity as a result of 5′-ATP utilization. Such adenosine could act presynaptically to reduce the rate of release of transmitter.

ACKNOWLEDGMENTS

This research was supported by the Medical Research Council of Canada.

REFERENCES

1. Alberty, R. A. (1968): *J. Biol. Chem.*, 243:1337–1343.
2. Berne, R. M., Rubio, R., and Curnish, R. (1974): *Cir. Res.*, 35:262–272.
3. Burnstock, G. (1972): *Pharmacol. Rev.*, 24:509–581.
4. Curtis, D. R., Phillis, J. W., and Watkins, J. C. (1961): *J. Physiol. (Lond.)*, 158:296–323.
5. Edstrom, J. P., and Phillis, J. W. (1976): *Can. J. Physiol. Pharmacol.*, 54:787–790.
6. Feldberg, W., and Sherwood, S. L. (1954): *J. Physiol. (Lond.)*, 123:148–167.
7. Galindo, A., Krnjević, K., and Schwartz, S. (1967): *J. Physiol. (Lond.)*, 192:359–377.
8. Ginsborg, B. L., and Hirst, G. D. S. (1972): *J. Physiol. (Lond.)*, 224:629–645.
9. Hedqvist, P., and Fredholm, B. B. (1976): *Arch. Pharmacol.*, 293:217–223.
10. Hoffer, B. J., Siggins, G. R., Oliver, A. P., and Bloom, F. E. (1971): *Ann. N. Y. Acad. Sci.*, 185:531–549.
11. Holton, F. A., and Holton, P. (1954): *J. Physiol. (Lond.)*, 126:124–140.
12. Holton, P. (1959): *J. Physiol. (Lond.)*, 145:494–504.
13. Huang, M., and Daly, J. W. (1974): *Life Sci.*, 14:489–503.
14. Kostopoulos, G. K., Limacher, J. J., and Phillis, J. W. (1975): *Brain Res.*, 88:162–165.
15. Kostopoulos, G. K., and Phillis, J. W. (1977): *Exp. Neurol.*, 55:719–724.
16. Kuroda, Y., and McIlwain, H. (1974): *J. Neurochem.*, 22:691–699.
17. Kuroda, Y., Saito, M., and Kobayashi, K. (1976): *Brain Res.*, 109:196–201.
18. Lekić, D. (1977): *Can. J. Physiol. Pharmacol.*, 55:1391–1393.
19. Mah, H. D., and Daly, J. W. (1976): *Pharmacol. Res. Commun.*, 8:65–79.
20. Phillis, J. W. (1974): *Adv. Behav. Biol.*, 10:57–77.
21. Phillis, J. W. (1977): *Can. J. Neurol. Sci.*, 4:151–195.
22. Phillis, J. W., and Kirkpatrick, J. R. (1978): *Gen. Pharmacol.*, 9:239–247.
23. Phillis, J. W., Kostopoulos, G. K., and Limacher, J. J. (1974): *Can. J. Physiol. Pharmacol.*, 52:1226–1229.
24. Pull, I., and McIlwain, H. (1972): *Biochem. J.*, 130:975–981.
25. Pull, I., and McIlwain, H. (1977): *Neurochem. Res.*, 2:203–216.
26. Sahyoun, N., Schmitges, C. J., Siegel, M. I., and Cuatrecasas, P. (1976): *Life Sci.*, 19:1971–1980.
27. Sattin, A., and Rall, T. W. (1970): *Mol. Pharmacol.*, 6:13–23.
28. Schubert, P., Rose, G., Lee, K., Lynch, G., and Kreutzberg, G. W. (1977): *Brain Res.*, 134:347–352.
29. Siggins, G. R., Hoffer, B. J., and Bloom, F. E. (1971): *Brain Res.*, 25:535–553.
30. Silinsky, E. M. (1975): *J. Physiol. (Lond.)*, 247:145–162.
31. Silinsky, E. M., and Hubbard, J. I. (1973): *Nature*, 243:404–405.
32. Stone, T. W., and Taylor, D. A. (1977): *J. Physiol. (Lond.)*, 266:523–543.
33. Su, C. (1978): *J. Pharmacol. Exp. Ther.*, 204:351–361.
34. Sulakhe, P. V., and Phillis, J. W. (1975): *Life Sci.*, 17:551–556.
35. Vizi, E. S., and Knoll, J. (1976): *Neuroscience*, 1:391–398.
36. White, T. D. (1978): *J. Neurochem.*, 30:329–336.
37. Wu, P. H., and Phillis, J. W. (1978): *Neurochem. Res.*, 3:563–571.
38. Yount, R. G. (1975): *Adv. Enzymol.*, 43:1–56.

Physiological and Regulatory Functions of Adenosine and Adenine Nucleotides, edited by H. P. Baer and G. I. Drummond. Raven Press, New York © 1979.

Adenosine and Its Mononucleotides as Regulatory and Adaptive Signals in the Brain

Henry McIlwain

Department of Biochemistry, Institute of Psychiatry, British Postgraduate Medical Federation, University of London, London SE5 8AF, United Kingdom

Adenine and mononucleotides are probably of prebiotic status: adenine is a pentamer of hydrocyanic acid and adenosine and phosphate derivatives are obtainable from it with minimal reagents in aqueous solutions (see ref. 16). Adenine nucleotide residues are present in the simplest known organisms, and in microorganisms they are already functioning as "indicator molecules," a type of bioseme (6,39). In this role they may serve as symbols for carbon- or nitrogen-source starvation, and in organisms generally may represent, intracellularly, a specific state of the internal or external environment, with the ability to transmit signals about that state to the protein-synthesizing apparatus of the cell and thereby to condition genetic expression.

SIGNALING: QUANTITIES AND SITES

One of the proposals of the present account is that such relationships can still be discerned in the behavior of adenosine and of metabolically related nucleotides in the mammalian central nervous system. Indeed, the signaling role of adenine derivatives may have undergone specific development in neural systems, of which signaling is a major function. Little will be said here about the important roles of adenosine triphosphate (ATP) in energy metabolism, except insofar as those roles result in the normal maintenance of a relatively stable and abundant supply of ATP at an intracellular concentration of about 3 μmoles/g tissue, much greater than those of the cerebral metabolites that derive from it. As summarized previously (18–20) and indicated by other contributions to this volume, adenosine, cyclic 3′,5′-adenosine monophosphate (cyclic AMP), and 5′-mononucleotides are neuroactive agents; adenine itself is little, if at all, active in this respect.

Normal cerebral concentrations of adenosine and cyclic AMP are shown by Table 1 to be about 0.1% of those of the ATP concentration quoted. Such source-signal relationships are emphasized as an essential requirement of metabolic regulatory systems (27), in addition to the metabolic mechanisms which rapidly destroy the signal to form inactive products that do not accumulate in the cell and which include phosphodiesterases, 5′-nucleotidases, and adenosine deaminase in neural tissue (31).

TABLE 1. *Cerebral concentrations of adenosine and of cyclic AMP*

Conditions	Content (nmoles/g fresh tissue)	
	Adenosine	Cyclic AMP
1. Rapidly fixed *in situ*	1.9 ± 0.3[a]	2.3 ± 0.2[a]
	2.3 ± 0.5	1.6 ± 0.3
2. Incubated and superfused in oxygenated glucose saline	2.0 ± 0.3	1.2 ± 0.2
3. As in condition 2, above, and medium also containing 100 μM noradrenaline	5.2 ± 0.8	4.8 ± 1.0
4. As in condition 2, above, and medium also containing 100 μM histamine	2.2 ± 0.8	5.0 ± 1.5
5. As in condition 2, above, and tissue electrically stimulated 4 min, 20 Hz	3.0 ± 0.4	5.0 ± 1.1
6. As in condition 5,above, but stimulation for 10 min, 50 Hz	5.0 ± 0.5	13.9 ± 1.8

Data (24) give mean values from three to 10 samples ± SEM and refer to guinea pig neocortex except those noted otherwise. Incubation was for 30 min at 38° C followed by superfusion and the treatments specified.

[a] Neocortex of rat.

Excitation of mammalian neocortical tissues is found to cause the extracellular appearance of increased concentrations of adenosine, cyclic AMP, and of 5′-nucleotides (30,32). When the nucleotides are added extracellularly to the tissues, they are converted to adenosine and the deaminated products inosine and hypoxanthine, of little or no neuroactivity; thus the extracellular appearance of adenosine and its mononucleotides is transitory, which is important to their signaling role.

INTRACELLULAR SITING AND TRANSLOCATION

The characteristic roles for cyclic AMP in higher organisms are, however, intracellular as "second messengers" controlled by extracellular "first messengers" (2,23,36). The formation of cyclic AMP has been found to be highly localized. By isotope labeling the cyclic AMP newly formed from ATP on excitation was found to derive from limited regions that held only some 5% of the tissue ATP content (4,17). Such sites were visualized immunohistologically in the cerebellar cortex and found to comprise the perikarya and dendrites of specific cell types, notably of Purkinje cells (2).

As is fitting in relation to "messenger" compounds, cyclic nucleotides do not remain at their sites of formation (1,25); translocation occurs (Fig. 1) in the periods of up to 30 min during which cyclic AMP and cyclic guanosine monophosphate have been observed to remain at elevated concentrations after their augmentation by stimulation of the brain *in situ* or of isolated cerebral tissues *in vitro* (13,26). Their action is thus an extended one in two senses, namely, in time and to sites other than those of their immediate generation. Indeed, migration of adenine nucleotides throughout the length of axons and dendrites of several cell types of the mammalian

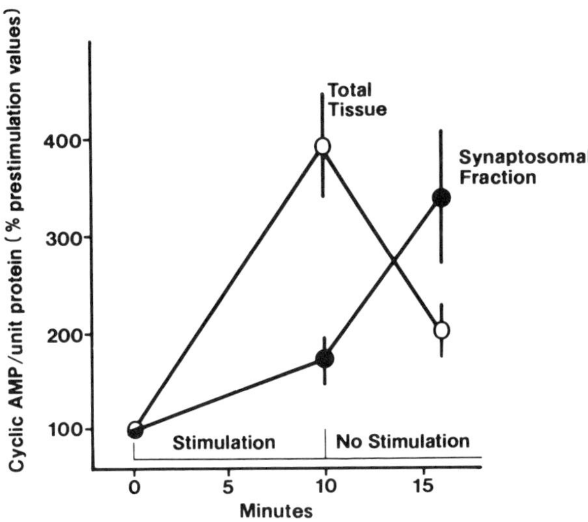

FIG. 1. Translocation of cyclic AMP following its generation by electrical excitation in guinea pig neocortical tissues. Values give mean ± SEM of four to seven samples incubated, superfused, and subsequently separated to subcellular fractions. Nuclear and mitochondrial fractions did not show the poststimulation increase observed in the synaptosomal fraction (25).

brain has been observed. This occurred at speeds of about 1 μm/sec (37,38); the maximum for this category of movement has been found to be about 5μm/sec (12,28).

This translocation may take place by more than one process but involves in part the mechanisms for active cytoplasmic movement which, understandably, are highly developed, in elongated cells such as neurons. Energy-requiring cytoplasmic transport movements are involved in the translocation of many substances and subcellular entities (e.g., proteins and mitochondria) other than the synaptic mediators, but can be recognized as having special significance in relation to these latter compounds which act as intracellular signals. The probable mechanism of active cytoplasmic transport includes the microtubular system (28) which is highly developed in the central nervous system: for example, the characteristic constituent of the microtubules, the protein tubulin, constitutes 10% of the total protein of rat brain (5).

A CONSEQUENCE OF TRANSLOCATION IN DENDRITES

Also highly developed in the central nervous system is cell connectivity: the multiplicity of terminals of a given neuron and multiplicity of sites for reception of terminals from other neurons. The human neocortical ratio of synapses/neuron is about 4×10^4 (3). There is, however, much order in this multiplicity. Over half the synapses in the mammalian cerebral cortex were found to have dendrites as their postsynaptic elements (29), and these dendrites are highly elongated and branched.

Thus there is a strong linear element in the postsynaptic component, and this is to a considerable extent independent of the relative positions of the cells between which the synapses are formed. Indeed, one presynaptic cell may make several postsynaptic contacts [the terminals of several branches of its axon(s)] at quite separate points in the dendrites of a second neuron. Instances in the olfactory bulb derived from Ramon y Cajal's description have been quoted (20,35).

In these and other instances, when the neurotransmitter and neurotransmitter mediators at several synapses of a series are likely to be similar, persistence and translocation of the postsynaptic mediator take on a special significance. It can result in a new type of signaling in which morphological and chemical characteristics of a relatively large unit—a dendritic branch—become expressed in modified cell firing. In specified cerebral systems, data collated (20) from many investigations showed that a single brief period of stimulation could inhibit, augment, or modify in a fluctuating fashion responses during subsequent periods of several minutes or hours. The durations of the modified responses were comparable with times required for cytoplasmic movements at observed speeds. It was therefore suggested (18,20) that the cell firing changes resulted from the excitation of synapses some of which operated by neurotransmitters and synaptic mediators at dendritic arrays, and that the arrival of a synaptic mediator at a given postsynaptic site by translocation produced the same effect there as did the locally generated mediator.

In interpreting such phenomena, the term rheosome was defined as a type of bioseme (6) (see first paragraph of this chapter) whose action as a carrier of molecular signification included a flow mechanism. The hypothesis was proposed that rheosemes that modify the cell firing of the neurons in whose dendrites they move do so in temporal patterns that reflect the spatial separation of rheoseme-sensitive postsynaptic regions, and that this action has formal similarities to that occurring in other instances of linear coding including that of the genetic system.

CONSEQUENCES OF THE PROPOSED LINEAR ELEMENT IN CODING

Some properties and data correlative to this hypothesis may now be considered. The linear element in genetic systems, and as proposed in the specified neural systems, stands in a defined relationship to a three-dimensional functioning component. Proteins synthesized ribosomally act as effectors, e.g., enzymes, after adopting particular spatial conformations; the rheosemes act by causing changes in cell firing equivalent to those produced by a three-dimensional pattern of cells that synapse with dendrite and involve the rheoseme, e.g., as postsynaptic mediator. A given linear array of nucleotide residues or of terminals can be read in one or more sequences and can contribute to an output of numerous replicated three-dimensional components. With rheoseme travel at 5μm/sec (28), terminal spacing is such that passage between adjacent postsynaptic regions would occupy 0.6 sec; along a dendritic spine, up to 4 sec; and to the maximal radius of a dendritic field in Golgi neurons of the dog forebrain, 230 sec; these times have been correlated with cell firing characteristics (20). Neither readout process disrupts the linear coding component, which is therefore repeatedly reusable. More than one linear unit is involved in characterizing or producing one unit of the three-dimensional component. Corresponding to the nucleotide triplet that specifies the insertion of a given amino acid residue is the necessity for a number of rheoseme-sensitive terminals to yield a characteristic temporal and spatial sequence of cell firing changes.

Transfer and transformation of genetic and neural signals can be expressed as shown in Fig. 2. This describes three-component systems in which a given component acts by two or more processes distinguished by numbered lines. Each process is complex in molecular terms and employs considerable structural and enzymic resources. In genetic and in neural systems, component A carries the endogenous record and acts as template for its readout; C receives materials and/or signals from the internal and external environments, and B is involved in amplification, control, and transmission of effects. In neural systems the major electrical signaling is by process 6, Fig. 2; neurotransmission involves processes 2–5, or 4 and 5, according to whether or not synaptic mediators are involved. Only the cytoplasmically mobile mediators involve process 1; rheosome readout to modify electrical signaling then proceeds by processes 1 and 7.

CONCLUDING NOTE ON NEURAL SIGNALING

Synapses on dendrites, noted above as very prominent features in central neural systems, are already considered to be major components in neuronal integration (see for example, refs. 11 and 34). Hitherto, these and other aspects of neural signaling have been considered to operate largely (14) by electrical means, with neurotransmitters as intermediaries. The view summarized in Fig. 2, dealing with interactions between different types of coding, has further characteristics that merit comment. Components A, B, and C in neural systems have distinct domains (intracellular, extracellular, and outer cell membrane) and their operative time scales shorten from A to C. Increased concentrations of cyclic nucleotides, when generated by a brief stimulus in the mammalian brain or in derived preparations

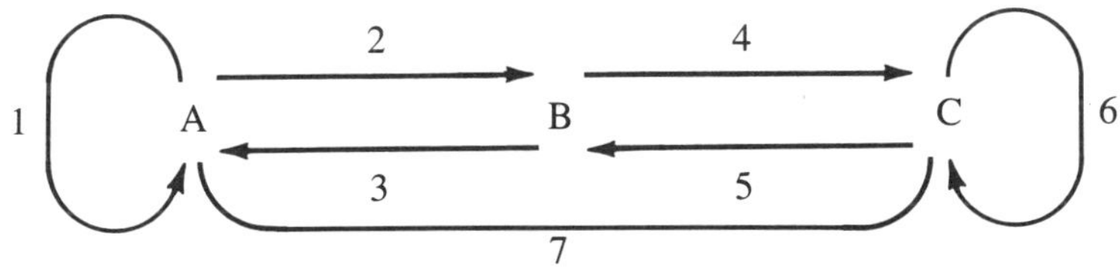

FIG. 2. Components on molecular signaling in genetic systems and in neural systems that include rheoseme-activated postsynaptic site arrays. In genetic systems **A** represents deoxyribonucleic acid; **B,** ribonucleic acid species; **C,** proteins. Arrows 1 to 4 give the "central dogma" and other lines the influence of protein components in controlling the expression of the genome (see refs. 10 and 40). In the neural systems **A** represents the rheoseme-postsynaptic site array; **B,** rheoseme-producing neurotransmitters; and **C,** electrical phenomena of cell polarization and the nerve impulse. The numbered processes are described in the text.

were noted above to persist for some minutes (Fig. 1), and thus component A of neural expression is an understandable source of slower cerebral signals that have been ascribed to sustained electrical activity.

The protein tubulin, the major component of the microtubules that appear to be largely responsible for axonal and dendritic transport of materials, including their "routing" to particular intraneuronal regions (28), is specifically present at the postsynaptic lattice of axodendritic synapses of the rat brain (15); some of these synapses have been shown to be sites of cyclic nucleotide generation (2). It is already considered that microtubule structure and functioning may be modified by changes in cyclic nucleotide metabolism (9) and that components of cyclic nucleotide systems can condition genetic expression (7,8,19). Common factors in neural and genetic expression are indicated, therefore, that concern the generation of the rheoseme-sensitive synaptic arrays in patterns that have significance in relation to sensory input. Situations have been examined (33) in which the relative permanence of synaptic connections can be expressed as a half-life of some days, with factors regulating their rates of formation and termination. Adaptive phenomena at this level have received comment elsewhere (22).

It should also be noted that the title compounds of this contribution and of this volume exemplify two of the three signaling categories of Fig. 2: adenosine of mainly extracellular action and cyclic AMP with mainly intracellular action. These actions, together with the metabolic interrelations, in which bound forms of cyclic AMP must be included (26), make them unusually versatile metabolic signals, meriting still further investigation in a variety of biological systems.

ACKNOWLEDGMENT

I am endebted to the Medical Research Council for support of this research, and to the Wellcome Trust for a Travel Grant.

REFERENCES

1. Barberis, C., McIlwain, H., and Newman, M. E. (1976): *Biochem. Soc. Trans.*, 4:748–749.
2. Bloom, F. E. (1975): *Rev. Physiol. Biochem. Pharmacol.*, 74:1–105.
3. Cragg, B. G. (1975): *Brain,* 98:81–90.
4. Daly, J. (1977): *Cyclic Nucleotides in the Nervous System.* Plenum, New York.
5. Eipper, B. A. (1975): *Ann. N. Y. Acad. Sci.*, 253:239–246.
6. Florkin, M. (1974): Concepts of molecular biosemiotics and of molecular evolution. In: *Comprehensive Biochemistry,* edited by M. Florkin and E. H. Stotz, pp. 1–123. Elsevier, Amsterdam.
7. Greengard, O. (1971): *Essays Biochem.*, 7:159–205.
8. Jungmann, R. A., and Russell, D. H. (1977): *Life Sci.*, 20:1787–1798.
9. Lagnado, J., Tan, L. P., and Reddington, M. (1975): *Ann. N.Y. Acad. Sci.*, 253:577–597.
10. Lehninger, A. L. (1975): *Biochemistry: The Molecular Basis of Cell Structure and Functioning,* pp. 853–1008. Worth, New York.
11. Llinas, R. (1975): *Adv. Neurol.*, 12:1–13.
12. Lubinska, L. (1975): *Int. Rev. Neurobiol.*, 17:241–296.
13. Lust, W. D., Goldberg, N. D., and Passonneau, J. V. (1976): *J. Neurochem.*, 26:5–10.
14. MacGregor, R. J., and Lewis, E. R. (1977): *Neural Modeling. Electrical Signal Processing in the Nervous System.* Plenum, New York.

15. Matus, A. I., Walters, B. B., and Mugal, S. (1975): *J. Neurocytol.*, 4:733–744.
16. McIlwain, H. (1972): *Biochem. Soc. Symp.*, 36:69–85.
17. McIlwain, H. (1974): *Biochem. Soc. Trans.*, 2:379–382.
18. McIlwain, H. (1976): *FEBS Lett.*, 64:271–273.
19. McIlwain, H. (1976): *Neurochem. Res.*, 1:351–368.
20. McIlwain, H. (1977): *Neuroscience*, 2:357–372.
21. McIlwain, H. (1978): *Prog. Neurobiol.*, 11:189–203.
22. McIlwain, H. (1978): In: *IBRO Monograph Series*, Vol. 4, Raven Press, New York.
23. Nathanson, J. A. (1977): *Physiol. Rev.*, 57:157–256.
24. Newman, M., and McIlwain, H. (1977): *Biochem. J.*, 164:131–137.
25. Newman, M., and McIlwain, H. (1977): *Biochem. Soc. Trans.*, 5:1074–1075.
26. Newman, M., and McIlwain, H. (1978): *Biochem. J.*, 170:73–79.
27. Newsholme, E. A., and Start, C. (1973): *Regulation in Metabolism*. Wiley and Sons, London.
28. Ochs, S. (1974): *Ann. N.Y. Acad. Sci.*, 228:202–223.
29. Pappas, G. D., and Waxman, S. G. (1972): In: *Structure and Function of Synapses*, edited by G. D. Pappas and D. P. Purpura, pp. 1–46. Raven Press, New York.
30. Pull, I., and McIlwain, H. (1972): *Biochem. J.*, 130:975–981.
31. Pull, I., and McIlwain, H. (1974): *Biochem. J.*, 144:37–41.
32. Pull, I., and McIlwain, H. (1976): *Neurochem. Res.*, 2:203–216.
33. Puro, D. G., De Mello, F. G., and Nirenberg, M. (1977): *Proc. Natl. Acad. Sci. U.S.A.*, 74:4977–4981.
34. Rall, W. (1970): In: *The Neurosciences Second Study Program*, edited by F. O. Schmitt, pp. 552–565. Rockfeller University Press, New York.
35. Reese, T. S., and Shepherd, G. M. (1972): In: *Structure and Function of Synapses*, edited by G. Pappas and D. P. Purpura, pp. 121–136. Raven Press, New York.
36. Robison, G. A., Butcher, R. W., and Sutherland, E. W. (1971): *Cyclic AMP*. Academic Press, New York.
37. Schubert, P. (1977): *Proc. Int. Soc. Neurochem.*, 6:56,80.
38. Schubert, P., and Kreutzberg, G. (1975): *Brain Res.*, 90:319–323.
39. Tomkins, G. M. (1975): *Science, N. Y.*, 188:760–763.
40. Watts, R. L., and Watts, D. C. (1975): In: *Synthesis of Amino Acids and Proteins*, edited by H. R. V. Arnstein. Butterworths, London.

Physiological and Regulatory Functions of Adenosine and Adenine Nucleotides, edited by H. P. Baer and G. I. Drummond. Raven Press, New York © 1979.

Effects of Purinergic Compounds on Excitable Membranes

L. P. Jager

Department of Pharmacology, Faculty of Pharmacy, State University Utrecht, Catharÿnesingel 60, 3511 GH, Utrecht, The Netherlands

Eighty years ago Langley (40) concluded that excitatory and inhibitory fibers were present in the vagus, innervating the gut. Based on his further investigations and those of Bayliss and Starling (4–6) he argued in 1922 (41) that the efferent enteric neural system was built from three functionally different intramural neurons, namely, excitatory, inhibitory, and associative neurons. Together with Magnus (42) he presented evidence that these intramural inhibitory nerves did not belong to the sympathetic system. Only in the 1960s were further details concerning the influence of these neurons on smooth cells reported (10,13,14,20,22,23). It was shown (11,13,14,29) that they caused a transient hyperpolarization of the smooth muscle cell membrane, inhibiting the generation of action potentials, and thereby inducing a relaxation of the contracted smooth muscle. Relaxations were often followed by a rebound contracture (10,27,34). Recent research has been aimed at identification and pharmacological characterization of the transmitter of these nonadrenergic, noncholinergic inhibitory (NAI) nerves, by studying the relaxing response of smooth muscle using mainly the guinea pig taenia caecum [by misnomer often referred to as taenia coli (23)].

METHODOLOGICAL ASPECTS

Bülbring (17,18) showed that the force exerted by the guinea pig taenia caecum, measured isometrically, increases with the rate of occurrence of muscle action potentials which, in turn, are inversely related to the potential difference across the cell membrane. The amplitudes of membrane hyperpolarizations and of muscle relaxations were found to be closely correlated (Fig. 1).

Although measurements of contractility are easily performed on smooth muscle preparations, this form of study has its pitfalls. For instance, the potentiating effect of dipyridamole on NAI nerve–induced relaxations (50) has been used as an indication that a purine compound such as adenosine triphosphate (ATP) may play the role of a transmitter in these nerves ("purinergic nerve hypothesis") (9,26,50), presumably because of the inhibitory effect of dipyridamole on adenosine uptake. However, the inhibitory junction potential (IJP) was found to be unaffected by dipyridamole (Fig. 2) (37). Furthermore, dipyridamole reduced the contraction and

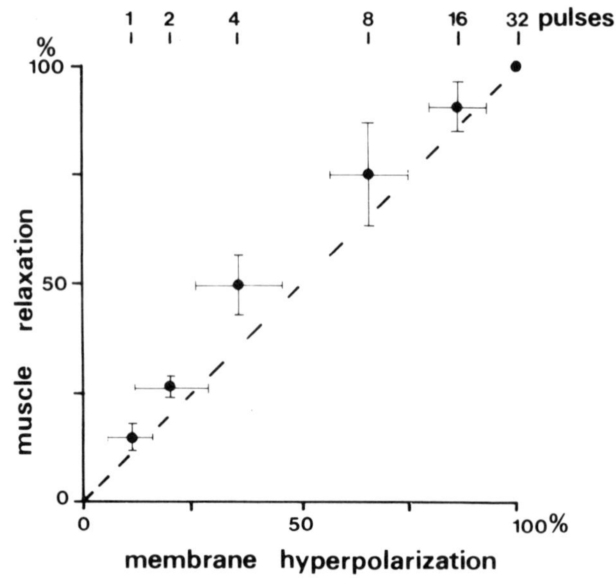

FIG. 1. Relation between the amplitude of the IJP and the amplitude of the relaxation following it. Measurements were made simultaneously with sucrose-gap apparatus and isotronic force-transducer, preload 10^{-2} N. The IJP was elicited by field stimulation [pulse rate, 30 pulse/sec; pulse duration, 0.3 msec; pulse amplitude, supramaximal. The IJP amplitude was varied by changing the duration of the pulse train (15 to 1,500 msec)]. 100% = 14.9 ± 3.1 mV and 100% = 1.9 ± 0.3 mN, N = 4, mean ± SEM.

the spike discharge but did not change the membrane depolarization induced with carbachol, indicating a dissociation between membrane potential and spike discharge by dipyridamole (3,9,37). Therefore, pros and cons of the purinergic nerve hypothesis (24) based on measurements of contractility alone should be regarded

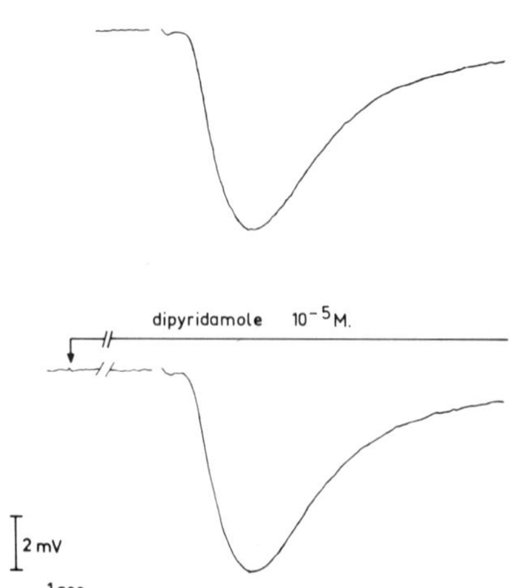

FIG. 2. Effect of dipyridamole on the IJP. **Upper trace:** Control response. **Lower trace:** Inhibitory junction potential evoked in the presence of dipyridamole (15 min). Traces are interrupted by the stimulus artifact. Field stimulation for 0.1 sec was used to elicit the IJP. Note the unchanged membrane potential and IJP in the presence of the drug. (From ref. 37, with permission.)

with caution. Illustrative of this point are the references given by Weisenthal et al. (57) as compatible with an adrenergic transmitter of NAI nerves, which were exactly identical with those given by Burnstock (21) as indicating a nonadrenergic transmitter.

So far, the sucrose gap method (12,15,25,43) seems to be the ideal compromise between rather laborious measurements of membrane potentials with microelectrodes and measurement of muscle contractions. Besides changes in contractility, changes in membrane potential and membrane resistance are measured simultaneously and easily. Johnson and Lieberman (38,47) showed, however, that this method also has its limitations and needs calibration with microelectrode measurements (1,29,45).

STUDIES ON THE TRANSMITTER OF NAI NERVES AND EFFECTS OF PURINE COMPOUNDS

Electrophysiological Studies with Intestinal Smooth Muscle Cell Membranes

Although the presence of intramural NAI nerves in the gut had been established beyond doubt in the 1960s, a hypothesis concerning the identity of their transmitter was put forward only at the end of that decade. It was suggested by Satchell et al. (49) that ATP or a related purine compound served the role of a chemical transmitter in these nerves. Proof of this "purinergic nerve hypothesis" would have to be provided on various levels of experimentation, including electrophysiological studies of membrane function involved in the control of muscle contractility.

Bennett et al. (13) suggested that the IJP evoked by stimulation of NAI nerves was caused by an increase in potassium permeability of the smooth muscle cell membrane. This hypothesis was further corroborated (28,55) by the following observations: (a) The membrane resistance of smooth muscle cells decreased transiently during the IJP. (b) The amplitude of IJP decreased with increasing membrane potential, and a reversal potential was observed. (c) The reversal potential of the IJP varied with the extracellular potassium concentration in a manner predicted by the Goldman equation for a selective increase in the membrane permeability for potassium (Fig. 3). (d) During the IJP the efflux of potassium from the smooth muscle increased (Fig. 4) whereas that of chloride and sodium hardly changed.

In conclusion, the following sequence of events has emerged: The transmitter of NAI nerves somehow induces a selective increase in membrane permeability for potassium ions; the resulting efflux of these ions increases the membrane potential; this hyperpolarization will, if the muscle cells are generating action potentials, inhibit this activity and thus evoke a relaxation of the contracted smooth muscle.

Sufficient evidence has been reported (7,32,35,39,46,53) to render this representation of the effects of the NAI transmitter on cell membranes valid for the longitudinal as well as the circular muscle layer of the mammalian gut.

Regarding other organs (e.g., blood vessels, trachea, bladder) in which NAI nerves seem to be present and those intestinal preparations in which an excitatory

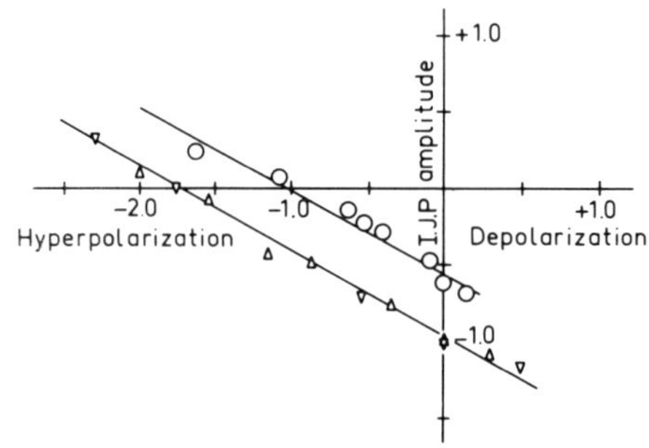

FIG. 3. The effect of potassium on the relationship between IJP amplitude and membrane polarization. The IJP amplitude was measured at different membrane polarizations in 2 mM (∇), in 12 mM (○), and recovery in 2 mM (Δ) extracellular potassium. Inhibitory junction potential amplitude *(ordinate)* and membrane polarization *(abscissa)* are given relative to the IJP amplitude in 2 mM potassium without membrane polarization (1.0 = 6.4 mV).

NAI innervation was found, no observations on membrane potentials have been reported. The only exception (8) indicates that in the guinea pig uterine artery, NAI nerve-induced relaxation is also mediated via a selective increase of membrane permeability for potassium ions.

Effects of Exogenous Purine Compounds on Intestinal Smooth Muscle Cell Membrane

Although numerous observations of effects of purine nucleotides on contractility of different kinds of smooth muscle have been reported, only five studies included observations of effects on smooth muscle cell membranes (2,19,36,48,56). In the guinea pig taenia caecum and the circular muscle of the frog stomach, purine nucleotides induce a decrease of resistance and a hyperpolarization of the smooth muscle cell membrane (Fig. 5). The potency of the different purine nucleotides to induce these changes was found to decrease in the following order: ATP>adenosine diphosphate (ADP)≫adenosine monophosphate (AMP)>adenosine, which is similar to that observed for producing relaxations (24). Since changes in resistance and in membrane potential were reported to be insensitive to changes in the extracellular concentration of chloride and sodium ions, an increase in the membrane permeability for potassium seemed a likely cause of the hyperpolarization. In accordance, the potassium efflux was found to be increased in the presence of ATP, whereas chloride and sodium efflux was unchanged (Fig. 6). Thus the same sequence of events as observed with NAI nerve stimulation appears to be valid for the action of these nucleotides.

However, relatively high concentrations (above 100 μM) of these compounds are required to obtain membrane responses of a magnitude comparable to the IJP, whereas muscle relaxations are observed with much lower concentrations. This discrepancy was noted by Tomita and Watanabe (56) and compared with the difference between α- and β-adrenoceptors. Thus an explanation for the different

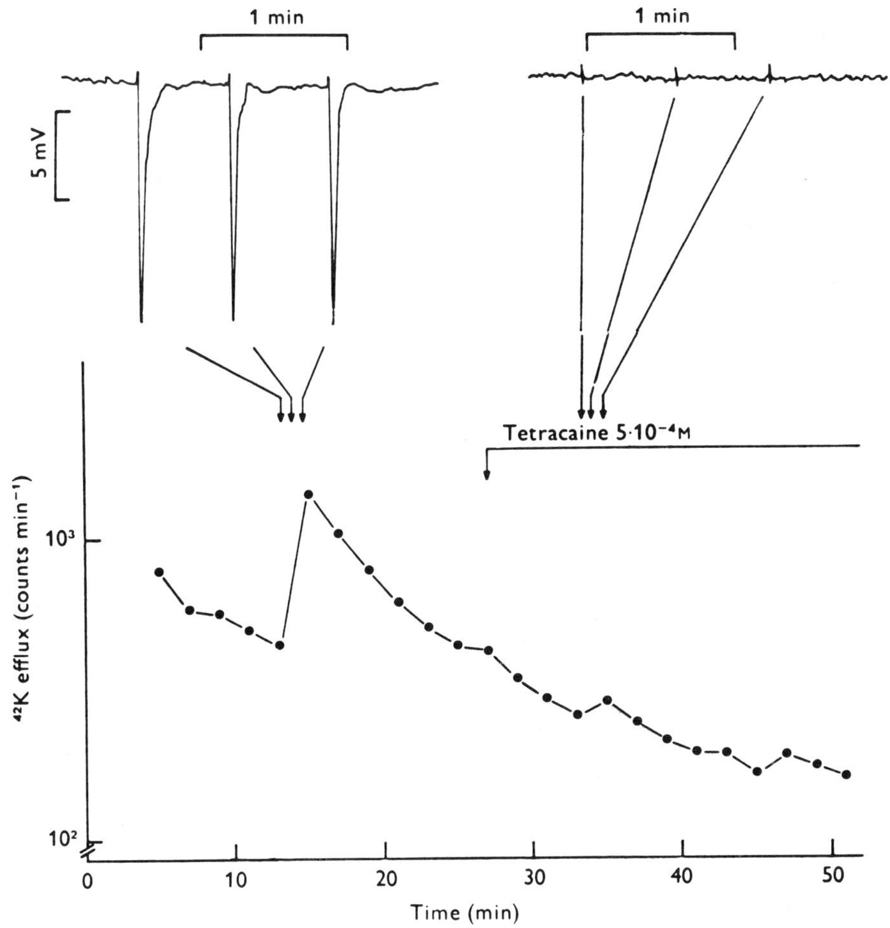

FIG. 4. Inhibitory potentials evoked by field stimulation *(trace)* and efflux of ^{42}K from the guinea pig taenia caecum measured simultaneously *(graph)*. Inhibitory junction potential and efflux increase are abolished in the presence of tetracaine. The activity of the effluent was counted during 4 min. (From ref. 28, with permission.)

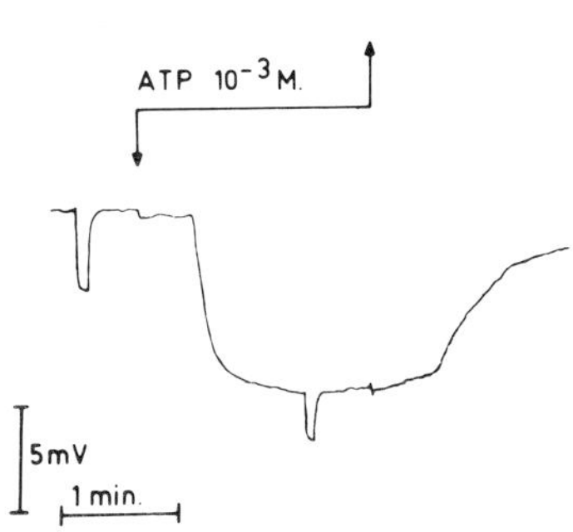

FIG. 5. Effect of exogenously applied ATP on membrane potential and membrane resistance of smooth muscle cells in the guinea pig taenia caecum. (From ref. 36, with permission.)

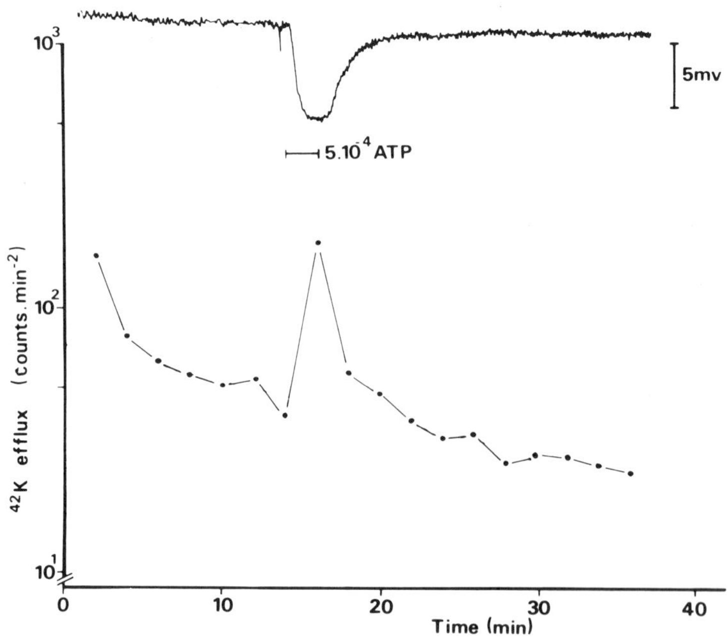

FIG. 6. Effect of exogenously applied ATP on the membrane potential *(trace)* and on the ^{42}K efflux from the guinea pig taenia caecum *(graph)*.

dose levels of nucleotides required for changes in membrane responses and relaxation might be the existence of different purine receptors. Receptors for ATP have been suggested to be distinct from those for adenosine on the basis of selective blockade of ATP action by 2,2′-pyridylisatogen tosylate in the guinea pig taenia caecum (52). Possibly, the ATP receptor is not readily accessible to exogenous agonist, being localized at the nerve-muscle junction and shielded anatomically in analogy to adrenergic nerve termini in vas deferens (30,54), thus requiring high concentrations of exogenous agonist to initiate muscle relaxation. A further type of purine receptor, which is also sensitive to adenosine, may exist on muscle cell membranes, thus being readily accessible to exogenous agonists. Stimulation of this latter receptor would produce relaxation by a mechanism other than alteration of membrane potential.

Effect of Purines on Atrial Muscle Membrane

Regarding the effects of purine nucleotides on other excitable membranes, only reports on membranes of heart muscle cells are available. Generally, an increase of a potassium efflux in the presence of nucleotides has been observed in mammalian (16,33) and amphibian (31) atrial muscle. This enhancement of the delayed outward current, carried mostly by potassium ions, is manifested as a negative inotropic effect. The order of potency is again ATP>ADP>AMP>adenosine. In addition, effects on calcium influx by ATP and ADP, but not by adenosine, were reported (31,51). Based on structural requirements of adenine nucleotides affecting contractility of atria (44), dual sites of action are evident, as suggested above for intestinal smooth muscle. Thus a receptor for ATP or ADP and a receptor sensitive to these nucleotides as well as to adenosine seem to exist. However, contrasting with the intestinal system, the receptor sensitive to both adenosine and nucleotides should be linked to an increase in membrane permeability for potassium ions.

SUMMARY

Although the transmitter substance of NAI nerves has not yet been identified, a good case for ATP or a related nucleotide has been presented by Burnstock in the form of the purinergic nerve hypothesis. To the extent that this hypothesis rests on the use of ATP as an exogenous agonist and the similarities of response profiles to nerve stimulation and ATP, it must be noted that the IJP and associated membrane potential changes caused by potassium efflux, which are responsible for the NAI nerve–mediated relaxation response in intestinal muscle, are not evident with doses of ATP causing relaxation. Only at high doses above 100 μM are membrane potential changes comparable to those caused by the physiological NAI transmitter obtained.

In view of this discrepancy, the purinergic nerve hypothesis can be considered further only by assuming that the postulated nerve-related purine (ATP) receptor sites are shielded to exogenous agonist such that high concentrations are required to affect the IJP. The relaxant response obtained with doses of agonists not affecting the IJP must then involve an additional purine receptor on muscle cells that is also sensitive to adenosine in some tissues. This latter adenosine receptor may have significance for autoregulatory functions independent of membrane potential changes. It should be noted, however, that to date nerve-muscle structures that would support the idea of anatomical shielding to exogenous agonist have been described only for vas deferens. Thus the hypothesis of ATP as a neurotransmitter in NAI nerves remains seriously questioned by the available electrophysiological evidence.

ACKNOWLEDGMENTS

I am greatly indebted to Dr. A. den Hertog and Mr. J. A. M. Schevers for their cooperation. I also wish to thank Mrs. G. C. van den Bogaard and Mr. P. W. van Dorp van Vliet for their assistance in the presentation of this chapter. The foundation for Medical Research FUNGO is acknowledged for financial support.

REFERENCES

1. Abe, Y., and Tomita, (1968): *J. Physiol.*, 196:87–100.
2. Axelsson, J., and Holmberg, B. (1969): *Acta Physiol. Scand.*, 75:149–156.
3. Baer, H. P., Frew, R., and Burnstock, G. (1977): *Can. J. Physiol. Pharmacol.*, 55:394–398.
4. Bayliss, W. M., and Starling, E. H. (1899): *J. Physiol.*, 24:99–143.

5. Bayliss, W. M., and Starling, E. H. (1900): *J. Physiol.,* 26:107–118.
6. Bayliss, W. M., and Starling, E. H. (1901): *J. Physiol.,* 26:125–138.
7. Beani, L., Bianchi, C., and Crema, A. (1971): *J. Physiol.,* 217:259–279.
8. Bell, C. (1969): *J. Physiol.,* 205:695–708.
9. Bell, C. (1974): *Br. J. Pharmacol.,* 52:13–17.
10. Bennett, M. R. (1966): *J. Physiol.,* 185:124–131.
11. Bennett, M. R. (1966): *J. Physiol.,* 185:132–147.
12. Bennett, M. R., and Burnstock, G. (1966): *J. Physiol.,* 183:637–648.
13. Bennett, M. R., Burnstock, G., and Holman, M. E. (1963): *J. Physiol.,* 169:33–34P.
14. Bennett, M. R., Burnstock, G., and Holman, M. E. (1966): *J. Physiol.,* 182:541–558.
15. Berger, W. (1963): *Pfluegers Arch.,* 277:570–576.
16. Bertelli, A., Bianchi, C., and Beani, L. (1972): *Eur. J. Pharmacol.,* 19:130–133.
17. Bülbring, E. (1954): *J. Physiol.,* 125:302–315.
18. Bülbring, E. (1955): *J. Physiol.,* 128:200–221.
19. Bülbring, E., and Hardman, J. H. (1976): In: *Smooth Muscle Pharmacology and Physiology,* edited by M. Worcel and G. Vassort, pp. 125–134. INSERM, Paris.
20. Bülbring, E., Kuriyama, H., and Tomita, T. (1965): *J. Physiol.,* 181:8–10P.
21. Burnstock, G. (1972): *Pharmacol. Rev.,* 24:509–581.
22. Burnstock, G., Campbell, G., Bennett, M., and Holman, M. E. (1964): *Int. J. Neuropharmacol.,* 3:163–166.
23. Burnstock, G., Campbell, G., and Holman, M. E. (1966): *J. Physiol.,* 182:504–526.
24. Burnstock, G., Campbell, G., Satchell, D., and Smythe, A. (1970): *Br. J. Pharmacol.,* 40:668–688.
25. Burnstock, G., and Straub, R. W. (1958): *J. Physiol.,* 140:156–167.
26. Coleman, R. H., and Levy, G. P. (1974): *Br. J. Pharmacol.,* 52:167–174.
27. Day, M. D., and Warren, P. R. (1968): *Br. J. Pharmacol. Chemother.,* 32:227–240.
28. Den Hertog, A., and Jager, L. P. (1975): *J. Physiol.,* 250:681–691.
29. Furness, J. B. (1969): *J. Physiol.,* 205:549–562.
30. Furness, J. B. (1974): *Br. J. Pharmacol. Chemother.,* 50:63–68.
31. Goto, M., Yatani, A., and Tsuda, Y. (1977): *Jpn. J. Physiol.,* 27:81–94.
32. Hirst, G. D. S., and McKirdy, H. C. (1974): *J. Physiol.,* 238:129–143.
33. Hollander, P. B., and Webb, J. L. (1957): *Circ. Res.,* 5:349–353.
34. Holman, M. E., and Hughes, J. R. (1965): *Aust. J. Exp. Biol. Med. Sci.,* 43:277–290.
35. Ito, Y., and Kuriyama, H. (1973): *J. Physiol.,* 231:455–470.
36. Jager, L. P. (1974): *Eur. J. Pharmacol.,* 25:372–382.
37. Jager, L. P. (1976): *Arch. Int. Pharmacodyn.,* 221:40–53.
38. Johnson, E. A., and Lieberman, M. (1971): *Annu. Rev. Physiol.,* 33:479–532.
39. Kinoshita, S. (1973): *Tohoku J. Exp. Med.,* 110:119–138.
40. Langley, J. N. (1898): *J. Physiol.,* 23:407–414.
41. Langley, J. N. (1922): *J. Physiol.,* 56:39P.
42. Langley, J. N., and Magnus, R. (1905): *J. Physiol.,* 33:34–51.
43. McCann, F. V., Stibitz, G. R., and Huguenin, J. (1973): *Biophys. J.,* 13:1183–1199.
44. Meinertz, T., Nawrath, H., and Scholz, H. (1973): *Naunyn Schmiedebergs Arch. Pharmacol.,* 278:165–178.
45. Ohba, M., Sakamoto, Y., Tokuno, H., and Tomita, T. (1976): *J. Physiol.,* 256:527–540.
46. Ohga, A., and Taneike, T. (1977): *Br. J. Pharmacol.,* 60:221–231.
47. Poindessault, J. P., Duval, A., and Léoty, C. (1976): *Biophys. J.,* 16:105–120.
48. Poskonova, M. A., and Malchikova, L. S. (1970): *Dokl. Akad. Nauk. SSSR,* 190:1494–1497.
49. Satchell, D. G., Burnstock, G., and Campbell, G. (1969): *Aust. J. Exp. Biol. Med. Sci.,* 47:24P.
50. Satchell, D. G., Lynch, A., Bourke, P. M., and Burnstock, G. (1972): *Eur. J. Pharmacol.,* 19:343–350.
51. Schneider, J. A., and Sperelakis, N. (1975): *J. Mol. Cell. Cardiol.,* 7:249–273.
52. Spedding, M., and Weetman, D. F. (1976): *Br. J. Pharmacol.,* 57:305–310.
53. Suzuki, H., and Kuriyama, H. (1975): *Jpn. J. Physiol.,* 25:759–773.
54. Swedin, G. (1971): *Acta Physiol. Scand. [Suppl.],* 369.
55. Tomita, T. (1972): *J. Physiol.,* 225:693–703.
56. Tomita, T., and Watanabe, H. (1973): *J. Physiol.,* 231:167–177.
57. Weisenthal, L. M., Hug, C. C., Weissbrodt, N. W., and Bass, P. (1971): *J. Pharmacol. Exp. Ther.,* 178:497–508.

Fat Cells, Immuno System, and Protaglandin Involvement

Physiological and Regulatory Functions of Adenosine and Adenine Nucleotides, edited by H. P. Baer and G. I. Drummond. Raven Press, New York © 1979.

Adenosine Effects in Adipose Tissue: Effect of Adenosine on the Cyclic Adenosine Monophosphate System and Lipolysis of Fat Cells

Thomas Trost and Klaus Stock

Institut für Pharmakologie und Toxikologie, Abteilung II, Medizinische Hochschule Hannover, Karl-Wiechert-Allee 9, D-3000 Hannover 61, Federal Republic of Germany

The role of cyclic adenosine monophosphate (cyclic AMP) as mediator of lipolysis in adipose tissue has been studied and reviewed extensively (2,7,9,26). Concerning the action of adenosine, Dole (4) first reported in 1961 that adenosine and related compounds like adenosine triphosphate, adenosine monophosphate, and cyclic AMP reduced the adrenaline-stimulated release of free fatty acids from isolated fat pads. Later, several other investigators (1,14,19,20) confirmed the results of Dolc (4). In 1968, Davies (3) demonstrated that adenosine not only inhibited hormone-induced glycerol output, but also antagonized lipolysis evoked by theophylline. Similar effects of adenosine on lipolysis were obtained with isolated adipocytes (5) and, in 1972, Fain et al. (10) found adenosine to be inhibitory on noradrenaline-stimulated cyclic AMP accumulation in the intact fat cell.

All these data indicated that adenosine may play a physiologically important role as a regulator of the cyclic AMP system in fat cells (10). Further evidence for this hypothesis was reported by Schwabe et al. (22,23) who found that the effect of lipolytic stimulants on cyclic AMP levels and glycerol production was strongly dependent on the concentration of fat cells in the incubation medium. The increased effectiveness of hormones in diluted cell suspensions suggested the existence of an inhibitory substance produced by fat cells (22), as already supposed by Manganiello et al. (17) and by Ho and Sutherland (13). Schwabe et al. (22) were able to prove that the inhibitor released from isolated fat cells was adenosine. However, the release of adenosine into the incubation medium was not dependent on the presence of noradrenaline as stimulator of cyclic AMP accumulation or lipolysis (22). Accumulation of considerable amounts of adenosine was dependent only on time and cell concentration, which put into question the role of adenosine as a feedback regulator (22).

In further studies efforts were made to identify the site of adenosine action. From experiments with dipyridamole, which strongly decreased the uptake of adenosine into fat cells, Ebert and Schwabe (6) first assumed that the site of action of adenosine is localized on the outer surface of the cell membrane. Dipyridamole did not counteract the antilipolytic action of adenosine. Similar results were obtained by Fain (8) in isolated fat cells: Adenosine was inhibitory on noradrenaline-induced cyclic AMP accumulation in the presence or absence of dipyridamole. Further evidence that adenosine is likely to act on the outer surface of the fat cell was obtained with adenosine deaminase. Addition of the enzyme, which is unlikely to penetrate cell membranes, to incubation media of fat cells produced a strong lipolytic response accompanied by a rise in cyclic AMP levels (11,21). Striking similarities to the effects of adenosine deaminase were obtained with theophylline (21). Both agents produced only a relatively small increase in cyclic AMP levels but a maximal lipolytic response and acted synergistically with noradrenaline on cyclic AMP accumulation (8,21). These findings suggested that theophylline, apart from inhibiting phosphodiesterase, may exert its lipolytic effect by antagonizing the action of adenosine on the cell membrane (6,21). This hypothesis is in good agreement with the finding that adenosine competitively inhibited the lipolytic action of theophylline (6). The site at which adenosine acted on the lipolytic system showed a high degree of specificity and sensitivity toward the nucleoside. Among several nucleotides and nucleosides, such as ATP, adenosine diphosphate, 5′-AMP, inosine, and hypoxanthine, adenosine proved to be the most potent inhibitor of noradrenaline-induced cyclic AMP accumulation (22). In the presence of only 10 nM adenosine, more than 50% inhibition of noradrenaline-stimulated cyclic AMP accumulation was observed (22). Furthermore, the onset of inhibition of noradrenaline-stimulated cyclic AMP accumulation occurred within seconds (8). Adenosine appears not to affect other components of the adenylate cyclase system, e.g., increasing the activity of phosphodiesterase in fat cells. The low K_m phosphodiesterase activity of fat cells was only reduced to about 80% by 70 μM adenosine, which completely inhibited hormone-induced cyclic AMP levels (10). Lipolysis induced by dibutyryl cyclic AMP, which is believed to circumvent the adenylate cyclase (12), was not inhibited by adenosine up to 1 mM (6).

All these results strongly support the hypothesis that adenosine acts on an extracellular site of the fat cell plasma membrane. If a compound is believed to act on a site at the cell membrane and exerts an inhibitory effect on adenylate cyclase more or less directly, it should be possible to demonstrate this inhibitory effect in a cell-free system. Indeed, adenosine has been found to be an inhibitor of adenylate cyclase activity of fat cell ghosts (10,18,23). However, 1,000-fold higher concentrations of adenosine were required in the ghost preparation than in the intact cell to obtain significant inhibitory effects (23). This was thought to be the result of the high activity of the adenosine kinase, which has free access to the nucleoside only in the broken cell system (23).

The inhibitory mechanism of adenosine on cyclic AMP and lipolysis in fat cells appeared to be almost completely understood, but one point remained unclear. In 1969 Westermann et al. (27) reported that the N^6-monosubstituted adenosine derivative N^6-phenylisopropyladenosine (PIA) exhibited strong antilipolytic effects *in vivo* and *in vitro*. Using isolated fat cells, Fain (8) found that PIA was a more potent inhibitor of noradrenaline-stimulated cyclic AMP accumulation than adenosine. In contrast to the isolated cell, PIA was completely inactive on adenylate cyclase in a cell-free system, whereas adenosine was inhibitory in both intact cells and broken cells (10).

To clarify this discrepancy, a number of N^6-monosubstituted adenosine derivatives were studied with respect to their effects on cyclic AMP accumulation and lipolysis in the isolated fat cell and, on the other hand, with respect to their effects on adenylate cyclase activity of fat cell plasma membranes (25).

TWO DIFFERENT SITES OF ADENOSINE ON ADENYLATE CYCLASE OF FAT CELLS

Effects of Adenosine and N^6-Substituted Adenosine Derivatives

Figure 1 shows the inhibitory effects of N^6-cyclohexyladenosine, adenosine, and N^6-methyladenosine on noradrenaline-stimulated cyclic AMP accumulation in isolated rat fat cells. These two N^6-substituted derivatives may be considered to be representative of a whole series of compounds tested: N^6-cyclohexyladenosine was more potent and N^6-methyladenosine less potent than adenosine. Both derivatives, as well as adenosine, inhibited cyclic AMP accumulation in a concentration-dependent manner and almost completely. The same order of potency was found when the compounds were tested against lipolysis stimulated submaximally by 0.3 μM noradrenaline (Fig. 2). However, inhibition of lipolysis by the nucleosides was found only when submaximal concentrations of noradrenaline were used. Even under these conditions inhibition of noradrenaline-induced glycerol production by adenosine and N^6-substituted derivatives never exceeded 50%.

When 1 mM theophylline was used to maximally stimulate lipolysis, adenosine and all N^6-substituted derivatives tested were able to suppress lipolysis almost completely (25). Again, N^6-cyclohexyladenosine was more potent and N^6-methyladenosine less potent than adenosine (25). Table 1 shows the 50% inhibitory concentrations (IC_{50}) and the relative potencies of all N^6-inhibited derivatives versus adenosine. In the presence of about 8 nM adenosine, a 50% inhibition of noradrenaline-induced cyclic AMP accumulation was seen in isolated fat cells. All N^6-substituted derivatives tested were inhibitors, being either more or less potent than adenosine. A similar pattern of relative potency as shown in Table 1 was found when lipolysis was induced by theophylline or noradrenaline (25). The N^6-substituted derivatives showed similar range of potency on lipolysis induced by adenosine deaminase with the exception of N^6-methyladenosine, which was probably hydrolyzed by the enzyme. N^6-Cyclohexyladenosine was the most potent inhibitor, whereas N^6-benzyladenosine proved to be about 400 times less active (T. Trost and K. Stock, *unpublished data*). The results with adenosine deaminase indicate that a fat cell remains sensitive against the N^6-substituted derivatives when adenosine is removed from the medium and thus appears to act directly and not via endogenous adenosine.

Concerning the structure-activity relationship of the N^6-substituted compounds,

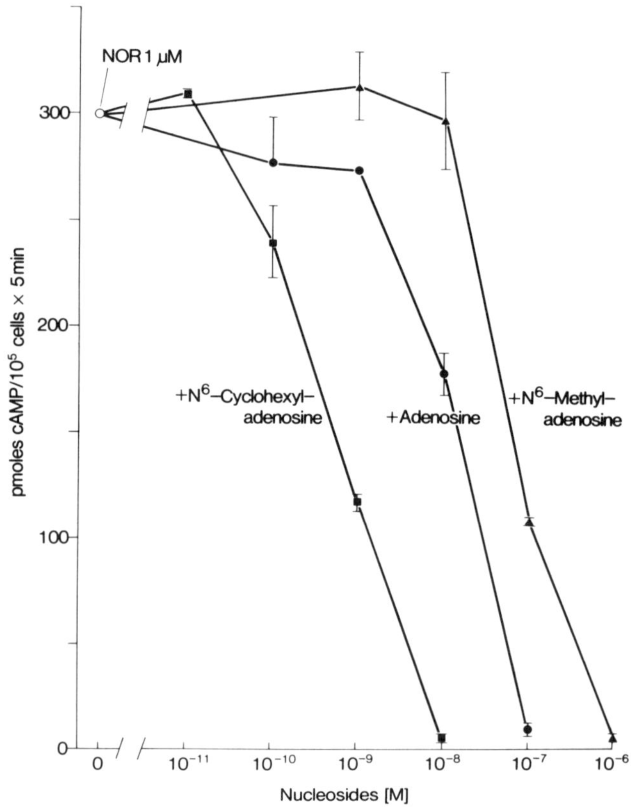

FIG. 1. Concentration-inhibition curves of N⁶-cyclohexyladenosine, adenosine, and N⁶-methyladenosine on cyclic AMP accumulation in isolated rat fat cells, stimulated by 1 μM noradrenaline (NOR). Values are means of duplicate or triplicate (± SEM) incubates from a single cell preparation. (Data from Trost and Stock, ref. 25.)

no clear-cut correlations existed between inhibitory potency and the lipophilic nature of the substituents in the N⁶ position, since N⁶-xylyladenosine, for instance, is more lipophilic than N⁶-phenyladenosine but significantly less active as an inhibitor of stimulated cyclic AMP accumulation or lipolysis. Moreover, the dextroisomer of PIA was around 70 times less active than the levoisomer (27), although they possess the same lipophilic properties.

In further experiments we studied the effects of adenosine and N⁶-methyladenosine on adenylate cyclase activity in fat cell plasma membranes. Adenosine was found to be inhibitory on noradrenaline-stimulated adenylate cyclase activity in a concentration-dependent way, whereas addition of N⁶-methyladenosine in concentrations up to 1 mM did not affect the hormone-stimulated enzyme activity (Fig. 3, left panel). No inhibitory activity of N⁶-methyladenosine was detected even at low concentrations of noradrenaline (Fig. 3, right panel). Similar results were

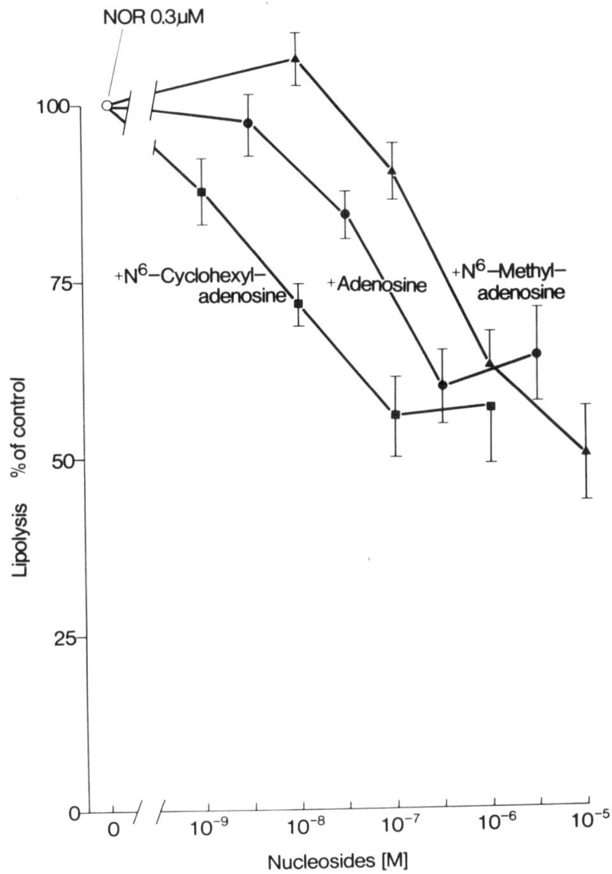

FIG. 2. Concentration-inhibition curves of N^6-cyclohexyladenosine, adenosine, and N^6-methyladenosine on lipolysis in isolated rat fat cells stimulated by 0.3 μM noradenaline (NOR). Mean of controls (effect of noradrenaline without nucleosides) was used as 100% (absolute value: 67.2 $\pm$ 9.5 nmoles glycerol/10^5 cells $\times$ 60 min). Values are means $\pm$ SEM of percent change from controls and were obtained from three to four different cell preparations with triplicate incubates. (Data from Trost and Stock, ref. 25.)

obtained for all N^6-substituted adenosine derivatives tested, irrespective of the degree of inhibitory potency of the compound on the level of the intact cell.

Effects of Adenosine and Deoxyadenosine Derivatives

In further experiments, we studied the effects of adenosine derivatives altered in the ribose moiety of the molecule. Fain et al. (10) first reported that 2′,5′-dideoxyadenosine was more potent than adenosine as an inhibitor of the adenylate cyclase of fat cell ghosts. On the other hand, this compound was found to be relatively ineffective as an inhibitor of cyclic AMP levels in intact fat cells (8).

TABLE 1. *IC_{50} values and relative potencies of N^6-substituted derivatives versus adenosine*

Nucleosides	IC_{50} (nM)	Relative potency
N^6-Cyclohexyladenosine	0.63	12.6
N^6-n-Propyladenosine	2.0	4.0
N^6-Phenyladenosine	3.2	2.5
Adenosine	7.9	1.0
N^6-Methyladenosine	63.1	0.13
N^6-o-Xylyladenosine	50.1	0.16
N^6-Benzyladenosine	125.9	0.06

IC_{50} values were obtained from inhibition curves of noradrenaline-stimulated (1 μM) cyclic AMP accumulation in isolated rat fat cells.
Data from ref. 25, with permission.

Figure 4 shows the effects of 2′,5′-dideoxyadenosine and adenosine on cyclic AMP levels in intact cells and on adenylate cyclase of fat cell plasma membranes. In isolated fat cells a 50% inhibition of noradrenaline-stimulated cyclic AMP accumulation by 2′,5′-dideoxyadenosine was found at a concentration of 10 μM, indicating

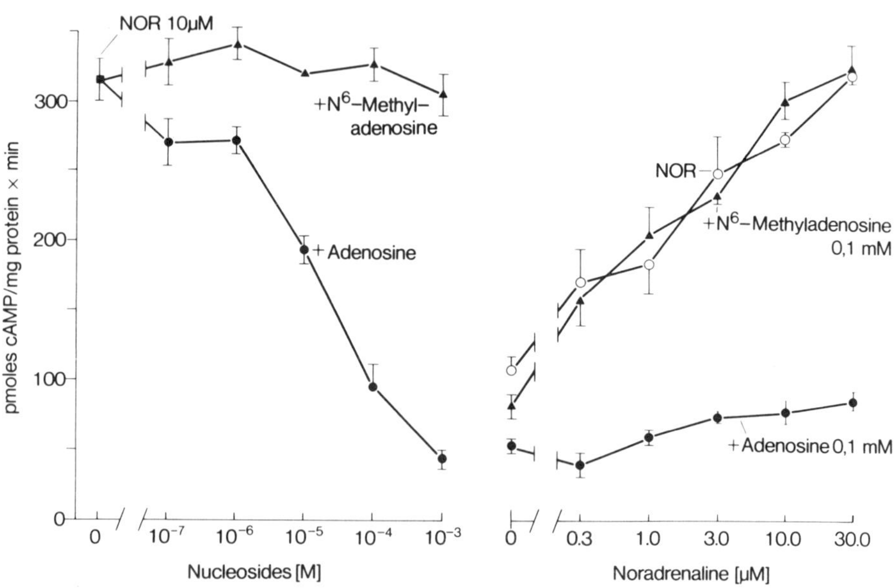

FIG. 3. Left panel: Effects of N^6-methyladenosine and adenosine on adenylate cyclase activity of fat cell plasma membranes, stimulated by 10 μM noradrenaline (NOR). **Right panel:** Concentration-response curves of noradrenaline on adenylate cyclase activity of fat cell plasma membranes in the absence of nucleosides, in the presence of 0.1 mM N^6-methyladenosine, and in the presence of 0.1 mM adenosine. Values are means ± SEM (N = 3) from a single membrane preparation. (Data from Trost and Stock, ref. 25.)

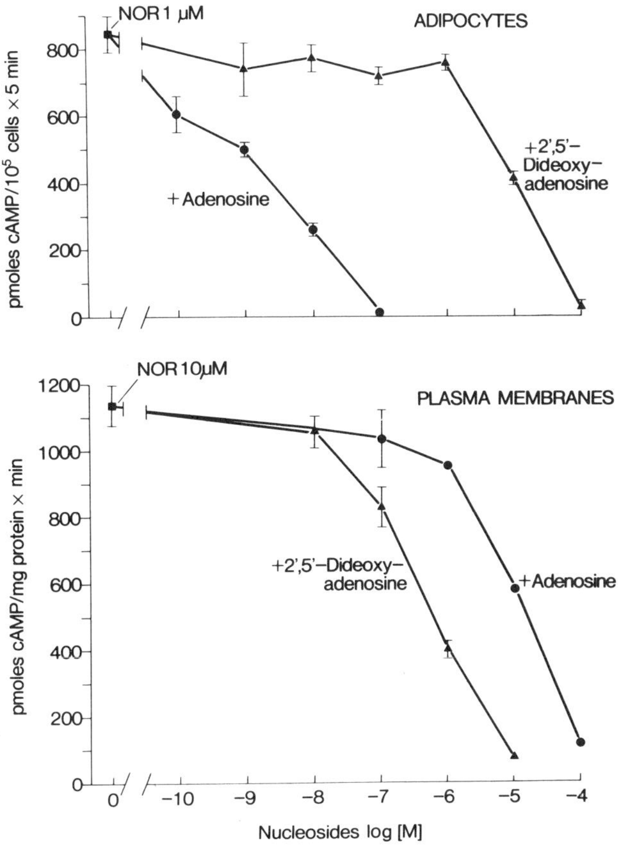

FIG. 4. Upper panel: Concentration-inhibition curves of 2′,5′-dideoxyadenosine and adenosine on cyclic AMP accumulation in isolated rat adipocytes stimulated by 1 μM noradrenaline (NOR). Values are means ± SEM *(N* = 3 to 4) from a single cell preparation. **Lower panel:** Concentration-inhibition curves of 2′,5′-dideoxyadenosine and adenosine on adenylate cyclase activity of adipocyte plasma membranes stimulated by 10 μM noradrenaline. Values are means of duplicate or triplicate (± SEM) incubates from a single membrane preparation. (Data from Trost and Stock, ref. 25.)

that this compound is about 1,000 times less active than adenosine (Fig. 4, upper panel). However, in plasma membranes 2′,5′-dideoxyadenosine was found to be about 40 times more active than adenosine as an inhibitor of noradrenaline-stimulated adenylate cyclase activity. In general, all deoxyadenosine derivatives tested were considerably less potent than adenosine as inhibitors of noradrenaline-stimulated cyclic AMP accumulation in isolated fat cells (Table 2).

Contrary to these findings with isolated fat cells, in fat cell plasma membranes a concentration of adenosine 1,000-fold higher was necessary to obtain 50% inhibition of noradrenaline-stimulated adenylate cyclase activity and all deoxyadenosine

TABLE 2. *IC_{50} values and relative potencies of deoxy derivatives versus adenosine*

Nucleosides	Adipocytes		Plasma membranes	
	IC_{50} (μM)	Relative potency	IC_{50} (μM)	Relative potency
Adenosine	0.008	1.0	12.6	1.0
2′,5′-Dideoxyadenosine	10.0	0.0008	0.3	39.8
3′-Deoxyadenosine [a]	0.4	0.02	1.6	8.0
2′-Deoxyadenosine	1.0	0.008	6.3	2.0

IC_{50} values were obtained from inhibition curves of noradrenaline-stimulated (1 μM) cyclic AMP accumulation in isolated rat fat cells and of noradrenaline-stimulated (10 μM) adenylate cyclase activity of fat cell plasma membranes.

Data from ref. 25, with permission.

[a] Unpublished values.

derivatives tested proved to be nearly equipotent or more potent than adenosine as inhibitors of plasma membrane adenylate cyclase (Table 2). From these data it could be concluded that adenosine has two different sites of action on fat cell adenylate cyclase. One site on the outer cell surface, the "outer site," appears to be the important physiological regulator site. Removal of adenosine from this site results in an activation of lipolysis via the cyclic AMP system. Adenosine derivatives, substituted in the N^6 position of the molecule, are agonists of the "outer site." These agonists show varying potency depending on the structure of the substituent in the N^6 position. Alterations of adenosine in the ribose moiety strongly reduced the affinity to the "outer site."

The other site, the "inner site," appears to be accessible only in the broken cell preparation. Its physiological and regulatory importance remains to be clarified. Action of adenosine at this site results in an inhibition of adenylate cyclase activity. In contrast to the "outer site," a free N^6 amino group appears to be an absolute requirement for action of adenosine at the "inner site," whereas alterations of adenosine in the ribose moiety can even enhance the affinity of adenosine to the "inner site." Two distinct adenosine-sensitive sites were also found by Londos and Wolff (15) for adenylate cyclase from other tissues. In their experiments two adenosine-reactive sites have been identified: (a) the "R" site, occupancy of which usually leads to activation of adenylate cyclase and which requires integrity of the ribose ring for activity, and (b) the "P" site, which mediates inhibition and requires integrity of the purine ring for activity.

From our data and those of Londos and Wolff (15), we assume that the "R" site is identical with the "outer site," occupancy of which leads to inhibition of adenylate cyclase in fat cells. The "P" site appears to be identical with the direct inhibitory "inner site" of fat cell adenylate cyclase.

Effects of Theophylline on Cyclic AMP Accumulation in Isolated Fat Cells and on Adenylate Cyclase Activity of Fat Cell Plasma Membranes

To further characterize the two adenosine sites with respect to the known antagonism between theophylline and adenosine in isolated cells, we compared the effects of theophylline on cyclic AMP levels in isolated cells and on adenylate cyclase of plasma membranes (T. Trost and K. Stock, *unpublished data*). Figure 5 shows that the addition of theophylline to isolated fat cells increased the noradrenaline-stimulated cyclic AMP level in a concentration-dependent manner, reaching a plateau at 0.1 mM theophylline. In the presence of 10 μM adenosine the noradrenaline-stimulated cyclic AMP accumulation was inhibited to nearly basal values and addition of theophylline raised the cyclic AMP level in a concentration-dependent manner. The inhibitory effect of adenosine was almost overcome at a concentration of 10 mM theophylline. These data show that a 1,000-fold excess of theophylline over adenosine was necessary to overcome the inhibitory effect of the nucleoside in isolated fat cells.

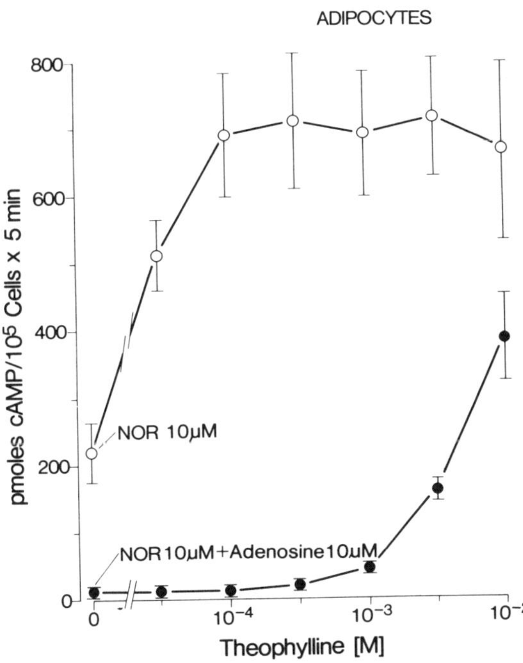

FIG. 5. Effect of theophylline on cyclic AMP accumulation in isolated rat adipocytes, stimulated by 10 μM noradrenaline (NOR) in the absence and in the presence of adenosine (10 μM). Values are means ± SEM (N = 3 to 6) obtained from three to six different cell preparations with duplicate or triplicate incubates for each concentration. Basal value without any additions (not shown) was 7.9 ± 3.1 pmoles cyclic AMP/10^5 cells × 5 min (N = 6). For methods see Trost and Stock (25). [Data from Trost and Stock *(unpublished data)*.]

Addition of theophylline to plasma membranes did not enhance noradrenaline-stimulated adenylate cyclase activity (Fig. 6) and, in contrast to isolated fat cells, high concentrations of theophylline even caused a considerable inhibition. The inhibitory action of adenosine on the noradrenaline-stimulated enzyme was not influenced by theophylline even at concentrations of 3 to 10 mM. In isolated fat cells, such concentrations of theophylline clearly antagonized the inhibitory action of adenosine (Fig. 5).

These results show that theophylline is an antagonist of adenosine action at the "outer site." Thus theophylline appears to be a tool for demonstrating adenosine effects at the "outer site" since it has no effect on the "inner site."

Recently, Malbon et al. (16) presented data on [^{3}H]adenosine binding to fat cell membranes. They found two populations of adenosine binding sites, possessing different affinities for adenosine. The apparent dissociation constants (K_d) were about 10 μM for the high- and about 1 mM for the low-affinity adenosine binding sites. However, even 100 μM PIA did not affect binding of [^{3}H]adenosine to fat cell plasma membranes, whereas in the presence of 100 μM 2′,5′-dideoxyadenosine a 54% inhibition of [^{3}H]adenosine binding occurred. From our point of view, the

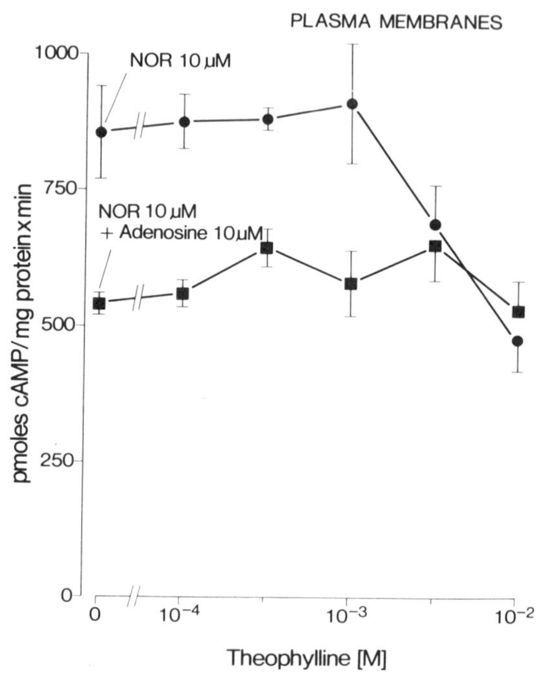

FIG. 6. Effect of theophylline on adenylate cyclase activity of fat cell plasma membranes stimulated by 10 μM noradrenaline (NOR) in the absence and presence of adenosine (10 μM). Values are means ± SEM (N = 3), obtained from three different experiments with duplicate or triplicate incubates for each concentration. Basal value without any additions (not shown) was 297.9 ± 4.3 pmoles cyclic AMP/mg protein × min (N = 3). For methods see Trost and Stock (25). [Data from Trost and Stock *(unpublished data).]*

high-affinity site characterized by these binding studies thus appears to be identical with the "inner site."

With fat cells, the "outer site" has not yet been demonstrated in broken cell preparations, neither by [^{3}H]adenosine binding studies nor by measurement of adenylate cyclase activity. It is unlikely that the "outer site" is lost during the preparation of plasma membranes (11). We assume that the conditions of the *in vitro* adenylate cyclase assay interfere with the effects of N^6-substituted adenosine derivatives (8).

DISSOCIATION BETWEEN CYCLIC AMP AND LIPOLYSIS BY PIA

The dissociation between cyclic AMP and lipolysis by PIA has already been well studied by Stock and Prilop (24). Similar dissociation between the lipolytic response and cyclic AMP levels were also reported for adenosine (9,10,22). More recently, Wiedbrauck and Stock *(unpublished results)* resumed investigations on the "dissociation phenomenon" by simultaneous measurement of cyclic AMP levels, protein kinase activity, and glycerol release in rat fat cells from the same cell preparation. These studies revealed some new aspects of the physiological importance of adenosine action in adipose tissue.

As shown in Fig. 7, addition of 1 mM theophylline to isolated fat cells resulted in a sixfold increase of cyclic AMP level. This increase was sufficient for a complete activation of the protein kinase activity and lipolysis. Addition of PIA resulted in a concentration-dependent decrease of cyclic AMP, protein kinase activity, and lipolysis to basal values. If theophylline and adenosine compete for a common receptor site located on the cell membrane (6) ("outer site"), these data suggest a physiological role for adenosine in the regulation of lipolysis. Adenosine deficiency may result in an uncontrolled, continuous lipolysis. Only in the presence of adenosine is lipolysis under control. In this model adenosine does not play the role of a classical feedback regulator (22).

When noradrenaline was used as the lipolytic agent results were quite different from those with theophylline. Figure 8 shows that at maximal stimulation of cyclic AMP accumulation by 3 μM noradrenaline, protein kinase activity as well as lipolysis was also activated maximally. In contrast to results with theophylline, a level of cyclic AMP about 11-fold higher was obtained. Furthermore, addition of PIA decreased cyclic AMP to almost basal values. Protein kinase activity and lipolysis remained unaffected.

This dissociation between lipolysis and cyclic AMP accumulation in fat cells proved to be rather complex. Using maximal concentrations of noradrenaline, no inhibition of lipolysis by adenosine or N^6-substituted derivatives occurred, whereas cyclic AMP levels decreased to almost basal values in the presence of these nucleosides (T. Trost and K. Stock, *unpublished data)*. At submaximal concentrations of noradrenaline, inhibition of lipolysis by adenosine or N^6-substituted derivatives occurred (Fig. 2). Even under these conditions maximal inhibition was less than 50% (Fig. 2). From these and other results (8,24) it was hypothesized that

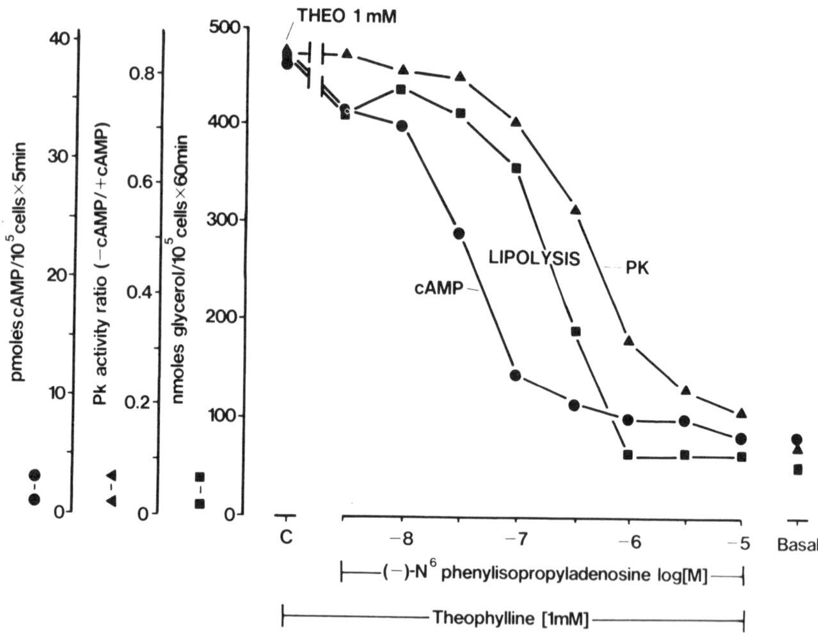

FIG. 7. Effect of PIA on cyclic AMP, protein kinase (PK) activity ratio (−cAMP/+cAMP), and lipolysis of rat fat cells stimulated maximally by theophylline (THEO). **Basal:** No additions; PIA alone had no significant effect on any parameter. **C:** Effect of 1 mM theophylline in the absence of PIA. Maximal protein kinase activity (+cAMP) remained unaffected by either theophylline or PIA. Values are means *(N* = 4) obtained from four different cell preparations with one to two incubates for each concentration. [Data from Wiedbrauck and Stock *(to be published).]*

unknown factors other than cyclic AMP were involved in maintenance of the lipolytic action of catecholamines. However, the data presented in Fig. 9 argue against the hypothesis of unknown factors. In this experiment protein kinase activity and lipolysis were maximally activated by 3μM noradrenaline in the presence of 1μM PIA, which was maximally inhibitory at this concentration with respect to cyclic AMP accumulation. From the data presented in Fig. 8 it can be concluded that cyclic AMP levels were reduced to almost basal value. However, the decrease in cyclic AMP effected by PIA does not represent the basal level. As shown in Fig. 9, the cyclic AMP level in the presence of PIA is still slightly elevated and is further decreased to basal level by addition of a β-blocker. Displacement of noradrenaline from its receptor by the β-blocking agent bupranolol results in a suppression of the residual cyclic AMP level to its basal value, accompanied by a simultaneous decline of protein kinase activity and lipolysis.

From these data one can assume the existence of two states of activity of the fat cell adenylate cyclase: (a) an "adenosine-sensitive" state and (b) an "adenosine-insensitive" state. A 100% "adenosine-sensitive" state of adenylate cyclase activity is represented in the absence of adenosine. A small increase of cyclic AMP

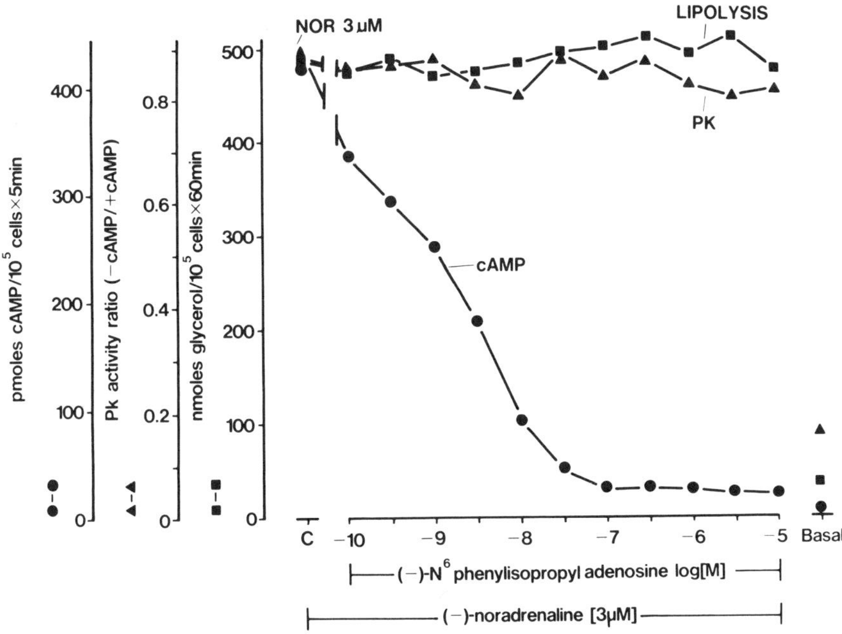

FIG. 8. Effect of PIA on cyclic AMP, protein kinase (PK) activity ratio (−cAMP/+cAMP), and lipolysis of rat fat cells stimulated maximally by noradrenaline (NOR). **Basal:** No additions; none of the basal values was significantly affected by PIA at the concentrations used. **C:** Effect of 3 μM noradrenaline in the absence of PIA. Maximal protein kinase activity (+cAMP) remained unaffected by either noradrenaline or PIA. Values are means (*N*=3) obtained from three different cell preparations with one to two incubates for each concentration. [Data from Wiedbrauck and Stock *(to be published).]*

levels can occur at this state, sufficient to completely activate protein kinase activity and lipolysis. In the "adenosine-sensitive" state the adenylate cyclase shows a high degree of sensitivity toward hormones, e.g., at maximally active concentrations of a hormone a powerful amplification of the enzyme activity can occur, reflected in high levels of cyclic AMP. However, on addition of adenosine and in the presence of a hormone the adenylate cyclase can alter its state of activity, and at maximally active concentrations of a hormone *plus* maximally active concentrations of adenosine a 100% "adenosine-insensitive" state of adenylate cyclase activity is obtained. At this state a small increase of cyclic AMP level can be seen, sufficient to completely activate protein kinase activity and lipolysis.

Both states of adenylate cyclase activity can exist at the same time. Therefore, the model of two states of adenylate cyclase activity can be used to explain the partial inhibition of lipolysis by adenosine after submaximal stimulation with noradrenaline (Fig. 2); the adenylate cyclase is partly in the "adenosine-sensitive" state and partly in the "adenosine-insensitive" state. Thus the total amount of cyclic AMP formed in the presence of submaximal concentrations of noradrenaline derives from "two"

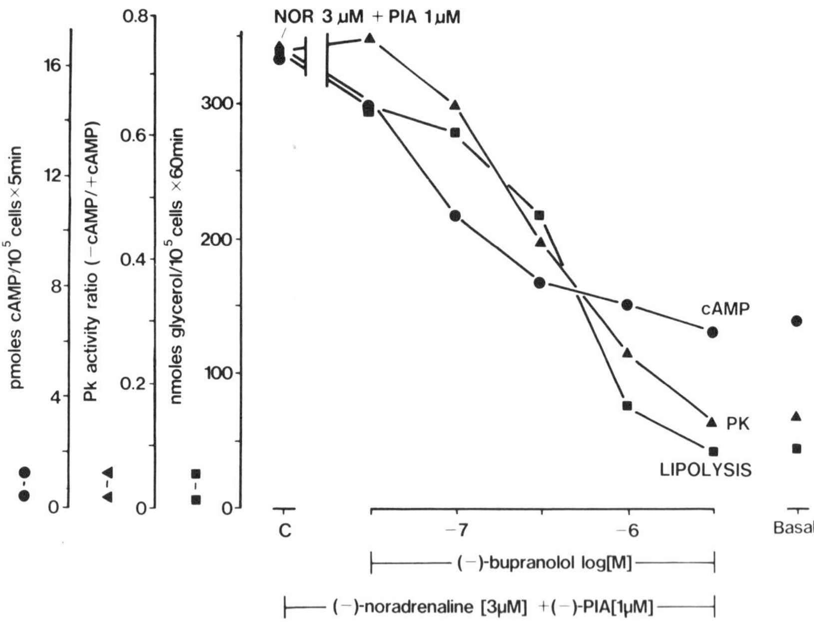

FIG. 9. Effect of the β-adrenolytic (−)-bupranolol on cyclic AMP, protein kinase (PK) activity ratio (−cAMP/+cAMP), and lipolysis of rat fat cells in the presence of the maximally active concentrations of 3 μM noradrenaline (NOR) plus 1 μM PIA. **Basal:** No additions; none of the basal values was significantly affected by (−)-bupranolol or PIA alone. **C:** Effect of 3 μM noradrenaline plus 1 μM PIA in the absence of (−)-bupranolol; note numerical values for cyclic AMP on ordinate. Maximal protein kinase activity (+cAMP) remained unaffected by (−)-bupranolol, noradrenaline, or PIA. Values are means (N = 2) obtained from two different cell preparations with two to three incubates for each concentration. [Data from Wiedbrauck and Stock *(to be published).*]

sources, being responsible for a submaximal lipolysis. On addition of adenosine the lipolytic response can be further decreased via inhibition of the "adenosine-sensitive" state of adenylate cyclase activity. The remaining lipolysis, which is not suppressible by the action of adenosine, reflects a submaximal activity of the "adenosine-insensitive" adenylate cyclase. However, all these assumptions are only speculative at this time and further studies are necessary.

Nevertheless, adenosine undoubtedly is a local regulatory agent in adipose tissue. In the case of hormone-stimulated lipolysis, adenosine appears to prevent excessive energy expenditure. The enormous increase of cyclic AMP owing to hormones in the absence of adenosine is unnecessary for a maximal lipolytic response, because even in the presence of adenosine, i.e., at low cyclic AMP levels, a regulation of lipolysis by hormones occurs.

ACKNOWLEDGMENTS

These studies were supported by the Deutsche Forschungsgemeinschaft (Sto 80/7). We thank Mrs. Karin Meyer and Mrs. Gabriele Rachow for their excellent technical assistance. Some of the data presented are part of the doctoral thesis of T. T. (Fachbereich Pharmazie, Universität Tübingen).

REFERENCES

1. Blackard, W. G., and Cameron, T. (1967): *Metabolism,* 16:91–95.
2. Butcher, R. W. (1971): *Naunyn Schmiedebergs Arch. Pharmacol.,* 269:358–372.
3. Davies, J. I. (1968): *Nature,* 218:349–352.
4. Dole, V. P. (1961): *J. Biol. Chem.,* 236:3125–3130.
5. Ebert, R., and Schwabe, U. (1971): *Acta Pharmacol. Toxicol.,* 29(Suppl. 4):10.
6. Ebert, R., and Schwabe, U. (1973): *Naunyn Schmiedebergs Arch. Pharmacol.,* 278:247–259.
7. Fain, J. N. (1973): *Pharmacol. Rev.,* 25:67–118.
8. Fain, J. N. (1973): *Mol. Pharmacol.,* 9:595–604.
9. Fain, J. N. (1977): In: *Cyclic 3',5'-Nucleotides: Mechanisms of Action,* edited by H. Cramer and J. Schultz, pp. 207–228. Wiley and Sons, London-New York-Sydney-Toronto.
10. Fain, J. N., Pointer, R. H., and Ward, W. F. (1972): *J. Biol. Chem.,* 247:6866–6872.
11. Fain, J. N., and Wieser, P. B. (1975): *J. Biol. Chem.,* 250:1027–1034.
12. Henion, W. F., Sutherland, E. W., and Posternak, T. (1967): *Biochim. Biophys. Acta,* 148:106–114.
13. Ho, R. J., and Sutherland, E. W. (1971): *J. Biol. Chem.,* 246:6822–6827.
14. Kappeler, H. (1966): *Diabetologia,* 2:52–61.
15. Londos, C., and Wolff, J. (1977): *Proc. Natl. Acad. Sci. U.S.A.,* 74:5482–5486.
16. Malbon, C. C., Hert, R. C., and Fain, J. N. (1978): *J. Biol. Chem.,* 253:3114–3122.
17. Manganiello, V. C., Murad, F., and Vaughan, M. (1971): *J. Biol. Chem.,* 246:2195–2202.
18. McKenzie, S. G., and Baer, H. P. (1973): *Can. J. Physiol. Pharmacol.,* 51:190–196.
19. Pereira, J. N., and Holland, G. F. (1966): *Experientia,* 22:658–659.
20. Raben, M. S., and Matsuzaki, F. (1966): *J. Biol. Chem.,* 241:4781–4786.
21. Schwabe, U., and Ebert, R. (1974): *Naunyn Schmiedebergs Arch. Pharmacol.,* 282:33–44.
22. Schwabe, U., Ebert, R., and Erbler, H. C. (1973): *Naunyn Schmiedebergs Arch. Pharmacol.,* 276:133–148.
23. Schwabe, U., Ebert, R., and Erbler, H. C. (1975): In: *Advances in Cyclic Nucleotide Research,* Vol. 5, edited by G. I. Drummond, P. Greengard, and G. A. Robinson, pp. 569–584. Raven Press, New York.
24. Stock, K., and Prilop, M. (1974): *Naunyn Schmiedebergs Arch. Pharmacol.,* 282:15–31.
25. Trost, T., and Stock, K. (1977): *Naunyn Schmiedebergs Arch. Pharmacol.,* 299:33–40.
26. Westermann, E. (1967): In: *12. Symposium der Deutschen Gesellschaft für Endokrinologie,* pp. 154–173. Springer-Verlag, Berlin-Heidelberg-New York.
27. Westermann, E., Stock, K., and Bieck, P. (1969): In: *Fettstoffwechsel,* vol. 5, edited by L. Demling and G. Berg, pp. 68–73. Pallas Verlag, Lochham-München.

Physiological and Regulatory Functions of Adenosine and Adenine Nucleotides, edited by H. P. Baer and G. I. Drummond. Raven Press, New York © 1979.

Abnormalities of Purine Metabolism in Human Immunodeficiency Diseases

J. Edwin Seegmiller

Department of Medicine, University of California San Diego, La Jolla, California 92093

The inborn errors of human metabolism have long been regarded as "experiments of nature" and their detailed investigation has provided valuable insight into a wide range of biochemical and clinical problems (19,64). To date, nearly 200 genetically determined defects in enzymes or other proteins have been identified in the human species (40,64). In the last six years the abnormal gene products responsible for the loss of function of portions of the immune system have begun to be identified including deficiencies of any one of three sequential enzyme systems of purine nucleotide degradation that have been found in association with an aberrant function of the immune system (Fig. 1).

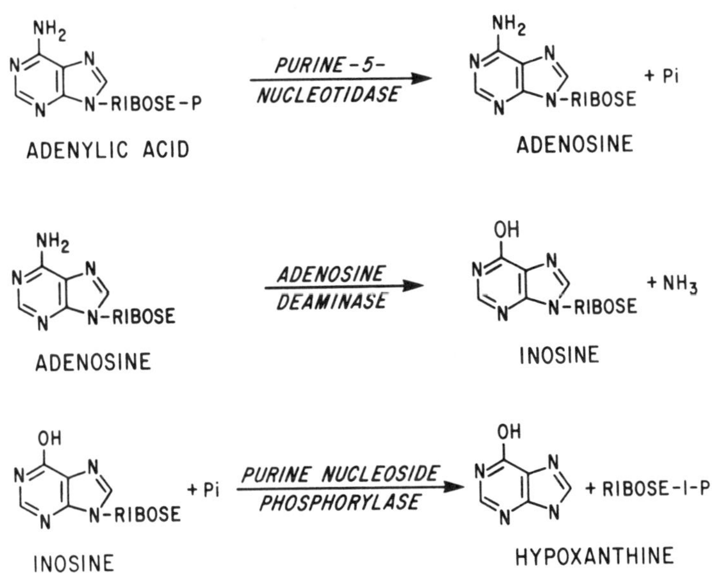

FIG. 1. Enzyme defects of purine nucleotide catabolism associated with various immunodeficiency diseases. Purine 5′-nucleotidase deficiency has been reported in patients with defective B cell function, both X linked (16) and adult onset agammaglobulinemia (14,70). Adenosine deaminase deficiency is found in some patients with severe combined immunodeficiency disease with dysfunction of T cells and, to a lesser extent, B cells. Purine nucleoside phosphorylase deficiency is found in patients with an isolated T cell function impairment.

ENZYME DEFECTS IN PURINE METABOLISM ASSOCIATED WITH IMMUNODEFICIENCY DISEASE

Adenosine Deaminase Deficiency

The first enzyme defect to be identified was a gross deficiency of adenosine deaminase (ADA) found by Giblett et al. in 1972 (22) in two unrelated children with severe combined immunodeficiency disease, the most severe type of disorder of the immune system affecting both T and B cell function (25,32). Well over two dozen families have since been described in which this association has been found, and ADA deficiency accounts for up to one-half of the patients with the autosomal recessive forms of the severe combined immunodeficiency diseases (32).

Purine Nucleoside Phosphorylase Deficiency

Three years later, Giblett et al. (21) reported a child with isolated T cell dysfunction and severe deficiency of purine nucleoside phosphorylase. A total of 7 patients have since been found with this type of association in four different families all showing autosomal recessive inheritance (57). Affected children are particularly susceptible to viral illnesses and overwhelming vaccinia infection has taken the life of two such patients (57). Two siblings in one of the families have an incomplete deficiency of the enzyme associated with a clinically attenuated form of the immunological disease (16).

Ecto-Purine 5′-Nucleotidase Deficiency

Edwards et al. (14) have reported a gross deficiency of ecto-purine 5′-nucleotidase of the lymphocyte membrane to one-third of control values associated with congenital agammaglobulinemia in males. Affected males are unable to make gammaglobulins as a result of a B cell dysfunction and therefore are particularly susceptible to recurrent bacterial infections. This enzyme defect is not in agreement with the observations of Webster et al. (70) who have reported that lymphocytes from patients with Bruton's agammaglobulinemia show activities for this enzyme in the lower one-third of the range found in cells from normal individuals. A gross deficiency of this enzyme has also been reported in some patients with nonfamilial agammaglobulinemia of adult onset (35,70).

PURINE METABOLISM IN THE NORMAL CELL

Role in Cellular Proliferation

The need for an abundant supply of purine and pyrimidine nucleotides for ribonucleic acid (RNA) required for protein synthesis and deoxyribonucleic acid (DNA) for replication during cellular proliferation has long been realized and provides the basis for the large investments in programs to develop purine and pyrimidine analogs as possible cancer chemotherapeutic agents (48). A complication arising during the therapeutic evaluation of many of these agents in treatment of cancer was the development of suppression of the immune system.

Roles for ADA

The association of severe combined immunodeficiency with a gross deficiency of the enzyme ADA came as something of a surprise to most biochemists and geneticists. In retrospect, this need not have been such a surprise if we consider the distribution of ADA. It is most abundant in the thymus and next most abundant in the spleen and lymph nodes (1,7).

The enzyme would be expected to be of functional significance in organs and tissues in which physiological roles of adenosine are likely or have been proposed, including fat cells (15,68), "purinergic" nerves (6), and blood vessels (20). A role for the "adenosine cycle" as a major pathway of purine degradation and metabolism was proposed by Green and Ishii (27), but no evidence that this cycle plays a major role in purine metabolism has been found by subsequent work of Brox and Henderson (5).

Metabolic Fates of Adenosine

Adenosine has three possible metabolic fates: (a) It can be reconverted into adenylic acid through the action of adenosine kinase which shows a high affinity for adenosine with a K_m of 2.0 μM (52). (b) The enzyme ADA, by contrast, has a higher K_m of 20 to 40 μM (3,47), and shows a substantially higher capacity (V_{max}). (c) A third fate for adenosine is to react with homocysteine to form S-adenosylhomocysteine with a K_m of about 1.0 μM that could well inhibit RNA methylation reactions required for RNA and protein synthesis (29,36,37).

The kinetic properties of ADA seem well suited for the rapid removal of any excessive quantities of adenosine that might be presented to the cell, either from extracellular sources or from increased breakdown of adenosine 5′-monophosphate or of S-adenosylhomocysteine. Presumably, the principal function of ADA in the cell is to provide a "spillway" for disposing of increased amounts of adenosine.

Metabolism of Deoxyadenosine

Deoxyadenosine is an even better substrate for ADA than adenosine (3,61). It can also be converted to a nucleotide by a kinase; however, the K_m of the kinase is around 400 μM, making the pathway by ADA the preferential route of metabolism (7,17). The deoxyadenylic acid formed by the kinase is rapidly phosphorylated further to deoxy-adenosine triphosphate (deoxy-ATP).

Role of Purine Nucleoside Phosphorylase

The enzyme purine nucleoside phosphorylase (PNP) is responsible for converting both deoxyribonucleosides and ribonucleosides of guanine and hypoxanthine in the presence of inorganic phosphate to free bases and pentose-1-phosphate (71). It shows little or no detectable action with adenine nucleosides. The reaction is readily reversible but the absence of a kinase for inosine or guanosine in human tissues (17) makes it unlikely that the reverse reaction is of any importance in the regeneration of nucleotides.

Role of Ecto-Purine 5′-Nucleotidase

The enzyme purine 5′-nucleotidase converts purine nucleotides to the purine nucleosides (18). It is a sparingly soluble enzyme when associated with the cell wall and must be distinguished from similar enzyme activities found in the cell lysate that are not altered in patients with agammaglobulinemia (14,35,70). If the nucleotides were normally most abundant in B cells, the seeming deficiency in lymphocytes of patients with agammaglobulinemia could merely reflect the low B cell count characteristic of this disease.

METABOLIC CONSEQUENCE OF ENZYME DEFECTS

Ecto-Purine 5′-Nucleotidase Deficiency

A decreased conversion of adenylic acid to adenosine would be expected in patients with deficiency of this enzyme, but no detailed studies have been performed so far. The extent to which intracellular purine nucleotides may be increased in lymphocytes of affected patients also remains to be evaluated.

Adenosine Deaminase Deficiency

The study of the deficiency of this enzyme has been greatly aided by the availability of a number of inhibitors of ADA, the most potent of which is deoxycoformycin with a K_I of 2.5 pM (2,34) and erythro-9-(2-hydroxy-3-nonyl) adenine (EHNA) which has a K_I of 1.3 nM (49). These agents have been added with adenosine to provide a model of ADA deficiency in a variety of cell systems.

The lethality of adenosine for cultured mammalian cells of lymphoid origin reported by Ishii and Green (33) was associated with a gross deficiency of pyrimidine nucleotides and was reversed by uridine suggesting that this was the main action for adenosine toxicity (26). A similar type of adenosine toxicity in human lymphoblast lines treated with EHNA was reversed only in part by uridine (62). These observations suggested that ADA deficiency led to adenosine toxicity which could be reversed by uridine in affected children. However, when added to their cells *in vitro,* uridine appeared to be of no value in restoring the ability of their lymphocytes to respond to mitogens (44,45). Also, direct treatment of affected children with uridine was unsuccessful (32). One might expect an accumulation of adenosine to be demonstrable in serum or urine of affected children but only minimal accumulations have been reported and only in serum (42).

Presumably, the alternate route of adenosine metabolism via adenosine kinase could lead to an increase in the cellular adenine nucleotides in affected patients. Such an increase has been reported in red cells (42) and in lymphocytes (24,43,51). The elevation of ATP in lymphocytes was corrected by a form of enzyme-replacement therapy consisting of transfusion of irradiated erythrocytes that concurrently restored the immune function in an affected child (43). The latter approach was prompted by noting an increase in responsiveness of lymphocytes of an ADA-deficient child to phytohemagglutinin (PHA) stimulation by addition of exogenous ADA to the culture medium (45). Unfortunately, the red cell transfusion has not been successful in all patients with ADA deficiency (50). Furthermore, the identity of the ATP was not firmly established and it now appears that it may well have been deoxy-ATP (see below).

An increased concentration of adenine in the serum and urine of affected children has been reported (42). Subsequent studies of the same patient showed no evidence of abnormal amounts of adenine or adenosine excretion in the urine but showed a 6,000-fold increase in concentration of deoxyadenosine (24,38,60). The possibility that the free adenine may have arisen during sample preparation from hydrolysis of deoxyadenosine seems very likely (24).

Purine Nucleoside Phosphorylase Deficiency

Children with the most extreme degrees of PNP deficiency have shown a profound hypouricemia and excretion of very little uric acid in the urine as might be expected from the position of the enzyme defect (Fig. 1) (10,65). Instead, they excrete in the urine inosine and guanosine and lesser amounts of deoxyinosine and deoxyguanosine. On a molar basis, the total amount of purine excretion is quite comparable to the amount of uric acid excreted by children with Lesch-Nyhan disease who are deficient in the enzyme hypoxanthine-guanine phosphoribosyltransferase (HPRT) (39,55,58).

A common denominator to account for the excessive purine production in these two conditions is the absence of hypoxanthine reutilization to inosinic acid in both. It results from the virtual absence of the enzyme HPRT required for hypoxanthine reutilization in children with Lesch-Nyhan disease and from the inability to make hypoxanthine in children with PNP deficiency (Fig. 2). Studies of human cells cultured *in vitro* show the rate of synthesis of purines *de novo* in normal cells can be accelerated to values comparable to that seen in cells from children with Lesch-Nyhan disease by removing all hypoxanthine from the medium (30,67).

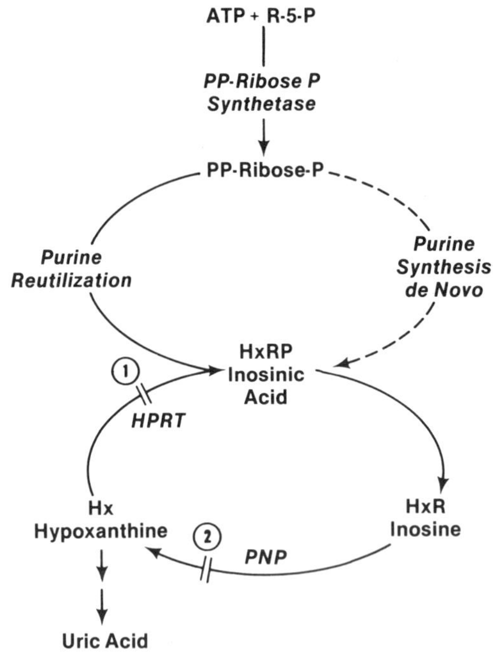

FIG. 2. Interruption of hypoxanthine reutilization cycle as a proposed mechanism of purine overproduction in **(1)** Lesch-Nyhan syndrome (HPRT deficiency), and **(2)** PNP deficiency. (From Seegmiller, ref. 56, with permission.)

POSSIBLE MECHANISMS OF IMMUNOSUPPRESSION IN ADA AND PNP DEFICIENCY

A wide variety of mechanisms have been proposed to account for the immunodeficiency associated with ADA deficiency (10,24,41,54–57,59,69).

Chromosomal Deletion

The possibility that ADA deficiency is an incidental accompaniment of a chromosomal deletion involving primarily the immune response genes and nearby locus for ADA was one of the first proposals (22). However, a chromosomal deletion is a relatively rare occurrence and the subsequent finding of this same correlation of ADA deficiency and the clinical findings of severe combined immunodeficiency disease being present in children in many different families made this hypothesis highly unlikely. The subsequent identification of the histocompatibility gene locus on chromosome 6 and ADA on chromosome 20 completely eliminated this mechanism from consideration (40).

Adenosine Versus Deoxyadenosine as a Possible Mediator

Most of the other mechanisms proposed have postulated a toxic action on lymphocytes or lymphocyte precursors from accumulation of an unreacted substrate of the missing enzyme rather than from a deficiency of a product of the reaction. Inosine, a major product of ADA, can be formed by two alternate pathways: (a) from a reversal of the reaction catalyzed by purine nucleoside phosphorylase and (b) from the dephosphorylation of inosinic acid by purine 5′-nucleotidase. The same two enzymes provide an alternative route that is probably of greater importance than ADA for the synthesis of hypoxanthine as well (5).

The above considerations led most investigators to consider adenosine the culprit since it was already known to be toxic and to cause a depletion of pyrimidine nucleotides (26).

The impairment of the proliferative response of normal lymphocytes to mitogens produced by adenosine and its potentiation by EHNA or deoxycoformycin (9,63) and the ability of these same agents to inhibit the production of antibodies *in vitro* (59) was taken as additional evidence for a possible role for adenosine as a mediator of the imunosuppression.

Very modest elevations of the plasma concentration of adenosine were accompanied by substantially elevated concentrations of ATP in erythrocytes (42) and even more elevated concentrations in lymphocytes (43,51). One might expect such an increase in ATP from the preferential action of adenosine kinase. The correlation of a lowering of lymphocyte ATP with restoration of immune response after enzyme replacement by administration of a transfusion of irradiated erythrocytes occurred even though the lymphocytes showed no increase in ADA. It was also compatible with the view that adenosine might be mediating the toxicity (43). The report of an increased ATP concentration in lymphocytes from other patients with combined immunodeficiency disease and normal ADA and PNP suggested that such an accumulation might well be the final common mediator of the immunosuppression in other metabolic defects as well (23).

However, a number of observations emerged that threw into question the view that adenosine was the toxic mediator. Adenosine was found to be toxic in cultured lymphoblasts deficient in adenosine kinase (31). The failure to find adenosine in the urine (42) was also troublesome with respect to this concept. Furthermore, the report of free adenine in plasma and urine (42) of an ADA-deficient patient was not in keeping with the concepts and suggested the possibility that adenine might be an artifact of sample preparation.

The reason most investigators ignored the possibility that deoxyadenosine accumulation might be a significant mediator of toxicity reflected the general lack of appreciation of the magnitude and importance of this metabolic pathway. The surprisingly large amount of deoxynucleosides excreted in the urine of children with PNP deficiency demonstrated for the first time the turnover in the body of substantial amounts of these nucleosides and focused attention on their metabolic origin. This, in turn, led to the realization that some deoxyinosine could well originate from deoxyadenosine through the action of ADA and that the quantities of deoxyadenosine processed by this enzyme could be substantial.

These considerations led Carson et al. (7) to compare the relative toxicity of ribonucleosides and deoxynucleosides for growth of human lymphoblasts in culture and their effect on mitogen stimulation of peripheral blood lymphocytes. When HPRT-deficient lymphoblasts were used to eliminate toxic effects from guanine accumulation, deoxyinosine and deoxyguanosine showed substantially higher toxicity for growth than did the corresponding ribonucleoside. Likewise, the toxicity of deoxyadenosine in PHA-stimulated lymphocytes was enhanced over 200-fold by the simultaneous addition of the ADA inhibitor EHNA. The toxicity was not affected significantly by addition of uridine, but addition of deoxycytidine restored growth to 70% of control values (7). A possible theoretical basis for the greater toxicity of the deoxynucleosides to T cells was provided by the demonstration of that deoxynucleoside kinase is most abundant in human thymus, even though its K_m is quite high at 400 μM (7). Subsequent work suggested that the greater effect on T cells *in vivo* may reflect a fundamental difference in their metabolism (8). Permanent lymphoblast lines of T cell origin showed a substantially greater sensitivity to growth inhibition by deoxyadenosine in the presence of the ADA inhibitor deoxycoformycin than did similar lines derived from B cells despite their carrying similar activities of deoxynucleoside kinase (8).

The results suggested the possibility that one or more of these deoxynucleosides might mediate the immunosuppression of T cells in children with PNP deficiency. The possibility that deoxyadenosine might be a toxic agent in mediating immunosuppression in ADA deficiency would provide a unifying concept for relating the immunosuppression in both PNP and ADA deficiency. Indeed, deoxyadenosine proved to be more effective in suppressing cell proliferation than adenosine in a variety of cultured cell systems with potentiation of the toxic effect by addition of deoxycoformycin to inhibit ADA (4). Deoxyadenosine was about 10-fold more effective than adenosine in inhibiting growth of ADA-deficient human fibroblasts, as shown in Fig. 3, and in inhibiting proliferation of PHA-stimulated lymphocytes when ADA was inhibited by deoxycoformycin (Fig. 4) (4). Even in the absence of ADA inhibitors, deoxyadenosine was a more potent inhibitor of the mitogen response of human lymphocytes than adenosine (66) and was greatly potentiated by ADA inhibitors (28,60).

Deoxy-Adenosine Triphosphate Accumulation

The demonstration of an accumulation of deoxy-ATP in erythrocytes of children deficient in ADA provided an additional line of evidence in favor of a possible pathogenetic role of deoxyadenosine in this disease. Cohen et al. (12) found a more than 100-fold elevation in the concentration of deoxy-ATP of an ADA-deficient patient, and Coleman (13) has confirmed these findings with a demonstration of over 1,500 times the normal amount of deoxy-ATP and over 260 times the normal amount of deoxy-adenosine diphosphate in erythrocytes of an affected child. The

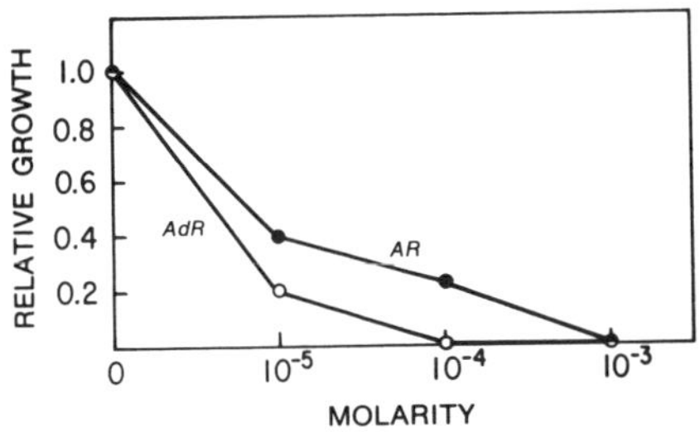

FIG. 3. Comparison of inhibition produced by 2′-deoxyadenosine (AdR) and by adenosine (AR) on growth of human fibroblasts from a patient with ADA deficiency. (From Bluestein et al. ref. 4, with permission.)

identity of the deoxynucleotide triphosphate was confirmed in each case by reaction with DNA polymerase as well as by gas chromatography of the trimethylsilyl derivatives and mass spectrometry (13). The presence of increased quantities of deoxy-ATP in erythrocytes of ADA-deficient children suggests that the 10-fold increase in "ATP" of their lymphocytes (24,43,51) may actually have been deoxy-ATP. No elevation of deoxy-ATP has yet been identified in lymphocytes of affected children. However, in a model system consisting of PHA-stimulated lymphocytes with ADA inhibited by deoxycoformycin, deoxyadenosine treatment produced a substantial increase in deoxy-ATP as shown in Fig. 5 (4).

Deoxy-Guanosine Triphosphate Accumulation in PNP Deficiency

An accumulation of deoxy-guanosine triphosphate (deoxy-GTP) has been found in erythrocytes of a child with PNP deficiency quite comparable in magnitude to the deoxy-ATP accumulation in ADA deficiency (11).

Correction of an *In Vitro* Model of ADA Deficiency by Balanced Deoxynucleoside Addition

A fresh approach to possible treatment of ADA and PNP deficiency is provided by studies of a model system for ADA deficiency. In the presence of the ADA inhibitor deoxycoformycin the proliferative response of phytohemagglutinin-stimulated lymphoblasts, as measured by incorporation of [^{14}C]leucine into protein,

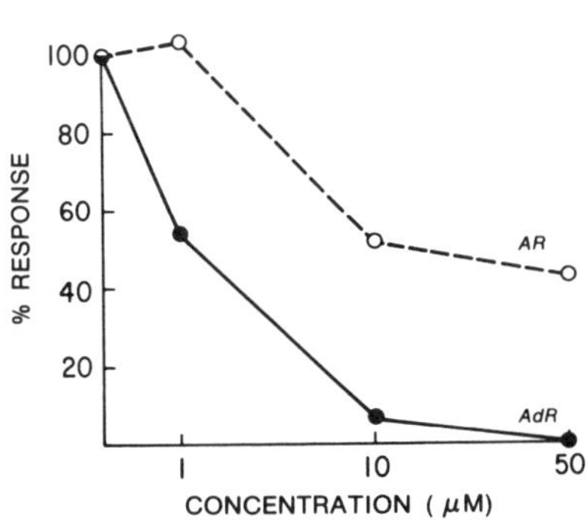

FIG. 4. Inhibition of phytohemagglutinin-induced proliferation of human peripheral blood lymphocytes by 2′-deoxyadenosine (AdR) and by adenosine (AR). Incorporation of [^{14}C]leucine into cell protein was used as an index of the proliferative response. All cultures contained 1 μM 2′-deoxycoformycin to inhibit adenosine deaminase. (From Bluestein et al. ref. 4, with permission.)

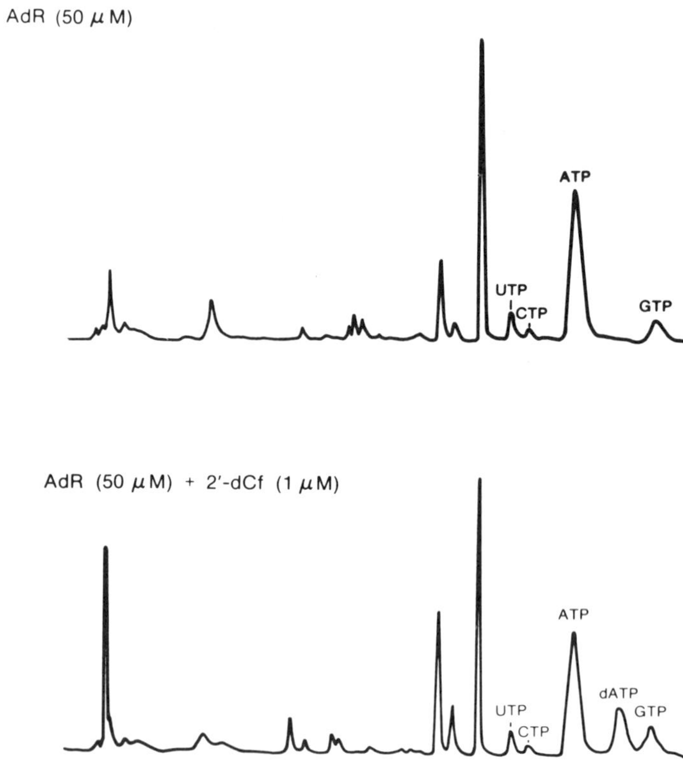

FIG. 5. A model system of ADA deficiency; accumulation of deoxy-ATP in phytohemagglutinin-stimulated lymphocytes from human peripheral blood by treatment with 50 μM 2′-deoxy-adenosine and 1μM 2′-deoxycoformycin to inhibit ADA. (From Bluestein et al. ref. 4, with permission.)

was inhibited to 10% of control values by addition of deoxyadenosine as shown in Fig. 6. This inhibition was substantially overcome by addition of various other single deoxynucleosides, and a combination of either thymidine or deoxycytosine with deoxyguanosine completely overcame the inhibition (4).

HYPOTHETICAL MECHANISMS

Possible Role of Increased Cyclic-3′,5′-Adenylic Acid in ADA Deficiency

The possibility that the defect in purine metabolism in ADA deficiency might be mediated through raising the intracellular concentrations of cyclic-3′,5′-adenylic acid (cAMP) has been a recurring hypothesis (41). Ullman et al. (69) presented arguments against it by showing equal sensitivity to toxic effects of adenosine in mutant mouse lymphoma cell lines in which receptors for cAMP were present or deficient. On the other hand, Goldblum et al. (24) have reported a two- to threefold

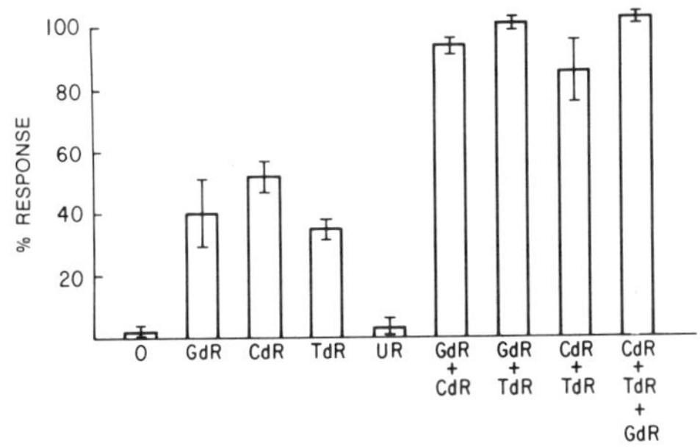

FIG. 6. Inhibition of phytohemagglutinin-induced proliferation of human peripheral blood lymphocytes by 10 μM deoxyadenosine and 1 μM deoxycoformycin and reversal of the inhibition by addition of deoxyguanosine (GdR), deoxycytosine (CdR), thymidine (TdR), and uridine (UR), and combinations of nucleosides. (From Bluestein et al. ref. 4, with permission.)

increase in concentration of cAMP in ADA-deficient lymphocytes which was decreased 21% by incubation with exogenous ADA. Other, less direct, evidence for a role for cAMP has come from the demonstrations by Polmar et al. (44) of a further decrease in mitogen response of ADA-deficient lymphocytes with little effect on normal lymphocytes by a variety of agents, such as theophylline, prostaglandins E_1 and E_2, and norepinephrine, that are known to raise the intracellular cAMP concentration, whereas lithium chloride, an inhibitor of cAMP synthesis, enhanced the mitogenic response. The latter observation suggests another possible pharmacologic approach to therapy. The possibility that adenosine receptors on the lymphocyte may mediate the increase in cAMP was suggested by Schwartz et al. (53).

Immunosuppression Through Purine Deoxynucleotide Accumulation in Both ADA and PNP Deficiency

An imbalance of any one of the deoxynucleotide triphosphates is known to exert an allosteric inhibition on ribonucleoside diphosphate reductase, an enzyme required for the synthesis of deoxynucleotides from ribonucleotides (46). An excess of deoxy-ATP such as would accumulate in ADA deficiency results in a general allosteric inhibition of reduction of all substrate ribonucleotides resulting in a cessation of DNA synthesis. An excess of deoxy-GTP such as would accumulate in PNP deficiency produces an allosteric inhibition principally of the synthesis of deoxy-CTP. In either case, DNA synthesis would be interrupted by the lack of a properly balanced supply of deoxynucleoside triphosphates as shown in Fig. 7. Conceivably, deoxycytosine could provide a specific correction of any replicating process interrupted by a lack of deoxy-CTP in PNP deficiency.

The fact that the inhibition in this model system of ADA deficiency (4) can be countered completely by providing a more balanced supply of deoxynucleosides is quite compatible with three quite different mechanisms. The additional deoxynucleosides could directly restore the balance of intracellular deoxynucleotide triphos-

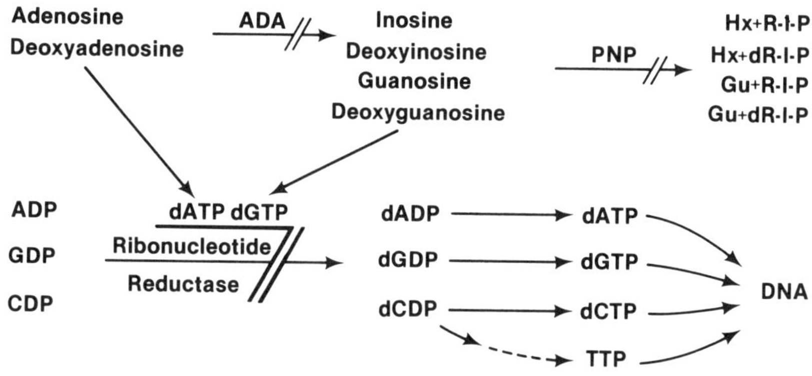

FIG. 7. Unitary hypothesis for mechanism of immunodeficiency through inhibition of lymphocyte proliferation in ADA and PNP deficiency. The accumulated deoxynucleosides (deoxyadenosine in ADA deficiency and deoxyguanosine in PNP deficiency) are converted to their respective triphosphates. These in turn produce an allosteric inhibition of ribonucleotide reductase that shuts off production of deoxynucleotides and thereby blocks DNA synthesis. Reversal of the inhibition by other deoxynucleosides could be from providing precursors for synthesis of a balanced complement of deoxynucleotide triphosphates needed for DNA synthesis, or from competition of added deoxynucleosides to reduce deoxyadenosine transport or phosphorylation by deoxynucleoside kinase.

phates needed for DNA synthesis. Alternative explanations might be an interference by the additional deoxynucleosides with either the action of the deoxynucleoside kinase used by all of the deoxynucleosides or in a transport process of nucleosides into the cells shared by all of the deoxynucleosides. In any event, the success of the added deoxynucleosides in overcoming the inhibition in the model system gives promise of a fresh approach to possible clinical treatment of these disorders *in vivo*.

ACKNOWLEDGMENTS

This work was supported in part by grants AM-13622 and GM-17702 from the United States Public Health Service and grants from the Kroc Foundation.

REFERENCES

1. Adams, A., and Harkness, R. A. (1976): *Clin. Exp. Immunol.,* 26:647–649.
2. Adamson, R. H., Zaharevitz, D. W., and Johns, D. G. (1977): *Pharmacology,* 15:84–89.
3. Agarwal, R. P., Sagar, S. M., and Parks, R. E., Jr. (1975): *Biochem. Pharmacol.,* 24:693–701.
4. Bluestein, H. G., Willis, R. C., Thompson, L. F., Matsumoto, S., and Seegmiller, J. E. (1978): *Trans. Assoc. Am. Phys. (in press).*
5. Brox, L. W., and Henderson, J. F. (1976): *Can. J. Biochem.,* 54:200–202.
6. Burnstock, G. (1978): Past and current evidence for the purinergic nerve hypothesis. *(This volume.)*
7. Carson, D. A., Kaye, J., and Seegmiller, J. E. (1977): *Proc. Natl. Acad. Sci. U.S.A.,* 74:5677–5681.
8. Carson, D. A., Kaye, J., and Seegmiller, J. E. (1978): *J. Immunol.,* 121:1726–1731.
9. Carson, D. A., and Seegmiller, J. E. (1976): *J. Clin. Invest.,* 57:274–282.
10. Cohen, A., Doyle, D., Martin, D. W., Jr., and Ammann, A. J. (1976): *N. Engl. J. Med.,* 295:1449–1454.

11. Cohen, A., Gudas, L. J., Ammann, A. J., Staal, G. E. J., and Martin, D. W. (1978): *J. Clin. Invest.*, 61:1405–1409.
12. Cohen, A., Hirschhorn, R., Horowitz, S. D., Rubinstein, A., Polmar, S. H., Hong, R., and Martin, D. W., Jr. (1978): *Proc. Natl. Acad. Sci. U.S.A.*, 75:472–476.
13. Coleman, M. S., Donofrio, J., Hutton, J. J., Hahn, L., Daoud, A., Lampkin, B., and Dyminski, J. (1978): *J. Biol. Chem.*, 253:1619–1626.
14. Edwards, N. L., Magilavy, D. B., Cassidy, J. T., Fox, I. H. (1978): *Science*, 201:628–630.
15. Fain, J. N., and Wieser, P. B. (1975): *J. Biol. Chem.*, 250:1027–1034.
16. Fox, I. H., Andres, C. M., Gelfand, E. W., and Biggar, D. (1977): *Science*, 197:1084–1086.
17. Fox, I. H., and Kelley, W. N. (1978): *Annu. Rev. Biochem.*, 47:655–686.
18. Fox, I. H., and Marchant, P. J. (1976): *Can. J. Biochem.*, 54:462–469.
19. Garrod, A. E. (1923): *Inborn Errors of Metabolism*. Oxford, London.
20. Gerlach, E., Deuticke, B., and Dreisbach, R. H. (1963): *Naturwissenschaften*, 6:228–229.
21. Giblett, E. R., Ammann, A. J., Wara, D. W., Sandman, R., and Diamond, L. K. (1975): *Lancet*, 1:1010–1013.
22. Giblett, E. R., Anderson, J. E., Cohen, F., Pollara, B., and Meuwissen, H. J. (1972): *Lancet*, 2:1067–1069.
23. Goldblum, R. C., Schmalstieg, F. C., Goldman, A. S., Nelson, J. A., and Monahan, T. M. (1976): *Pediatr. Res.*, 10:387.
24. Goldblum, R. M., Schmalstieg, F. C., Nelson, J. A., and Mills, G. C. (1978): *Proc. Birth Defects Conf.*, Memphis, Tennessee *(in press)*.
25. Good, R. A., and Hansen, M. A. (1976): *Adv. Exp. Med. Biol.*, 73:155–178.
26. Green, H. (1975): In: *Combined Immunodeficiency Disease and Adenosine Deaminase Deficiency. A Molecular Defect*, edited by H. J. Meuwissen, R. J. Pickering, B. Pollara, and I. H. Porter, pp. 141–155. Academic Press, New York.
27. Green, H., and Ishii, K. (1972): *J. Cell. Sci.*, 11:173–177.
28. Harrap, K. R., and Paine, R. M. (1976): In: *Advances in Enzyme Regulation*, edited by G. Weber, pp. 169–193. Pergammon Press, New York.
29. Hershfield, M. S. (1978): *Fed. Proc.*, 37:1466.
30. Hershfield, M. S., and Seegmiller, J. E. (1977): *J. Biol. Chem.*, 252:6002–6010.
31. Hershfield, M. S., Snyder, F., and Seegmiller, J. E. (1977): *Science*, 197:1284–1286.
32. Hirschhorn, R. (1977): In: *Progress in Clinical Immunology*, edited by R. S. Schwartz, pp. 67–83. Grune & Stratton, San Francisco.
33. Ishii, K., and Green, H. (1973): *J. Cell. Sci.*, 13:429–439.
34. Johns, D. B., and Adamson, R. H. (1976): *Biochem. Pharmacol.*, 25:1441–1444.
35. Johnson, S. M., Asherson, G. L., Watts, R. W. F., North, M. E., Allsop, J., and Webster, A. D. B. (1977): *Lancet*, 1:168–170.
36. Kredich, N. M., and Hershfield, M. S. (1978): *Fed. Proc.*, 37:1465.
37. Kredich, N. M., and Martin, D. W., Jr. (1977): *Cell*, 12:931–938.
38. Kuttesch, J. F., Schmalstieg, F. C., and Nelson, J. A. (1978): *J. Liq. Chromatogr.*, 1:97–109.
39. Lesch, M., and Nyhan, W. L. (1964): *Am. J. Med.*, 36:561–570.
40. McKusick, V. A. (1975): *Mendelian Inheritance in Man*. The Johns Hopkins University Press, Baltimore.
41. Meuwissen, H. J., Pollara, B., Pickering, R. J., Porter, I. H., Hook, E. B., and Kelly, S. (1975): *Combined Immunodeficiency Disease and Adenosine Deaminase Deficiency*. Academic Press, San Francisco.
42. Mills, G. C., Schmalstieg, F. C., Trimmer, K. B., Goldman, A. S., and Goldblum, R. M. (1976): *Proc. Natl. Acad. Sci. U.S.A.*, 73:2867–2871.
43. Polmar, S. H., Stern, R. C., Schwartz, A. L., Wetzler, E. M., Chase, P. A., and Hirschhorn, R. (1976): *N. Engl. J. Med.*, 295:1337–1343.
44. Polmar, S. H., Wetzler, E. M., and Stern, R. C. (1978): In: *Inborn Errors of Immunity and Phagocytes*, edited by E. Wamberg and E. Guttler. NTP Press, Lancaster, England *(in press)*.
45. Polmar, S. H., Wetzler, E. M., Stern, R. C., and Hirschhorn, R., (1975): *Lancet*, 2:743–746.
46. Reichard, P. (1972): *Adv. Enzyme Regul.*, 10:3–16.
47. Rossi, C. A., Lucacchini, A., Montali, U., and Ronca, G. (1975): *Int. J. Pept. Protein Res.*, 7:81–89.
48. Sartorelli, A. C., and Johns, D. G. (editors) (1975): *Antineoplastic and Immunosuppressive Agents I and II*. Springer-Verlag, New York-Heidelberg-Berlin.
49. Schaeffer, H. J., and Schwende, G. F. (1974): *J. Med. Chem.*, 17:6–8.

50. Schmalstieg, F. C., Mills, G. C., Nelson, J. A., May, L. T., Goldman, A. S., and Goldblum, R. M. (1978): *J. Pediatr. (Submitted for publication.)*
51. Schmalstieg, F. C., Nelson, J. A., Mills, G. C., Monahan, T. M., Goldman, A. S., and Goldblum, R. M. (1977): *J. Pediatr.,* 91:48–51.
52. Schnebli, H. P., Hill, D. L., and Bennett, L. L., Jr. (1967): *J. Biol. Chem.,* 242:1997–2004.
53. Schwartz, A. L., Stern, R. C., and Polmar, S. H. (1978): *Clin. Immunol. Immunopathol. (in press).*
54. Seegmiller, J. E. (1976): In: *The Year in Metabolism 1975–1976,* edited by N. Freinkel, pp. 213–258. Plenum, New York.
55. Seegmiller, J. E. (1976): In: *Advances in Human Genetics,* edited by K. Hirschhorn, pp. 75–163. Plenum, New York.
56. Seegmiller, J. E. (1978): In: *The Year in Metabolism—1977,* edited by N. Freinkel, pp. 253–325. Plenum, New York.
57. Seegmiller, J. E. (1979): In: *Contemporary Metabolism: Analytical Reviews of Basic and Clinical Research,* edited by N. Freinkel, pp. 1–85. Plenum, New York.
58. Seegmiller, J. E., Rosenbloom, F. M., and Kelley, W. N. (1967): *Science,* 155:1682–1684.
59. Seegmiller, J. E., Watanabe, T., and Schreier, M. H. (1977): *Ciba Found. Symp. on Purine and Pyrimidine Metabolism,* pp. 249–266. Elsevier, Amsterdam.
60. Simmonds, H. A., Panayi, G. S., and Corrigal, V. (1978): *Lancet,* 1:60–63.
61. Snyder, F. F., and Henderson, J. F. (1973): *J. Biol. Chem.,* 248:5899–5904.
62. Snyder, F. F., Hershfield, M. S., and Seegmiller, J. E. (1978): *Cancer Res.,* 38:2357–2362.
63. Snyder, F. F., Mendelsohn, J. and Seegmiller, J. E. (1976): *J. Clin. Invest.,* 58:654–666.
64. Stanbury, J. B., Wyngaarden, J. B., and Frederickson, D. S. (editors) (1978): *The Metabolic Basis of Inherited Disease.* McGraw-Hill, San Francisco.
65. Stoop, J. W., Zegers, B. J. M., Hendricks, G. F. M., Siegenbeek van Heukelom, H. L., Staal, G. E. J., deBree, P. K., Wadman, S. K., and Ballieux, R. E. (1977): *N. Engl. J. Med.,* 296:651–655.
66. Tattersall, M. H. N., Ganeshaguru, K., and Hoffbrand, A. V. (1975): *Biochem. Pharmacol.,* 24:1495–1498.
67. Thompson, L. F., Willis, R. C., Stoop, J. W., and Seegmiller, J. E. (1978): *Proc. Natl. Acad. Sci. U.S.A.,* 75:3722–3726.
68. Turpin, B. P., Duckworth, W. C., and Solomon, S. S. (1977): *J. Clin. Invest,* 60:442–448.
69. Ullman, B., Cohen, A., and Martin, D. W., Jr. (1976): *Cell,* 9:205–211.
70. Webster, A. D. B., North, M., Allsop, J., Asherson, G. L., and Watts, R. W. E. (1978): *Clin. Exp. Immunol.,* 31:456–463.
71. Zannis, V., Doyle, D., and Martin, D. W., Jr. (1978): *J. Biol. Chem.,* 253:504–510.

Physiological and Regulatory Functions of Adenosine and Adenine Nucleotides, edited by H. P. Baer and G. I. Drummond. Raven Press, New York © 1979.

Interrelationship Between Adenine Nucleotides and Prostaglandins

T. Kevin Sweeney and Philip Needleman

Department of Pharmacology, Washington University Medical School, St. Louis, Missouri 63110

PROSTAGLANDIN MEDIATION OF VASCULAR RESPONSIVENESS TO ADENINE NUCLEOTIDES

The administration of exogenous adenosine triphosphate (ATP) to an isolated perfused rabbit kidney caused a modest increase in renal resistance and resulted in the appearance of a substance in the renal venous effluent that had prostaglandin (PG) -like activity (37). The released material was characterized by its ability to contract PGE_2-sensitive assay organs (chick rectum, rat colon) in a superfusion cascade that was continuously bathed with the renal venous effluent. A bolus injection of 0.5 μmoles of ATP resulted in the release of 25 ng of PG (37). There are neither preformed stores of PG nor available precursor arachidonic acid (43). Thus ATP must have stimulated renal PG biosynthesis by activation of phospholipase A_2. Adenosine triphosphate and adenosine diphosphate (ADP) were equipotent as stimulators of PG biosynthesis, whereas adenosine 5′-monophosphate (AMP), adenosine, cyclic 3′,5′-adenosine monophosphate (cAMP), and guanosine monophosphate were inactive (37). Adenosine triphosphate and ADP (50 nmoles of each) stimulated PG biosynthesis in a wide variety of isolated, Krebs-perfused rabbit organs including kidney, heart, spleen, vascular smooth muscle, liver, and lung (37).

Burnstock (5) has suggested that ATP is the transmitter released from nonadrenergic inhibitory (purinergic) neurons in the gut. Dipyridamole was claimed to potentiate the effects of both exogenously applied ATP and stimulation of nonadrenergic inhibitory fibers in guinea pig tenia coli (50), whereas quinidine antagonized these actions (6). We found that neither dipyridamole nor quinidine blocked the ATP- or ADP-induced renal PG release (37). Indomethacin, however, abolished the adenine nucleotide–induced PG synthesis and release.

The actual PG released by stimulation with exogenous adenine nucleotides seems to be dependent on the tissue. When isolated perfused rabbit hearts are slowly infused with [^{14}C]arachidonic acid in the absence of albumin the label is quantitatively (~90%) incorporated into the tissue phospholipid (23). Stimulation of such prelabeled hearts with ATP resulted in the release of radioactive PG into the venous effluent. The release of PG by ATP was temporally correlated with a decrease in coronary resistance; both were abolished by pretreatment with indomethacin. The predominant peak of radioactivity obtained by acid lipid extraction of the perfusion media after such stimulation comigrated with standard PGE_2 in a solvent system (system C) consisting of chloroform:methanol:acetic acid:water, 90:8:1:0.8, without demonstrable $PGF_{2\alpha}$ present (22). However, when similarly obtained samples were chromatographed in system BDA (benzene:dioxane:acetic acid, 60:30:3), two peaks were observed, one of which had chromatographic mobility similar to that of $PGF_{2\alpha}$ (22). Sodium borohydride reduction, alkaline and acid hydrolysis, and chromatographic mobility of the novel radioactive substance led to the identification of 6-keto-$PGF_{1\alpha}$ as the primary cardiac arachidonate metabolite generated by exogenous ATP or bradykinin (22). Subsequent experiments proved this compound to be the stable metabolite of prostacyclin (PGI_2) (25,34,44). On the other hand, in the normal kidney the primary PG produced by ATP proved to be authentic PGE_2 (39). However, in the isolated perfused hydronephrotic (ureter obstructed) rabbit kidney, ATP also caused the production of thromboxane (36).

It has previously been shown with bradykinin that the primary receptor event in heart and kidney leads to activation of phospholipase A_2 to generate free arachidonic acid independent of its subsequent oxidation (21). Thus ATP, like bradykinin, apparently stimulates tissue lipase activity and the subsequent fate of the released arachidonate is dependent on the complement of endoperoxide isomerases available in the given tissue.

A typical experiment is shown in Fig. 1, illustrating renal and myocardial

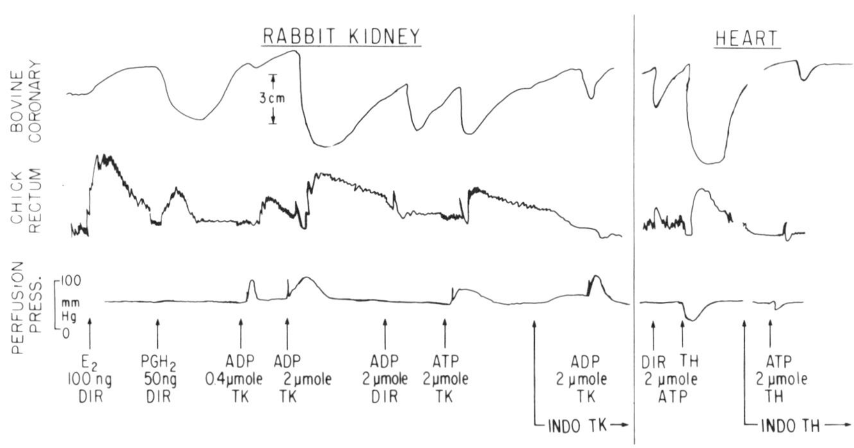

FIG. 1. The effect of PGE_2 (E_2), PGH_2, ATP, and ADP on perfused rabbit heart and kidney. Organs were perfused with a physiological salt solution *ex vivo* at a constant flow rate (kidney 10 ml/min, heart 20 ml/min) and the effluent superfused over a cascade including a bovine coronary artery strip and a chick rectum (21). Samples were injected through the kidney (TK), through the heart (TH), or directly over the superfusion cascade (DIR). At the end of each recording, indomethacin was perfused through the organs as indicated by arrows at a rate of 10 μg/min. Organ perfusion pressure is measured as mm Hg and reflects vasoactivity. Bovine coronary artery relaxation and chick rectum contraction are measured in cm of recording pen deflection.

vascular responsiveness to exogenous adenine nucleotides with coincident PG synthesis. Intrarenal injection of ADP or ATP produced a concentration-dependent renal vasoconstriction (shown as an increase in perfusion pressure) simultaneous with the release of a PGE_2-like substance (i.e., contraction of the chick rectum) and a prostacyclin (PGI_2) -like substance (relaxation of the bovine coronary artery assay strip) into the renal venous effluent (Fig. 1). In the heart, ATP or ADP produced a reduction in cardiac perfusion pressure associated with a profound release of PGI_2. Indomethacin abolished the PG release in both the kidney and heart produced by the adenine nucleotides. The ATP-induced PG release in heart must be primarily in the vasculature since indomethacin also abolished the coronary vasodilation induced by the adenine nucleotides. On the other hand, the ADP-induced renal PG biosynthesis does not appear to be in the region of the renal resistance determinants because indomethacin did not appear to influence the renal vasoconstriction.

ADENINE NUCLEOTIDE-PG INTERACTIONS IN PLATELETS

Adenosine diphosphate administered exogenously to human or rabbit (47), but not rat *(data to be published),* platelet-rich plasma causes release of arachidonic acid from endogenous phopholipids and subsequent synthesis of arachidonic acid derivatives (38). This phenomenon occurs only at doses of ADP sufficiently large to cause the second, irreversible, phase of ADP-induced platelet aggregation (1) (see Fig. 2). The second phase of ADP-induced aggregation is caused, at least in part, by endoperoxide synthesis from arachidonic acid. Cyclooxygenase inhibitors like aspirin given prior to phlebotomy or added *in vitro* to platelet-rich plasma abolish both the second phase of ADP-induced aggregation (41) and the associated release of

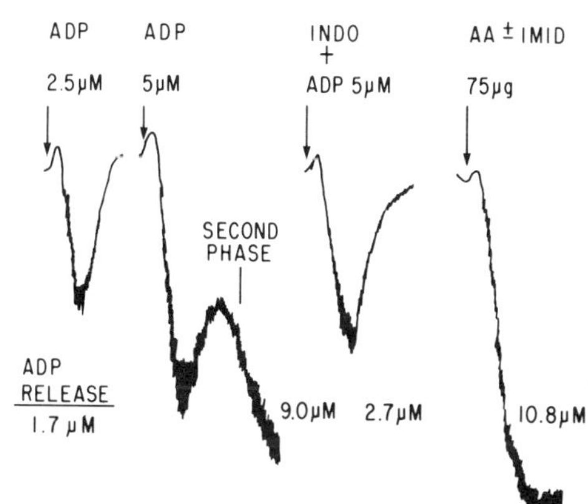

FIG. 2. The effect of ADP and arachidonic acid, added at the times indicated by the arrows, on human platelets in the presence and absence of 10 μg/ml indomethacin or 5 mM imidazole. Platelet aggregation is measured as an increase in light transmission (deflection downward). Release of ADP into extracellular fluid is measured in μM. As a time reference, the reversible aggregation produced by 2.5 μM ADP took 5 min.

granular ADP, ATP, serotonin, and calcium (13) as shown in Fig. 2. In the presence of aspirin or indomethacin, high-dose ADP will cause reversible human platelet aggregation, behaving like untreated rat platelets which aggregate reversibly in response to any dose of ADP (28). A patient with cyclooxygenase deficiency has been described whose platelets respond to high-dose ADP by aggregating reversibly without a significant release reaction (30).

In contrast to tissues such as heart and kidney, ATP has not been reported to induce synthesis of arachidonic acid derivatives in platelets. Rather, ATP inhibits ADP-induced platelet aggregation (1,4) and thrombin-induced platelet aggregation (40). Adenosine monophosphate and adenosine have also been shown to inhibit ADP-induced platelet aggregation (1,3), thrombin-induced platelet aggregation (18,27), and collagen-induced platelet aggregation (49) in man, rabbits, and hamsters (11), but not in rats (10,46), cats (2), or guinea pigs (9). Perhaps the most important inhibitory nucleotide is cAMP because of its key role in platelet function (45). Platelet cAMP levels increased by exogenous addition of cAMP or by stimulation of adenylate cyclase caused inhibition of platelet aggregation and release induced by such agonists as arachidonic acid, PG endoperoxides, and thrombin (33). When added to untreated platelets, agonists causing platelet aggregation lower basal cAMP levels, although this effect is weak *in vitro*. Elevated cyclic nucleotide levels suppress phospholipase A_2 but not cyclooxygenase activity in human platelet suspensions (33).

Arachidonic acid added *in vitro* to platelets can be converted to metabolites that influence platelet aggregation and release. In human washed platelets, most exogenous arachidonic acid is converted to 12-hydroxy-arachidonic acid (HETE), HHT, PG endoperoxides, and thromboxane B_2, the inactive metabolite of thromboxane A_2 (15). Thromboxane A_2, PGG_2, and PGH_2 cause platelet aggregation and release of dense granule components, including ADP (16,17). Thromboxane A_2 synthesis is not necessary for arachidonic acid–induced platelet aggregation since imidazole blocks thromboxane A_2 synthesis but does not affect arachidonic acid–induced platelet aggregation or release (38) (see Fig. 2). Since thromboxane A_2 is a very unstable molecule [$T_{1/2}$ in aqueous solution is about 30 sec (16)], it is difficult to isolate, and thus properties of this metabolite are largely inferred. The role of HETE and HHT in platelet aggregation and release is still unclear. Platelets given exogenous arachidonic acid also synthesize small amounts of PGE_2, $PGF_{2\alpha}$, and PGD_2. Since PGD_2 is an inhibitor of human platelet aggregation and release (29) (see Fig. 3), it has been postulated to act as a feedback inhibitor of arachidonic acid–induced platelet aggregation and release (42). Prostaglandin E_2 facilitates platelet aggregation (29,53). The molecule PGI_2 has been shown to be a very potent inhibitor of platelet aggregation and release (35), but has not been demonstrated to be synthesized by platelets. Finally, PGE_1 is another inhibitor of platelet aggregation and release *in vitro* (48). Prostaglandin E_1, PGD_2, and PGI_2 probably exert their inhibitory action by increasing platelet cAMP concentration (14,51).

Arachidonic acid–induced aggregation of human platelets is correlated in time with ADP secretion at all doses (7). Figure 3 demonstrates this direct temporal

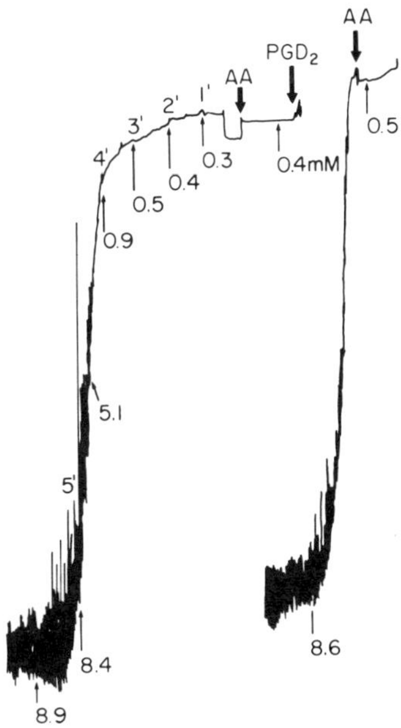

FIG. 3. The effect of exogenous PGD_2 on aggregation and ADP release by arachidonic acid (AA) in human platelet-rich plasma. Fifteen nanograms PGD_2 was added 2 min prior to 125 μg arachidonic acid to 0.4 ml of platelet-rich plasma, and aliquots were withdrawn for ADP determination every 30 sec until aggregation (shown moving from right to left) was complete. The numbers shown in the figure are ADP concentrations expressed in μM. Unstimulated platelet-poor plasma contained only 0.09 μM ADP. The numbers above the aggregation curve indicate the time in minutes at which the aliquots were taken. Thus arachidonic acid added after PGD_2 caused a complete irreversible aggregation by 5 min (5′).

relationship expressed as ADP concentrations in platelet-rich plasma supernatant continuously sampled during PGD_2-delayed, arachidonic acid–induced platelet aggregation. Aggregation curves such as that shown in Fig. 3 support the notion that ADP release is essential for arachidonic acid–induced platelet aggregation. Adenosine diphosphate scavengers such as phosphoenolpyruvate plus pyruvate kinase or creatine phosphate plus creatine kinase added to platelets prior to arachidonic acid or PGG_2 have been reported to inhibit aggregation (8,52). Although this too would seem to require an essential role for released ADP in arachidonic acid–induced platelet aggregation, ATP formed by the scavengers from ADP can directly inhibit platelet aggregation, as already noted. Another type of specific inhibitor of ADP-induced aggregation is furosemide which antagonizes PGG_2-induced platelet aggregation (30). Consequently, ADP release has been implicated as a causative factor for platelet aggregation following exposure to arachidonic acid or PGG_2 by several investigators.

More information concerning the requirement of ADP release for arachidonic acid– or PG endoperoxide–induced platelet aggregation has been obtained by studying platelets drawn from patients with storage pool disease which is a hematologic abnormality in which patients have bleeding times increased above control, and their platelets have markedly decreased concentrations of ADP, ATP, serotonin, and calcium (19,20,26). Within this entity, there appear to be subgroups as described below. Platelets from patients with one subgroup of storage pool disease aggregate with only one phase after addition of ADP or epinephrine and do not aggregate following collagen administration *in vitro*. Adenosine diphosphate and ATP release is small under these circumstances (19), which is explained by the paucity of ADP-rich dense granules seen electron microscopically within these platelets (54). Arachidonic acid or PGG_2 aggregates the platelets of these patients only very weakly even though thromboxane B_2 synthesis is normal (31). This combination of subnormal aggregation in response to PGG_2 and depleted ADP stores again infers that ADP is necessary for arachidonic acid– or PGG_2-induced platelet aggregation.

Platelets from a second subgroup of storage pool disease patients aggregate in a normal fashion to arachidonic acid but not to ADP, epinephrine, or collagen (32). In spite of markedly decreased ADP concentrations both within the platelets and extracellularly after arachidonic acid stimulation, these platelets aggregate vigorously in response to arachidonic acid. The differences between storage pool disease subgroups will have to be elucidated in order to ascertain under what circumstances ADP-depleted platelets will aggregate in the presence of arachidonic acid. This question assumes significance because ADP scavengers have been reported not to inhibit PGG_2-induced platelet aggregation (12) and to only partially inhibit arachidonic acid–induced aggregation (24). The contradictory data obtained by different researchers with ADP scavengers is still unexplained but may be related to ATP generation, as mentioned above. Additional evidence against an essential role for ADP release in arachidonic acid–induced platelet aggregation is available. Arachidonic acid causes aggregation even in thrombin-treated platelets deficient in ADP (24). Finally, low doses of PG endoperoxides cause reversible aggregation without measurable ATP release (7).

In summary, ADP causes synthesis of arachidonic acid derivatives in whole tissues and in a homogeneous platelet suspension. The subsequent synthesis of certain PGs causes effects not resulting from ADP alone, since aspirin-like inhibitors of PG synthesis will block these effects. Likewise, ATP causes PG-dependent effects in whole tissues but not in platelets. Cyclic adenosine monophosphate affects synthesis and expression of arachidonic acid in platelets and probably has a crucial role in tissue regulation of PGs also. Arachidonic acid and its metabolites, in turn, cause platelet aggregation that may be mediated via ADP release. Prostaglandins, ATP, and ADP interact in tissues and single cells in a complex manner that is only now beginning to be understood.

ACKNOWLEDGMENTS

This research was supported by National Institutes of Health Medical Health Training Grant 5 T32 GMO 7200 04 and by HL-20787.

REFERENCES

1. Born, G. V. R. (1962): *Nature,* 194:927–929.
2. Born, G. V. R., and Cross, M. J. (1963): *Nature,* 197:974–976.
3. Born, G. V. R., and Cross, M. J. (1963): *J. Physiol.,* 168:178–195.
4. Born, G. V. R., and Foulks, J. G. (1977): *Br. J. Pharmacol.,* 61:87–89.

5. Burnstock, G. (1972): *Pharmacol. Rev.*, 24:509–581.
6. Burnstock, G., Campbell, G., Stachell, D. G., and Smythe, A. (1970): *Br. J. Pharmacol.*, 40:668–688.
7. Charo, I. F., Feinman, R. D., Detwiler, T. C., Smith, J. B., Ingerman, C. M., and Silver, M. J. (1977): *Nature*, 269:66–69.
8. Claesson, H. E., and Malmsten, C. (1977): *Eur. J. Biochem.*, 76:277–284.
9. Constantine, J. W. (1966): *Nature*, 210:162–164.
10. Constantine, J. W. (1967): *Nature*, 214:1084–1086.
11. Cuthbertson, W. F. J., and Mills, D. C. B. (1963): *J. Physiol.*, 169:9P–10P.
12. Gerrard, J. M., Townsend, D., Stoddard, S., Witkop, C. J., and White, J. G., (1977): *Am. J. Pathol.*, 86:99–112.
13. Gerrard, J. M., White, J. G., Rao, G. H. R., Krivit, W., and Witkop, C. J., Jr. (1975): *Br. J. Haematol.*, 29:657–665.
14. Gorman, R. R., Bunting, S., and Miller, O. V. (1977): *Prostaglandins*, 13:377–388.
15. Hamberg, M., Svenssen, J., and Samuelsson, B. (1974): *Proc. Natl. Acad. Sci.*, 71:3400–3404.
16. Hamberg, M., Svensson, J., and Samuelsson, B. (1975): *Proc. Natl. Acad. Sci. U.S.A.*, 72:2994–2998.
17. Hamberg, M., Svensson, J., Wakabayashi, T., and Samuelsson, B. (1974): *Proc. Natl. Acad. Sci. U.S.A.*, 71:345–349.
18. Haslam, R. J. (1964): *Nature*, 202:765–768.
19. Holmsen, H., and Weiss, H. J. (1970): *Br. J. Haematol.*, 19:643–649.
20. Holmsen, H., and Weiss, H. J. (1972): *Blood*, 39:197–209.
21. Isakson, P. C., Raz, A., Denny, S. E., Wyche, A., and Needleman, P. (1977): *Prostaglandins*, 14:853–871.
22. Isakson, P. C., Raz, A., Denny, S. E., Pure, E., and Needleman, P. (1977): *Proc. Natl. Acad. Sci. U.S.A.*, 74:101–105.
23. Isakson, P. C., Raz, A., and Needleman, P. (1976): *Prostaglandins*, 12:739–748.
24. Kinlough-Rathbone, R. L., Reimers, H. J., Mustard, J. F., and Packham, M. A. (1976): *Science*, 192:1011–1012.
25. Johnson, R. A., Morton, D. R., Kinner, J. H., Gorman, R. R., McGuire, J. C., Sun, F. F., Whittaker, N., Bunting, S., Salmon, J., Moncada, S., and Vane, J. R. (1976): *Prostaglandins*, 12:915–928.
26. Lages, B., Scrutton, M. C., Holmsen, H., Day, J., and Weiss, H. J. (1975): *Blood*, 46:119–130.
27. MacMillan, D. C. (1966): *Nature*, 211:140–144.
28. MacIntyre, D. E., and Gordon, J. L. (1974): *Biochem. Soc. Trans.*, 2:1265–1269.
29. MacIntyre, D. E., and Gordon, J. L. (1975): *Nature*, 258:337–339.
30. Malmsten, C., Hamburg, M., Svensson, J., and Samuelsson, B. (1975): *Proc. Natl. Acad. Sci. U.S.A.*, 72:1446–1450.
31. Malmsten, C., Kindahl, H., Samuelsson, B., Levy-Toledano, S., Tobelem, G., and Caen, J. P. (1977): *Br. J. Haematol.*, 35:511–520.
32. Minkes, M. S., Joist, J. H., and Needleman, P. (1979): *(Submitted for publication.)*
33. Minkes, M., Stanford, N., Chi, M. M-Y., Roth, G. J., Raz, A., Needleman, P., and Majerus, P. (1977): *J. Clin. Invest.*, 59:449–454.
34. Moncada, S., Gryglewski, R., Bunting, S., and Vane, J. R. (1976): *Nature*, 263:663–665.
35. Moncada, S., Gryglewski, R. J., Bunting, S., and Vane, J. R. (1976): *Prostaglandins*, 12:715–737.
36. Morrison, A. R., Nishikawa, K., and Needleman, P. (1978): *J. Pharmacol. Exp. Ther.*, 205:1–8.
37. Needleman, P., Minkes, M. S., and Douglas, J. R. (1974): *Circ. Res.*, 34:455–460.
38. Needleman, P., Raz, A., Ferrendelli, J. A., and Minkes, M. (1977): *Proc. Natl. Acad. Sci. U.S.A.*, 74:1716–1720.
39. Nishikawa, K., Morrison, A., and Needleman, P. (1977): *J. Clin. Invest.*, 59:1143–1150.
40. O'Brien, J. R. (1963): *J. Clin. Pathol.*, 16:223.
41. O'Brien, J. R. (1968): *Lancet*, i:779–783.
42. Oelz, O., Oelz, R., and Knapp, H. R. (1977): *Prostaglandins*, 13:225–234.
43. Piper, P. J., and Vane, J. R. (1971): *Ann. N.Y. Acad. Sci.*, 180:363–383.
44. Raz, A., Isakson, P. C., Minkes, M. S., and Needleman, P. (1977): *J. Biol. Chem.*, 252:1123–1126.
45. Salzman, E. W., and Weisenberger, H. (1972): *Adv. Cyclic Nucleotide Res.*, 1:231–247.
46. Sinakos, Z., and Caen, J. (1966): *Thromb. Diath. Haemorrhag.*, 16:163–184.
47. Smith, J. B., Ingerman, C., Koisis, J. J., and Silver, M. J. (1973): *J. Clin. Invest.*, 52:965–969.

48. Sinha, A. K., and Colman, R. W. (1978): *Science,* 200:202–203.
49. Spaet, T. H., and Zucker, M. B. (1964): *Am. J. Physiol.,* 206:1267–1274.
50. Stachell, D. G., Lynch, A., Bourke, P. M., and Burnstock, G. (1972): *Eur. J. Pharmacol.,* 19:343–350.
51. Tateson, J. E., Moncada, S., and Vane, J. R. (1977): *Prostaglandins,* 13:389–397.
52. Vargaftig, B. B., and Zirinis, P. (1973): *Nature New Biol.,* 244:114–116.
53. Weiss, H. J., Willis, A. L., Kuhn, D., and Brand, H. (1976): *Br. J. Haematol.,* 32:257–272.
54. White, J. G., Edson, J. R., Desnick, S. J., and Witkop, C. J. (1971): *Am. J. Pathol.,* 63:319–330.

Physiological and Regulatory Functions of Adenosine and Adenine Nucleotides, edited by H. P. Baer and G. I. Drummond. Raven Press, New York © 1979.

Analysis of Adenosine and Adenine Nucleotides in Severe Combined Immunodeficiency Disease

*J. Arly Nelson, *John F. Kuttesch, **Randall M. Goldblum, **Armond S. Goldman, and **Frank C. Schmalstieg

**Department of Pharmacology and Toxicology and **Department of Pediatrics, University of Texas Medical Branch, Galveston, Texas 77550*

In this chapter I review some aspects of analytical procedures for the measurement of adenosine and present some of our recent observations concerning levels of purines and pyrimidines in patients with severe combined immunodeficiency (SCID). The many and complex functions of nucleosides in biological systems have been the topic of a previous conference and discussion (6,23).

ANALYSIS OF ADENOSINE

Several methods for the analysis of adenosine have been reported (Table 1). The classic spectrophotometric procedure (11) utilizes adenosine deaminase (ADA) alone (16) or coupled with purine nucleoside phosphorylase and xanthine oxidase (13). The latter two enzymes serve to convert the product of deamination, inosine, to hypoxanthine and eventually to uric acid. The increase in absorbance at 293 nm is indicative of uric acid formation, and interference by other components is minimized at this wavelength. The molar absorptivity change for adenosine → inosine is −8,100 at 265 nm (16) whereas uric acid has a value of 12,600 at 293 nm (25). In either case, a realistic limit of detection can be estimated from these values, and this limit is generally about 1 nmole. The exact limit will depend on the cuvette and spectrophotometer used. The steady state level of adenosine in normal oxygenated tissues and plasma is in the range of 0.1 to 1.0 nmole/g. Consequently, the amount of tissue extract required may be excessive for many applications. Other limitations of the spectrophotometric methods relate to the specificity of the enzymes used and to the presence of ADA in highly purified preparations of purine nucleoside phosphorylase (13). Fox et al. (9) used this methodology to detect the release of adenosine by the human heart, and overcame the problem of sensitivity by using a large volume (15 ml) of blood from which adenosine was extracted and concentrated prior to analysis.

Namm and Leader (17) reported a kinetic radioisotopic method for measuring adenosine in tissues using partially purified adenosine kinase from the rat heart. Adenosine kinase has a higher affinity ($K_m \sim 1\ \mu M$) for adenosine than ADA ($K_m \sim$

TABLE 1. *Methods used to measure adenosine in biological tissues*

Method	References	Remarks[a]
Spectrophotometric, colorimetric	9,11,13,16,29	Sensitivity generally limited, 0.1 to 1 nmole.
Adenosine kinase, radioisotope dilution	17,20	Good sensitivity (i.e., 20 pmoles); enzyme not commercially available; minimal sample preparation.
Microbiological	5	Must obtain or derive auxotrophs; may have to remove adenine and deoxy-adenosine; good sensitivity (20 pmoles or less).
Radioligand binding	21	Must prepare binding protein; extensive cleanup required; sensitive (~ 1 pmole).
Radioisotope, chemical	22	High sensitivity; separation required before or after tritium derivatization; biological applications to tRNA analysis reported.
Fluorescence after column or thin-layer chromatography	3,12,15	Derivatization step introduced; technique required cleanup; good sensitivity (~ 1 pmole).

[a] Levels of sensitivity are approximations from the cited references.

25 μM) (2). In the adenosine kinase assay, nonsaturating levels of radioactive adenosine are used and the amount of labeled adenosine monophosphate (AMP) is measured on addition of nonlabeled adenosine (such as that present in sample extracts). Namm and Leader (17) illustrated the utility of this technique to measure adenosine in heart tissue extracts. Newman and McIlwain (20) applied this method to the measurement of adenosine in guinea pig brain. An advantage of this technique is that extensive prior treatment of tissue extracts is not needed owing to the specificity of adenosine kinase. A commercial source of stable adenosine kinase would facilitate its use in other laboratories.

Berlin and Oliver (5) described a novel assay for adenine and adenosine using a double auxotroph of *Escherichia coli* that required both uracil and adenine. The adenine-dependent radiolabeled uracil incorporation represents the basis for the sensitive (~20 pmoles) bioassay. As described in their paper, the theoretical limit for detection by this method is lower since the specific activity of the uracil and volumes of solution can be increased. Availability and stability of the mutant cells limit the widespread use of this technique. With the adenine-uracil double auxotroph, adenosine must first be hydrolyzed to adenine which requires measurement of adenosine by difference (i.e., with and without hydrolysis) or after prior separation from adenine. Furthermore, deoxyadenosine would also have to be separated from adenosine, a task that could be accomplished easily (28).

Recently, Olsson et al. (21) used an adenine analog–binding protein from rabbit erythrocytes to measure adenosine. The sensitivity (1 pmole) and precision of the method are excellent. Extensive prior removal of adenine and adenine nucleotides is required. Because of the popularity of radioimmunoassays, this relatively simple procedure may soon be used in many laboratories.

A procedure that has tremendous potential for the analysis of all nucleosides at an extremely high level of sensitivity is that devised by Randerath et al. for base analysis of transfer ribonucleic acid (tRNA) (22). Specifically, ribonucleosides are converted to nucleoside dialdehydes via oxidation with periodate. The dialdehydes are then labeled with [^{3}H]potassium borohydride to form radioactive trialcohols that are separated by thin-layer chromatography and counted by liquid scintillation spectrometry. Although an initial investment in time may be required to set up this procedure, it is probably the best because of its versatility, availability, and low cost.

Mills et al. (15) used chloroacetaldehyde to convert adenosine to a fluorescent compound, 1,N^6-ethenoadenosine (Fig. 1), to enhance the sensitivity for detection after column chromatography. This procedure is an application of that described by Avigad and Damle (3) and Barrio et al. (4). The condensation of chloroacetaldehyde with adenine and cytosine compounds occurs at mild conditions of pH and temperature (4,15). Since the fluorescent properties of the ethenoadenine and ethenocytosine compounds differ, this reaction will rather specifically yield fluorescent derivatives of adenine-containing compounds. We used fluorescence detection for analysis of 1,N^6-ethenoadenosine by high-pressure liquid chromatography (HPLC) (12), an extension of the above-mentioned efforts of others. Detection of adenosine in plasma by HPLC with usual ultraviolet absorbance (10) is limited by sensitivity (~50 pmoles), similar to the limitations in the spectrophotometric-enzyme methods described above. For the fluorescence assay using HPLC (12), partial purification of acid extracts is necessary for most tissues or fluids. We have used batch and column procedures with Dowex anion exchange resins for urine, lymphocyte, and erythrocyte extracts (12,14). Extracts of plasma do not require the Dowex pretreatment,

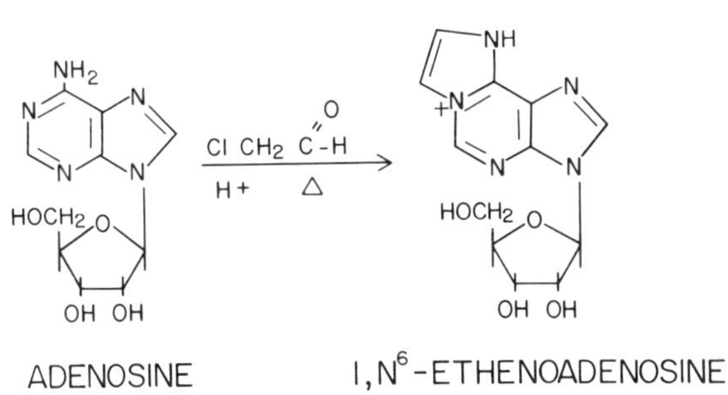

FIG. 1. Conversion of adenosine to the fluorescent, 1,N^6-etheno derivative (4).

and the limit of sensitivity is about 10 nM. This procedure is limited by the availability of the instrumentation; however, many investigators are adding HPLC to their laboratories because of its versatility.

PURINES AND PYRIMIDINES IN SCID

Our primary concern with methods allowing efficient and sensitive measurement of adenosine arose during the study of a patient at our institution with SCID associated with ADA deficiency (SCID:ADA$^-$). Seegmiller *(this volume)* has reviewed aspects of this disease that relate to the theme of this volume. Increased amounts of ADA substrates, i.e., adenosine and deoxyadenosine, may play a role in the etiology of this disease. Our studies of this SCID:ADA$^-$ child have been described in several recent reports (15,26,27). When we used HPLC with fluorescence detection to measure adenosine in the child's urine (Fig. 2), we observed a high level of deoxyadenosine (12). The level of deoxyadenosine in plasma was below the limit of detection (Table 2). Therefore, it appears that deoxyadenosine is either produced or rapidly secreted by the kidneys of this patient. It is also interesting to note (Table 2) that, unlike many tissues, very high levels of adenosine ($\sim$30 μM) are present in lymphocytes. Results of an extensive study of the urinary excretion of purines, purine nucleosides, and pseudouridine in this patient have recently been summarized (14). This study indicated that deoxyadenosine was the major urinary purine excreted in abnormal amounts and that the excretion was reduced following transfusion of the patient with red blood cells. In addition, a decrease in the excretion of 7-methylguanine after initiation of transfusion therapy suggests that nucleic acid methylation may be altered in ADA deficiency. This effect may be caused by alterations produced by adenosine on the S-adenosylmethionine metabolic pathway (14). In 1974, Dr. Lee Bennett, Jr., at Southern Research Institute, remarked in a personal communication that deoxyadenosine is a likely candidate for a major role in this disease for the reasons discussed by Seegmiller *(this volume)*. Recent reports (7,8) of elevated deoxy-adenosine triphosphate (dATP) in the erythrocytes of 3 patients with SCID:ADA$^-$ led us to consider the possibility that the elevated ATP we reported earlier in blood lymphocytes (18,27) and erythrocytes (1,15,18) may have been caused by dATP rather than ATP. We have used a Partisil-SAX column for analysis of ribonucleotides (18,19) which will readily separate ATP and dATP (Fig. 3) as well as arabinosyladenine 5′-triphosphate (araATP) (24). Actually, araATP elutes between ATP and dATP. We have used this mode of separation to determine adenine nucleotide levels in 2 SCID:ADA$^-$ children. In one case, measurement of the nucleotide pools in erythrocytes was made following a fetal liver transplantation (18), and in the other patient, measurement of lymphocyte nucleotide pools was performed before and after transfusion with normal red blood cells (26,27). At the time these measurements were reported (1,18,26), the cells examined did not contain levels of dATP that approximated those reported recently in erythrocytes of other SCID:ADA$^-$ patients (7,8). However, when the red blood cell transfusion therapy was withheld recently

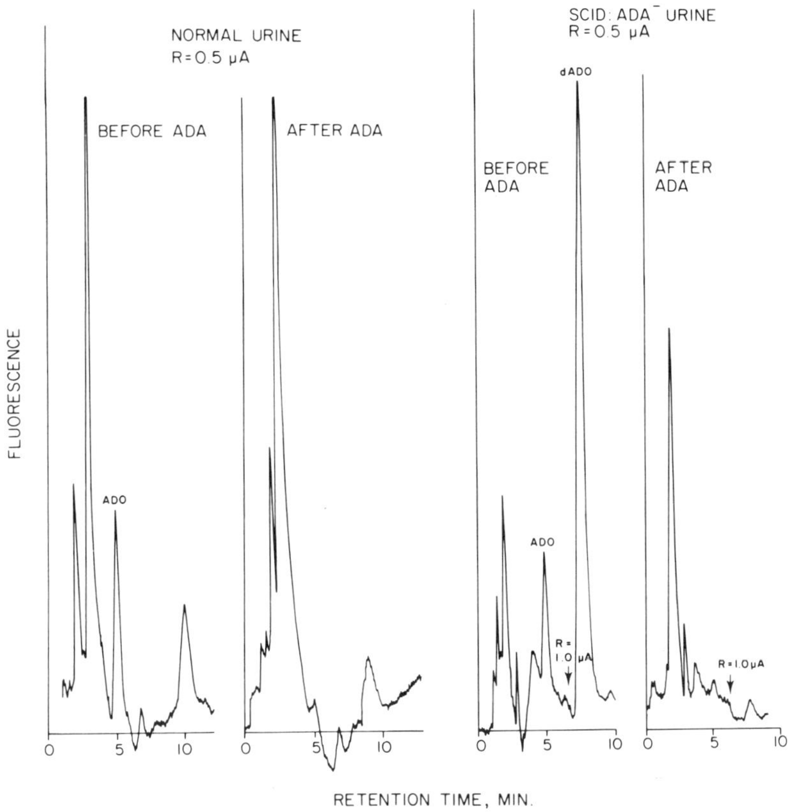

FIG. 2. Detection of adenosine and deoxyadenosine in human urine using HPLC with fluorescence detection. Adenine was separated from adenosine and deoxyadenosine prior to derivatization or other treatment (12,14). The 1,N^6-etheno derivatization was performed without prior treatment with ADA (BEFORE ADA) or after treatment with ADA (AFTER ADA). Isocratic elution of a cation exchange column (Partisil-SCX, Whatman, Inc., Clifton, N.J.) was performed as described previously (12). ADO = 1,N^6-ethenoadenosine; dADO = 1,N^6-ethenodeoxyadenosine. (From ref. 12; with permission.)

for a period of 4 to 6 months, the dATP level in the erythrocytes of the SCID:ADA^- child in Galveston increased from a virtually undetectable level (by the method described above) to 350 nmoles per ml packed cells. Thus, in the absence of transfusion therapy, this patient also exhibits marked elevation of dATP in erythrocytes. We are currently attempting to measure the deoxyribonucleotide levels in the affected cells (lymphocytes) using a more sensitive analytical method.

I would like to turn to another group of SCID patients we have studied during the past four years. This group of patients, referred to as SCID:ADA^+ in Fig. 4, presented with SCID but appeared to have normal activities of ADA and purine nucleoside phosphorylase. The levels of purine and pyrimidine ribonucleotides in the SCID:ADA^+ lymphocytes were elevated in the group as a whole (Fig. 4; Table 3).

TABLE 2. *Approximate levels of adenine compounds in normal adults and a SCID:ADA⁻ patient*

	Adenosine [a]		Deoxyadenosine [a]		Adenine [a]	
Source	Normal	SCID:ADA⁻	Normal	SCID:ADA⁻	Normal	SCID:ADA⁻
Lymphocyte	30	55	<25 [b]	<25	130	490
Erythrocyte	0.2	0.7	<0.2	<0.2	15	95
Plasma	0.1	1	<0.06	<0.06	—	—
Urine	5	5	<0.01	60	3	50

[a] Concentration, nmole/ml or nmole/10^9 cells.
[b] "<" indicates minimal detectable level for that tissue (12).

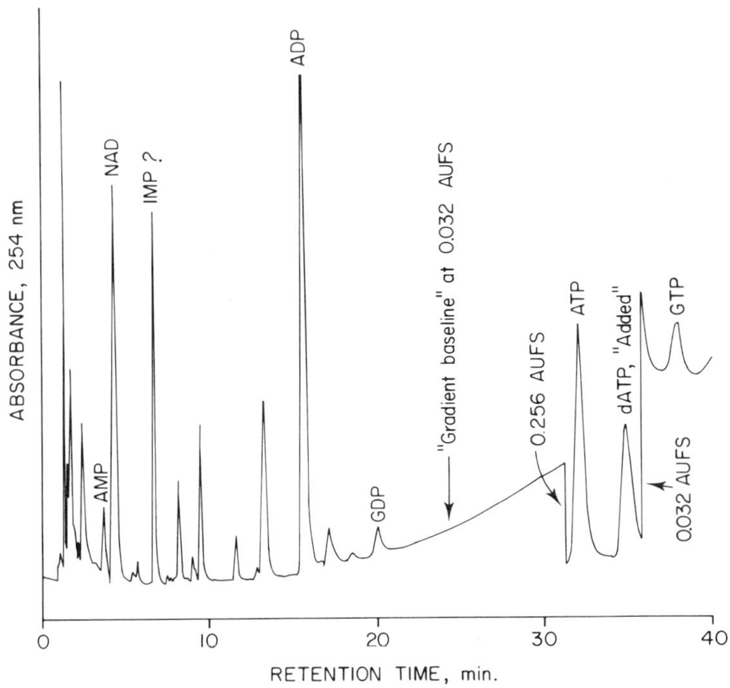

FIG. 3. Representative separation of ATP and dATP using HPLC with an anion exchange column (Partisil-SAX, Whatman, Inc., Clifton, N.J.). The dATP (25 nmoles) was injected simultaneously with a trichloroacetic acid extract representing approximately 12 mg of erythrocytes from a SCID:ADA⁻ patient (26). Separation and detection of components of the mixture were accomplished as described previously (19). There was no detectable dATP in the patient's extract (<20 μM) without added dATP.

When the absolute values for pool sizes of these compounds were compared to that of uridine triphosphate (UTP) (lower portion of Table 3), ATP in the SCID:ADA$^+$ patients was significantly increased. In one of the children (no. 3 in Fig. 4), the uptake of radioactive adenosine into red blood cell ATP was increased about twofold over that observed in normal controls (Fig. 5 shows one of two such experiments). Furthermore, erythrocyte ATP levels in 3 of these SCID:ADA$^+$ children ranged from 1,800 to 2,100 μM, values that exceed that of normals, i.e., 1,398 ± 145 μM, mean ± SD for 50 control samples (Gordon C. Mills, *unpublished observation*). Differences in plasma, urine, and erythrocyte levels of other adenine components were not obvious. Erythrocyte data for the remaining SCID:ADA$^+$ patient in Fig. 4 are not available. It appears, therefore, that there are abnormalities in purine metabolism in the lymphocytes and erythrocytes of SCID:ADA$^+$ as well as SCID:ADA$^-$ patients, particularly in the maintenance of high levels of ATP (18,26,27; *this chapter*). We do not know whether the increased ATP is caused by enhanced synthesis, reduced catabolism, or both. It seems plausible that a defect in regulation in an associated

TABLE 3. *Purine and pyrimidine ribonucleotide levels in peripheral lymphocytes of patients with SCID*

Sample	ATP	ADP	GTP (nmoles/10^9 cells)	GDP	CTP	UTP
Controls [a]	738 ± 258	268 ± 66	131 ± 17	58 ± 16	23 ± 3	109 ± 14
SCID:ADA+ [b]	3,054 ± 1675 [d]	852 ± 278 [d]	408 ± 119 [d]	137 ± 41 [d]	82 ± 26 [d]	252 ± 41 [d]
SCID:ADA− [c]	5,491	2,047	1,191	562	132	463
			Level relative to UTP			
Controls [a]	6.98 ± 0.48	2.23 ± 0.41	1.21 ± 0.08	0.47 ± 0.09	0.24 ± 0.03	1.00
SCID:ADA+ [b]	11.48 ± 1.47 [d]	3.34 ± 0.70	1.52 ± 0.25	0.54 ± 0.11	0.32 ± 0.07	1.00
SCID:ADA− [c]	11.86	4.42	2.57	1.21	0.28	1.00

ADP-adenosine diphosphate. GTP=guanosine triphosphate. GDP=guanosine diphosphate. CTP=cytidine triphosphate.

[a] $N=8$; seven normal adults and one nonSCID pediatric patient; mean values ± SEM are given.

[b] $N=4$; refer to Fig. 4.

[c] $N=1$; refer to Fig. 4; SCID:ADA− patient prior to transfusion therapy (27).

[d] Significantly different ($p < 0.05$) from control values.

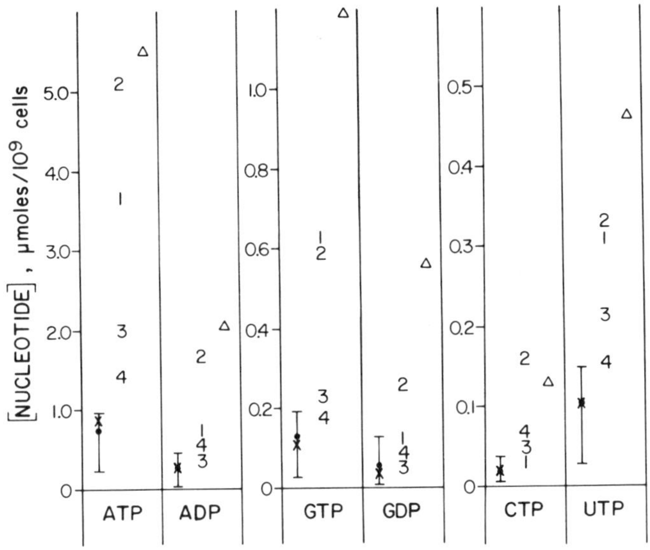

FIG. 4. Purine and pyrimidine ribonucleotide levels in isolated peripheral lymphocytes of control and SCID individuals. Determination of nucleotide pools was performed as described previously (19). ♦ = mean ± range of values for eight samples, i.e., 7 normal adults and 1 nonSCID pediatric patient represented by x; numbers 1, 2, 3, and 4 represent the nucleotide pool values for each of 4 SCID:ADA$^+$ patients; Δ = a SCID:ADA$^-$ child (27) prior to extensive transfusion therapy (26). ADP = adenosine diphosphate. GTP = guanosine triphosphate. GDP = guanosine diphosphate. CTP = cytidine triphosphate. UTP = uridine triphosphate.

pathway, for example, at the level of phosphofructokinase, exists in SCID:ADA$^+$ patients such that the higher steady state pool of ATP is maintained.

SUMMARY

Development of technology to permit the rapid analysis of purines and pyrimidines has led to a better descriptive understanding of SCID. Elevated ATP in lymphocytes and erythrocytes of SCID:ADA$^-$ and SCID:ADA$^+$ patients appears to be a common abnormality. The relevance of this observation to the immunodeficiency remains obscure. An important role for adenosine and adenine nucleotides in the etiology of these diseases and in normal immune function appears highly likely and is the subject of continuing investigation.

ACKNOWLEDGMENTS

The original work reported herein has been aided in part by Grants from the American Cancer Society, CH-88, and from the National Foundation March of Dimes, 6-130. Funds for the major equipment were provided by Department of Health, Education, and Welfare Capitation Grant 065000-170. Clinical support was

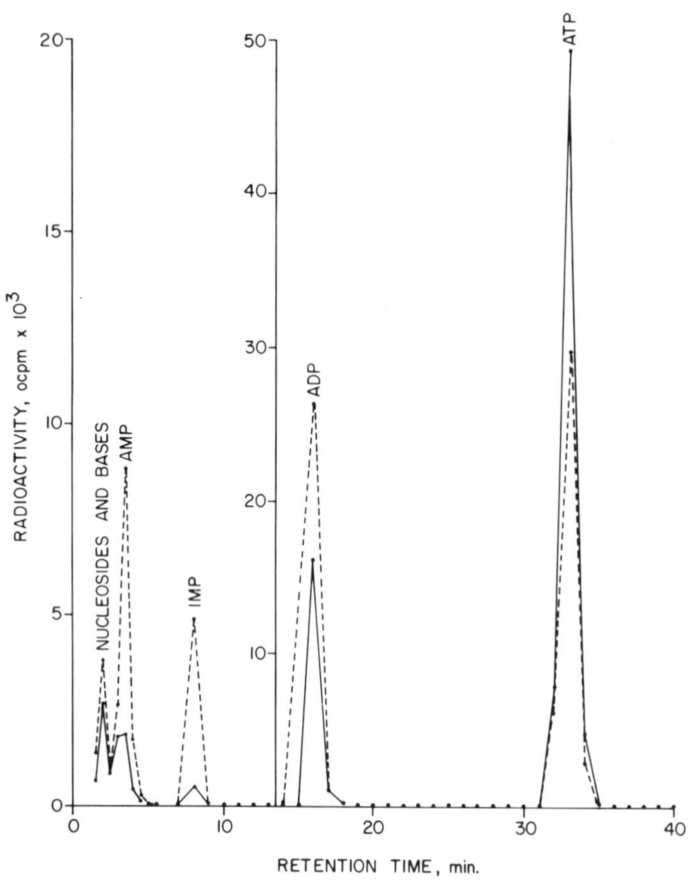

FIG. 5. Incorporation of radioactive adenosine into nucleotides of human red blood cells. Red blood cells (15%, vol/vol) from a normal adult and patient no. 3 of Fig. 4, were incubated for 30 min with 3.7 μM [8-^{14}C]adenosine in the presence of 2.2 μM erythro-9-(2-hydroxy-3-nonyl)adenine. After addition of excess (100 μM) cold adenosine, the cells were isolated and extracted with cold trichloroacetic acid. After neutralization by washing with ether, the extract was applied to a Partisil-SAX column for nucleotide analysis as described (19). Aliquots (1 min) of the eluent were collected and radioactivity was determined by liquid scintillation spectrometry (18). (•----•), Control; (•—•), SCID:ADA$^+$. IMP = inosine monophosphate. ADP = adenosine diphosphate.

provided in part by Grant RR-00073-14, General Clinical Research Centers Branch, Division of Research Facilities and Resources, NIH, DHEW. J. F. Kuttesch is a recipient of a McLaughlin Predoctoral Fellowship, and F. C. Schmalstieg is a recipient of a Young Investigator Research Grant Award, AI-14245, National Institute of Allergy and Infectious Diseases, NIH.

REFERENCES

1. Agarwal, R. P., Crabtree, G. W., Parks, R. E., Jr., Nelson, J. A., Keightley, R., Parkman, R., Rosen, F. S., Stern, R. C., and Polmar, S. H. (1976): *J. Clin. Invest.*, 57:1025–1035.

2. Agarwal, R. P., Sagar, S. M., and Parks, R. E., Jr. (1975): *Biochem. Pharmacol.*, 24:693–701.
3. Avigad, G., and Damle, S. (1972): *Anal. Biochem.*, pp. 321–323.
4. Barrio, J. R., Secrist, J. A., III, and Leonard, N. J. (1972): *Biochem. Biophys. Res. Commun.*, 46:597–604.
5. Berlin, R. D., and Oliver, J. M. (1973): *Anal. Biochem.*, 53:21–27.
6. Bloch, A. (1975): *Chemistry, Biology, and Clinical Uses of Nucleoside Analogs, Ann. N.Y. Acad. Sci.*, Vol. 255.
7. Cohen, A., Hirschhorn, R., Horowitz, S. D., Rubinstein, A., Polmar, S. H., Hong, R., and Martin, D. W., Jr. (1978): *Proc. Natl. Acad. Sci. U.S.A.*, 75:472–476.
8. Coleman, M. S., Donofrio, J., Hutton, J. J., and Hahn, L. (1978): *J. Biol. Chem.*, 253:1619–1626.
9. Fox, A. C., Reed, G. E., Glassman, E., Kaltman, A. J., and Silk, B. B. (1974): *J. Clin. Invest.*, 53:1447–1457.
10. Hartwick, R. A., and Brown, P. R. (1977): *J. Chromatogr.*, 143:383–389.
11. Kalckar, H. M. (1947): *J. Biol. Chem.*, 167:445–459.
12. Kuttesch, J. F., Schmalstieg, F. C., and Nelson, J. A. (1978): *J. Liq. Chromatogr.*, 1:97–109.
13. Manoukian, E., and Fawaz, G. (1971): *Hoppe Seylers Z. Physiol. Chem.*, 352:1631–1634.
14. Mills, G. C., Goldblum, R. M., Newkirk, K. E., and Schmalstieg, F. C. (1978): *Biochem. Med.*, 20:180–199.
15. Mills, G. C., Schmalstieg, F. C., Trimmer, K. B., Goldman, A. S., and Goldblum, R. M. (1976): *Proc. Natl. Acad. Sci. U.S.A.*, 73:2867–2871.
16. Mollering, H., and Bergmeyer, H. U. (1974): In: *Methods of Enzymatic Analysis,* Vol. 4, edited by H. U. Bergmeyer, pp. 1919–1922. Academic Press, New York.
17. Namm, D. H., and Leader, J. P. (1974): *Anal. Biochem.*, 58:511–524.
18. Nelson, J. A. (1977): *Adv. Chromatogr.*, 15:273–305.
19. Nelson, J. A., Rose, L. M., and Bennett, L. L., Jr. (1976): *Cancer Res.*, 36:1375–1378.
20. Newman, M., and McIlwain, H. (1977): *Biochem. J.*, 164:131–137.
21. Olsson, R. A., Davis, C. J., Gentry, M. K., and Vomacka, R. B. (1978): *Anal. Biochem.*, 85:132–138.
22. Randerath, E., Yu, C. T., and Randerath, K. (1972): *Anal. Biochem.*, 48:172–198.
23. Robins, R. K. (1975): *Ann. N.Y. Acad. Sci.*, 255:597–610.
24. Rose, L. M., and Brockman, R. W. (1977): *J. Chromatogr.*, 133:335–343.
25. Scheibe, P., Bernt, E., and Bergmeyer, H. U. (1974): In: *Methods of Enzymatic Analysis,* Vol. 4, edited by H. U. Bergmeyer, pp. 1951–1958. Academic Press, New York.
26. Schmalstieg, F. C., Mills, G. C. Nelson, J. A., May, L. T., Goldman, A. S., and Goldblum, R. M. (1978): *J. Pediatr.*, 93:597–603.
27. Schmalstieg, F. C., Nelson, J. A., Mills, G. C., Monahan, T. M., Goldman, A. S., and Goldblum, R. M. (1977): *J. Pediatr.*, 91:48–51.
28. Uziel, M., Smith, L. H., and Taylor, S. A. (1976): *Clin. Chem.*, 22:1451–1455.
29. Van Belle, H. (1969): *Anal. Biochem.*, 32:238–250.

Subject Index

Acetylcholine (ACh), 9
Acetylcholinesterase, 323
Actin, platelet, 338
Actomyosin ATPase, 333
Adenine
  analog-binding protein, 419
  analogs, effect of, 247
  derivatives on CNS, actions of, 355-357
  radioactive, 222
Adenine nucleotide(s)
  actions of adenosine and, on cerebral cortical neurons, 343-348
  analogs, phosphate-modified, 36-40
  compartmentation of, 130-133, 319
  and magnesium ion, role of balance of, 172-174
  and PG, 409-416
  receptor sites, specificity of, 33-43
  role of, 177
Adenosine
  accumulation, 320
  action
    on adenylate cyclase, two sites for, 272-274
    on adipocyte adenylate cyclase, 276-279
    inner site of, 386
    mechanisms of, 220-225
    outer site of, 386
    and release on cardiac function, 172-179
  adenine nucleotides and, 71, 343-348, 417-427
  adenylate cyclase and, 262-263, 271-281, 381-389
  and β-adrenergic stimulants, 138
  and 5′-AMP, potentiation of, 353-354
  analogs, 233-234, 309-310
    hemodynamics and metabolic actions of, 158-165
    synthesis of, 183
  antilipolytic action of, 380
  attenuation by, 137-139
  binding
    activity of heart muscle, 298
    proteins, 245-248, 297-304
    rate of, 299
  and biogenic amines, 231-232
  biphasic effect of, 190-191, 264
  in brain, 120, 237
  in brain, release, and metabolism, 230-231
  in brain cAMP, 217-227
  and cAMP, 229-241, 362-363
  5′-carboxylates and carboxamides, 183-187
  and cardiac adenylate cyclase system, 140-142
  in cardiac tissues, electrophysiological effects of, 167-172
  catecholamine-antagonistic effects of, 142-144
  cleavage, 316-317
  coronary activity of, 155-156
  in CSF, 122
  cyclic, monophosphate, *see* Cyclic adenosine monophosphate
  deaminase, *see* Deaminase, adenosine
  deamination of, 308
  deficiency, 389
  versus deoxyadenosine as mediator, 401-402
  deoxyadenosine derivatives, effects of, 383-387
  depression of central neurons, 343-359
  derivative, 133
    on slow atrial action potential, effects of, 169-170
  detection of P and R, sites, 274-276
  effects in adipose tissue, 379-393
  D- and L-enantiomers of, 234
  endogenous, 318-322
  epicardial, 118
  -5′ethylcarboxamide (744-96), 160
  excitant actions and neurons, 352-353
  exogenous, 315-318
  external, receptor, 221
  extracellular metabolism of, 315
  formation, functions of pathways of, 321-322
  free, 319
  in heart, 117-120, 127-136
  hypothetical scheme of effects of, 211
  inactivation by deamination, 353
  -induced relaxations, effects of PIT on, 89-90
  influent, 308
  inhibited, deaminase, 237
  intracellular, 297, 303, 315

Adenosine (*contd.*)
and isoprenaline, 206-212
kinase, *see* Kinase, adenosine
membrane effects and control, 167-182
metabolism
enzymes involved in, 244
fate of, 397
in heart, 127
regulation of, 315-322
and methylxanthine antagonism, 354-355
muscle, content, 147
myocardial, 118
and neural signaling, 361-367
and neuroblastoma cyclase, 249-257
and neuron
depressant activity, 348-352
depression, 343-345
excitation, 345-348
penetration, dissociation of, 221
in pericardial infusates, 123
pharmacological actions of, 358
5'-phosphate, X-ray coordinates of, 284
and phosphodiesterase inhibitors, 232-233
and PIA, effects of, 381-383
potency of, 73
presynaptic inhibition by, 71-73, 343
production, 145-153
purine-modified, analogs, 272
radioactive, 418
rates of, 315, 317
and reactive hyperemia, 156-158
receptor, *see* Receptor(s), adenosine
regional and species variation in response to, 220
regulation in intact cells, 317-318
relaxation
effect of agonist dose in, 96-97
of VSM, mechanisms for, 101
release, possible mechanisms controlling, 172-177
site of action, 133, 211
on slow atrial action potential, effect of, 167-169
on slow calcium-dependent action potential, effect of, 170-172
and smooth muscle adenylate cyclase, 205-213
specific, phosphorylase, 317
steady state versus transient release of, 177-179
-stimulated cAMP accumulation, 218-220
stimulation of adenylate cyclase, evidence for, 221-223
stoichiometric production of extracellular, 336
storage of, 230
studies with intact cells, 249-252
synergistic responses to, 219
synthesis, pathways of, 319
tetraphosphate, 40
threshold concentration for, 356
toxicity, 398
transport, 305-313
triphosphate (ATP), 3
analogs, 41
antagonists, 55-57
blocking agents for, 14
calcium, 11-12, 375
concentration-response curves to, 88
desensitization to, 57
effects of, 87-88
inactivation of, 8-9
-induced relaxation by PIT, antagonism of, 88-89
and intramural nerve stimulation, potentiators of, 54-55
lack of mimicking action of, 24
low potency to, 23
and magnesium, effect of, 262-265
mimicry by, 53
potassium-induced release of, 13
by pyridylisatogen, antagonism of, 85-93
receptor, 374
release of, 6-7
sources of, 75
storage, 6, 11-13
tachyphylaxis to, 24
time course of conversion of, 325-326
following transmural stimulation release of, 51-53
tritiated, interaction of L, S, and, 299-300
uptake, rate of, 308
on vascular smooth muscle, action of, 95-102
with vasodilators, interaction of, 98-101
in ventricular myocardium, 137-145
S-Adenosylhomocysteine, 319
Adenylate
cyclase, *see* Cyclase, adenylate
kinase, 13
Adenylic acid, 397
Adipocyte(s), 271, 333
ADP, putative, receptor, 189
α-Adrenoceptors, 51, 219, 223
β-Adrenoceptors, 51
α, β-Adrenoceptors, 69-70
Agammaglobulinemia, congenital, 396
Agonist(s)
β-adrenergic, 218

Agonist(s) (*contd.*)
  dose in adenosine relaxation, 96-97
  muscarinic, 69, 70
  type, 97
Alimentary canal, 15
Alkylxanthine, 232
Amine(s)
  adenosine and biogenic, 230-232
  on cAMP, effect of biogenic, 243
  receptors, desensitization of, 239
Aminophylline, 98, 99, 355
AMP
  deaminase and 5'-nucleotidase activities, 149-152
  degradation, pathways of, 145
  phosphodiesterase, absence of inhibitors of cyclic, 196-197
  platelet cyclic
    mechanism of action of, 200-201
    and platelet aggregation, 198-199
Amplitude, IJP, 45
Analogs
  biological effects of nucleoside, 284
  conformational distribution of purine, 283
Analysis, pharmacological, in neurogenic excitation, 50-57
Angiotensin, 69
Anti conformation and anti orientation, 286, 291
Anti preference, 287
Anti versus, 286
Arachidonic acid (AA), 65, 67, 411, 412
Arcs, reflex, 50
Arterioles, pial, 121
ATP, *see* Adenosine triphosphate
ATPase
  actomyosin, 327, 333
  plasma, 337
Autoradiography, electronmicroscopic, 11
Axon profiles, 4

B cell dysfunction, 396
Barrier, endothelial cell, 121
Basal activity, 277
Bayliss effect, 158
Blockade, adrenergic β-receptor, 207
Blood vessels, 17-18
Bradykinin, 410
Brain
  adenosine and cAMP formation in, 229-230
  adenosine and cAMP-generating systems in, 229-241
  adenosine in, 120
  adenosine receptors in, 243-248
  adenylate cyclase, 221
  alterations in cAMP-generating systems in, 240
  basal levels of cAMP in, 234
  cAMP, adenosine actions on, 217-227
  cells, studies with perinatal, 218-219
  homogenate preparation, 259-260
  Krebs-Ringer homogenates of, 230
  purinergic nerve mechanism in, 244
  release, metabolism, and action of adenosine in, 230-231
α, β-Bungarotoxin, 108, 109
Bupranolol, 389

Caffeine, 277
Calcium
  availability, 127
  and cAMP in brain, 237-239
  complexation of, 39
  deprivation, 224
  -free medium, 238
  influx by ATP and ADP, 375
  reduction of, 238
  slow, channels, 169
  transmembranal, movement, 127, 137
cAMP, *see* Cyclic adenosine monophosphate
Carboxylates, synthesis of, 138
Catecholamine(s)
  -antagonistic effects of adenosine, 142-144
  endogenous, 139
  lipolytic action of, 390
CBF, *see* Coronary blood flow
Cell(s)
  eukaryotic, 200
  regulation of adenosine in intact, 317-318
  sonicated, preparations, 334
Central nervous system (CNS), 18
  actions of adenine derivative on, 355-357
  adenylate cyclase in, 259-269
  ecto-enzymes, 335-337
Cerebrospinal fluid (CSF), 121, 122
2-Chloroadenosine, 189, 191, 233
Choline acetyl transferase activity, 218
Cholinoceptors, nicotinic, 48
Circular dichroism studies, 291
Circulation, coronary, 98
Clefts, synaptic, 9
CNS, *see* Central nervous system
CoA, *see* Coenzyme A
Cocaine, 169, 243
Coding, consequences of proposed linear element in, 364-365
Coenzyme A (CoA), 64, 104, 110
Conductance changes, 13

Conformation
  and modifications at 8 position of purines, 291-292
  and N3 electronic properties, 287-291
Conformational
  effects of molecular modifications, 287-292
  energy calculations, 284-287
  energy profile of inosine, 288
  ranges, definition of glycosidic, 284
Contractions
  atropine-insensitive "off," 62
  atropine-sensitive "on," 61
Coronary blood flow (CBF), 117, 128
Cotransmitter, 19-21
Creatine phosphate
  in control of 5'-nucleotidase, role of, 174
  free magnesium ion effect on, 174-177
  role of, 177
CSF, *see* Cerebrospinal fluid
Curve(s)
  concentration-effect, of methylene isosteres, 35-36
  cumulative dose-response, 104
  distinctive dose-effect, 42
  dose-inhibition, 96
Cyclase, adenylate
  activation of, 230, 275
  activity
    effects of adenosine on, 190-195, 271-281
    effects of theophylline on cAMP and on, 387-389
    in membrane, 275
  adenosine and
    on adipocyte, 276-279
    blood platelet, 189-204
    neuroblastoma, 249-257
    smooth muscle, 205-213
  adipocyte, 274
  assay, 260, 262-263
  basal activity of liver, 275
  and cAMP, 14
  in CNS, 259-269
  concentrations of, 320
  coupled receptor, 243
  dephosphorylation of, 319-320
  distribution of β-adrenergic sensitive, 267
  effect of solubilization of, 210-212
  evidence for adenosine stimulation of, 221-223
  fat cell, 277, 390
  hormone-sensitive, 274
  inhibition of, 358
    fat cell, 279
    in intact platelets, 199-200
    by methylxanthines and adenosine deaminase, 260-261
  metabolism, alternative pathways of, 320-321
  neuroblastoma, 221
  platelet, 189
  receptor-mediated stimulation of, 212
  regional distribution of, 267
  striatal, 264
  subcellular distribution of, 265-266
  system, adenosine and, 140-142
  topographical distribution of, 260, 265-268
  two sites for adenosine on, 272-274, 381-389
Cyclic adenosine monophosphate (cAMP), 205
  accumulation, 218-220, 226
  adenosine action on brain, 217-227
  adenosine and
    formation in brain, 229-230
    -generating systems in brain, 229-241
    intracellular siting and translocation, 362-363
  adenylate cyclase and, 14
  in brain, basal levels of, 234
  calcium and, in brain, 237-239
  dissociation between, and lipolysis by PIA, 389-392
  effects of theophylline on, accumulation and on adenylate cyclase activity, 387-389
  of fat cells, 276
  formation, synergistic stimulation of, 223-225
  -generating systems in brain, alterations in, 240
  glucagon-stimulated accumulation of, 212
  neuronal, metabolism, 249
  platelet, 412
  residual, 390
Cyclic-3', 5'-adenylic acid in ADA deficiency, 404-405
Cyclic guanosine monophosphate (cyclic GMP), 230
Cytosine, 419

Deaminase, adenosine (ADA), 169, 230, 235-237, 396
  deficiency, 396
    correction of, by deoxynucleoside, 403-404
  cyclic-3', 5' adenylic acid in, 404-405
  inhibition of adenylate cyclase by methylxanthines and, 260-261
  inhibitor, 117
  roles for, 397
  theophylline and, 193-195

Deamination
  of adenosine, rates of, 315
  adenosine inactivation by, 353
  ratio of phosphorylation to, 317
Deficiency
  ADA, 396, 403-405
  deoxy-guanosine triphosphate accumulation in PNP, 403
  ecto-purine 5'-nucleotidase, 396
  immunosuppression in ADA and PNP, 404-406
  of pyrimidine nucleotide, 398
Deletion, chromosomal, 400
Dendrites, consequence of translocation in, 363-364
Deoxyadenosine
  versus adenosine as mediator, 401-402
  derivatives, effects of adenosine and, 383-387
  metabolism of, 397
  triphosphate accumulation, 402-403
Deoxycytidine kinase, 290
Deoxy-guanosine triphosphate accumulation in PNP deficiency, 403
Deoxynucleoside, correction of ADA deficiency by, 403-404
Dephosphorylation of adenylate, 319-320
Depolarization
  membrane, 252
  neuronal, 236
  potassium, 96
Depolarizing conditions, 235
Diazepam, 353
Digitonin solubilization and phosphatidylserine, influence of, 210-211
Dipyridamole, 23, 58, 311-312
  effect of, 40, 85, 86, 98, 99
  inhibition, 311
Distortion, steric, 39
Dopamine, 69

Ecto-AMPase, 327, 335
Ecto-enzymes
  biological significance of, 337-338
  CNS, 335-337
  mammary gland and tumor, 335
  muscle, 325-333
  in nucleotide metabolism, 323-339
Ecto-purine 5'-nucleotidase deficiency, 396
Eicosatetraynoic acid (ETA), 65, 66, 67
EJP (excitatory junction potentials), 111
Electrochemical studies, 81-84
Electrodegradation and multireceptors, 79-84
Electrophoresis, slab gel, 201
Endocardium, 118
Endoperoxide(s)
  stable, 66
  synthesis, 411
Enhancement, postjunctional, 21
Enzyme defects, metabolic consequences of, 398-400
Epicardium, 118
Erythrocyte(s)
  nucleoside transport by, 305
  proteins, 302
ETA (eicosatetraynoic acid), 65, 66, 67
Excitation, rebound, 8, 47
Excitatory junction potentials (EJPs), 111
External membrane, 324

Falck technique, 63
Fiber(s)
  fast-twitch, 146
  highly oxidative "red,"146
  low oxidative "white," 146
  quinacrine-positive, 11
  slow-twitch, 146
  types, characteristics of three major, 146
Fluorescence
  detection, 419, 420
  quinacrine, 24
Fluoride sensitivity, 211
Frequency, electrical pulse, 46

Ganglion
  bilateral stellate, stimulation, 121
  graded, stimulation, 119
  peripheral, 18
Gastrointestinal tract, mammalian, 103
Glucagon
  and insulin secretion, 16
  plasma, and insulin, 163
  -stimulated accumulation of cAMP, 212
Glutamate on cAMP, effect of, 243
Guanethidine, 90, 104

Heart
  adenosine action in, 127-136
  embryonic chick, cells, 118
  muscle, adenosine binding activity of, 298
HeLa cell(s)
  adenosine transport by, 308
  monolayers, 311
Histamine (HIST), 90, 127, 219, 333
  release, 333
Homogenate(s)
  brain, preparation, 259-260
  studies on cell-free NS20, 253-256
Hydrolysis, rapid, 71

Hyoscine, 90
Hypercapnia, 122
Hyperemia, reactive
  adenosine and, 156-158
  mediator of, 157
Hyperpolarization
  of intestinal muscle, 53
  membrane, 47, 48, 101
Hypocapnia, 122
Hypothesis
  mobile receptor, 211
  one-receptor, 357
  purinergic nerve, 3-32
Hypoxanthine, 117
Hypoxia, 98

IDM (indomethacin), 65
IJP, *see* Inhibitory junction potential
Immunosuppression in ADA and PNP deficiency, 400-406
IMP, *see* Inosine monophosphate
Indicator molecules, 361
Indomethacin (IDM), 65
Inhibition
  of adenylate cyclase by methylxanthines and adenosine deaminase, 260-261
  chemical stimulation in neurogenic, 48, 49
  competitive, by theophylline, 254
  dipyridamole, 311
  electrical stimulation in neurogenic, 45-48
  of fat cell adenylate cyclase, 279
  intramural, in intestine, 45-60
  mechanical stimulation in neurogenic, 49-50
  of nucleoside transport, 308-309
  prejunctional, 21
  presynaptic, of adrenergic neurotransmission, 69-77
  primary, 47
  transynaptic, 230
Inhibitory junction potential (IJP), 45, 111
Inosine, 117
  conformational energy profile of, 288
  monophosphate (IMP), 145, 287-288
  nucleotides, 356
Inotropism, negative, 171
Insulin and glucagon, 16
Interneurons, 49
Ionophores, blocking action of, 252
Ischemia, myocardial, 118
Isoelectric focusing, 298
Isoprenaline, 205
  effects of adenosine and
    on adenylate cyclase, 209-212
    on coronary strips, 206-209
Isoproterenol, 223
Isosteres, methylene-substituted, 110
Iterative extended Huckel theory (IEHT), 284

Kinase, adenosine, 118, 230, 244, 288-289
Krebs-Henseleit solution, 127
Krebs-Ringer homogenates of brain, 230

Lactate, muscle, content, 147
Lactate dehydrogenase, retention of, 333
Leakage, enzyme, 332
Leukocytes and lymphocytes, 333-335
Lidoflazine, 353
Linear elements, consequences of proposed in coding, 364-365
Lipolysis
  catecholamine-induced, 19
  in fat cells, 337
  hormone-stimulated, 392
  mediator of, 379
  by PIA, dissociation between cAMP and, 389-392
Liver cyclase, activation of, 275
Lymphoblast lines, permanent, 402
Lymphocytes, stimulated, and unstimulated, 317
Lymphocytes and leukocytes, 333-335

Magnesium (Mg)
  concentration of, 173
  role of free, 177
Mast cells and adipocytes, 333
Mechanisms, noradrenergic and nonadrenergic inhibitory, 51
Mediator(s)
  adenosine versus deoxyadenosine as, 401-402
  endogenous purine nucleotide, 104
  of lipolysis, 379
  NAI, 63
  neurotransmitter, 364
  purinergic, source and identity, 103-113
  putative, 64
  of reactive hyperemia, 156
  of relaxation, 65
Membrane(s)
  effects of
    exogenous purine compounds on intestinal smooth muscle cell, 372-374
    purinergic compounds on excitable, 369-376
    purines on atrial muscle, 374-375

Mediator(s) (*contd.*)
electrophysiological studies with intestinal smooth muscle cell, 371-372
hyperpolarization, 101
potential, 101
resistance, 54
Meromyosin, 338
Messengers, first and second, 362
Metabolism
alternative pathways of adenylate, 320-321
of deoxyadenosine, 397
ecto-enzymes in nucleotide, 323-339
enzymes involved in adenosine, 244
extracellular, of adenosine, 315
myocardial, 119
neuronal cAMP, 249
purine, in normal cell, 396-398
Methylene isosteres, 34, 352
concentration-effect curves of, 35-36
Methylxanthine(s)
antagonism and adenosine, 354-355
inhibition of adenylate cyclase by, and adenosine deaminase, 260-261
MgATPase of oligomycin, 334
Michaelis constants, 315, 316
Michaelis-Menten kinetics *in situ*, 336
Microsacs, 230
Microtubules, 366
Mitochondrial subfraction, 265
Modulation
of adrenergic activity, 21-22
of cholinergic nerve activity, 22
Moiety, catalytic, 211
Molecules, indicator, 361
Monocarboxylic amino acid, 344
Multireceptors and electrodegradation, 79-84
Muscle
action of adenosine on vascular smooth, 95-102
adenosine and smooth, adenylate cyclase 205-213
adenosine content, 147
analyses for sodium and potassium, 327
cell membrane
effects of exogenous purine compounds on intestinal smooth, 372-374
electrophysiological studies with intestinal smooth, 371-372
ecto-enzymes, 325-333
hyperpolarization of intestinal, 53
lactate content, 147
membrane, effects of purines on atrial, 374-375
responses, 22
sarcolemma, 332
skeletal, adenosine production in, 145-153
Mycoplasm infection, 317
Myosin, myofibrillar, 332

Nerve(s)
adrenergic, 19
calcium-dependent release of ATP from, rather than muscle, 11-12
cholinergic, 20, 112
distribution
of functional NAI, 61-62
and physiological roles of purinergic, 15-19
purinergic, 25
excitatory, 5
guanethidine-resistant adrenergic, 22
intramural inhibitory, 369
modulation of cholinergci, activity, 22
NAI, *see* Nonadrenergic, noncholinergic inhibitory nerves
nonadrenergic inhibitory, 4, 61-68, 85
noncholinergic, 22
noncholinergic inhibitory, 4
peptidergic, 25
periarterial, stimulation, 57
purinergic, 23-24, 30-32, 244
sensory, 20
stimulation, effects of PIT on responses to inhibitory, 90-92
storage, release, and inactivation of ATP from purinergic, 11-13
varicosities, distribution of, 62-65
Neuroblastoma
adenylate cyclase, 221
cells, 336
cyclase, adenosine and, 249-257
Neuroma cells, studies with, 218
Neuromodulation, 21-22
Neuron(s)
adenosine excitant actions and, 352-353
basal firing frequency of, 353
in brain, 20
depressant activity and adenosine, 348-352
depression and adenosine, 343-345
dissociated, 336
excitation and adenosine, 345-348
intracellular records from corticospinal, 346
response characteristics of, 219
role of noradrenergic, 50-57
spontaneous discharge frequency of, 353
Neuronal
firing, purinergic depression of, 355
-glial interaction, 336
Neurotransmission, presynaptic inhibition of adrenergic, 69-77
Neurotransmitter(s), 5, 6

Nitrobenzylthioinosine inhibitor, 306
Nonadrenergic, noncholinergic inhibitory (NAI), 61
-mediated responses, role of PG's in, 65-68
mediator, 63
nerves, 61-62, 369, 371-375
Noradrenaline (NA), 7, 19
Norepinephrine (NE), 217
Nucleoside(s)
analogs, biological effects of, 284
conformation, 283-294
conformational properties of, 283
cytotoxic, 310
interchangeability of purine and pyrimidine, 288
permeants, 306
purine, phosphorylase, 289-290
transport
by erythrocytes, 305
inhibition of, 308-309
nitrobenzylthioinosine as inhibitor of, 306
uptake process, 306-307
Nucleotidase, lysosomal, 320
5'-Nucleotidase, 118
AMP deaminase and, activities, 149-152
localization of, 133
role of creatine phosphate in control of, 174
Nucleotide(s)
inosine, 356
metabolism, ecto-enzymes in, 323-339
modified, 33
radiolabeled, 51
role of adenine, 51-57

Oligomycin MgATPase, 334
Ontogenetic studies, 219
Opiates, 69
Optimal inhibitory activity, 42
Ouabain, 325, 327

Papaverine, studies in presence of, 197-198
PG, *see* Prostaglandin
PGI_2, *see* Prostacyclin
N^6-Phenylisopropyladenosine (PIA), 380
and adenosine, effects of, 381-383
Phosphatase
acid and alkaline, 320
nonspecific, activity, 324
Phosphatidylserine, 211
influence of digitonin solubilization and, 210-211
Phosphodiesterase, 380
adenosine and, inhibitors, 232-233
Phosphofructokinase, 425
Phosphorylase
nucleoside, 118, 123, 289-290
specific adenosine, 317
Phosphorylation
of adenosine, rates of, 315
and deamination, 317
ratio of, to deamination, 317
uncoupler of oxidative, 117
PIA, *see* N^6-Phenylisopropyladenosine
PIT, *see* 2'-2'-Pyridylisatogen tosylate
Plasma membrane (PM) enzymes, 323-324
Platelet(s)
actin, 338
adenine nucleotide-PG interactions in, 411-414
aggregation, 191, 413
cAMP, 412
ecto-ATPase, 338
Plexus, quinacrine-positive perivascular, 18
PM, *see* Plasma membrane
PNP, *see* Purine nucleoside phosphorylase
Polyxyethylene ether, 211
Potassium
depolarization, 96
efflux, increase of, 374
Potential
effects of adenosine derivatives on slow atrial action potential, 169-170
effects of adenosine on slow atrial action, 167-169
effects of adenosine on slow calcium-dependent action, 170-172
inhibitory postjunctional, 45
membrane, 101
recordings of, 54
Prostacyclin (PGI_2), 65
Prostaglandin (PG), 61
and adenine nucleotides, 409-416
interactions in platelets, 411-414
E_1, 189
exogenous, 251
F_2, 66
mediation and adenine nucleotides, 409-411
in NAI-mediated responses, role of, 65-68
receptors, 66
synthesis of, 14
Protein(s)
adenine analog-binding, 419
adenosine-binding, 254, 297-304
aggregation, 301
erythrocyte, 302
kinase activity, 389
kinase II, 291
tubulin, 363
Pseudouridine, 420
Pulse, electrical, frequency, 46

Purine(s)
analogs, conformational distribution of, 283
conformation and modifications at 8 position of, 291-292
metabolic defects, 395-408
metabolism in normal cells, 396-398
nucleoside phosphorylase (PNP), 289-290, 396
and ADA deficiency, immunosuppression in, 405-406
deficiency, 396
deficiency, deoxy-guanosine triphosphate accumulation in, 403
role of, 398
pyrimidine nucleosides, interchangeability of, 288
and pyrimidines in SCID, 420-425
role in cellular proliferation, 396-397
Purkinje cells, cerebellar, 355
2′-2′-Pyridylisatogen tosylate (PIT), 85, 89-90, 92-93
Pyrimidine(s)
nucleotide, deficiency of, 398
and purines in SCID, 420-425

Quinacrine
fluorescence, 24
localization of, 11
-positive
fibers, 11
perivascular plexus, 18

Radioactivity, analysis of, 6
Rayleigh light scattering, 310-312
Reaction velocity, 263
Receptor(s)
activation, 7-8
adenosine, 42
in brain, 243-248
desensitization of, 239
evidence for, 220-221
external, 221
interaction, 223
in neural tissue, 217
α-adrenergic, 219, 223
β-adrenergic blockade, 207
ATP, 374
block of postsynaptic, 345
cyclase-coupled, 243
desensitization of amine, 239
extracellular location of, 357
-mediated stimulation of adenylate cyclase, 212
muscarinic cholinergic, 252
PG, 66, 289-304
populations for AMP, ADP, and ATP, 36
prejunctional, 21
purinergic, 13-15, 73-75
putative ADP, 133-135, 189, 261
stimulation of extracellular adenosine, 357
studies, 80-81
Record, endogenous, 365
Relaxation
maximum, 42
mediator of, 65
Resistance, membrane, 54
Respiratory system, 17
Response(s)
characteristics of intramural inhibitory, 45-50
inhibitory, tachyphylaxis of, 107
role of PG's in NAI-mediated, 65-68
supramaximal, 42
Reticulum, cardiac sarcoplasmic, 201
Rheoseme, 364
-sensitive postsynaptic regions, 364
Ribavirin, 287
RNA
hydrolysis of, 319
terminal groups, cleavage of, 319

Scattered light, angular dependence of, 302
SCID, *see* Severe combined immunodeficiency disease
Secretion, glucagon and insulin, 16
Severe combined immunodeficiency disease (SCID), 417
purines and pyrimidines in, 420-425
Signaling, neural, 365-366
and adenosine, 361-367
of quantities and sites, 361-362
Snake venom, 174, 175
Sotalol, 210
Stimulation
β-adrenergic, of heart, 129
antidromic, 3, 7
bilateral stellate, 121
chemical, in neurogenic inhibition, 48, 49
effects of PIT on responses to inhibitory nerve, 90-92
electrical, in neurogenic inhibition, 45-48
graded ganglion, 119
mechanical, in neurogenic inhibition, 45-50
nonadrenergic inhibitory nerve, 7
periarterial nerve, 57
release of ATP following transmural, 51-53
Subfraction, mitochondrial, 265
Substance P, 64
Sucrose gap method, 371
Superfusion, convexity-cisternal, 122
Synaptosomes, brain, 75
Synconformation, 287, 290

T cell dysfunction, 396
T tubules, 33
Tachyphylaxis, 9
  to ATP, 24
  of inhibitory responses, 107
Technique
  of Falck, 63
  of Veech, 123
Theophylline, 56
  and adenosine deaminase, 193-195
  blockade, 79, 81
  competitive inhibition by, 254
  effect of, on cAMP accumulation, 387-389
  influence of, 207-209
  inhibition, 251
  -sensitive sites, 93
Thromboasthenin, 338
Thromboxane A_2, 412
Thymidine, 305
Tissue
  adenosine
    effects in adipose, 379-393
    receptors in neural, 217
  cerebellar, 220
  excitation of mammalian neocortical, 362
  levels
    of adenosine, 179
    of hypoxanthine, 179
    of inosine, 179
Tone, coronary vascular, 155
Translocation, consequence of, in dendrites, 363-364
Transmission
  auriculoventricular (AV), 170
  inhibitory effect on adrenergic, 69-70
  mechanism of inhibitory action on adrenergic, 70-71
  presynaptic inhibition of adrenergic, 75
  purinergic, 54
    motor, 23
  synaptic, 345
Transmitter, putative, 53
Transport, axonal and dendritic, 366
Tritium-labeled compounds, release of, 23-24
Tubercidin, 310
Tubulin, protein, 363
Tyramine, 139

Ultraviolet
  absorbance, 419
  light, 14
Uridine, 305, 398
  triphosphate (UTP), 423
Urogenital system, 16-17

Varicosities, distribution of nerve, 62-65
Vascular smooth muscle (VSM), 95
  adenosine relaxation of various, 95-96
  mechanisms for adenosine relaxation of, 101
Vasodilatation, cerebral, 121
Velocity, reaction, 263
Vesicles
  adrenergic, 75
  cholinergic, 111
  large granular, 62
  opaque (LOVs), 5
VSM, *see* Vascular smooth muscle